The Stainless Banner
Volume 3

Edited by C.L. Gray

The Stainless Banner Publishing Company

Copyright © 2014 by C.L. Gray
All Rights Reserved
This edition is published by The Stainless Banner Publications, LLC, Fairfield, Ohio.

ISBN: 978-1-62752-010-2

All rights reserved. No part of this publication may be reproduced, stored in a retrieval system, or transmitted in any form or by any means, electronic, mechanical, recording or otherwise, without the prior written permission of the publisher.

Cover design by Kristin Luther/Luther Multimedia (www.luthermultimedia.com)

Printed in the United States of America.

www.thestainlessbanner.com

For those who did not return home.

The Stainless Banner

An e-zine dedicated to the armies of the Confederacy

Issue 3, Volume 1
January 2012

THE FIRST DOMINO FALLS!
THE LOSS OF FORT HENRY

Columbus, Kentucky! The river city seated on high bluffs above the Mississippi River beckoned Confederate General Leonidas Polk. It was the perfect place from which to defend the Mississippi River from Union gunboats steaming from Cairo, and he desired it above all things. The problem was Richmond had forbidden Polk to enter Kentucky and violate the State's neutrality. Orders or the consequences Cumberland and Tennessee rivers from Union gunboats.

To compound his error, Polk vacillated on whether or not he should occupy Paducah on the mouth of the Tennessee River. Ulysses S. Grant did not show such reticence. Paducah was occupied; Columbus was flanked; and Kentucky's neutrality was flaunted for no reason.

All this served as the background for Albert Sidney Johnston's arrival in Richmond. Jefferson Davis appointed Johnston as a full general and gave him command of the overall war effort in the West. Johnston's department included

for breaking Kentucky's neutrality were not enough to keep Polk from his prize. On September 3, 1861, he paraded his troops through the city and turned Columbus into a river fortress.

It was a grievous wound; a needless, self-inflicted wound. Kentucky provided a huge buffer between North and South from the Appalachian Mountains to the Mississippi River. More importantly, the State's neutrality protected the Tennessee, Arkansas, the western part of Mississippi, Kentucky, Missouri, Kansas, and the Indian Territory. The only thing Johnston did not command was the

What's Inside:	
Tilghman's Official Report	6
The Defense of Fort Henry	15
The Fall of Fort Henry	20
Gunboats at Fort Henry	24
Felix Zollicoffer	30
The Character of Robert E. Lee	32
Stonewall Jackson – Soldier	40

coastal defenses.

When Johnston arrived in Nashville, he immediately realized the loss of Kentucky meant his line would have to be expanded to include the Kentucky/Tennessee border. His immediate problem was the lack of manpower and weapons. Richmond said no to his repeated requests for both. Finally, Davis ordered him to do the best he could with what he had.

What Johnston had, he quickly redeployed. Simon Buckner and 4,000 men were reassigned to Bowling Green to defend the railroad that ran between Louisville and Nashville. William Hardee's small force in Arkansas was put on the train and sent to Bowling Green. Polk had 11,000 men in Columbus. Felix Zollicoffer was guarding the Cumberland Gap with another 5,400 men. All-in-all, Johnston had 40,000 men, while Union forces numbered 90,000 with reinforcements coming daily.

Against so great a force, Johnston used the only weapon he possessed. He bluffed with all the finesse of a professional gambler. So good was Johnston at the game that he bluffed William T. Sherman into a nervous breakdown. Sherman commanded the Union forces in Kentucky and as Johnston raided, demonstrated, and moved troops from one point to another and back again; Sherman became convinced he was grossly outnumbered by an ever growing Confederate hoard. Of course, nothing could have been further from the truth, but Johnston was not about to prematurely show his hand. Sherman cracked under the pressure and was out.

Work on Fort Henry

When work began on Henry, a neutral Kentucky influenced where the fort was situated. The engineer placed the works at the first viable location on the Tennessee side of the border.

Unfortunately, the first viable location meant Henry was in a flood plain. Across the river were high bluffs commanding the fort. Due to the sweep of the river, the bluffs were in Kentucky. Protection enough before Polk went into Columbus, but after Columbus, a huge liability.

Johnston's entire command had three engineers and all three had been shanghaied by Polk in his efforts to fortify Columbus. What Polk possessed, he was not giving up. Johnston ordered Polk to send Lieutenant Joseph Dixon to Fort Henry, but Polk delayed Dixon's departure before informing Johnston that Dixon could not be spared. It took two stern orders before Polk released the engineer.

Dixon inspected the works and sent a mixed assessment to Johnston. Henry might not be in the best defensive position, but that was not reason enough to abandon the work already done. The best thing to do would be to build

> *Johnston bluffed with all the finesse of a professional gambler. So good was Johnston at the game that he bluffed William T. Sherman into a nervous breakdown.*

fortifications on the bluffs opposite the fort.

For command of Henry and Donelson, Johnston requested Major A.P. Stewart. Secretary of War Benjamin Judah acting on behalf of Davis, who was acting on behalf of Polk, sent Lloyd Tilghman instead. There was nothing wrong with the choice. Tilghman was a West Point graduate, Mexican War veteran and an experienced engineer. He was conscientious enough to be concerned about the lack of progress on Henry and Donelson to write Davis directly, but his letters did not convey any sense of urgency. Therefore, Davis left the construction of the forts to Polk, since the region was in Polk's department.

But Polk had Columbus on his mind and gave very little thought to river forts. He left its construction to Tilghman, to Dixon and to the department's chief engineer. There were too many cooks in the kitchen, and the three men just got in to each other's way, which slowed the construction considerably. The work slowed even more when Gideon Pillow, in command of Tennessee's defense, stuck his nose into the matter. He countermanded orders and shuffled engineers from one post to another. The end result was that Henry and Fort Heiman, on the opposite bluffs, were not finished by time Grant began his attack on the Tennessee River.

Satisfied with the Columbus defenses, Polk informed Johnston that he would be dismissing some of his troops to the partisan bands engaged in a guerilla war against Union forces in Kentucky. Johnston was aghast! Manpower was too scarce for such a dubious deployment, so he requested the men be sent to Henry and Donelson instead.

Polk's reaction was typical Polk. At first, he ignored the request. When Johnston made it an order, Polk sent Davis his resignation. Davis begged Polk to reconsider. But, before Polk could do that, Union gunboats came steaming down the Mississippi. Polk was all aquiver. It was the battle he had been prophesying long before he entered Columbus.

The fight across the Mississippi at Belmont, Missouri, amounted to a skirmish and the Confederates forced Grant's small army back on their transports and up the river to Cairo. Polk crowed about his victory to anyone who would listen. He dispatched a message to Johnston that he would now need all his men in case the Federals returned.

Johnston did not see it that way and sent orders to dispatch 3,100 men to the river defenses. Polk promptly sent Johnston his resignation. Fed-up with trying to force Polk to do anything that resembled obeying orders, Johnston looked elsewhere for the necessary troops to garrison Henry and Donelson.

The First Push Against Johnston's Line

Not everyone in Tennessee was for secession, and when the state joined the Confederacy, the citizens in the eastern part of Tennessee threatened to secede from the state. Polk, who witnessed the tension firsthand on his way from Richmond to Memphis, wrote Davis that a strong general would be the remedy to anti-Confederate sentiments. Davis asked Governor Harris for suggestions. Harris happily complied, sending Davis a list that included both Whig and Democrat candidates. Harris preferred a Whig for the assignment to counterbalance the overabundance of Democratic generals that had been appointed by his office. At the top of the list, the top Whig – Felix Zollicoffer.

Zollicoffer was a politician and not a military man, not withstanding his brave service in the Seminole War. Davis did not see Zollicoffer's lack of experience as a deterrent. The Unionists were not all that organized and any trouble they may cause could be handled politically, or, if that failed, by strong police action. Besides, Kentucky was neutral. What kind of trouble could Zollicoffer possibly get into? Zollicoffer was commissioned a brigadier general and given command of the military district of Eastern Tennessee.

Johnston passed through the region right after Polk seized Columbus. An energized Zollicoffer, full of military strategies, advanced his plans to occupy the strategic Cumberland Gap. This gateway into Tennessee had to be protected, so Johnston gave permission. Zollicoffer assembled his troops and marched into Eastern Kentucky.

Abraham Lincoln was very interested in the strong pro-Union sentiment in east Tennessee. When the war began, he had hoped there would be enough Unionists in the Confederacy to prevent the war from becoming a long and drawn out affair, but that did not prove to be the case. East Tennessee and western Virginia were the only two regions where Unionists outnumbered Secessionists. Lincoln ordered Union troops into East Tennessee. The Cumberland Gap, where Zollicoffer was setting up defenses, was the preferred route.

Now that Kentucky was in play, Davis was having second thoughts about Zollicoffer's capabilities. What the new general needed was someone to keep a close eye on him and head off any trouble Zollicoffer might accidentally stumble into. It was a strategy that had worked before. At the beginning of the war, Davis had sent Polk to Tennessee to babysit Pillow. Why not send George Crittenden to East Tennessee to babysit Zollicoffer and, at the same time, go on the offense against Union forces in Kentucky.

> **The gateway to Tennessee was open, and Johnston was in real danger of being flanked out of Nashville and the rest of Tennessee.**

Crittenden came to Tennessee fully expecting to lead troops into Kentucky. In Richmond, Davis had promised him ten regiments to do just that. Johnston could have told Crittenden that there were not ten regiments sitting around Knoxville with nothing to do.

With no force to rally, Crittenden did not have any reason to leave the comfort of his Knoxville headquarters. He never went to check on his charge, which meant Zollicoffer was left to his own inexperienced devices. Satisfied with his fortifications of the Gap, Zollicoffer marched his forces closer to Bowling Green and to the enemy. His next step was to find a good area in which to go into winter quarters. He selected Mill Springs.

Mill Springs was on the southern side of the upper Cumberland River about 100 miles east of Bowling Green. It was a strong position for Zollicoffer had chosen a high bluff to plant his army. The surrounding countryside was full of supplies, and Zollicoffer's men would be able to pass the winter in comfort.

Except Zollicoffer was not satisfied. Peering down on the low, flat plain across the river, he thought it a more desirable than the heights. After all, his flanks would be protected by the river, as would his rear. Zollicoffer did not recognize that he was walking into a trap. Any enterprising Federal artillery captain could roll his guns to the edge of the bluff and blast Zollicoffer into surrender.

When Johnston received Zollicoffer's letter informing him of the movement, Johnston ordered him back to the bluffs. It was too late. The first reason was logistical. Zollicoffer could not locate enough boats to ferry his men across the river. The second reason was obduracy. He was perfectly content where he was, so why change?

Crittenden finally arrived at Zollicoffer's headquarters and found the small army in dire straits. Federal forces under the command of General George Thomas had encamped at Logan's Crossroads, six miles from Zollicoffer's camp. There was no way to affect a successful withdrawal now. Complicating matters, another Federal force was garrisoned at Somerset.

The two Union forces were separated by Fishings Creek. Winter rains made the creek unfordable. The thing to do would be to attack one force and defeat it, then wait for the water to recede and attack the other force. Crittenden chose Thomas' force at Logan's Crossroads to attack first.

The small Confederate army left their camps and marched through the mud during a night of torrential rain. The mud slowed their march and by time they arrived at Logan's Crossroads, Thomas was waiting for them. The rain had also dampened their powder, making their old flintlock muskets practically useless.

Still the Confederates pitched in with bravery and for a while were actually winning. Then disaster hit. Zollicoffer was killed, and the Confederate assault ground to a halt. The Union line stiffened and drove the Confederates back in disarray. Zollicoffer's army abandoned supplies, arms, and wagons and did not stop running until they reached Chestnut Mound, fifty miles east of Nashville.

The gateway to Tennessee was open and Johnston was in real danger of being flanked out of Nashville and the rest of Tennessee.

In the meantime, Grant had been agitating his commanding officer to strike both Forts Henry and Donelson. The opening of the Cumberland Gap made this possible.

Grant Moves to Take Fort Henry

February 6, 1862 – Grant disembarked 15,000 men from transports upriver from Fort Henry. Four squat, ugly things, barely recognizable as boats, chugged toward Henry. They were the ironclads and, throughout the war, the South could never find an answer for these armored boats.

Tilghman realized the unfinished fort could withstand neither navy nor infantry. He sent his men to Donelson and ordered Captain Jesse Taylor to hold off the Federals as long as possible to give his men time to escape. Taylor put his big guns into action, but they were only delaying the inevitable. Tilghman ordered the white flag raised. Henry was so flooded that the delegation from Flag Officer Foote rowed right through the gates.

Within days of Henry's fall, Foote's gunboats were steaming as far south as Florence, Alabama, destroying bridges, putting ashore raiding parties and spreading fear throughout the region.

The real damage was, of course, the loss of the Tennessee River, the severing of communications, and the splitting of Johnston's army in two.

The first domino had fallen.

GENERAL LLOYD TILGHMAN'S OFFICIAL REPORT ON THE LOSS OF FORT HENRY

February 12, 1862.

SIR: My communication of the 7th instant, sent from Fort Henry, having announced the fact of the surrender of that fort to Commodore Foote, of the Federal Navy, on the 6th instant, I have now the honor to submit the following report of the details of the action, together with the accompanying papers, containing a list of officers and men surrendered, together with casualties, &c.:

On Monday, February 3, in company with Major Gilmer of the Engineers, I completed the inspection of the main work as well as outworks at Fort Heiman, south of the Tennessee River, as far as I had been able to perfect them, and also the main work, entrenched camp, and exterior line of rifle pits at Fort Henry. At 10:00 a.m. on that morning (the pickets on both sides of the Tennessee River extended well in our front, having reported no appearance of the enemy), I left, in company with Major Gilmer, for Fort Donelson for the purpose of inspecting with him the defenses of that place.

Tuesday, the 4th instant, was spent in making a thorough examination of all the defenses at Fort Donelson. At noon heard heavy firing at Fort Henry for half an hour. At 4:00 p.m., a courier reached me from Colonel Heiman at Fort Henry informing me that the enemy was landing in strong force at Bailey's Ferry, three miles below and on the east bank of the river.

Delaying no longer than was necessary to give all proper orders for the arrangement of matters at Fort Donelson, I left with an escort of Tennessee cavalry, under command of Lieutenant Colonel Gantt, for Fort Henry accompanied by Major Gilmer. Reaching that place at 11.30 p.m., I soon became satisfied that the enemy was really in strong force at Bailey's Ferry, with every indication of reinforcements arriving constantly.

Colonel Heiman of the 10th Tennessee, commanding with most commendable alacrity and good judgment, had thrown forward to the outworks covering the Dover road two pieces of light artillery, supported by a detachment from the 4th Mississippi Regiment, under the command of Captain W.C. Red. Scouting parties of cavalry, operating on both sides of the river, had been pushed forward to within a very short distance of the enemy's lines. Without a moment's delay, after reaching the fort, I proceeded to arrange the available force to meet whatever contingency might arise.

The 1st Brigade under Colonel Heiman was composed of the 10th Tennessee, the 27th Alabama, the 48th Tennessee, light battery of four pieces, and the Tennessee Battalion of Cavalry. Total officers and men equaled 1,444.

The 2nd Brigade under Colonel Joseph Drake was composed of the 4th Mississippi, the 15th Arkansas, the 51st Tennessee, the Alabama Battalion, light battery of three pieces, and the Alabama Battalion of Cavalry. Total officers and men equaled 1,215. The heavy artillery, under command of Captain Taylor, numbering seventy-five men, was placed at the guns in Fort Henry.

As indicated some time since to the general commanding the department, I found it impossible to hold the commanding ground south of the Tennessee River with the small force of badly-armed men at my command. Notwithstanding the fact that all my defenses were commanded by the high ground on which I had commenced the construction of Fort Heiman, I deemed it proper to trust to the fact that the extremely bad roads leading to that point would prevent the movement of heavy guns by the enemy by which I might be annoyed. Leaving the Alabama Battalion of Cavalry and Captain Padgett's spy company on the western bank of the river, I transferred the force encamped on that side to the opposite bank.

At the time of receiving the first intimation of the approach of the enemy, the 48th and 51st Tennessee regiments having only just reported were encamped at Danville and at the mouth of Sandy (Creek) and had to be moved from five to twenty miles in order to reach Fort Henry. This movement, together with the transfer of the 27th Alabama and 15th Arkansas Regiments from Fort Heiman across the river, was all perfected by 5:00 a.m. on the morning of the 5th.

The Enemy is Sighted

Early on the morning of the 5th, the enemy was plainly to be seen at Bailey's Ferry, three miles below. The large number of heavy transports reported by

> *As indicated some time since to the general commanding the department, I found it impossible to hold the commanding ground south of the Tennessee River with the small force of badly-armed men at my command*

our scouts gave evidence of the fact that the enemy was there in force even at that time, and the arrival every hour of additional boats showed conclusively that I should be engaged with a heavy force by land, while the presence of seven gunboats, mounting fifty-four guns, indicated plainly that a joint attack was contemplated by land and water.

On leaving Fort Donelson, I ordered Colonel Head to hold his own and Colonel Sugg's regiments, Tennessee volunteers, with two pieces of artillery, ready to move at a moment's warning, with three days' cooked rations, and without camp equipage or wagon train of any kind, except enough to carry the surplus ammunition.

On the morning of the 5th, I ordered him, in case nothing more had been heard from the country below, on the Cumberland, at the time of the arrival of my messenger, indicating an intention on the part of the enemy to invest Fort Donelson, to move out with the two regiments and the two pieces of artillery and take position at the Furnace, half way on the Dover road to Fort Henry; the force embraced in this order was about 750 men, to act as circumstances might dictate.

Thus matters stood at 9:00 a.m. on the morning of the 5th. The wretched military position of Fort Henry and the small force at my disposal did not permit me to avail myself of the advantages to be derived from the system of outwork built with the hope of being reinforced in time, and compelled me to determine to concentrate my efforts by land within the rifle pits surrounding the camp of the 10th Tennessee and the 4th Mississippi regiments in case I deemed it possible to do more than operate solely against the attack by the river.

Accordingly, my entire command was paraded and placed in the rifle pits around the above camps, and minute instructions given, not only to brigades, but to regiments and companies, as to the exact ground each was to occupy. Seconded by the able assistance of Major Gilmer of whose valuable services I thus early take pleasure in speaking, and by Colonels Heiman and Drake, everything was arranged to make a formidable resistance against anything like fair odds.

It was known to me on the day before that the enemy had reconnoitered the roads leading to Fort Donelson from Bailey's Ferry by way of Iron Mountain Furnace, and at 10:00 a.m. on the 5th, I sent forward from Fort Henry a strong reconnoitering party of cavalry. They had not advanced more than one and a half miles in the direction of the enemy when they encountered their reconnoitering party. Our cavalry charged them in gallant style, upon which the enemy's cavalry fell back, with a loss of only one man on each side. Very soon the main body of the Federal advance guard composed of a regiment of infantry and a large force of cavalry was met upon which our cavalry retreated.

On receipt of this news I moved out in person with five companies of the 10th Tennessee, five companies of the 4th Mississippi, and fifty cavalry, ordering at the same time two additional companies of infantry to support Captain Red at the outworks. Upon advancing well to the front I found that the enemy had retired. I returned to camp at 5:00 p.m., leaving Captain Red reinforced at the outworks.

The enemy was again reinforced by the arrival of a number of large transports.

At night the pickets from the west bank reported the landing of troops on that side (opposite Bailey's Ferry), their advance picket having been met one and a half miles from the river. I, at once, ordered Captain Hubbard of the Alabama cavalry to take fifty men, and, if possible, surprise them. The inclemency of the weather, the rain having commenced to fall in torrents, prevented anything being accomplished.

Early on the morning of the 6th, Captain Padgett reported the arrival of five additional transports overnight and the landing of a large force on the west bank of the river at the point indicated above. From that time up to 9:00, it appeared as though the force on the east bank was again reinforced, which was subsequently proven to be true.

The movements of the fleet of gunboats at an early hour prevented any communication, except by a light barge, with the western bank, and by 10:00 a.m., it was plain that the boats intended to engage the fort with their entire force, aided by an attack on our right and left flanks from the two land forces in overwhelming numbers.

To understand properly the difficulties of my position, it is right that I should explain fully the unfortunate location of Fort Henry in reference to resistance by a small force against an attack by land cooperating with the gunboats, as well as its disadvantages in even an engagement with boats alone.

The entire fort, together with the entrenched camp spoken of, is enfiladed from three or four points on the opposite shore, while three points on the eastern bank completely command them both, all at easy cannon range.

At the same time the entrenched camp, arranged as it was in the best possible manner to meet the case, was two-thirds of it completely under the control of the fire of the gunboats. The history of military engineering records no parallel to this case. Points within a few miles of it, possessing great advantages and few disadvantages, were totally neglected, and a location fixed upon without one redeeming feature or filling one of the many requirements of a site for a work such as Fort Henry. The work itself was well built; it was completed long before I took command, but strengthened greatly by myself in building embrasures and impalements of sand bags. An enemy had but to use their most common sense in obtaining the advantage of high water, as was the case, to have complete and entire control of the position.

I am guilty of no act of injustice in this frank avowal of the opinions entertained by myself, as well as by all other officers who have become familiar with the location of Fort Henry; nor do I desire the defects of location to have an undue influence in directing public opinion in relation to the battle of the 6th instant. The fort was built when I took charge, and I had no time to build anew. With this seeming digression, rendered necessary to a correct understanding of

> *The fate of our right wing at Bowling Green depended upon a concentration of my entire division on Fort Donelson and the holding of that as long as possible.*

the whole affair, I will proceed with the details of the subsequent movements of the troops under my command.

The Battle Begins

By 10:00 a.m. on the 6th, the movements of the gunboats and land force indicated an immediate engagement, and in such force as gave me no room to change my previously conceived opinions as to what, under such circumstances, should be my course. The case stood thus: I had at my command a grand total of 2,610 men, of which only one-third had been at all disciplined or well armed. The high water in the river filling the sloughs gave me but one route by which to retire, if necessary, and that route for some distance in a direction at right angles to the line of approach of the enemy, and over roads well nigh impassable for artillery, cavalry, or infantry. The enemy had seven gunboats, with an armament of fifty-four guns, to engage the eleven guns at Fort Henry. General Grant was moving up the east bank of the river from his landing, three miles below, with a force of 12,000 men, verified afterwards by his own statement, while General Smith, with 6,000 men, was moving up the west bank, to take a position within 400 or 500 yards, which would enable him to enfilade my entire works.

The hopes (founded on a knowledge of the fact that the enemy had reconnoitered on the two previous days thoroughly the several roads leading to Fort Donelson) that a portion only of the land force would cooperate with the gunboats in an attack on the fort were dispelled, and but little time left me to meet this change in the circumstances which surrounded me. I argued thus: Fort Donelson might possibly be held, if properly reinforced, even though Fort Henry should fall; but the reverse of this proposition was not true. The force at Fort Henry was necessary to aid Fort Donelson either in making a successful defense or in holding it long enough to answer the purposes of a new disposition of the entire army from Bowling Green to Columbus, which would necessarily follow the breaking of our center, resting on Forts Donelson and Henry.

The latter alternative was all that I deemed possible. I knew that reinforcements were difficult to be had, and that unless sent in such force as to make the defense certain which I did not believe practicable, the fate of our right wing at Bowling Green depended upon a concentration of my entire division on Fort Donelson and the holding of that place as long as possible, trusting that the delay by an action at Fort Henry would give time for such reinforcements as might reasonably be expected to reach a point sufficiently near Fort Donelson to cooperate with my division, by getting to the rear and right flank of the enemy, and in such a position as to control the roads over which a safe retreat might be effected. I hesitated not a moment.

My infantry, artillery, and cavalry, removed of necessity to avoid the fire of the gunboats to the outworks, could not meet the enemy there; my only chance was to delay the enemy every moment possible and retire the command, now outside the main work, towards Fort Donelson, resolving to suffer as little loss as possible. I retained only the heavy artillery company to fight the guns, and gave the order to commence the movement at once.

At 10:15, Lieutenant Colonel MacGavock sent a messenger to me, stating that our pickets reported General Grant approaching rapidly and within half a mile of the advance work, and movements on the west bank indicated that General Smith was fast approaching also. The enemy, ignorant of any movement of my main body, but knowing that they could not engage them behind our entrenched camp until after the fort was reduced or the gunboats retired, without being themselves exposed to the fire of the latter, took a position north of the forks of the river road, in a dense wood (my order being to retreat by way of the Stewart road), to await the result.

At 11:00 a m., the flotilla assumed their line of battle. I had no hope of being able successfully to defend the fort against such overwhelming odds, both in point of numbers and in caliber of guns. My object was to save the main body by delaying matters as long as possible, and to this end I bent every effort.

At precisely 11:45 a.m., the enemy opened from their gunboats on the fort. I waited a few moments until the effects of the first shots of the enemy were fully appreciated. I then gave the order to return the fire, which was gallantly responded to by the brave little band under my command. The enemy, with great deliberation, steadily closed upon the fort, firing very wild until within 1,200 yards. The cool deliberation of our men told from the first shot fired with tremendous effect.

At 12:35 p.m., the bursting of our 24-pounder rifled gun disabled every man at the piece. This great loss was to us in a degree made up by our disabling entirely the *Essex* gunboat, which immediately floated downstream. Immediately after the loss of this valuable gun, we sustained another loss, still greater in the closing up of the vent of the 10-inch Columbiad, rendering that gun perfectly useless and defying all efforts to reopen it.

The fire on both sides was now perfectly terrific. The enemy's entire force was engaged, doing us but little harm, while our shot fell with unerring certainty upon them and with stunning effect.

At this time a question presented itself to me with no inconsiderable degree of embarrassment. The moment had arrived when I should join the main body of troops retiring toward Fort Donelson, the safety of which depended upon a protracted defense of the fort. It was equally plain that the gallant men working the batteries, for the first time under fire, with all their heroism, needed my presence. Colonel Heiman, the next in command, had returned to the fort for instructions. The men working the heavy guns were becoming exhausted with the rapid firing. Another gun became useless by an accident, and yet another by the explosion of a shell immediately after, striking the muzzle, involving the death of two men and disabling several others. The effect of my absence at such a critical moment would have been disastrous. At the earnest solicitation of many of my officers and men I determined to remain, and ordered Colonel Heiman to join his command and keep up the retreat in good order, while I should fight the guns as long as one man was left, and sacrifice myself to save the main body of my troops.

No sooner was this decision made known than new energy was infused. The enemy closed upon the fort to within 600 yards, improving very much in their fire, which now began to tell with great effect upon the parapets, while the fire from our guns (now reduced to seven) was returned with such deliberation and judgment that we scarcely missed a shot. A second one of the gunboats retired, but I believe was brought into action again.

At 1:10 p.m., so completely broken down were the men, that, but for the fact that four only of our guns were then really serviceable, I could not well have worked a greater number. The fire was still continued with great energy and tremendous effect upon the enemy's boats.

At 1:30 p.m., I took charge of one of the 32-pounders to relieve the chief of that piece, who had worked with great effect from the beginning of the action. I gave the flagship *Cincinnati* two shots, which had the effect to check a movement, intended to enfilade the only guns now left me.

It was now plain to be seen that the enemy were breaching the fort directly in front of our guns, and that I could not much longer sustain their fire without an unjustifiable exposure of the valuable lives of the men who had so nobly seconded me in this unequal struggle.

Surrender

Several of my officers, Major Gilmer among the number, now suggested to me the propriety of taking the subject of surrender into consideration. Every moment I knew was of vast importance to those retreating on Fort Donelson, and I declined, hoping to find men enough at hand to continue a while longer the fire now so destructive to the enemy. In this I was disappointed.

My next effort was to try the experiment of a flag of truce, which I waved from the parapets myself. This was precisely at 1:50 p.m. The flag was not noticed, I presume, from the dense smoke that enveloped it, and leaping again into the fort, I continued the fire for five minutes, when, with the advice of my brother officers, I ordered the flag to be lowered, and after an engagement of two hours and ten minutes with such an unequal force the surrender was made to Flag Officer Foote, represented by Captain Stembel, commanding gunboat *Cincinnati*, and was qualified by the single condition that all officers should retain their sidearms, that both officers and men should be treated with the highest consideration due prisoners of war, which was promptly and gracefully acceded to by Commodore Foote.

The retreat of the main body was effected in good order, though involving the loss of about twenty prisoners, who from sickness and other causes were unable to encounter the heavy roads. The rear of the army was overtaken at a distance of some three miles from Fort Henry by a body of the enemy's cavalry, but, on being engaged by a small body of our men, under Major Garvin, were repulsed and retired.

> *I should fight the guns as long as one man was left and sacrifice myself to save the main body of my troops.*

This fact alone shows the necessity of the policy pursued by me in protracting the defense of the fort as long as possible, which only could have been done by my consenting to stand by the brave little band. No loss was sustained by our troops in this affair with the enemy.

I have understood from the prisoners that several pieces of artillery also were lost, it being entirely impossible to move them over four or five miles with the indifferent teams attached to them.

The entire absence of transportation rendered any attempt to move the camp equipage of the regiments impossible. This may be regarded as fortunate, as the roads were utterly impassable, not only from the rains, but the backwater of the Tennessee River.

A small amount of quartermaster's and commissary stores, together with what was left of the ordnance stores, were lost to us also.

The tents of the Alabama Regiment were left on the west bank of the river, the gunboats preventing an opportunity to cross them over.

Confident of having performed my whole duty to my Government in the defense of Fort Henry, with the totally inadequate means at my disposal, I have but little to add in support of the views before expressed. The reasons for the line of policy pursued by me are to my mind convincing.

> *Confident of having performed my whole duty to my government in the defense of Fort Henry, with the totally inadequate means at my disposal, I have but little to add in support of the views before expressed. The reasons for the line of policy pursued by me are to my mind convincing.*

Against such overwhelming odds as 16,000 well-armed men (exclusive of the force on the gunboats) to 2,610 badly armed, in the field, and fifty-four heavy guns against eleven medium ones in the fort, no tactics or bravery could avail.

The rapid movements of the enemy, with every facility at their command, rendered the defense from the beginning a hopeless one.

I succeeded in doing even more than was to be hoped for at first. I not only saved my entire command outside of the fort, but damaged materially the flotilla of the enemy, demonstrating thoroughly a problem of infinite value to us in the future.

Had I been reinforced, so as to have justified my meeting the enemy at the advanced works, I might have made good the land defense on the east bank. I make no inquiry as to why I was not, for I have entire confidence in the judgment of my commanding general.

The elements even were against us and had the enemy delayed his attack a few days, with the river rising, one-third of the entire fortifications (already affected by it) would have been washed away, while the remaining portion of the works would have been untenable by reason of the depth of water over the whole interior portion.

The number of officers surrendered was twelve; the number of non-commissioned officers and privates in the fort at the time of the surrender was 66, while the number in the hospital-boat Patton was sixteen.

I take great pleasure in making honorable mention of all the officers and men under my command. To Captain Taylor, of the artillery, and the officers of his corps, Lieutenants Watts and Weller; to Captain G.R.G. Jones, in command of the right battery; to Captains Miller and Hayden, of the Engineers; to Acting Assistant Adjutant General McConnico; to Captain H.L. Jones, brigade quartermaster; to Captain McLaughlin, quartermaster of the 10th Tennessee, and to Surgeons Voorhies and Horton, of the 10th Tennessee, the thanks of the whole country are due for their consummate devotion to our high and holy cause. To Sergeants John Jones, Hallam, Cubine, and Silcurk, to Corporals Copass, Cavin, and Renfro, in charge of the guns, as well as to all the men, I feel a large debt is due for their bravery and efficiency in working the heavy guns so long and so efficiently.

Officers and men alike seemed actuated but by one spirit – that of devotion to a cause in which was involved life, liberty, and the pursuit of happiness. Every blow struck was aimed by cool heads, supported by strong arms and honest hearts.

I feel that it is a duty I owe to Colonel A. Heiman, commanding the 10th Tennessee Regiment (Irish), to give this testimony of my high appreciation of him as a soldier and a man, due to his gallant regiment, both officers and men. I place them second to no regiment I have seen in the Army.

To Captain Dixon, of the Engineers, I owe (as does the whole country) my special acknowledgments of his ability and unceasing energies. Under his immediate eye were all the works proposed by myself at Fort Donelson and Heiman executed, while his fruitfulness in resources to meet the many disadvantages of position alone enabled us to combat its difficulties successfully.

To Lieutenant Watts, of the heavy artillery, as acting ordnance officer at Fort Henry, I owe this special notice of the admirable condition of the ordnance department at that post. Lieutenant Watts is the coolest officer under fire I ever met with.

I take pleasure in acknowledging the marked courtesy and consideration of Flag Officer Foote, of the Federal Navy; of Captain Stembel and the other naval officers to myself, officers, and men. Their gallant bearing during the action gave evidence of a brave and therefore generous foe.

Respectfully, your obedient servant,
Lloyd Tilghman
Brigadier General, Commanding.

The Defense of Fort Henry

By Captain Jesse Taylor. *Battles and Leaders, Volume 1*, pages 368-372.

About the 1st of September, 1861, while I was in command of a Confederate "camp of artillery instruction," near Nashville, Tennessee, I received a visit from Lieutenant-Colonel Milton A. Haynes of the 1st Regiment Tennessee Artillery, who informed me of the escape of a number of our steamers from the Ohio River into the Tennessee, and of their having sought refuge under the guns of Fort Henry; that a "cutting-out" expedition from Paducah was anticipated, and that as there was no experienced artillerist at the fort, the Governor (Isham G. Harris) was anxious that the deficiency should immediately be supplied; that he had no one at his disposal unless I would give up my light battery (subsequently Porter's and later still Morton's), and take command at Fort Henry. Anxious to be of service, and convinced that the first effort of the Federals would be to penetrate our lines by the way of the Tennessee River, I at once, in face of the loudly expressed disapproval and wonder of my friends, consented to make the exchange.

Arriving at the fort, I was convinced by a glance at its surroundings that extraordinarily bad judgment, or worse, had selected the site for its erection. I found it placed on the east bank of the river in a bottom commanded by high hills rising on either side of the river, and within good rifle range. This circumstance was at once reported to the proper military authorities of the State at Nashville, who replied that the selection had been made by competent engineers and with reference to mutual support with Fort Donelson on the Cumberland, twelve miles away; and knowing that the crude ideas of a sailor in the navy concerning fortifications would receive but little consideration when conflicting with those entertained by a "West Pointer," I resolved quietly to acquiesce, but the accidental observation of a water-mark left on a tree caused me to look carefully for this sign above, below, and in the rear of the fort; and my investigation convinced me that we had a more dangerous force to contend with than the Federals, namely, the river itself. Inquiry among old residents confirmed my fears that the fort was not only subject to overflow, but that the highest point within it would be – in an ordinary February rise – at least two feet under water. This alarming fact was also communicated to the State authorities, only to evoke the curt notification that the State forces had been transferred to the Confederacy, and that I should apply to

General Polk, then in command at Columbus, Kentucky. This suggestion was at once acted on, not once only, but with a frequency and urgency commensurate with its seeming importance, the result being that I was again referred, this time to General A.S. Johnston, who at once dispatched an engineer (Major Jeremy F. Gilmer) to investigate and remedy; but it was now too late to do so effectually, though an effort was made looking to that end, by beginning to fortify the heights on the west bank (Fort Heiman).

The armament of the fort at the time I assumed command consisted of six smooth-bore 32-pounders and one 6-pounder iron-gun; February 1st, 1862, by the persistent efforts of General Lloyd Tilghman and Colonel A. Haiman, this had been increased to eight 32 pounders, two 4 pounders, and one 128-pounders (Columbia), five 18 pound siege guns, all smooth-bore, and one 6-inch rifle. We also had six 12 pounders, which looked so much like pot metal that it was deemed best to subject them to a test, and as two of them burst with an ordinary charge, the others were set aside as useless encumbrances.

The powder supplied was mostly of a very inferior quality, so much so that it was deemed necessary to adopt the dangerous expedient of adding to each charge a proportion of quick burning powder. That this was necessary will, I think, be admitted when it is understood that with the original charge it was almost impossible to obtain a random shot of a little over one mile (that being the distance to a small island below the fort).

Union Gunboats Try the Defenses

During the winter of 1861 and 1862, the Federal gunboats, notably the *Lexington* and *Conestoga*, made frequently appearances in the Tennessee, and coming up under the cover of this island would favor the fort with an hour or more of shot and shell, but, as their object was to draw our fire and thus obtain the position of our guns, we, though often sorely tempted by the accuracy of their fire, deemed it best not to gratify them.

On the 4th of February, the Federal fleet of gunboats, followed by countless transports, appeared below the fort. Far as eye could see, the course of the river could be traced by the dense volumes of smoke issuing from the flotilla indicating that the long-threatened attempt to break our lines was to be made in earnest. The gunboats took up a position about three miles below and opened a brisk fire, at the same time shelling the woods on the east bank of the river, thus covering the debarkation of their army.

The 5th was a day of unwonted animation on the hitherto quiet waters of the Tennessee; all day long the floodtide of arriving and the ebb of returning transports continued ceaselessly. Late in the afternoon, three of the gunboats, two on the west side and one on the east at the foot of the island, took position and opened a vigorous and well directed fire, which was received in silence until the killing of one man and the wounding of three provoked an order to open with the Columbiad and the rifle. Six shots were fired in return – three from each piece – and with such effect that the gunboats dropped out of range and ceased firing.

At night, General Tilghman called his leading officers in consultation – Colonels Heiman, Forrest, and Drake are all that I can now recall as having been present. The Federal forces were variously estimated by us, as being 25,000 at the least. To oppose this force, General Tilghman had less than four thousand men – mostly raw regiments armed with shotguns and hunting rifles. In fact, the best equipped regiment of his command, the 10th Tennessee, was armed with old flint-lock "Tower of London" muskets that had "done the state some service" in the war of 1812. The general opinion and final decision was that successful resistance to such an overwhelming force was impossible, that the army must fall back and unite with Pillow and Buckner at Fort Donelson. General Tilghman, recognizing the difficulty of withdrawing undisciplined troops from the front of an active and superior opponent, turned to me with the question, "Can you hold out for one hour against a determined attack?" I replied that I could. "Well, then, gentlemen, rejoin your commands and hold them in readiness for instant motion."

The garrison left at the fort to cover the withdrawal consisted of part of Company B, 1st Tennessee Artillery, Lieutenant Watts, and fifty-four men.

> **The general opinion and final decision was that successful resistance to such an overwhelming force was impossible, that the army must fall back and unite with Pillow and Buckner at Fort Donelson.**

The Battle Begins

The forenoon of February 6th was spent by both sides in making needful preparations for the approaching struggle. The gunboats formed line of battle abreast under the cover of the island. The *Essex*, the *Cincinnati*, the *Carondelet*, and the *St. Louis*, the first with four and the others each with thirteen guns, formed the van; the *Tyler*, *Conestoga*, and *Lexington*, with fifteen guns in all, formed the second or rear line. Seeing the formation of battle, I assigned to each gun a particular vessel to which it was to pay its especial compliments and directed that the guns be kept constantly trained on the approaching boats. Accepting the volunteered services of Captain Hayden (of the engineers) to assist at the Columbiad, I took personal supervision of the rifle.

When the gunboats were out of cover of the island they opened fire, and as they advanced they increased the rapidity of their fire, until as they swung into the main channel above the island, they showed one broad and leaping sheet off lame. At this point, the van being a mile distant, the command was given to commence firing from the fort. Let me say that it was as pretty and as simultaneous a "broadside" was delivered as I ever saw flash from the sides of a frigate.

The action now became general, and for the next twenty or thirty minutes was, on both sides, as determined, rapid, and accurate as a heart could wish, and apparently inclined in favor of the fort. The ironclad *Essex*, disabled by a shot through her boiler, dropped out of line. The fleet seemed to hesitate, when a succession of untoward and unavoidable accidents happened in the fort; thereupon the flotilla continued to advance.

First, the rifle gun, from which I had just been called, burst, not only with destructive effect to those working it, but with a disabling effect on those in its immediate vicinity. Going to the Columbiad as the only really effective gun left, I met General Tilghman, who I supposed was with the retreating army. While consulting with him, a sudden exclamation drew me to the Columbiad, which I found spiked with its own priming wire, completely disabled for the day at least.

The Federal commander, observing the silence of these two heavy guns, renewed his advance with increased precision of fire. Two of the 32 pounders were struck almost at the same instant, and the flying fragments of the shattered guns and bursted shells disabled every man at the two guns. His rifle shot and shell penetrated the earthworks as readily as a ball from a navy Colt would pierce a pine board, and soon so disabled other guns as to leave us but four capable of being served.

General Tilghman now consulted with Major Gilmer and myself as to the situation, and the decision was that further resistance would only entail a useless loss of life. He therefore ordered me to strike the colors, now a dangerous as well as a painful duty. The flag mast, which had been the center of fire, had been struck many times; the top mast hung so far out of the perpendicular that it seemed likely to fall at any moment; the flag halyards had been cut by shot, but had fortunately become "foul" at the cross trees.

I beckoned – for it was useless to call amid the din – to Orderly Sergeant Jones, an old "men-of-war's men," to come to my assistance, and we ran across to the flagstaff and up the lower rigging to the cross trees, and by our united efforts succeeded in clearing the halyards and lowering the flag.

The view from that elevated position at the time was grand, exciting, and striking. At our feet, the fort with her few remaining guns was sullenly hurling her harmless shot against the sides of the gunboats, which, now apparently within two hundred yards of the fort, were, in perfect security, and with the coolness and precision of target practice, sweeping the entire fort. To the north and west, on both sides of the river, were the hosts of "blue coats," anxious and interested spectators, while to the east the feeble forces of the Confederacy could be seen making their weary way toward Donelson.

On the morning of the attack, we were sure that the February rise of the Tennessee had come. When the action began, the lower part of the fort was already flooded; and when the colors were hauled down, the water was waist deep there; and when the cutter came with the officers to receive the formal surrender, she pulled into the "sally port" between the fort and the position which had been occupied by the infantry support was a sheet of water a quarter of a mile or more wide, and "running like a mill-race." If the attack had been delayed forty-eight hours, there would hardly have been a hostile shot fired; the Tennessee would have accomplished the work by drowning the magazine.

The Surrender

The fight was over; the little garrison were prisoners; but our army had been saved. We had been required to hold out an hour; we had held out for over two.

We went into the fight with nine guns bearing on the river approach – we had two more forty-two pounders, but neither shot nor shell for them. Of these all were disabled but four. Of the fifty-four men who went into action, five were killed, eleven wounded or disabled, and five missing.

When the *Essex* dropped out of the fight, I could see her men wildly throwing themselves into the swollen river. Admiral Foote reported that his flagship was struck thirty-eight times, and the commanding officers of gunboats (with several of whom I had enjoyed a personal acquaintance) complimented me highly on what they termed the extraordinary accuracy of the fire. I believe that with effective guns the same precision of fire would have sunk or driven back the flotilla.

The formal surrender was made to the naval forces; Lieutenant Commander Phelps acting for Flag Officer Foote, and me, representing General Tilghman. The number captured, including Tilghman and staff, hospital attendants and some stragglers from the infantry, amounted to about seventy.

During the evening a large number of army officers came into the fort, to whom I was introduced by my old messmates, Lieutenant Commanders Gwin and Shirk. Here I first saw General Grant, who impressed me, at the time, as a modest, amiable, kind hearted but resolute man.

While we were at headquarters, an officer came in to report that he had not as yet found any papers giving information of our forces, and, to save him further looking, I informed him that I had destroyed all the papers bearing on the subject, at which he seemed very wrath, fussily demanding, "By what authority?" Did I not know that I laid myself open to punishment, etc., etc. Before I could reply fully, General Grant quietly broke in with, "I would be very much surprised and mortified if one of my subordinate officers should allow information which he could destroy to fall into the hands of the enemy."

We were detained for several days at the fort and were confined to the same steamer on which General Grant had established his headquarters, and as the officers, Confederate and Federal, messed together, I saw much of the general during that time. We were treated with every courtesy; so our confinement was less irksome than we had anticipated and was only marred by one incident.

Two of the younger Confederate officers having obtained liquor became vociferous. At dinner General Grant did not take his seat with the rest, and this restraint being removed, the young men, despite frowns and nudges, persisted in discussing politics, military men, and movements, etc. While they were thus engaged, General Grant, unobserved by them, entered, took his seat, and dined without appearing to notice their conversation, but when the youngsters left the table, they were dumbfounded to meet a corporal and file of men, who ceremoniously conducted them to the "nursery" and left them under guard, where I shortly visited them.

At last I promised to intercede, which I did, carrying with me regrets, explanations, and apologies. The general smiled and said that he had confined them partly for their own sakes, lest they might fall in with some of his own men in a similar condition; that he did not believe the young men knew of his presence, and that he would order their release so soon as they became sober, which he did.

THE FALL OF FORT HENRY

By Ulysses S. Grant. *The Personal Memoirs of Ulysses S. Grant*, (New York: Charles L. Webster, 1885).

The enemy at this time occupied a line running from the Mississippi River at Columbus to Bowling Green and Mill Springs, Kentucky. Each of these positions was strongly fortified, as were also points on the Tennessee and Cumberland rivers near the Tennessee state line. The works on the Tennessee were called Fort Heiman and Fort Henry, and that on the Cumberland was Fort Donelson. At these points, the two rivers approached within eleven miles of each other.

The lines of rifle pits at each place extended back from the water at least two miles, so that the garrisons were in reality only seven miles apart. These positions were of immense importance to the enemy; and of course correspondingly important for us to possess ourselves of.

With Fort Henry in our hands we had a navigable stream open to us up to Muscle Shoals, Alabama. The Memphis and Charleston Railroad strikes the Tennessee at Eastport, Mississippi, and follows close to the banks of the river up to the shoals. This road, of vast importance to the enemy, would cease to be of use to them for through traffic the moment Fort Henry became ours.

Fort Donelson was the gate to Nashville – a place of great military and political importance – and to a rich country extending far east in Kentucky. These two points in our possession the enemy would necessarily be thrown back to the Memphis and Charleston road, or to the boundary of the cotton states, and, as before stated, that road would be lost to them for through communication.

The designation of my command had been changed after Halleck's arrival, from the District of Southeast Missouri to the District of Cairo, and the small district

> *With Fort Henry in our hands, we had a navigable stream open to us up to Muscle Shoals, Alabama.*

commanded by General C.F. Smith, embracing the mouths of the Tennessee and Cumberland rivers, had been added to my jurisdiction.

Early in January, 1862, I was directed by General McClellan, through my department commander, to make a reconnaissance in favor of Brigadier General Don Carlos Buell, who commanded the Department of the Ohio, with headquarters at Louisville, and who was confronting General S.B. Buckner with a larger Confederate force at Bowling Green. It was supposed that Buell was about to make some move against the enemy, and my demonstration was intended to prevent the sending of troops from Columbus, Fort Henry or Donelson to Buckner.

I at once ordered General Smith to send a force up the west bank of the Tennessee to threaten Forts Heiman and Henry; McClernand at the same time with a force of 6,000 men was sent out into west Kentucky, threatening Columbus with one column and the Tennessee River with another. I went with McClernand's command.

The weather was very bad; snow and rain fell; the roads, never good in that section, were intolerable. We were out more than a week splashing through the mud, snow and rain, the men suffering very much. The object of the expedition was accomplished. The enemy did not send reinforcements to Bowling Green, and General George H. Thomas fought and won the battle of Mill Springs before we returned.

As a result of this expedition, General Smith reported that he thought it practicable to capture Fort Heiman. This fort stood on high ground, completely commanding Fort Henry on the opposite side of the river, and its possession by us, with the aid of our gunboats, would insure the capture of Fort Henry. This report of Smith's confirmed views I had previously held, that the true line of operations for us was up the Tennessee and Cumberland rivers. With us there, the enemy would be compelled to fall back on the east and west entirely out of the State of Kentucky.

On the 6th of January, before receiving orders for this expedition, I had asked permission of the general commanding the department to go to see him at St. Louis. My object was to lay this plan of campaign before him. Now that my views had been confirmed by so able a general as Smith, I renewed my request to go to St. Louis on what I deemed important military business. The leave was granted, but not graciously. I had known General Halleck but very slightly in the old army, not having met him either at West Point or during the Mexican war. I was received with so little cordiality that I perhaps stated the object of my visit with less clearness than I might have done, and I had not uttered many sentences before I was cut short as if my plan was preposterous. I returned to Cairo very much crestfallen.

Flag Officer Foote commanded the little fleet of gunboats then in the neighborhood of Cairo and, though in another branch of the service, was subject to the command of General Halleck. He and I consulted freely upon military matters and he agreed with me perfectly as to the feasibility of the campaign up the Tennessee. Notwithstanding the rebuff I had received from my immediate chief, I therefore, on the 28th of January, renewed the suggestion by telegraph that "if permitted, I could take and hold Fort Henry on the Tennessee." This time I was backed by Flag Officer Foote, who sent a similar dispatch.

On the 29th, I wrote fully in support of the proposition. On the 1st of February I received full instructions from department headquarters to move upon Fort Henry. On the 2nd, the expedition started.

In February, 1862, there were quite a good many steamers laid up at Cairo for want of employment, the Mississippi River being closed against navigation below that point. There were also many men in the town whose occupation had been following the river in various capacities, from captain down to deck hand. But there were not enough of either boats or men to move at one time the 17,000 men I proposed to take with me up the Tennessee.

I loaded the boats with more than half the force, however, and sent General McClernand in command. I followed with one of the later boats and found McClernand had stopped, very properly, nine miles below Fort Henry. Seven gunboats under Flag Officer Foote had accompanied the advance. The transports we had with us had to return to Paducah to bring up a division from there, with General C.F. Smith in command.

Before sending the boats back, I wanted to get the troops as near to the enemy as I could without coming within range of their guns. There was a stream emptying into the Tennessee on the east side, apparently at about long range distance below the fort. On account of the narrow watershed separating the Tennessee and Cumberland rivers at that point, the stream must be insignificant at ordinary stages, but when we were there, in February, it was a torrent. It would facilitate the investment of Fort Henry materially if the troops could be landed south of that stream.

To test whether this could be done, I boarded the gunboat *Essex* and requested Captain William Porter to approach the fort to draw its fire. After we had gone some distance past the mouth of the stream, we drew the fire of the fort, which fell much short of us. In consequence, I had made up my mind to return and bring the troops to the upper side of the creek, when the enemy opened upon us with a rifled gun that sent shot far beyond us and beyond the stream. One shot passed very near where Captain Porter and I were standing, struck the deck near the stern, penetrated and passed through the cabin and so out into the river. We immediately turned back, and the troops were debarked below the mouth of the creek.

When the landing was completed, I returned with the transports to Paducah to hasten up the balance of the troops. I got back on the 5th with the advance the remainder following as rapidly as the steamers could carry them.

At ten o'clock at night, on the 5th, the whole command was not yet up. Being

> *The weather was very bad; snow and rain fell; the roads, never good in that section, were intolerable. We were out more than a week splashing through the mud, snow, and rain, the men suffering very much.*

anxious to commence operations as soon as possible before the enemy could reinforce heavily, I issued my orders for an advance at 11:00 a.m. on the 6th. I felt sure that all the troops would be up by that time.

Fort Henry occupies a bend in the river which gave the guns in the water battery a direct fire down the stream. The camp outside the fort was entrenched with rifle pits and outworks two miles back on the road to Donelson and Dover. The garrison of the fort and camp was about 2,800, with strong reinforcements from Donelson halted some miles out. There were seventeen heavy guns in the fort. The river was very high, the banks being overflowed except where the bluffs come to the water's edge. A portion of the ground on which Fort Henry stood was two feet deep in water. Below, the water extended into the woods several hundred yards back from the bank on the east side. On the west bank, Fort Heiman stood on high ground completely commanding Fort Henry.

The distance from Fort Henry to Donelson is but eleven miles. The two positions were so important to the enemy, *as he saw his interest*, that it was natural to suppose that reinforcements would come from every quarter from which they could be got. Prompt action on our part was imperative.

The plan was for the troops and gunboats to start at the same moment. The troops were to invest the garrison and the gunboats to attack the fort at close quarters. General Smith was to land a brigade of his division on the west bank during the night of the 5th and get it in rear of Heiman.

At the hour designated the troops and gunboats started. General Smith found Fort Heiman had been evacuated before his men arrived. The gunboats soon engaged the water batteries at very close quarters, but the troops which were to invest Fort Henry were delayed for want of roads, as well as by the dense forest and the high water in what would in dry weather have been unimportant beds of streams. This delay made no difference in the result.

On our first appearance, Tilghman had sent his entire command, with the exception of about one hundred men left to man the guns in the fort, to the outworks on the road to Dover and Donelson, so as to have them out of range of the guns of our navy; and before any attack on the 6th he had ordered them to retreat on Donelson. He stated in his subsequent report that the defense was intended solely to give his troops time to make their escape.

Tilghman was captured with his staff and ninety men, as well as the armament of the fort, the ammunition and whatever stores were there. Our cavalry pursued the retreating column towards Donelson and picked up two guns and a few stragglers; but the enemy had so much the start, that the pursuing force did not get in sight of any except the stragglers.

All the gunboats engaged were hit many times. The damage, however, beyond what could be repaired by a small expenditure of money, was slight, except to the *Essex*. A shell penetrated the boiler of that vessel and exploded it, killing and wounding forty-eight men, nineteen of whom were soldiers who had been detailed to act with the navy. On several occasions during the war such details were made when the complement of men with the navy was insufficient for the duty before them.

After the fall of Fort Henry Captain Phelps, commanding the ironclad *Carondelet*, at my request ascended the Tennessee River and thoroughly destroyed the bridge of the Memphis and Ohio Railroad.

★ ★ ★

GUNBOATS AT THE FALL OF FORT HENRY

By Rear Admiral Henry Walke, U.S.N. *Battles and Leaders*, Volume 1, pages 358-367.

At the beginning of the war, the army and navy were mostly employed in protecting the loyal people who resided on the borders of the disaffected States, and in reconciling those whose sympathies were opposed. But the defeat at Manassas and other reverses convinced the Government of the serious character of the contest, and of the necessity of more vigorous and extensive preparations for war. Our navy yards were soon filled with workmen; recruiting stations for unemployed seamen were established, and we soon had more sailors than were required for the ships that could be fitted for service. Artillerymen for the defenses of Washington being scarce, five hundred of these sailors, with a battalion of marines (for guard duty), were sent to occupy the forts on Shuter's Hill, near Alexandria. The Pensacola and the Potomac flotilla and the seaboard navy yards required nearly all of the remaining unemployed seamen.

> *Only after a most determined resistance and after all his heavy guns had been silenced, did General Tilghman lower his flag.*

While Foote was improvising a flotilla for the western rivers, he was making urgent appeals to the government for seamen. Finally some one at the Navy Department thought of the five hundred tars stranded on Shuter's Hill and obtained an order for their transfer to Cairo, where they were placed on the receiving ship *Maria Denning*. There they met freshwater sailors from our great lakes and steamboat hands from the western rivers. Of the seamen from the East, there were Maine lumbermen, New Bedford whalers, New York liners, and Philadelphia sea lawyers. The foreigners enlisted were mostly Irish, with a few English and Scotch, French, Germans, Swedes, Norwegians, and Danes. The Northmen, considered the hardiest race in the world, melted away in the Southern sun with surprising rapidity.

On my gunboat, the *Carondelet*, were more young men perhaps than on any other vessel in the fleet. Philadelphians were in the majority; Bostonians came next, with a sprinkling from other cities and just enough men-of-war men to leaven the lump with naval discipline.

The *De Kalb* had more than its share of men-of-war men, Lieutenant Commander Leonard Paulding having had the first choice of a full crew, and having secured all the frigate *Sabine's* reenlisted men who had been sent West.

During the spring and summer of 1861, Commander John Rodgers purchased, and he, with Commander Roger N. Stembel, Lieutenant S. L. Phelps, and Mr. Eads, altered, equipped, and manned for immediate service on the Ohio and Mississippi rivers, three wooden gunboats: the *Tyler*, of six 8-inch shell guns and two 32-pounders; the *Lexington*, of four 8-inch shell guns and two 32-pounders, and the *Conestoga*, of four 32-pounder guns. This nucleus of the Mississippi flotilla (like the fleets of Perry, Macdonough, and Chauncey in the war of 1812) was completed with great skill and dispatch; they soon had full possession of the western rivers above Columbus, Kentucky, and rendered more important service than as many regiments could have done.

On October 12th, 1861, the *St. Louis*, afterward known as the *De Kalb*, the first of the seven ironclad gunboats ordered of Mr. Eads by the Government, was launched at Carondelet, near St. Louis. The other ironclads, the *Cincinnati*, *Carondelet*, *Louisville*, *Mound City*, *Cairo*, and *Pittsburgh*, where launched soon after the *St. Louis*, Mr. Eads having pushed forward the work with most commendable zeal and energy. Three of these were built at Mound City, Illinois. To the fleet of ironclads above named were added the *Benton* (the largest and best vessel of the Western flotilla), the *Essex*, and a few smaller and partly armored gunboats.

Flag-Officer Foote arrived in St. Louis on September 6th, and assumed command of the Western flotilla. He had been my fellow midshipman in 1827, on board the United States ship *Natchez*, of the West Indian squadron and, was then a promising young officer. He was transferred to the *Hornet*, of the same squadron, and was appointed her sailing master. After he left the *Natchez*, we never met again until February, 1861, at the Brooklyn Navy Yard, where he was the executive officer. Foote, Schenck, and myself were then the only survivors of the midshipmen of the *Natchez*, in her cruise of 1827, and now I am the only officer left.

Foote arrived at Cairo September 12th, and relieved Commander John Rodgers of the command of the station. The first operations of the Western flotilla consisted chiefly of reconnaissances on the Mississippi, Ohio, Cumberland, and Tennessee rivers. At this time it was under the control of the War Department and acting in cooperation with the army under General Grant, whose headquarters were at Cairo.

The Battle of Belmont

On the evening of the 6th of November, 1861, I received instructions from General Grant to proceed down the Mississippi with the wooden gunboats *Tyler* and *Lexington* on a reconnaissance, and as convoy to some half-dozen transport steamers; but I did not know the character of the service expected of me until I anchored for the night, seven or eight miles below Cairo. Early the next morning, while the troops were being landed near Belmont, Missouri, opposite Columbus, Kentucky, I attacked the Confederate batteries, at the request of General Grant, as a diversion, which was done with some effect.

But the superiority of the enemy's batteries on the bluffs at Columbus, both in the number and the quality of his guns, was so great that it would have been too hazardous to have remained long under his fire with such frail vessels as the *Tyler* and *Lexington*, which were only expected to protect the land forces in case of a repulse. Having accomplished the object of the attack, the gunboats withdrew, but returned twice during the day and renewed the contest.

During the last of these engagements, a cannon ball passed obliquely through the side, deck, and scantling of the *Tyler*, killing one man and wounding others. This convinced me of the necessity of withdrawing my vessels, which had been moving in a circle to confuse the enemy's gunners. We fired a few more broadsides, therefore, and, perceiving that the firing had ceased at Belmont, an ominous circumstance, I returned to the landing, to protect the army and transports. In fact, the destruction of the gunboats would have involved the loss of our army and our depot at Cairo, the most important one in the West.

Soon after we returned to the landing place, our troops began to appear, and the officers of the gunboats were warned by General McClernand of the approach of the enemy. The Confederates came en masse through a cornfield, and opened with musketry and light artillery upon the transports, which were filled or being filled with our retreating soldiers. A well-directed fire from the gunboats made the enemy fly in the greatest confusion.

Foote was at St. Louis when the battle of Belmont was fought, and made a report to the Secretary of the Navy of the part which the gunboats took in the action, forwarding my official report to the Navy Department. The officers of the vessels were highly complimented by General Grant for the important aid they rendered in this battle; and in his second official report of the action he made references to my report. It was impossible for me to inform the flag officer of the general's intentions, which were kept perfectly secret.

Battle of Fort Henry

During the winter of 1861-62, an expedition was planned by Flag Officer Foote and Generals Grant and McClernand against Fort Henry, situated on the eastern bank of the Tennessee River, a short distance south of the line between Kentucky and Tennessee. In January the ironclads were brought down to Cairo, and great efforts were made to prepare them for immediate service, but only four of the ironclads could be made ready as soon as required.

On the morning of the 2nd of February, the flag officer left Cairo with the four armored vessels above named, and the wooden gunboats *Tyler*, *Lexington*, and *Conestoga*, and in the evening reached the Tennessee River. On the 4th, the fleet anchored six miles below Fort Henry. The next day, while reconnoitering, the *Essex* received a shot which passed through the pantry and the officers' quarters and visited the steerage. On the 5th, the flag officer inspected the officers and crew at quarters, addressed them, and offered a prayer.

Heavy rains had been falling, and the river had risen rapidly to an unusual height; the swift current brought down an immense quantity of heavy driftwood, lumber, fences, and large trees, and it required all the steam-power of the *Carondelet*, with both anchors down, and the most strenuous exertions of the officers and crew, working day and night, to prevent the boat from being dragged down stream.

This adversity appeared to dampen the ardor of our crew, but when the next morning they saw a large number of white objects, which through the fog looked like polar bears, coming down the stream, and ascertained that they were the enemy's torpedoes forced from their moorings by the powerful current, they took heart, regarding the fresher as providential and as a presage of victory. The overflowing river, which opposed our progress, swept away in broad daylight this hidden peril; for if the torpedoes had not been disturbed, or had broken loose at night while we were shoving the driftwood from our bows, some of them would surely have exploded near or under our vessels.

The 6th dawned mild and cheering with a light breeze, sufficient to clear away the smoke. At 10:20 the flag-officer made the signal to prepare for battle, and at 10:50 came the order to get under way and steam up to Panther Island, about two miles below Fort Henry. At 11:35, having passed the foot of the island, we formed in line and approached the fort four abreast: the *Essex* on the right, then the *Cincinnati*, *Carondelet*, and *St. Louis*. For want of room, the last two were interlocked, and remained so during the fight.

As we slowly passing up this narrow stream, not a sound could be heard nor a moving object seen in the dense woods which overhung the dark and swollen river. The gun crews of the *Carondelet* stood silent at their posts, impressed with the serious and important character of the service before them.

About noon the fort and the Confederate flag came suddenly into view, the barracks, the new earthworks, and the great guns well manned. The captains of our guns were men-of-war's men, good shots, and had their men well drilled.

The flag steamer, the *Cincinnati*, fired the first shot as the signal for the others to begin. At once the fort was ablaze with the flame of her eleven heavy guns. The wild whistle of their rifle shells was heard on every side of us. On the *Carondelet* not a word was spoken more than at ordinary drill, except when Matthew Arthur, captain of the starboard bowgun, asked permission to fire at one or two of the enemy's retreating vessels, as he could not at that time bring his gun to bear on the fort. He fired one shot, which passed through the upper cabin of a hospital boat, whose flag was not seen, but injured no one.

The *Carondelet* was struck in about thirty places by the enemy's heavy shot and shell. Eight struck within two feet of the bow ports, leading to the boilers, around which barricades had been built – a precaution which I always took before going into action, and which on several occasions prevented an explosion. The *Carondelet* fired 107 shell and solid shot; none of her officers or crew was killed or wounded.

The firing from the armored vessels was rapid and well sustained from the beginning of the attack, and seemingly accurate, as we could occasionally see the earth thrown in great heaps over the enemy's guns. Nor was the fire of the Confederates to be dispersed; their heavy shot broke and scattered our iron plating as if it had been putty, and often passed completely through the casemates. But our old men-of-war's men, captains of the guns, proud to show their worth in battle to their young comrades. When these experienced gunners saw a shot coming toward a port, they had the coolness and discretion to order their men to bow down, to save their heads.

After nearly an hour's hard fighting, the captain of the *Essex*, going below, and complimenting the First Division for their splendid execution, asked them if they did not want to rest and give three cheers, which were given with a will. But the feelings of joy on board the Essex were suddenly changed by a calamity when a shot from the enemy pierced the casement just above the porthole on the port side, then through the middle boiler, and opening a chasm for the escape of the scalding steam and water.

> **The Confederate soldiers fought as valiantly and as skillfully as the Union sailors.**

The *Essex* before the accident had fired seventy shots from her two 9-inch guns. A powder boy, Job Phillips, fourteen years of age, coolly marked down upon the casemate every shot his gun had fired, and his account was confirmed by the gunner in the magazine. The loss in killed, wounded, and missing was thirty-two.

The *St. Louis* was struck seven times. She fired 107 shots during the action. No one on board the vessel was killed or wounded.

Foote, during the action, was in the pilot house of the *Cincinnati*, which received 32 shots. Her chimneys, after cabin, and boats were completely riddled. Two of her guns were disabled. The only fatal shot she received passed through the larboard front, killing one man and wounding several others. I happened to be looking at the flag steamer when one of the enemy's heavy shot struck her. It had the effect, apparently, of a thunder bolt, ripping her side timbers and scattering the splinters over the vessel. She did not slacken her speed, but moved on as though nothing unexpected had happened.

From the number of times the gunboats were struck, it would appear that the Confederate artillery practice, at first, at least, was as good, if not better, than ours. This, however, was what might have been expected, as the Confederate gunners had the advantage of practicing on the ranges the gunboats would probably occupy as they approached the fort. The officers of the gunboats, on the contrary, with guns of different caliber and unknown range, and without practice, could not point their guns with as much accuracy. To counterbalance this advantage of the enemy, the gunboats were much better protected by their casemates for distant firing than the fort by its fresh earthworks.

The Confederate soldiers fought as valiantly and as skillfully as the Union sailors. Only after a most determined resistance, and after all his heavy guns had been silenced, did General Tilghman lower his flag. The Confederate loss, as reported, was five killed, eleven wounded, and five missing. The prisoners, including the general and his staff, numbered seventy-eight in the fort and sixteen in a hospital boat; the remainder of the garrison, a little less than 3,600, having escaped to Fort Donelson.

Surrender

Our gunboats continued to approach the fort until General Tilghman, with two or three of his staff, came off in a small boat to the *Cincinnati* and surrendered the fort to Flag Officer Foote, who sent for me, introduced me to General Tilghman, and gave me orders to take command of the fort and hold it until the arrival of General Grant.

General Tilghman was a soldierly looking man, a little above medium height, with piercing black eyes and a resolute, intelligent expression of countenance. He was dignified and courteous, and won the respect and sympathy of all who became acquainted with him. In his official report of the battle, he said that his officers and men fought with the greatest bravery until 1:50 p.m., when seven of his eleven guns were disabled; and, finding it impossible to defend the fort, and wishing to spare the lives of his gallant men, after consultation with his officers, he surrendered the fort.

It was reported at the time that, in surrendering to Flag Officer Foote, the Confederate general said, "I am glad to surrender to so gallant an officer," and that Foote replied, "You do perfectly right, sir, in surrendering, but you should have blown my boat out of the water before I would have surrendered to you." I was with Foote soon after the surrender, and I cannot believe that such a reply was made by him. He was too much of a gentleman to say anything calculated to wound the feelings of an officer who had defended his post with signal courage and fidelity, and whose spirits were clouded by the adverse fortunes of war.

When I took possession of the fort the Confederate surgeon was laboring with his coat off to relieve and save the wounded; and although the officers and crews of the gunboats gave three hearty cheers when the Confederate flag was hauled down, the first inside view of the fort sufficed to suppress every feeling of exultation and to excite our deepest pity. On every side the blood of the dead and wounded was intermingled with the earth and their implements of war. Their largest gun, a 128-pounder, was dismounted and filled with earth by the bursting of one of our shells near its muzzle; the carriage of another was broken to pieces, and two dead men lay near it, almost covered with heaps of earth; a rifled gun had burst, throwing its mangled gunners into the water. But few of the garrison escaped unhurt.

General Grant, with his staff, rode into the fort about 3:00 on the same day, and relieved me of the command. The general and staff then accompanied me on board the *Carondelet* (anchored near the fort), where he complimented the officers of the flotilla in the highest terms for the gallant manner in which they had captured Fort Henry. He had expected his troops to take part in a land attack, but the heavy rains had made the direct roads to the fort almost impassable.

The wooden gunboats *Conestoga* and *Lexington* engaged the enemy at a long range in the rear of the ironclads. After the battle, they pursued the enemy's transports up the river, and the *Conestoga* captured the steamer *Eastport*. The news of the capture of Fort Henry was received with great rejoicing all over the North.

Following upon the capture of Fort Henry (February 6th, 1862) and of Fort Donelson (February 16th), the fortifications of Columbus on the Mississippi were evacuated February 20th. In January General Halleck reached the conclusion that the object for which General Polk had labored in fortifying Columbus had been accomplished, for on the 20th he wrote General McClellan: "Columbus cannot be taken without an immense siege-train and a terrible loss of life. I have thoroughly studied its defenses-they are very strong; but it can be turned, paralyzed, and forced to surrender." In accordance with the idea suggested in their dispatch, the Federal movement upon Forts Henry and Donelson was decided upon.

HI-LIGHTS OF A HERO'S LIFE
FELIX ZOLLICOFFER

★ 1812 – (October 12) Born in Maury County, Tennessee.

★ 1828 – Became an apprentice printer in Paris, Tennessee.

★ 1831 – Moved to Knoxville and worked as a journeyman printer for the *Knoxville Register*.

★ 1834 – Became editor and part owner of the *Columbia Observer*.

★ 1835 – (September 24) Married Louisa Pocahontas Gordon.

★ 1836 – Volunteered for the army and served as a lieutenant in the Second Seminole War.

★ 1837 – Became the owner of the *Columbia Observer* and the *Southern Agriculturist*.

★ 1843 – Edited the *Republican Banner*, the state organ for the Whig Party.

- ★ 1845 – Named Comptroller of the State Treasury.

- ★ 1849 – Delegate in the State Senate.

- ★ 1852 – Delegate to the Whig National Convention and supported the candidacy of General Winfield Scott.

- ★ 1853 – Served in the House of Representatives for two terms.

- ★ 1858 – Retired to private life.

- ★ 1860 – Supported John Bell for president.

- ★ 1861 – Served as a member of the peace convention. Though he believed in states rights, he was not a proponent of secession.

- ★ 1861 – (May 9) Appointed brigadier general of the Provisional Army of Tennessee.

- ★ 1861 – (July 9) Transferred to the Confederate States Army as a brigadier general.

- ★ 1861 – (July 26) Harris ordered Zollicoffer's brigade to Knoxville to suppress East Tennessee's resistance to secession.

- ★ 1861 – (August 1) To help facilitate the suppression, Zollicoffer is given command of the Eastern District of Tennessee.

- ★ 1861 – (September 17) Led a force of 5,400 men from Tennessee through the Cumberland Gap to seize eastern Kentucky.

- ★ 1861 – Won the Battle of Barbourville.

- ★ 1861 – Defeated at the Battle of Wildcat Mountain and forced to retreat back to eastern Tennessee.

- ★ 1861 – (November) Returned to eastern Kentucky and took up a defensive position at Mill Springs.

- ★ 1861 – (December 8) George Crittenden assumed command of the department. Zollicoffer retained command of the 1st Brigade.

- ★ 1862 – (January 19) Killed at the Battle of Mill Springs. He was buried in Nashvillle.

THE LIFE AND CHARACTER OF ROBERT EDWARD LEE

By William Evelyn Cameron. *Southern Historical Society Papers, Volume 29*, (January 19, 1901), pages 82-99. (Edited by C.L. Gray)

Not unmindful of the magnitude of the task your partial judgment has assigned to me – diffident of my power to clothe your love and reverence for Robert Lee in adequate phrase – I have yet accepted your invitation as a command, to which neither inclination nor duty could remain irresponsive; and I throw myself upon your generous indulgence as in sober speech I try to portray to you "The man he was who held a nation's heart in thrall."

By time Virginia called in 1861, he was already a veteran in war, master of its theories, ripe in its practice, in the flush of health in mind and body. He was the center of expectancy and of confidence. In the old army he had won a reputation second to none. Scott, his old commander, had declared of him, in his stilted but sincere way, that he was the "the greatest military genius in America, the best soldier I ever saw in the field; and if opportunity should offer he will show himself the foremost captain of his time."

It was through the influence of this Virginian, then at the head of the United States army, that President Lincoln was induced to offer that high command to Colonel Lee. This tender so calculated to gratify an ordinary pride, and great enough to satisfy any ambition, came to a man who was controlled in every act of his existence by his desire to do the right. In all that memorable career there is not an act nor utterance which suggests a motive less noble than a sense of duty. His answer to the overture was a courteous negative, and forthwith he saw that the time had come to leave the service of the Union.

That his resignation from the United States army was a step taken in sorrow and after severe conflict of mind is not to be doubted by any who read the calm yet mournful letters in which at this juncture he announced his decision to his sister. He severed the ties and relinquished the aspirations of a lifetime to enter upon a contest which promised nothing but loss and danger to him and his. He relinquished high opportunity to embark fame and fortune upon a more than doubtful struggle. That his reluctance and regret were sincere none who knew the stern integrity of the man can doubt. He

> *Scott, his old commander, had declared of him, in his stilted but sincere way, that he was "the greatest military genius in America, the best soldier I ever saw in the field; and if opportunity should offer, he will show himself the foremost captain of his time."*

says that his heart bled within him at the prospect, and this is the deliberate statement of one to whom falsehood was impossible.

Entering the service of Virginia as Commander-in-Chief of her forces, for nearly a year he held no important command in the field, and this is another illustration of the entire freedom of the man from self-seeking. He was content to be of use; and while engaged in the essential work of organizing the troops as they arrived from the South, with headquarters at Richmond, he saw without regret and with no effort to assert his claim, the conduct of operations in the field entrusted to others.

It was not until the spring of 1862 that General Johnston, having been wounded at Seven Pines, the opportunity was born which gave to Lee an adequate field for the exercise of his abilities. Thenceforward until the closing scene at Appomattox he was never absent from that army with whose achievements his name is inseparably linked.

His face and figure were soon familiar to every man in the command. He was constantly on the lines, rarely attended by any escort save a single staff officer. An active and perfect horseman, of distinguished and handsome countenance, he looked every inch the gallant soldier

From the very first he inspired officers and men with a trusting affection which later grew into worship. He had none of the arts by which lovers of popularity seek to ingratiate themselves with their subordinates. In his intercourse with soldiers of whatever rank, so far as my knowledge goes, General Lee never unbent from the somewhat formal courtesy habitual to him. The magnetism was there though, if not perceptible, and it wrought devotion and implicit confidence in the hearts of the coldest.

Even before we met the enemy under the direction of that steady eye, he was all in all to us. After the first trial, when McClellan had been driven to the plains of Berkeley, the army of Virginia pinned its faith to him with a tenacity which no subsequent disaster was able to shake. And that mere corporal's guard of us who still survive, our ranks growing thinner hour by hour, despite the fact that the mechanic grasp of fate denied the victor's laurel to that brow, we who gloried the more in his initial triumphs because they were his, who felt the sting of final disaster more keenly because it pierced so cruelly that great heart, we believe in him still.

To resume for a moment the parallel previously drawn, I think that in the qualities of their military genius, Washington and Lee – I name them in the order of time – had many points in common. The characteristic of both was pugnacity, and the campaigns of Lee in Virginia, as those of Washington in the Jerseys, were superb examples of what is technically known as the offensive-defensive. The vigilance of both was sleepless; both were acute in penetrating the designs and anticipating the movements of the enemy; neither ever willingly neglected an opportunity to take the initiative.

From the swoop upon McClellan's right, through the campaigns against Pope, in the battles of 1863, in his manner of meeting Grant's advance through the Wilderness, and even after lines of circumvallation were drawn at

Petersburg, General Lee was constantly and consistently aggressive.

No finer example of this trait is known to military history than that given at Chancellorsville, where, with the swiftness of a practiced fencer, General Lee passed from the attitude of the assailed to that of the assailant, ere his antagonist had time to realize the changed conditions. To find Lee in line of battle parallel to his lines of communication was the first surprise which disconcerted the Federal commander; but even then he never dreamed of the prescient boldness that was to amuse Sedgwick with Early's handful, hold his own front against Hooker's main force, with barely eleven thousand men, while Jackson, with two-thirds of the Confederate troops, was sent across the front and well to the right and rear of an army of ninety-two thousand muskets.

The easy confidence with which Lee responded to a movement upon his flank of an overwhelming enemy, while at the same time another force nearly equal to his total strength was thundering in his rear, proved that from the very first he felt himself, despite the disparity in numbers, to be master of the situation. The only doubt he seems to have entertained after the first intelligence of Hooker's presence on the south side of the Rappahannock, was whether first to push Jackson against Sedgwick on the plains where Burnside met his crushing defeat. But his consideration of this plan was brief, though Jackson favored it, and instead he seized his right wing and hurled it in reverse, as an athlete might have slung a stone, over field and forest, upon the one vulnerable spot in the strong formation of his foe.

But uniformly his tactics and his "noble ire of battle" were alike the servants of that cool, clear judgment which seldom erred. Self-discipline with him had been brought to a science.

> **We believe in him still.**

I have used the term "combative by calculation," meaning by that the conviction of General Lee that the Confederate armies could not afford to conduct a purely defensive warfare – if in strategy, not in tactics. His greatest successes were won by aggressive operations.

So McClellan's grand army was pushed back upon its gunboats, the siege of Richmond raised, and an hundred thousand of the best troops of the Union paralyzed and neutralized, while the army of Northern Virginia first staggered Banks at Cedar Mountain and then drove Pope's legions in pell mell disorder back into the entrenchments around Washington. It was so, as has been said, that he compassed that victory at Chancellorsville, which is still the study and wonder of the military schools of the world. It was so that he freed the Valley of Virginia from invasion, sent Hooker back into Pennsylvania to defend his own; and it was so that the ark of Southern independence might have floated on the high tide of Gettysburg, but for contingencies, which as they are the subject of controversy, I shall not bring into formal discussion here.

If he erred in aggression there, the error was born of a noble confidence in that magnificent army which had so often under his leadership accomplished the improbable, that he had come to deem its valor invincible. Success held in its beckoning arms such glorious fruit for the cause he represented, that, in the light of all that failure cost us, I still hold from a soldier's point of view that the effort was justified by the prospect.

Our commander had reason to believe, which afterwards turned out to be true, that he had out-maneuvered Meade, and that his full concentration was confronted by only a portion of the latter's army. This was a situation which offensive operations alone could utilize. Whether the subsequent engagement was fought as he designed, it is a question which I believe will be answered by history in an emphatic negative. At least, the assaults in detail by fragments of corps, when whole divisions lay idle in our lines, bore no resemblance to any other attack delivered by Lee before or afterwards – for Malvern Hill, where Jackson was misled by his guides, and where D. H. Hill precipitated the action by misinterpretation of a signal, does not offer a proper basis of comparison.

Generally the instinct of an army may be trusted to adjudge responsibility for its reverses, after the event. In the case in hand there was no diminution in the affection or confidence of the army of Northern Virginia in its commander. Even the remnants of the brave divisions which gained the heights in vain, found voice when reeling back in bloody disarray, to give him greeting, and though he then and there avowed the blame with generous disregard of self, it was only as if he had said, "You were not at fault, you that came back from the heroic effort, or those whose bodies dot that deadly slope; you did all that human bravery could do."

The army took his grave, kind words as meaning that – no more nor less. Nor do I think at this late day the survivors will accept a version that would stamp their beloved leader as self-convicted of the blunders, or worse, of that ill-starred 3rd of July.

Illustrating Lee's offensive strategy is the movement by which, in the autumn of 1863, he flanked Meade out of his position at Culpeper, and forced him back into the lines at Centreville, and this, too, though his army had been depleted one-third by the dispatch of Longstreet to the west.

And when in December, Meade crossed the Rapidan and established himself across the roads leading from Orange Courthouse to Fredericksburg, not a step in retrograde did the Southern General take. He accepted the challenge from a superior force, marched promptly out with the corps of Ewell and Hill, planted himself on the ridges over Mine Run, and offered battle for two whole days. On the night of the third, he massed two divisions on his right to assault the left flank of the enemy, but in the morning, an advance in the gray light found only empty trenches.

The same movement essentially was repeated in the following spring when Grant came southward of the river. Here again, instead of retiring behind the North Anna as his antagonist presumed, Lee barred the path of invasion in the old battlefields of the Wilderness, and on the 6th of May, became the assailant after a vigorous fashion. Thereafter our commander proved the subordination of his temperament to his judgment by compelling battle from time to time on his own ground, giving his troops the advantage also of entrenchments.

If his military reputation should rest on this campaign alone, from the initial gun at the Wilderness to the passage of Grant's army to the south side of the James, Lee would deserve to rank among the few past masters in the art of war. From day to day he divined the movements of the enemy with an accuracy which was never at fault. At every successive point – Spotsylvania, Hanover Courthouse, Cold Harbor – Grant found his pathway barred by the grim veterans in gray.

Time and time again, exasperated by the consummate skill with which prompt check was given to his every maneuver, the Federal commander threw his bare breasted divisions against the works of Lee. As often the brave fellows recoiled with torn ranks from the desperate work, until at last, after the bloodiest of all bloody days, that at Cold Harbor, the bugles sounded the advance, the officers bared their swords and pointed the way, but the men with one accord stood motionless in their ranks – a silent, but effective protest against a further application of the policy of attrition.

On the 14th of June, the advance corps of the Army of the Potomac reached the pontoon bridge which was to bear them to the new scene of action at Petersburg. Since the 5th of May, their losses in killed, wounded and missing, according to the official returns of the Federal Surgeon-General, had been 67,000 — or 3,000 more than the number of men with which Lee had entered upon the campaign. Up to this time, including Smith's corps, Grant had received in reinforcements 51,000 muskets, Lee 14,000. These statistics are pregnant with testimony as to the skill of our commander and the efficient valor of his troops.

But the end was not yet. Once in front of the historic town on the Appomattox – where for the first and only time in the game of strategy, the Federal general fairly stole a march upon his opponent – but where Beauregard with a brilliant audacity, not yet sufficiently recognized, defended the position against great odds until the lost time was repaired – the situation seemed to Grant or Meade to justify a renewal of those clashes of solid lines upon well-manned earthworks to which the Federal army had already sacrificed so many lives and so much morale.

The result was disastrous as usual, and again the army and Northern public murmured at what they deemed a reckless expenditure of blood. And then the taciturn and persistent Union commander announced in general orders that no more assaults upon entrenched lines would be made. The engineers were brought up, the great guns were sent for, and the siege of Petersburg was set on foot.

> *That to decree the latter was the acceptance of a bitterness worse than death to the brave spirit upon whom the responsibility rested is only to say that he was a soldier and a Lee.*

The operations progressed with varying fortunes through the months of summer and autumn. Gradually the clasp of the besiegers grew closer and closer around the beleaguered army. There were some days of great glory for the Confederates. Longstreet held the north shore and the approaches to Richmond with a grip not to be shaken. Mahone and his division won fame in no scant measure at the Crater and on the Weldon road. Heth and Hampton broke through Hancock's ranks at Reams' Station and captured many prisoners, colors, and guns. The cavalry wrought wonders on the flanks. But further and further westward crept that fateful left flank of the Federal army. It was badly punished in each extension, but every inch of ground that General Warren gained he held.

Dark days were upon us. The shadow of the inevitable was beginning to obscure the bow of hope. It was as the winter fell that I first observed the deepened lines of care that not all the serenity of a soul at peace with God and itself could smooth from the countenance of General Lee. The raven hair of four years before was already bleached into silver, and though too thorough a gentleman to betray abstraction, his speech, except on business, was rare.

In fact, at this period the perils and privations of the troops were never absent from his thought. So patient of privation himself, he was indignant at what he believed to be the neglect of the supply department in furnishing clothing and provisions to the men. The Secretary of War made petulant inquiry of the General as to the cause of such frequent desertions from the ranks. His curt endorsement, amply justified by the facts, evinced his grave displeasure. "I suppose the causes to be the lack of food, fuel and clothing, and constant duty in the trenches."

As the winter waned his perplexities were redoubled. True, the wonderful resources of his genius, the magnetic influence which tied men to him as with links of steel – the influence of his goodness as well as his greatness – and the elastic vitality of his army had sufficed so far to hold intact the works around Petersburg and Richmond, and to preserve insecure communication between these positions and their nearer bases of supplies; but in other sections of the country reverse after reverse had overtaken the Southern arms.

For now the tale of ravaged lands and the wails of suffering wife and children – for Sherman's triumphal progress left desolation in its wake – come on the southern breeze to men whose cup of ills had already overflowed. There is – must be – some boundary to endurance, on touching which the staunchest heart must sue for truce.

For so it was. Night by night brought darkness, and each recurring morning showed the vacant places of some who dreamed of ruined homes and unprotected dear ones, and waked to yield to an unconquerable yearning to fly to their relief. And thus one enemy, so long repelled with scorn, had gained a foothold in our camp at last.

It has been said that Washington and Lee had kinship of most of the sublimest qualities of manhood but differed in fortune. I can picture to myself how the former bore himself during the trials of Valley Forge, by recalling the demeanor of Lee during that last terrible winter at Petersburg. Almost without hope; hampered by conditions over which he had no control; overwrought with duties not attaching to his position; denied by the narrow blindness of the government the only avenue of escape which remained to him; his heart bleeding for the sufferings of his faithful followers, and yearning more in sorrow than anger for those who found not the strength to endure to the end – yet was he patient, always striving, inventing *that* makeshift, urging *this* experiment, encouraging the officers, knocking constantly at the door of the government to better the condition of the men, stifling his own forebodings, careless of his own discomforts: the heart, the brain, the eyes of that brave, beset and beleaguered body of starving men.

He had a burden to bear which his great prototype was never called on to endure. Already he had reported to the War Department that except on certain conditions (which the Commissary General had declared to be impossible of fulfillment), he could neither hold his lines nor remove the army in safety from them. There remained for him the most exacting ordeal that can confront the commander of any army – to determine without reference to his feelings where the point of military honor ceases and where the duty to humanity begins – what protraction of a hopeless condition is justifiable. He must fight until the verdict of fate was plainly beyond his power to affect it. He must not anticipate that juncture, nor must he protract the struggle one hour beyond it.

When the time arrived for the rendering of that decision, General Lee was equal to it. Through no fault of his the retreat, begun, as he knew too late was interrupted by the fatal miscarriage of provisions ordered to meet the army en route. The delay so caused brought Meade upon his rear, and enabled Sheridan's hard riders to reach his flank. The disaster at Sailor's Creek, conclusive in its dimensions, brought the army, two days later, face to face with annihilation or surrender. That to decree the latter was the acceptance of a bitterness worse than death to the brave spirit upon whom the responsibility rested, is only to say that he was a soldier and a Lee.

But he met the crisis as he met all other demands upon his conscience – simply, promptly, and with a mien as calm as his soul was lofty. That he would have worn the crown of success without elation is as certain as that he rose superior to defeat. He never knew ambition in its vulgar sense.

He came not back, when his stainless sword was sheathed, to triumphal processions, civic honors, and ceremonial pomp. But the tears of the rugged soldiers who gathered around his horse at Appomattox and invoked the blessings of heaven on his honored head, was a

> *Patiently instilling the lessons of virtue into the minds of the Southern youth, presiding at the vestry meetings of his church, foremost in unheralded charities, so passed the few years that remained on earth to Robert E. Lee.*

tribute as precious as was ever offered at the shrine of human greatness.

Whether posterity will assign to General Lee the rank as a commander which the South claims for him is a question which need not be discussed here. The judgment of foreign critics of this generation places him high in the list of the born leaders of men. That he accomplished much with limited resources, that he elicited the best skill and valor of the Union by his persistent defense of Virginia, that he overmatched many generals and decimated several armies before his own succumbed, and that he finally gave to the victor a costly triumph, are facts not to be gainsaid.

He fought for the cause of his conscience until further contest would have been a useless and criminal sacrifice of life. He surrendered in good faith to a generous foe, and thereafter gave his example to the building up of substantial peace and a real Union. He laid aside his stainless sword as bravely as he had drawn it, and without repining for the past he turned to the duties of the present. Patiently instilling the lessons of virtue into the minds of the Southern youth, presiding at the vestry meetings of his church, foremost in unheralded charities, so passed the few years that remained on earth to Robert Lee.

He lived among us, to all appearances, absorbed and contented in the routine of educational work. If he repined under failure, he gave no sign. If he found the utter revolution in his life irksome to the spirit once "wrapped in high emprise," he uttered no complaint. If he felt anxiety as to the judgment of posterity upon his military career, he made no effort to place the records in evidence. In the controversial disputes among others of our military chieftains, which sprung up from the ashes of defeat as weeds from the wreck of some proud edifice, he took no part. He seemed to be content to leave his character and services in the keeping of his countrymen without a word of his own to prejudice their judgment.

It should also be recorded that he never spoke nor wrote a word which would prolong the bitterness of our ended strife, or reawaken sectional animosity. He seemed to have put the past behind him. It was only at the last when his mind wandered that the stirring memories of the old days triumphed over that strong will and asserted a momentary sway. The warrior in him awoke for one brief instant before the light of eternal peace cast all earth into shadow. "Bring up the troops," he said, "Let A.P. Hill prepare for action."

And so he passed away. And the world was poorer for his death.

STONEWALL JACKSON – SOLDIER

By Reverend Dr. J. Williams Jones. *Southern Historical Society Papers, Volume 35*, pages 79-98.

I used to hear the cadets of the Virginia Military Institute speak of a grim professor whom they called Old Jack, who was very eccentric and upon whom they delighted to play all sorts of pranks. Stories were told of his having greatly distinguished himself when serving in the regular army in the Mexican War, and of his steady promotion for gallantry and meritorious conduct from brevet second lieutenant to brevet major. But this gallant record had been overlooked or forgotten in the odd stories that were told of his conduct at the Institute, and when Governor Letcher, his neighbor and friend, nominated him as colonel in the Virginia volunteers in May, 1861, there was very general surprise, and many expressions of regret, especially among old cadets and people about Lexington who knew him. When his confirmation by the Virginia Convention was under consideration, a member arose and inquired, "Who is this Major Jackson anyway? And what are his qualifications for this important position?" It required all of the powers of the Lexington delegation and the influence of Governor Letcher to secure his confirmation by the convention.

He was soon sent to the command of Harper's Ferry, then popularly regarded as one of the strongholds of the Confederacy, and those of us who were stationed there eagerly inquired, "What is this newly made colonel?"

Some of the Lexington soldiers, and some of the old cadets, sneered at his appointment; made all manner of fun of him, and told various anecdotes of his career at the Virginia Military Institute to disparage him. I remember one of them said to me: "Governor Letcher has made a great mistake in promoting Old Jack. He is no soldier. If he wanted a real soldier, why did he not give the place to Major – "mentioning the name of a worthy gentleman, who afterwards served in the army, but made no reputation as a soldier.

But when Old Jack took command, we were soon made to see the difference between his rule and that of certain militia officers who had been commanding us, and were made to feel and know that a real soldier was now at our head. He soon reduced the high-

> *He will be known in history not by the name Thomas Jonathan Jackson, which his parents gave him, but as Stonewall Jackson. And yet the name was a misnomer. Thunderbolt, Tornado, or Cyclone would be more appropriate to Jackson's character as a soldier.*

spirited mob who rushed to the front at the first call of their native Virginia into the respectable Army of the Shenandoah, which he turned over to General Joseph E. Johnston when he came to take command of the department.

Jackson won some reputation in several skirmishes in the lower valley, and at this time very small affairs were magnified into brilliant victories.

Stonewall

But it was on the plains of first Manassas, July 21, 1861, that he first became famous.
General McDowell had ably and skillfully outgeneraled Beauregard, and crossing the upper fords of Bull Run, had moved down on the Confederate flank, driving before him the small Confederate force stationed there.

General Bee, in the agony of being driven back, galloped up to Jackson, who, in command of a Virginia brigade, was stationed on the Henry House Hill, and exclaimed: "General, they are beating us back!"

Jackson's eyes glittered beneath the rim of his old cadet cap as he almost fiercely replied: "Sir, we will not be beaten back. We will give them the bayonet."

Bee rushed to his own decimated ranks and rallied them by exclaiming: "Look! There stands Jackson like a stone wall! Rally on the Virginians! Let us determine to die here, and we will conquer!"

Jackson not only stood the shock of the heavy attack made on him, but did give them the bayonet, checked the onward tide of McDowell's victory, and held his position until Kirby Smith and Early came up on the flank. Jeb Stuart made a successful cavalry charge, Johnston and Beauregard had time to hurry up other troops, and a great Confederate victory was snatched from impending disaster.

The name which the gallant Bee, about to yield up his noble life, gave Jackson that day, clung to him ever afterwards, and he will be known in history not by the name Thomas Jonathan Jackson, which his parents gave him, but as Stonewall Jackson. And yet the name was a misnomer. Thunderbolt, Tornado or Cyclone would be more appropriate to Jackson's character as a soldier.

I cannot, within the proper limits of this paper, give even an outline of Jackson's subsequent career as a soldier – that would be to sketch the history of the Army of Northern Virginia, while he remained in it. But I propose rather to give and illustrate several salient points in his character as a soldier.

Rapidity of Movement

Nathan Bedford Forrest, the wizard of the saddle, when asked the secret of his wonderful success, replied: "I am there first with most men." Stonewall Jackson always got there first, and while his force was always inferior in numbers to the enemy, he not infrequently had the most men at the point of contact.

When General Banks reported that Jackson was in full retreat up the Valley, started a column to join McClellan east of the Blue Ridge, and was on his own way to report at Washington, Jackson (on a mistaken report of the number left in the Valley) suddenly wheeled, made a rapid march and struck at Kernstown a blow, which, while the only defeat he ever sustained, brought back the column which was crossing the mountains, and disarranged McClellan's plan of campaign.

He then moved up the Valley, took a strong position in Swift Run Gap, and after Ewell's Division joined him, he left Ewell to watch Banks, made a rapid march to unite with Edward Johnson, and sent (May the 9th) his famous dispatch: "God blessed our arms with victory at McDowell yesterday."

Ordering Ewell to join him at Luray, he pushed down the Valley, drove in Bank's flank at Front Royal, cut his retreating column at Middletown, marched all night by the light of the burning wagons of the enemy, and early the next morning drove Banks from Winchester and pursued him to the Potomac.

Learning that Shields, from McDowell's column at Fredericksburg, and Fremont, from the west, were hurrying to form a junction in his rear, he marched his old brigade thirty-five miles, and one of the regiments, the 2nd Virginia, forty-two miles a day, and safely passed the point of danger at Strasburg, carrying his immense wagon train loaded with captured stores, his prisoners and everything, not leaving behind so much as a broken wagon wheel. He then moved leisurely up the Valley until at Cross Keys and Port Republic he suffered himself to be caught, and proved beyond question that the man who caught Stonewall Jackson had indeed caught a Tartar.

> *"Look! There stands Jackson like a stone wall! Rally on the Virginians! Let us determine to die here, and we will conquer!"*

Penchant for Secrecy

Not long after the close of the Valley Campaign, when we were resting in the beautiful region around Port Republic, I got a short furlough to go to Nelson County to see my family and uncle. Colonel John Marshall Jones, Ewell's Chief of Staff, told me that if I would come by headquarters, he would ride with me as far as Staunton. Accordingly, I rode by Ewell's headquarters, and just before we left the grounds, General Ewell came out and said to us in a confidential tone: "If you gentlemen wish to stay a little longer than your leave it will make no difference, we are going to move down the Valley to beat up Banks' quarters again."

I did not overstay my brief furlough, for I was hurrying back in hope that our rest near Port Republic would give the chaplains especially good opportunities for preaching to the men, but when I reached Charlottesville, I found Jackson's troops marching through the town.

Jackson was anxious to be reinforced and move down the Valley again, but General Lee wrote him, "I would be glad for you to make that move, and will give you needed reinforcements; but you must first come down here and help me drive these people from before Richmond."

Reinforcements were sent Jackson, and pains taken to let the enemy know, and Jackson so completely deceived them as to his plans that at the time he was thundering on McClellan's flank before Richmond, they were entrenching at

Strasburg, some two hundred miles away, against an expected attack from him.

I remember that on this march we were in profound ignorance as to our destination. At Charlottesville, we expected to move into Madison County. At Gordonsville, we expected to move towards Washington. At Louisa, we expected to move on to Fredericksburg. At Hanover Junction, we expected to move up the railway to meet McDowell's Column. It was only on the afternoon of June 26th, when we heard A. P. Hill's guns at Mechanicsville that we fully realized where we were going.

In the second Manassas campaign, Jackson conducted his movements to Pope's flank and rear so secretly that just before he captured Manassas Junction, with its immense stores, Pope reported to Washington that Jackson was in full retreat to the mountains.

So at Chancellorsville he moved to Hooker's flank and rear so secretly that he struck Howard's corps entirely unprepared for his attack.

My accomplished friend, Reverend James Power Smith, D.D., the only surviving member of Jackson's staff, gave me an incident the other day illustrating how he concealed his plans from even his staff.

After the return of Lee from the first Maryland campaign, Jackson and his corps were left for a time in the Valley, while the rest of the army crossed the mountains to Eastern Virginia.
After lingering around Winchester for a time, Jackson's whole command was moved one day on Berryville, and it seemed very evident that they were about to ford the Shenandoah, and cross the mountains to join Lee.

Captain Smith went to his general and said: "As we are going to cross the mountains, general, I should like very much to ride back to Winchester to attend to some matters of importance to me personally, if you can give me a permit."

"Certainly I will give you the permit," was the reply, "and if we cross the mountains, you will be able to overtake us tomorrow."

Captain Smith rode into Winchester, and started early the next morning to overtake, as he supposed, the moving column. He had only ridden several miles when he met Jackson at the head of his corps moving back to Winchester and was greeted by the salutation, "I suppose Mr. Smith that you are on your way to cross the mountains."

It was then currently believed that Jackson would spend the winter in the Valley with headquarters at Winchester and a vacant house was selected for the general and his staff. After a day or two, Captain Smith and Colonel Pendleton, as a committee of the staff, waited on the general, and said: "As it is understood that we are to spend the winter here, we called to ask permission to get some necessary furniture."

"That would add very much to our comfort, but I think we had better wait until tomorrow, and decide definitely on what we need," was the reply. The next day Jackson started on his famous march to join Lee in time for the battle of Fredericksburg.

Stern Discipline

He put General Garnett under arrest at Kernstown for ordering a retreat of his brigade when they were out of ammunition and almost surrounded, saying, "He ought to have held his position with the bayonet."

Garnett was still under arrest when Jackson died. General Lee released him and put him in command of one of Pickett's Brigades, the gallant gentleman being killed in the charge at Gettysburg, while leading his men.

On the Valley campaign I chanced to witness a scene in which Jackson rode up to a gallant colonel, commanding a brigade, and said: "Colonel, the orders were for you to move in the rear of General – today.'

The colonel replied in a rather rollicking tone: 'Yes, I knew that General, but my fellows were ready to march and General – was not. I thought that it would make no difference which moved first, as we are not going to fight today. But if you prefer it, I can halt my brigade, and let General – pass us." Jackson replied, almost fiercely: 'How do you know that we are not going to fight today? Besides, colonel, I want you to distinctly understand that you must obey my orders first and reason about them afterwards. Consider yourself under arrest, sir, and march in the rear of your brigade."

In one of his battles, a brigadier rode up to him and asked: "General, did you order me to move my brigade across that plain, and charge that battery?" "Yes, sir, I sent you that order," said Jackson, "Have you obeyed it?"

"Why, no! General, the enemy's artillery will sweep that field, and my brigade would be literally annihilated if I move across it."

Jackson replied in tones not to be mistaken: "General, I always try to take care of my wounded and bury my dead. Obey that order, sir, and do it at once."

It is needless to add that the order was obeyed, and the battery captured.

> *"Have that horse shod immediately, or there will come an order down here from Old Jack wanting to know why the gray mare is allowed to go with a shoe off of her left hind foot."*

At one time he put every commander of a battery in A. P. Hill's Light Division under arrest for some slight disobedience of orders.

He put A. P. Hill under arrest several times, and there were charges and countercharges between these accomplished soldiers until General Lee intervened to effect a compromise.

But Jackson was always ready to obey himself orders from his superiors. General Lee once said of him: "I have only to intimate to him what I wish done, and he promptly obeys my wishes."

Dr. James Power Smith gives a striking incident illustrating this: General Lee sent Jackson, by Captain Smith, a message to the effect that he would be glad if he would call at his headquarters the first time he rode in that direction, but that it was a matter of no pressing importance, and he must not trouble himself about it.

When Jackson received this message he said: "I will go early in the morning, Captain Smith, I wish you to go with me."

The next morning when Captain Smith looked out he saw that a fearful snowstorm was raging, and took it for granted that Jackson would not undertake to ride fourteen miles to General Lee's quarters through that blizzard.

Very soon, however, Captain Smith's servant came to say, "The general done got his breakfast, and is almost ready to start."

Hurrying his preparations, the young aide galloped after his chief through the raging storm. On reaching Lee's quarters, the general greeted him with, "Why, what is the matter, general; have those people crossed the river again?"

"No, sir; but you sent me word that you wished to see me."

"But I hope that Captain Smith told you that I said it was not a matter of pressing importance, and that you must not trouble yourself about it. I had no idea of your coming in such weather as this."

Bowing his head, Jackson gave the emphatic reply: "General Lee's slightest wish is a supreme order to me, and I always try to obey it promptly."

Attention to Minute Details

He had an interview with his quartermaster, commissary, chief of ordnance, and surgeon-general every day, and kept minutely posted as to the condition of their departments. This was so well understood throughout the army, that I once heard a quartermaster say to his sergeant: "Have that horse shod immediately, or there will come an order down here from Old Jack wanting to know why the gray mare is allowed to go with a shoe off of her left hind foot."

He kept the most minute knowledge of the topography of the country in which he was campaigning and the roads over which he might move, and often when his men were asleep in their bivouac, he was riding to and fro inspecting the country and the roads.

But when he began to ask me which side of certain creeks was the highest, and whether there was not a blind road turning off at this point or that, and showed the most perfect familiarity with the country and the roads, I had to interrupt him by saying: "Excuse me, General, I thought I knew not only every road, but every footpath in that region, but I find that you really know more about them than I do, and I can give you no information that would be valuable to you." I can never forget another interview I had with him on the Second Manassas campaign. His corps had crossed the South Fork of the Rappahannock River, General Ewell's Division had been formed on the bank of the North Fork, and the rest of the corps were marching up between the two rivers to Warrenton White Sulphur Springs, where it was General Lee's purpose to cross his whole army and plant it in General Pope's rear at Warrenton. In bringing a wounded man of my regiment – the 13th Virginia – back from Ewell's Division to our surgeon, and returning, I saw a skirmish line of the boys in Blue who had crossed at the forks of the river below and were moving up in General Ewell's rear between him and the moving column of

Hill's Division. I waited to satisfy myself that they were real Blue Coats, and becoming fully satisfied by their firing at me, one of the bullets cutting off the extreme end of my horse's ear, I had, of course, important business elsewhere and was galloping to find General Hill, who commanded that part of our column, when I ran up against old Stonewall himself. I approached him, trying to be as calm as possible, and the following colloquy ensued: "General, are you aware that the enemy have crossed at the forks of the river and are now moving up in the rear of General Ewell, and between him and A. P. Hill's column?"

"No! Have they?"

"Yes, sir, I have seen them."

"Are you certain they are the enemy?"

"Yes, sir, I am."

"How close did you get to them?"

"I suppose about 1,000 yards. I could plainly see their blue uniforms and the United States flag which they carried. They shot at me and cut the ear of my horse, as you see, and then I got away from there as fast as my horse would bring me."

I expected that he would now send staff officers in every direction with orders to meet this new movement, but Jackson coolly replied: "1 am very much obliged to you, sir, for the information you have given me, but General Trimble will attend to them. I expected this movement, and ordered Trimble posted there to meet it."

> *Fitz Lee facetiously said that Hooker was in imminent peril when the "Blue-light Presbyterian" was praying on his flank and rear.*

He rode off, seemingly as unconcerned as if nothing had happened. Trimble did attend to them, and after a severe fight drove them back.

General Lee was prevented by a sudden rise of the river from a severe storm from crossing at Warrenton White Sulphur Springs, but the next day Jackson forded the river higher up and made his famous movement to Pope's flank and rear.

Turning to one of his staff, he said: "Gallop as hard as you can and tell Major Andrews to bring sixteen guns to bear on that battery and silence it immediately."

Soon Andrews was in position. His guns opened and before long the battery was silenced. When this was reported to Jackson, he said, with a quiet smile: "Now, tell General Ewell to drive them."

In the afternoon at Gaines Mill, June 27th, 1862, the progress seemed not to have been as rapid as he expected, as gallant Fitz John Porter made a heroic defense, and Jackson exclaimed to one of his staff: "This thing has hung fire too long; go rapidly to every brigade commander in my corps and tell him that if the enemy stands at sundown he must advance his brigade regardless of others, and sweep the field with the bayonet."

It was this order that won the day despite the gallant defense. I chanced to be near and heard the order he gave General Early at Cedar Run (Slaughter's Mountain) in the fight with our old friend, General Banks (Stonewall Jackson's quartermaster, our men facetiously called him), who commanded the advance of General Pope's Army. We had been skirmishing all of the morning, and Colonel Pendleton, of Jackson's staff, rode up to General Early and said quietly: "General Jackson's compliments to General Early, and says that he must advance on the enemy, and he will be supported by General Winder."

Grim old Early replied in his curtest tones: "Give my compliments to General Jackson, and tell him I will do it."

It was on this field that several of Jackson's Brigades were broken, and it looked as if Banks was about to win, when Jackson dashed in among them, and rallied the confused ranks by exclaiming, "Rally on your colors and let your general lead you to victory. Jackson will lead you." His presence acted like magic, the broken troops were rallied, the lines restored and the victory won.

At Chancellorsville, Fitz Lee discovered that from a certain hill a full view of Hooker's flank and rear could be seen. He galloped back until he met Jackson and conducted him to the spot, accompanied by a single courier. Jackson swept the scene with his glasses, decided at once that he should move further on the flank and rear than he had intended, and turning to his courier said: "Tell the head of my column to cross that road, and I'll meet them there."

Fitz Lee said that he made no reply to his remarks, but after gazing intently for a few moments longer at the enemy's exposed flank, he lifted his hand in that position which indicated, that he was engaged in prayer and then galloped rapidly down the hill to hurl his column like a thunderbolt on Hooker's flank and rear.

Fitz Lee facetiously said that Hooker was in imminent peril when the "Blue-light Presbyterian" was praying on his flank and rear.

Lee called Jackson his right arm, and wrote him when he was wounded at Chancellorsville:
"Could I have dictated events I should have chosen, for the good of the country, to have been disabled in your stead."

I had the privilege once of hearing General Lee in his office in Lexington pronounce a glowing eulogy on Jackson, in which he said with far more than his accustomed warmth of feeling: "He never failed me. Why, if I had had Stonewall Jackson at Gettysburg, I should have won that battle; and if I had won a decided victory there we would have established the independence of the Confederacy."

They were born, Lee on the 19th of January, and Jackson on the 21st of the same month. Cavalier and Puritan, but brothers in arms, brothers in faith, and brothers in glory, they will shine forever in the world's galaxy of true patriotism, stainless gentlemen, great soldiers and model Christians.

From all parts of the world pilgrims come to visit their tombs, and loving hands bring them fresh flowers, immortelles and evergreens, fit emblems of the fadeless wreaths which now deck their brows. The blue mountains of their loved Virginia sentinel their graves, and young men from every section throng the classic shades of Washington and Lee University and the Virginia Military Institute and delight to keep watch and ward at which flow along their emerald streams that seem to murmur their praise and roll on their fame to the ocean.

To subscribe, send an email to:
thestainlessbanner@gmail.com
or visit our website at www.thestainlessbanner.com

Subscription is free.

The Stainless Banner

An e-zine dedicated to the armies of the Confederacy

Volume 3, Issue 2
February 2012

THE SECOND DOMINO FALLS!
THE LOSS OF FORT DONELSON

Albert Sidney Johnston's immediate need was for more troops and weapons, but his urgent telegrams to Richmond had netted neither. What they did net, however, was the victor of Manassas – Pierre Gustave Toutant Beauregard. By sending Beauregard west, Richmond hoped his reputation would lift the morale of the troops and rally those men who had yet to join the colors.

Jefferson Davis' motives in sending Beauregard had less to do with Johnston's need than Davis' intense desire to rid himself of an unwanted burden. Since his victory at Manassas, Old Bory had become a political pain in the president's neck.

Beauregard had made numerous complaints about the lack of supplies flowing from Richmond to the front. Commissary General Lucius Northrop was a good friend of Davis, and the president did not like to hear criticism about his friends. Yet, for the good of the Cause, Davis swallowed his resentment and responded to Beauregard's complaints in a conciliatory manner.

Tensions mounted after Beauregard submitted his official report on the battle. In the report, he accused Davis of not allowing him to capture Baltimore and Washington before the Federals marched from the capital. There was some truth to the accusation. Beauregard was always full of plans to bring the war to an end in one fell swoop. But most of these plans were outlandish, pie-in-the-sky strategies, with no hope of succeeding. Davis would

What's Inside:	
Floyd's Official Report	6
Buckner's Official Report	11
Forrest's Statement	12
The Fall of Fort Donelson	13
River Battles	23
Nathan Bedford Forrest	39
Land Batteries versus Gunboats	41
Why We Fought	48

www.thestainlessbanner.com

have been derelict in his duty if he had not said no.

Beauregard sent the report to the War Department, but he also released a summary to the newspapers. The president dashed off a terse letter accurately pinpointing Beauregard's motives in publicly releasing the report. Beauregard did not have a defense, but, all the same, he resented the president's keen observations about his self-centered motivations.

Beauregard escalated the fight by mailing his reply, not to the president, but to a Richmond newspaper. With a melodramatic flair, he headed his letter: *Centreville – within earshot of enemy guns*. The letter was nothing more than a thinly veiled rant against Davis.

Beauregard then unleashed an all out assault on Davis. He sent letters to newspaper editors and politicians. Each one filled with more and more ridiculous strategies to end the war. After a congressman received a letter, he would hurry to the floor of the Congress and read it into the record as a perfectly legitimate way to win the war.

Davis had had enough. It was time to get Beauregard and his letter-writing campaign as far from Richmond as possible.

But Beauregard did not want to go any where. He was more than happy holding court at the front. So, he sat down and wrote one more letter to an influential congressman and laid down the law. He would go to Nashville, but under one condition: Once he achieved victory in the west, he would be recalled to Virginia so he could win the war in the east also. The congressman assured him that he would bring the terms to Davis, but he never did. Beauregard, none the wiser, set off for Tennessee.

His arrival had a ripple effect throughout the theater. Generals Halleck and Grant saw it as a harbinger of an upcoming Confederate offensive. Grant had been chomping at the bit to be allowed to attack both Forts Henry and Donelson. Halleck, worried about Beauregard's presence, gave his consent.

> *The loss of Nashville would be devastating. The river city was one of the two main supply depots in the South, and, more importantly, an industrial center that produced gun powder and processed food rations.*

On February 6th, Flag Officer Foote's gunboats silenced Confederate cannon at Fort Henry.

Johnston received the news with great alarm. If Union gunboats were really that powerful and effective, then Fort Donelson would not be able to stand. And when Donelson fell, the gunboats would be able to sail up the Cumberland River and into Nashville.

The loss of Nashville would be devastating. The river city was one of the two main supply depots in the South and, more importantly, an industrial center that produced gun powder and processed food rations. The loss of the Cumberland would trap Johnston's army north of the river, leaving the Deep South vulnerable to attack.

Johnston decided to make his stand at Donelson. He sent 12,000 reinforcements to the fort. With the reinforcements, he also sent orders for Donelson to be evacuated when it could no longer be held. Johnston believed that Union gunboats were more than capable of taking the fort, but he needed time to get his army over the Cumberland.

Johnston sent Beauregard to take command of the western wing of the army now cut off by the capture of Fort Henry and the loss of the Tennessee River. Word was sent to Polk to abandon his prized fortress city and head south. Johnston planned to delay the Federal gunboats along the Mississippi at New Madrid, Island Number 10, Fort Pillow, and at Memphis. The rest of the army would head south as soon as possible.

Richmond suddenly woke to the peril their policies had wrought. Orders were sent to the coastal defenses to get the men up and on the road to Fort Donelson. It was too little too late. Before the troops could reach Donelson, the fort would fall and the Cumberland River would be in Federal hands.

The Battle for the Fort

After General Tilghman was captured at Fort Henry, command at Donelson devolved upon Simon Buckner – a good and competent soldier. Johnston's reinforcements brought Gideon Pillow and John Floyd to the fort. It could not have been a more inept combination.

Pillow was such a bad soldier that Davis had felt it necessary to send Polk to the theater to babysit him. But Floyd was the worst. The former United States Secretary of War was a political appointment and his lack of military experience was evident. Proven to be totally incompetent in the campaign in western Virginia (not even Robert E. Lee could overcome Floyd's shortcoming), Davis unloaded him and sent him west, with no idea that Floyd would wind up in command of such vital ground. Floyd had seniority, so he was in command with Pillow next in charge, then Buckner.

On February 13, Federal infantry arrived at Fort Donelson. Grant placed his soldiers around the earthworks and waited for the gunboats to begin their attack the next morning. Grant's strategy was simple. The gunboats would bombard the fort into submission; then he would capture the garrison.

The moment the gunboats opened fire, Floyd wilted. He sent a frantic telegram to Johnston prophesying doom and despair. But Donelson was not Henry, and Confederate and Union soldiers learned this truth together: Land batteries can defend a well established defensive position against naval forces. The Union gunboats received the worse end of the beating.

The sight of gunboats floating helplessly down the river stiffened Floyd's spine. He sent another telegram to Johnston proclaiming a great victory.

Unfortunately, when Grant had come up, he had been allowed to surround the fort without a fight, trapping the Confederates inside. Floyd ordered a dawn attack to drive the Federals back and open up an escape route.

The next day's fighting succeeded and several escape routes were opened. Buckner hurried to take advantage, but Pillow ordered Buckner's men back into the fort. Buckner protested such an idiotic move. Floyd happened on the scene and Buckner informed him about Pillow's order. Floyd agreed with Buckner. The fort would be evacuated. Then Pillow had his say. He stressed to Floyd that the men were tired and hungry and, besides, their blankets were in the fort and the weather had turned freezing. Just like the politician he was, Floyd flipped-flopped and ordered Buckner into the fort.

A war council was called at the Old Dover Inn. Pillow arrived first and entertained Floyd with stories of Buckner's incompetence. Buckner arrived and his mood was pessimistic. He informed the two generals that the time had come to surrender. Suddenly, Pillow was full of fight and told Buckner that the men could hold against the Federals for a couple of days. This would give the Confederates time to figure another way out of the Federal trap.

Any suggestion Pillow made was tainted in Buckner's eyes. They had served together in Mexico and Buckner thought Pillow a fool then. Nothing in the last few days had changed his opinion. Despite Pillow's grandiose claims, Buckner insisted that his men could not hold, and any attempt to break out of Grant's trap would result in seventy-five percent casualties. The only right course of action would be to surrender the fort.

Colonel Nathan Bedford Forrest was present at the meeting, and Forrest was not the surrendering type. He had been fighting against Grant's advance all day. When he arrived at the inn, the three generals informed him that Grant had landed eleven boatloads of reinforcements. Not only that, but the Federals had reoccupied the ground on the Confederate left. Forrest did not believe them. He had just come from that portion of the field, and it was firmly in Confederate hands.

> *Buckner found Grant's terms of unconditional surrender less than chivalrous. After all, Grant had been penniless when he arrived in New York after resigning from the army due to rumors of drunkenness. Buckner had lent Grant the money to get home to Missouri. Unconditional surrender was no way to repay the loan, and he told Grant so.*

Forrest sent two scouts to check out the report. The scouts crawled on their bellies through freezing backwaters to where they could see the campfires of the Union pickets. They returned and told Forrest that they had not encountered the enemy, and the picket fires were in the same location as the evening before.

When the generals insisted that escape was impossible, Forrest volunteered to place his cavalry at any point in the line and cover the retreat or even to cut his way out if necessary. He left the room to let the generals deliberate. He returned and was informed that the garrison would surrender.

A defiant Forrest told them that he would not surrender his command under any circumstance. Furthermore, he would take any soldier who wanted to leave with him. Forrest was given permission to try.

Floyd was also not up for surrendering either. He was a wanted man in the North. As Secretary of War in the Buchanan administration, he had been accused (falsely) of transferring ordnance supplies to southern states in anticipation of the war. There was talk of hanging, and Floyd was not about to voluntarily hand himself over to the hangman. He turned command over to Pillow.

Pillow did not want command. This defeatist talk had not originated with him, and he was not going to surrender. He passed command to Buckner. Buckner took it and began to make plans to surrender the garrison.

Floyd wanted to leave Donelson as soon as possible. A steamship arrived in the morning; Floyd commandeered it, loaded his Virginian troops on board, and sailed out of Federal reach. He also sailed away from his Mississippi regiments, who were abandoned on the landing.

Pillow was the next to leave – in a rowboat.

Buckner raised the white flag and sought terms from his old friend Grant. He found Grant's terms of unconditional surrender less than chivalrous. After all, Grant had been penniless when he arrived in New York after resigning from the army due to rumors of drunkenness. Buckner had lent Grant the money to get home to Missouri. Unconditional surrender was no way to repay the loan, and he told Grant so.

Grant was unmoved by such sentiment. His terms remained the same. Buckner surrendered over 12,000 men.

The second domino had fallen. Nashville would have to be evacuated. Johnston's troops headed toward Mississippi to rendezvous at a steamboat stop called Pittsburgh Landing.

Boys, these people are talking about surrendering, and I am going out of this place before they do or bust hell wide open.

Nathan Bedford Forrest at Fort Donelson

OFFICIAL REPORT OF GENERAL FLOYD

CAMP NEAR MURFREESBOROUGH, TENNESSEE
February 27, 1862

SIR: Your order of the 12th of this month, transmitted to me from Bowling Green by telegraph to Cumberland City, reached me the same evening. It directed me to repair at once, with what force I could command, to the support of the garrison at Fort Donelson. I immediately prepared for my departure, and effected it in time to reach Fort Donelson the next morning (13th) before daylight. Measures had been already taken by Brigadier General Pillow, then in command, to render our resistance to the attack of the enemy as effectual as possible. He had, with activity and industry, pushed forward the defensive works towards completion.

These defenses consisted in an earthwork in Fort Donelson, in which were mounted guns of different calibers to the number of thirteen. A field work, intended for the infantry support, was constructed immediately behind the battery and upon the summit of the hill in rear. Sweeping away from this field work eastward, to the extent of nearly two miles in its windings, was a line of entrenchments, defended on the outside at some points with abatis. These entrenchments were occupied by the troops already there and by the addition of those which came upon the field with me.

The position of the fort, which was established by the Tennessee authorities, was by no means commanding, nor was the least military significance attached to the position. The entrenchments, afterwards hastily made, in many places were injudiciously constructed, because of the distance they were placed from the brow of the hill, subjecting the men to a heavy fire from the enemy's sharpshooters opposite as they advanced to or retired from the entrenchments.

Soon after my arrival the entrenchments were fully occupied from one end to the other, and just as the sun rose the cannonade from one of the enemy's gunboats announced the opening of the conflict, which was destined to continue for three days and nights. In a very short time the fire became general along our whole lines, and the enemy, who had already planted batteries at several points around the whole circuit of our entrenchments, as shown by a diagram herewith sent, opened a general and active fire from all

> *The position of the fort, which was established by the Tennessee authorities, was by no means commanding, nor was the least military significance attached to the position.*

arms upon our trenches, which continued until darkness put an end to the conflict.

They charged with uncommon spirit at several points along the line, but most particularly at a point undefended by entrenchments, down a hollow, which separated the right wing, under the command of Brigadier General Buckner, from the right of the center, commanded by Colonel Heiman. This charge was prosecuted with uncommon vigor, but was met with a determined spirit of resistance – a cool, deliberate courage – both by the troops of Brigadier General Buckner and Colonel Heiman, which drove the enemy, discomfited and cut to pieces, back upon the position he had assumed in the morning. Too high praise cannot be bestowed upon the battery of Captain Porter for their participation in the rout of the enemy in this assault. My position was immediately in front of the point of attack, and I was thus enabled to witness more distinctly the incidents of it.

The enemy continued their fire upon different parts of our entrenchments throughout the night, which deprived our men of any opportunity to sleep. We lay that night upon our arms in the trenches. We confidently expected at the dawn of day a more vigorous attack than ever; but in this we were entirely mistaken. The day advanced and no preparations seemed to be making for a general onset; but an extremely annoying fire was kept up from the enemy's sharpshooters throughout the whole length of the entrenchments from their long-range rifles. While this mode of attack was not attended with any consider able loss, it nevertheless confined the men to their trenches and prevented their taking their usual rest.

So stood the affairs of the field until about 3:00 p.m., when the fleet of gunboats in full force advanced upon the fort and opened fire. They advanced in the shape of a crescent, and kept up a constant and incessant fire for one hour and a half, which was replied to with uncommon spirit and vigor by the fort. Once the boats reached a point within a few hundred yards of the fort, at which time it was that three of their boats sustained serious injuries from our batteries and were compelled to fall back. The line was broken and the enemy discomfited on the water, giving up the fight entirely, which he never afterwards renewed.

I was satisfied from the incidents of the last two days that the enemy did not intend again to give us battle in our trenches. They had been fairly repulsed with very heavy slaughter upon every effort to storm our position, and it was but fair to infer that they would not again renew the unavailing attempt at our dislodgment when certain means to effect the same end without loss were perfectly at their command. We were aware of the fact that extremely heavy reinforcements had been continually arriving day and night for three days and nights, and I had no doubt whatever that their whole available force on the Western waters could and would be concentrated here if it was deemed necessary to reduce our position.

I had already seen the impossibility of holding out for any length of time with our inadequate numbers and indefensible position. There was no place within our entrenchments but could be reached by the enemy's artillery from their boats or their batteries.

It was but fair to infer that while they kept up a sufficient fire upon our entrenchments to keep our men from sleep and prevent repose, their object was merely to give time to pass a column above us on the river, both on the right and the left banks, and thus to cut off all our communication and to prevent the possibility of egress. I then saw clearly that but one course was left by which a rational hope could be entertained of saving the garrison or a part of it – that was to dislodge the enemy from his position on our left, and thus to pass our people into the open country lying southward towards Nashville.

> *The question then arose whether, in point of humanity and a sound military policy, a course should be adopted from which the probabilities were that the larger proportion of the command would be cut to pieces in an availing fight against overwhelming numbers.*

I called for a consultation of the officers of divisions and brigades to take place after dark, when this plan was laid before them, approved, and adopted, and at which it was determined to move from the trenches at an early hour on the next morning and attack the enemy in his positions.

It was agreed that the attack should commence upon our extreme left, and this duty was assigned Brigadier General Pillow, assisted by Brigadier General Johnson, having also under his command commanders of brigades Colonel Baldwin, commanding Mississippi and Tennessee troops, and Colonel Wharton and Colonel McCausland, commanding Virginians. To Brigadier General Buckner was assigned the duty of making the attack from near the center of our lines upon the enemy's forces upon the Wynn's Ferry road.

The attack on the left was delayed longer than I expected, and consequently, the enemy was found in position when our troops advanced. The attack, however, on our part was extremely spirited; and although the resistance of the enemy was obstinate, and their numbers far exceeded ours, our people succeeded in driving them, discomfited and terribly cut to pieces, from the entire left.

The Kentucky troops, under Brigadier General Buckner, advanced from their position behind the entrenchments upon the Wynn's Ferry road, but not until the enemy had been driven in a great measure from the position he occupied in the morning.

I had ordered on the night before that the two regiments stationed in Fort Donelson should occupy the trenches vacated by Brigadier General Buckner's forces, which, together with the men whom he detached to assist in this purpose, I thought sufficient to hold them. My intention was to hold with Brigadier General Buckner's command the Wynn's Ferry road, and thus to prevent the enemy during the night from occupying the position on our left which he occupied in the morning. I gave him orders upon the field to that effect. Leaving him in position, I started for the right of our command to see that all was secure there, my intention being, if things could be held in the condition that they then were, to move the whole army, if possible, to the open country lying southward beyond the Randolph Forge.

During my absence and from some misapprehension I presume of the previous order given, Brigadier General Pillow ordered Brigadier General Buckner to leave his position on the Wynn's Ferry road and to resume his place in his trenches on the right. This movement was nearly executed before I was aware of it. As the enemy was pressing upon the trenches, I deemed that the execution of this last order was all that was left to be done. The enemy, in fact, succeeded in occupying an angle of the trenches on the extreme right of Brigadier General Buckner's command; and, as the fresh forces of the enemy had begun already to move towards our left to occupy the position they held in the morning and as we had no force adequate to oppose their progress, we had to submit to the mortification of seeing the ground which we had won by such a severe conflict in the morning reoccupied by the enemy before midnight.

The enemy had been landing reinforcements throughout the day. His numbers had been augmented to eighty-three regiments. Our troops were completely exhausted by four days and nights of continued conflict. To renew it, with any hope of successful result was obviously vain, and such I understand to be the unanimous opinion of all the officers present at the council called to consider what was best to be done.

I thought and so announced that a desperate onset upon the right of the enemy's forces on the ground where we had attacked them in the morning might result in the extricating of a considerable proportion of the command from the position we were in, and this opinion I understood to be concurred in by all who were present; but it was likewise agreed, with the same unanimity, that it would result in the slaughter of nearly all who did not succeed in effecting their escape.

The question then arose whether, in point of humanity and a sound military policy, a course should be adopted from which the probabilities were that the larger proportion of the command would be cut to pieces in an unavailing fight against overwhelming numbers. I understood the general sentiment to be averse to the proposition, I felt that in this contingency while it might be questioned whether I should, as commander of the army, lead it to certain destruction in an unavailing fight, I had a right individually to determine that I would not survive a surrender there.

To satisfy both propositions, I agreed to hand over the command to Brigadier General Buckner through Brigadier General Pillow and to make an effort for my own extrication by any and every means that might present themselves to me. I therefore directed Colonel Forrest, a daring and determined officer, at the head of an efficient regiment of cavalry to be present for the purpose of accompanying me in what I supposed would be an effort to pass through the enemy's lines.

I announced the fact upon turning the command over to Brigadier General Buckner that I would bring away with me by any means I could command my own particular brigade, the propriety of which was acquiesced in on all hands. This by various modes I succeeded in accomplishing to a great extent and would have brought off my whole command in one way or another if I had had the assistance of the field officers who were absent from several of the regiments.

The command was turned over to Brigadier General Buckner, who at once opened negotiations with the enemy, which resulted in the surrender of the place.

Thus ended the conflict running through four days and four nights a large portion of which time it was maintained with the greatest fierceness and obstinacy in which we, with a force not exceeding 13,000, a large part of whom were ill armed, succeeded in resisting and driving back with discomfiture an army consisting of more than 50,000 men.

I have no means of accurately estimating the loss of the enemy. From what I saw upon the battlefield; from what I witnessed throughout the whole period of the conflict; from what I was able to learn from sources of information deemed by me worthy of credit, I have no doubt that the enemy's loss in killed and wounded reached a number beyond 5,000. Our own losses were extremely heavy, but for want of exact returns I am unable to state precise numbers. I think there will not be far from 1,500 killed and wounded.

Nothing could exceed the coolness and determined spirit of resistance which animated the men in this long and perilous conflict; nothing could exceed the determined courage which characterized them throughout this terrible struggle; and nothing could be more admirable than the steadiness which they exhibited until nature itself was exhausted in what they knew to be a desperate fight against a foe very many times their superior in numbers. I cannot particularize in this report to you the numberless instances of heroic daring performed by both officers and men, but must content myself for the present by saying in my judgment they all deserve well of their country.

I have the honor to be, very respectfully, your obedient servant,
John B. Floyd,
Brigadier General, Commanding

> *The command was turned over to Brigadier General Buckner, who at once opened negotiations with the enemy, which resulted in the surrender of the place.*

GENERAL BUCKNER'S OFFICIAL REPORT ON FORT DONELSON

Headquarters, Army of Cumberland,
Dover, Tennessee
February 18, 1862.

Col. W.W. MACKALL,
Assistant Adjutant General, Nashville, Tennessee

SIR: It becomes my duty to report that the remains of this army, after winning some brilliant successes both in repulsing the assaults of the enemy and in sallying successfully through their lines have been reduced to the necessity of surrender.

At the earliest practicable day I will send a detailed report of its operations. I can only say now that, after the battle of the 15th instant had been won and my division of the army was being established in position to cover the retreat of the army, the plan of battle seemed to have been changed and the troops were ordered back to the trenches. Before my own division returned to their works on the extreme right, the lines were assailed at that point and my extreme right was occupied by a large force of the enemy, but I successfully repelled their further assaults.

It was the purpose of General Floyd to effect the retreat of the army over the ground which had been won in the morning, and the troops moved from their works with that view; but before any movement for that purpose was organized, a reconnaissance showed that the ground was occupied by the enemy in great strength.

General Floyd then determined to retreat across the river with such force as could escape; but as there were no boats until nearly daylight on the 16th, he left with some regiments of Virginia troops about daylight and was accompanied by Brigadier General Pillow.

I was thus left in command of the remnant of the army, which had been placed in movement for a retreat which was discovered to be impracticable. My men were in a state of complete exhaustion from extreme suffering from cold and fatigue. The supply of ammunition, especially for the artillery, was being rapidly exhausted; the army was to a great extent demoralized by the retrograde movement.

On being placed in command, I ordered such troops as could not cross the river to return to their entrenchments, to make at the last moment such resistance as was possible to the overwhelming force of the enemy. But a small portion of the forces had returned to the lines when I received from General Grant a reply to my proposal to negotiate for terms of surrender. To have refused his terms would, in the condition of the army at that time, have led to the massacre of my troops without any advantage resulting from the sacrifice. I therefore felt it my highest duty to these brave men, whose conduct had been so brilliant and whose sufferings had been so intense, to accept the ungenerous terms proposed by the Federal commander who overcame us solely by overwhelming superiority of numbers.

This army is accordingly prisoners of war, the officers retaining their sidearms and private property and the soldiers their clothing and blankets. I regret to state, however, that, notwithstanding the earnest efforts of General Grant and many of his officers to prevent it, our camps have been a scene of almost indiscriminate pillage by the Federal troops.

In conclusion, I request at the earliest time practicable, a court of inquiry to examine into the causes of the surrender of this army.

I am, sir, very respectfully, your obedient servant,
S.B. BUCKNER,
Brigadier General, C.S. Army

Colonel Forrest's Statement on Fort Donelson

March 15, 1862

Between 1:00 and 2:00 on Sunday morning, February 16, being sent for, I arrived at General Pillow's headquarters, and found him, General Floyd, and General Buckner in conversation. General Pillow told me that they had received information that the enemy was again occupying the same ground they had occupied the morning before. I told him I did not believe it, as I had left that part of the field, on our left, late the evening before. He told me he had sent out scouts, who reported a large force of the enemy moving around to our left.

He instructed me to go immediately and send two reliable men to ascertain the condition of a road running near the river bank and between the enemy's right and the river, and also to ascertain the position of the enemy. I obeyed his instructions and awaited the return of the scouts.

They stated that they saw no enemy, but could see their fires in the same place where they were Friday night; that from their examination and information obtained from a citizen living on the river road, the water was about to the saddle skirts, and the mud about half-leg deep in the bottom where it had been overflowed. The bottom was about a quarter of a mile wide and the water then about 100 yards wide.

During the conversation that then ensued among the general officers, General Pillow was in favor of trying to cut our way out. General Buckner said that he could not hold his position over half an hour in the morning, and that if he attempted to take his force out it would be seen by the enemy (who held part of his entrenchments) and be followed and cut to pieces. I told him that I would take my cavalry around there and he could draw out under cover of them. He said that an attempt to cut our way out would involve the loss of three fourths of the men.

General Floyd said our force was so demoralized as to cause him to agree with General Buckner as to our probable loss in attempting to cut our way out. I said that I would agree to cut my way through the enemy's lines at any point the general might designate and stated that I could keep back their cavalry, which General Buckner thought would greatly harass our infantry in retreat.

General Buckner or General Floyd said that they (the enemy) would bring their artillery to bear on us. I went out of the room, and when I returned, General Floyd said he could not and would not surrender himself. I then asked if they were going to surrender the command. General Buckner remarked that they were. I then stated that I had not come out for the purpose of surrendering my command and would not do it if they would follow me out; that I intended to go out if I saved but one man; and then turning to General Pillow I asked him what I should do. He replied, "Cut your way out."

I immediately left the house and sent for all the officers under my command, and stated to them the facts that had occurred, stated my determination to leave and remarked that all who wanted to go could follow me, and those who wished to stay and take the consequences might remain in camp. All of my own regiment and Captain Williams, of Helm's Kentucky regiment, said they would go with me if the last man fell. Colonel Gantt was sent for and urged to get out his battalion as often as three times, but he and two Kentucky companies (Captains Wilcox and Huey) refused to come. I marched out the remainder of my command, with Captain Porter's artillery horses, and about 200 men of different commands up the river road and across the overflow, which I found to be about saddle skirt deep. The weather was intensely cold; a great many of the men were already frostbitten, and it was the opinion of the generals that the infantry could not have passed through the water and have survived it.

N. B. FORREST
Colonel, Commanding Forrest's Regiment of Cavalry

THE FALL OF FORT DONELSON

Ulysses S. Grant. *Personal Memoirs of U.S. Grant*, (New York: Charles L. Webster & Company, 1885).

I informed the department commander of our success at Fort Henry and that on the 8th, I would take Fort Donelson. But the rain continued to fall so heavily that the roads became impassable for artillery and wagon trains. Then, too, it would not have been prudent to proceed without the gunboats. At least it would have been leaving behind a valuable part of our available force.

On the 7th, the day after the fall of Fort Henry, I took my staff and the cavalry – a part of one regiment – and made a reconnaissance to within about a mile of the outer line of works at Donelson. I had known General Pillow in

Mexico and judged that with any force, no matter how small, I could march up to within gunshot of any entrenchments he was given to hold. I said this to the officers of my staff at the time.

I knew that Floyd was in command, but he was no soldier, and I judged that he would yield to Pillow's pretensions. I met, as I expected, no opposition in making the reconnaissance and, besides learning the topography of the country on the way and around Fort Donelson, found that there were two roads available for marching; one leading to the village of Dover, the other to Donelson.

Fort Donelson is two miles north, or down the river, from Dover. The fort, as it stood in 1861, embraced about one hundred acres of land. On the east it fronted the Cumberland; to the north it faced Hickman Creek, a small stream which at that time was deep and wide because of the backwater from the river; on the south was another small stream, or rather a ravine, opening into the Cumberland. This also was filled with backwater from the river. The fort stood on high ground, some of it as much as a hundred feet above the Cumberland.

Strong protection to the heavy guns in the water batteries had been obtained by cutting away places for them in the bluff. To the west there was a line of rifle pits some two miles back from the river at the farthest point. This line ran generally along the crest of high ground, but in one place crossed a ravine which opens into the river between the village and the fort.

The ground inside and outside of this entrenched line was very broken and generally wooded. The trees outside of the rifle pits had been cut down for a considerable way out, and had been felled so that their tops lay outwards from the entrenchments. The limbs had been trimmed and pointed and thus formed an abatis in front of the greater part of the line. Outside of this entrenched line and extending about half the entire length of it, is a ravine running north and south and opening into Hickman Creek at a point north of the fort. The entire side of this ravine next to the works was one long abatis.

I knew that Floyd was in command, but he was no soldier, and I judged that he would yield to Pillow's pretensions.

General Halleck commenced his efforts in all quarters to get reinforcements forwarded to me immediately on my departure from Cairo. General Hunter sent men freely from Kansas and a large division under General Nelson, from Buell's army, was also dispatched. Orders went out from the War Department to consolidate fragments of companies that were being recruited in the Western States so as to make full companies and to consolidate companies into regiments.

General Halleck did not approve or disapprove of my going to Fort Donelson. He said nothing whatever to me on the subject. He informed Buell on the 7th that I would march against Fort Donelson the next day; but on the 10th he directed me to fortify Fort Henry strongly, particularly to the land side, saying that he forwarded me entrenching tools for that purpose. I received this dispatch in front of Fort Donelson.

Movement against Donelson

I was very impatient to get to Fort Donelson because I knew the importance of the place to the enemy and supposed he would reinforce it rapidly. I felt that 15,000 men on the 8th would be more effective than 50,000 a month later. I asked Flag Officer Foote, therefore, to order his gunboats still about Cairo to proceed up the Cumberland River and not to wait for those gone to Eastport and Florence; but the others got back in time and we started on the 12th. I had moved McClernand out a few miles the night before so as to leave the road as free as possible.

Just as we were about to start, the first reinforcement reached me on transports. It was a brigade composed of six full regiments commanded by Colonel Thayer of Nebraska. As the gunboats were going around to Donelson by the Tennessee, Ohio, and Cumberland rivers, I directed Thayer to turn about and go under their convoy.

I started from Fort Henry with 15,000 men, including eight batteries and part of a regiment of cavalry, and, meeting with no obstruction to detain us, the advance arrived in front of the enemy by noon. That afternoon and the next day were spent in taking up ground to make the investment as complete as possible.

General Smith had been directed to leave a portion of his division behind to guard forts Henry and Heiman. He left General Lew Wallace with 2,500 men. With the remainder of his division he occupied our left extending to Hickman Creek. McClernand was on the right and covered the roads running south and southwest from Dover. His right extended to the backwater up the ravine opening into the Cumberland south of the village. The troops were not entrenched, but the nature of the ground was such that they were just as well protected from the fire of the enemy as if rifle pits had been thrown up. Our line was generally along the crest of ridges. The artillery was protected by being sunk in the ground. The men who were not serving the guns were perfectly covered from fire on taking position a little back from the crest.

The greatest suffering was from want of shelter. It was midwinter and during the siege we had rain and snow, thawing and freezing alternately. It would not do to allow campfires except far down the hill out of sight of the enemy, and it would not do to allow many of the troops to remain there at the same time. In the march over from Fort Henry numbers of the men had thrown away their blankets and overcoats. There was therefore much discomfort and absolute suffering.

During the 12th and 13th and until the arrival of Wallace and Thayer on the 14th, the National forces, composed of but 15,000 men, without entrenchments, confronted an entrenched army of 21,000, without conflict further than what was brought on by ourselves. Only one gunboat had arrived. There was a little skirmishing each day, brought on by the movement of our troops in securing commanding positions; but there was no actual fighting during this time except once, on the 13th, in front of McClernand's command.

That general had undertaken to capture a battery of the enemy which was annoying his men. Without orders or authority he sent three regiments to make the assault. The battery was in the main line of the enemy, which was defended by his whole army present. Of course the assault was a failure, and, of course, the loss on our side was great for the number of men engaged. In this assault Colonel William Morrison fell badly wounded.

Up to this time, the surgeons with the army had no difficulty in finding room in the houses near our line for all the sick and wounded; but now hospitals were overcrowded. Owing, however, to the energy and skill of the surgeons, the suffering was not so great as it might have been. The hospital arrangements at Fort Donelson were as complete as it was possible to make them, considering the inclemency of the weather and the lack of tents, in a sparsely settled country where the houses were generally of but one or two rooms.

The Gunboats Advance

On the return of Captain Walke to Fort Henry on the 10th, I had requested him to take the vessels that had accompanied him on his expedition up the Tennessee and get possession of the Cumberland as far up towards Donelson as possible. He started without delay, taking, however, only his own gunboat, the *Carondelet*, towed by the steamer *Alps*. Captain Walke arrived a few miles below Donelson on the 12th, a little after noon. About the time the advance of troops reached a point within gunshot of the fort on the land side, he engaged the water batteries at long range.

On the 13th, I informed him of my arrival the day before and of the establishment of most of our batteries, requesting him at the same time to attack again that day so that I might take advantage of any diversion. The attack was made and many shots fell within the fort, creating some consternation, as we now know. The investment on the land side was made as complete as the number of troops engaged would admit of.

During the night of the 13th, Flag Officer Foote arrived with the ironclads *St. Louis*, *Louisville*, and *Pittsburgh* and the wooden gunboats *Tyler* and *Conestoga*, convoying Thayer's brigade. On the morning of the 14th Thayer was landed. Wallace, whom I had ordered over from Fort Henry, also arrived about the same time. Up to this time he had been commanding a brigade belonging to the division of General C.F. Smith. These troops were now restored to the division they belonged to, and General Lew Wallace was assigned to the command of a division composed of the brigade of Colonel Thayer and other reinforcements that arrived the same day. This new division was assigned to the center, giving the two flanking divisions an opportunity to close up and form a stronger line.

The plan was for the troops to hold the enemy within his lines, while the gunboats should attack the water batteries at close quarters and silence his guns if possible. Some of the gunboats were to run the batteries, get above the fort and above the village of Dover. I had ordered a reconnaissance made with the view of getting troops to the river above

Dover in case they should be needed there. That position attained by the gunboats it would have been but a question of time – and a very short time, too – when the garrison would have been compelled to surrender.

By three in the afternoon of the 14th, Flag Officer Foote was ready and advanced upon the water batteries with his entire fleet. After coming in range of the batteries of the enemy, the advance was slow, but a constant fire was delivered from every gun that could be brought to bear upon the fort. I occupied a position on shore from which I could see the advancing navy. The leading boat got within a very short distance of the water battery, not further off I think than two hundred yards, and I soon saw one and then another of them dropping down the river, visibly disabled. Then the whole fleet followed and the engagement closed for the day.

The gunboat which Flag Officer Foote was on, besides having been hit about sixty times, several of the shots passing through near the waterline, had a shot enter the pilot house which killed the pilot, carried away the wheel and wounded the flag officer himself. The tiller ropes of another vessel were carried away and she, too, dropped helplessly back. Two others had their pilot houses so injured that they scarcely formed a protection to the men at the wheel.

The enemy had evidently been much demoralized by the assault, but they were jubilant when they saw the disabled vessels dropping down the river entirely out of the control of the men on board. Of course, I only witnessed the falling back of our gunboats and felt sad enough at the time over the repulse. Subsequent reports, now published, show that the enemy telegraphed a great victory to Richmond.

The Battle Not Decided

The sun went down on the night of the February, 14, 1862, leaving the army confronting Fort Donelson anything but comforted over the prospects. The weather had turned intensely cold; the men were without tents and could not keep up fires where most of them had to stay, and, as previously stated, many had thrown away their overcoats and blankets. Two of the strongest of our gunboats had been disabled, presumably beyond the possibility of rendering any present assistance. I retired this night not knowing but that I would have to entrench my position and bring up tents for the men or build huts under the cover of the hills.

On the morning of the 15th, before it was yet broad day, a messenger from Flag Officer Foote handed me a note, expressing a desire to see me on the flagship and saying that he had been injured the day before so much that he could not come himself to me. I at once made my preparations for starting. I directed my adjutant general to notify each of the division commanders of my absence and instruct them to do nothing to bring on an engagement until they received further orders, but to hold their positions.

From the heavy rains that had fallen for days and weeks preceding and from the constant use of the roads between the troops and the landing four to seven miles below, these roads had become cut up so as to be hardly passable. The intense cold of the night of the 14th/15th had frozen the ground solid. This made travel on horseback even slower than through the mud; but I went as fast as the roads would allow.

When I reached the fleet I found the flagship was anchored out in the stream. A small boat, however, awaited my arrival and I was soon on board with the flag officer. He explained to me in short the condition in which he was left by the engagement of the evening before and suggested that I should entrench while he returned to Mound City with his disabled boats, expressing at the time the belief that he could have the necessary repairs made and be back in ten days. I saw the absolute necessity of his gunboats going into hospital and did not know but I should be forced to the alternative of going through a siege. But the enemy relieved me from this necessity.

When I left the National line to visit Flag Officer Foote, I had no idea that there would be any engagement on land unless I brought it on myself. The conditions for battle were much more favorable to us than they had been for the first two days of the investment. From the 12th to the 14th, we had but 15,000 men of all arms and no gunboats.

Now we had been reinforced by a fleet of six naval vessels, a large division of troops under General Lew Wallace and 2,500 men brought over from Fort Henry belonging to the division of C.F. Smith. The enemy, however, had taken the initiative.

The Confederates Counter Attack

Just as I landed I met Captain Hillyer of my staff, white with fear, not for his personal safety, but for the safety of the National troops. He said the enemy had come out of his lines in full force and attacked and scattered McClernand's division, which was in full retreat. The roads, as I have said, were unfit for making fast time, but I got to my command as soon as possible. The attack had been made on the National right. I was some four or five miles north of our left. The line was about three miles long.

In reaching the point where the disaster had occurred, I had to pass the divisions of Smith and Wallace. I saw no sign of excitement on the portion of the line held by Smith; Wallace was nearer the scene of conflict and had taken part in it. He had, at an opportune time, sent Thayer's brigade to the support of McClernand and thereby contributed to hold the enemy within his lines.

I saw everything favorable for us along the line of our left and center. When I came to the right appearances were different. The enemy had come out in full force to cut his way out and make his escape. McClernand's division had to bear the brunt of the attack from this combined force. His men had stood up

> *In the course of our conversation, which was very friendly, he said to me that if he had been in command, I would not have got up to Donelson as easily as I did. I told him that if he had been in command, I should not have tried in the way I did.*

gallantly until the ammunition in their cartridge boxes gave out. There was abundance of ammunition near by lying on the ground in boxes, but at that stage of the war, it was not all of our commanders of regiments, brigades, or even divisions, who had been educated up to the point of seeing that their men were constantly supplied with ammunition during an engagement.

When the men found themselves without ammunition, they could not stand up against troops who seemed to have plenty of it. The division broke and a portion fled, but most of the men, as they were not pursued, only fell back out of range of the fire of the enemy. It must have been about this time that Thayer pushed his brigade in between the enemy and those of our troops that were without ammunition. At all events the enemy fell back within his entrenchments and was there when I got on the field.

I saw the men standing in knots talking in the most excited manner. No officer seemed to be giving any directions. The soldiers had their muskets, but no ammunition, while there were tons of it close at hand. I heard some of the men say that the enemy had come out with knapsacks and haversacks filled with rations. They seemed to think this indicated a determination on his part to stay out and fight just as long as the provisions held out.

I turned to Colonel J.D. Webster, of my staff, who was with me, and said: "Some of our men are pretty badly demoralized, but the enemy must be more so, for he has attempted to force his way out, but has fallen back. The one who attacks first now will be victorious and the enemy will have to be in a hurry if he gets ahead of me."

Grant Renews the Attack

I determined to make the assault at once on our left. It was clear to my mind that the enemy had started to march out with his entire force, except a few pickets, and if our attack could be made on the left before the enemy could redistribute his forces along the line, we would find but little opposition except from the intervening abatis. I directed Colonel Webster to ride with me and call out to the men as we passed: "Fill your cartridge boxes, quick, and get into line; the enemy is trying to escape and he must not be permitted to do so." This acted like a charm. The men only wanted some one to give them a command.

We rode rapidly to Smith's quarters, when I explained the situation to him and directed him to charge the enemy's works in his front with his whole division, saying at the same time that he would find nothing but a very thin line to contend with. The general was off in an incredibly short time, going in advance himself to keep his men from firing while they were working their way through the abatis intervening between them and the enemy.

The outer line of rifle pits was passed, and the night of the 15th, General Smith, with much of his division, bivouacked within the lines of the enemy. There was now no doubt but that the Confederates must surrender or be captured the next day.

There seems from subsequent accounts to have been much consternation, particularly among the officers of high rank, in Dover during the night of the 15th. General Floyd, the commanding officer, who was a man of talent enough for any civil position, was no soldier and possibly did not possess the elements of one. He was further unfitted for command, for the reason that his conscience must have troubled him and made him afraid. As Secretary of War he had taken a solemn oath to maintain the Constitution of the United States and to uphold the same against all its enemies. He had betrayed that trust. As Secretary of War he was reported through the northern press to have scattered the little army the country had so that the most of it could be picked up in detail when secession occurred. About a year before leaving the Cabinet he had removed arms from northern to southern arsenals. He continued in the Cabinet of President Buchanan until about the 1st of January, 1861, while he was working vigilantly for the establishment of a confederacy made out of United States territory. Well may he have been afraid to fall into the hands of National troops. He would no doubt have been tried for misappropriating public property, if not for treason, had he been captured.

General Pillow, next in command, was conceited, and prided himself much on his services in the Mexican war. He telegraphed to General Johnston at Nashville, after our men were within the rebel rifle pits, and almost on the eve of his making his escape, that the Southern troops had had great success all day. Johnston forwarded the dispatch to Richmond. While the authorities at the capital were reading it, Floyd and Pillow were fugitives.

Confederate Council of War

A council of war was held by the enemy at which all agreed that it would be impossible to hold out longer. General Buckner, who was third in rank in the garrison but much the most capable soldier, seems to have regarded it a duty to hold the fort until the general commanding the department, Albert Sidney Johnston, should get back to his headquarters at Nashville. Buckner's report shows, however, that he considered Donelson lost and that any attempt to hold the place longer would be at the sacrifice of the command. Being assured that Johnston was already in Nashville, Buckner too agreed that surrender was the proper thing. Floyd turned over the command to Pillow, who declined it. It then devolved upon Buckner, who accepted the responsibility of the position. Floyd and Pillow took possession of all the river transports at Dover and before morning both were on their way to Nashville, with the brigade formerly commanded by Floyd and some other troops, in all about 3,000.

Some marched up the east bank of the Cumberland; others went on the steamers. During the night Forrest also, with his cavalry and some other troops about a thousand in all, made their way out, passing between our right and the river. They had to ford or swim over the backwater in the little creek just south of Dover.

Before daylight General Smith brought to me the following letter from General Buckner:

Sir:

In consideration of all the circumstances governing the present situation of affairs at this station, I propose to the Commanding Officer of the Federal forces the appointment of Commissioners to agree upon terms of capitulation of the forces and fort under my command, and in that view suggest an armistice until 12:00 today.
I am, sir, very respectfully,
Your obedient servant, S. B. Buckner,
Brigadier General C. S. A.

To this I responded as follows:

Sir:
Yours of this date, proposing armistice and appointment of Commissioners to settle terms of capitulation, is just received. No terms except an unconditional and immediate surrender can be accepted. I propose to move immediately upon your works.
I am, sir, very respectfully,
Your obedient servant
U. S. Grant,
Brigadier General

To this I received the following reply:

Sir:
The distribution of the forces under my command, incident to an unexpected change of commanders, and the overwhelming force under your command, compel me, notwithstanding the brilliant success of the Confederate arms yesterday, to accept the ungenerous and unchivalrous terms which you propose.
I am, sir,
Your very obedient servant,
S. B. Buckner
Brigadier General C. S. A.

General Buckner, as soon as he had dispatched the first of the above letters, sent word to his different commanders on the line of rifle pits, notifying them that he had made a proposition looking to the surrender of the garrison, and directing them to notify National troops in their front so that all fighting might be prevented. White flags were stuck at intervals along the line of rifle-pits, but none over the fort.

Buckner Surrenders

As soon as the last letter from Buckner was received, I mounted my horse and rode to Dover. General Wallace, I found, had preceded me an hour or more. I presume that, seeing white flags exposed in his front, he rode up to see what they meant and, not being fired upon or halted, he kept on until he found himself at the headquarters of General Buckner.

I had been at West Point three years with Buckner and afterwards served with him in the army, so that we were quite well acquainted. In the course of our conversation, which was very friendly, he said to me that if he had been in command I would not have got up to Donelson as easily as I did. I told him that if he had been in command I should not have tried in the way I did: I had invested their lines with a smaller force than they had to defend them, and at the same time had sent a brigade full 5,000 strong, around by water; I had relied very much upon their commander to allow me to come safely up to the outside of their works.

I asked General Buckner about what force he had to surrender. He replied that he could not tell with any degree of accuracy; that all the sick and weak had been sent to Nashville while we were about Fort Henry; that Floyd and Pillow had left during the night, taking many men with them; and that Forrest, and probably others, had also escaped during the preceding night: the number of casualties he could not tell; but he said I would not find fewer than 12,000, nor more than 15,000.

He asked permission to send parties outside of the lines to bury his dead, who had fallen on the 15th when they tried to get out. I gave directions that his permit to pass our limits should be recognized. I have no reason to believe that this privilege was abused, but it familiarized our guards so much with the sight of Confederates passing to and fro that I have no doubt many got beyond our pickets unobserved and went on. The most of the men who went in that way no doubt thought they had had war enough, and left with the intention of remaining out of the army. Some came to me and asked permission to go, saying that they were tired of the war and would not be caught in the ranks again, and I bade them go.

The actual number of Confederates at Fort Donelson can never be given with entire accuracy. The largest number admitted by any writer on the Southern side, is by Colonel Preston Johnston. He gives the number at 17,000. But this must be an underestimate. The commissary general of prisoners reported having issued rations to 14,623 Fort Donelson prisoners at Cairo, as they passed that point.

General Pillow reported the killed and wounded at 2,000; but he had less opportunity of knowing the actual numbers than the officers of McClernand's division, for most of the killed and wounded fell outside their works, in front of that division, and were buried or cared for by Buckner after the surrender and when Pillow was a fugitive. It is known that Floyd and Pillow escaped during the night of the 15th, taking with them not less than 3,000 men. Forrest escaped with about 1,000 and others were leaving singly and in squads all night. It is probable that the Confederate force at Donelson, on the 15th of February, 1862, was 21,000 in round numbers.

On the day Fort Donelson fell, I had 27,000 men to confront the Confederate lines and guard the road four or five miles to the left, over which all our supplies had to be drawn on wagons. During the 16th, after the surrender, additional reinforcements arrived.

During the siege General Sherman had been sent to Smithland, at the mouth of the Cumberland River, to forward reinforcements and supplies to me. At that time he was my senior in rank and there was no authority of law to assign a junior to command a senior of the same grade. But every boat that came up with supplies or reinforcements brought a note of encouragement from Sherman, asking me to call upon him for any assistance he could render and saying that if he could be of service at the front, I might send for him and he would waive rank.

The Stainless Banner
February 2012

BATTLES AT FORT DONELSON, ISLAND NUMBER TEN, FORT PILLOW AND MEMPHIS

By Rear Admiral Henry Walke. *Battles and Leaders, Volume 1*, pages 430-452..

On the 7th of February, the day after the capture of Fort Henry, I received on board the *Carondelet* Colonels Webster, Rawlins, and McPherson, with a company of troops, and under instructions from General Grant proceeded up the Tennessee River and completed the destruction of the bridge of the Memphis and Bowling Green Railroad.

On returning from that expedition, General Grant requested me to hasten to Fort Donelson with the *Carondelet*, *Tyler*, and *Lexington*, and announce my arrival by firing signal guns. The object of this movement was to take possession of the river as soon as possible, to engage the enemy's attention by making formidable demonstrations before the fort, and to prevent it from being reinforced. On February 10th, the *Carondelet* alone (towed by the transport *Alps*) proceeded up the Cumberland River and on the 12th arrived a few miles below the fort.

Fort Donelson occupied one of the best defensive positions on the river. It was built on a bold bluff about 120 feet in height, on the west side of the river, where it makes a slight bend to the eastward. It had three batteries, mounting in all 15 guns: the lower, about twenty feet above the water; the second, about fifty feet above the water; the third, on the summit.

When the *Carondelet*, her tow being cast off, came in sight of the fort and proceeded up to within long range of the batteries, not a living creature could be seen. The hills and woods on the west side of the river hid part of the enemy's formidable defenses, which were lightly covered with snow; but the black rows of heavy guns, pointing down on us, reminded me of the dismal looking sepulchers cut in the rocky cliffs near Jerusalem, but far more repulsive.

At 12:50 p.m. to unmask the silent enemy and to announce my arrival to General Grant, I ordered the bow guns to be fired at the fort. Only one shell fell short. There was no response except the echo from the hills. The fort appeared to have been evacuated. After firing ten shells into it, the *Carondelet* dropped down the river about three miles and anchored. But the sound of her guns aroused our soldiers on the southern side of the fort into action; one report says that when they heard the guns of the avant courier of the fleet, they gave cheer upon cheer, and rather than permit the sailors to get ahead of them again, they engaged in skirmishes with the enemy, and began the battle of the three days following. On the *Carondelet*, we were isolated and beset with dangers from the enemy's lurking sharpshooters.

On the 13th, a dispatch was received from General Grant informing me that he had arrived the day before and had succeeded in getting his army in position, almost entirely investing the enemy's works. "Most of our batteries," he said, "are established and the remainder soon will be. If you will advance with your gunboat at 10:00 in the morning, we will be ready to take advantage of any diversion in our favor."

I immediately complied with these instructions, and at 9:05 a.m., with the *Carondelet* alone and under cover of a heavily wooded point, fired 139 70-pound and 64-pound shells at the fort. We received in return the fire of all the enemy's guns that could be brought to bear on the *Carondelet*, which sustained but little damage except from two shots.

One, a 128-pound solid, at 11:30 a.m., struck the corner of our port broadside casemate, passed through it and in its progress toward the center of our boilers glanced over the temporary barricade in front of the boilers. It then passed over the steam drum, struck the beams of the upper deck, carried away the railing around the engine room and burst the steam heater, and, glancing back into the engine room, "seemed to bound after the men," as one of the engineers said, "like a wild beast pursuing its prey."

I have preserved this ball as a souvenir of the fight at Fort Donelson. When it burst through the side of the *Carondelet*, it knocked down and wounded a dozen men, seven of them severely. An immense quantity of splinters was blown through the vessel. Some of them, as fine as needles, shot through the clothes of the men like arrows. Several of the wounded were so much excited by the suddenness of the event and the sufferings of their comrades that they were not aware that they themselves had been struck until they felt the blood running into their shoes. Upon receiving this shot we ceased firing for a while.

After dinner we sent the wounded on board the *Alps*, repaired damages, and not expecting any assistance, at 12:15 p.m., we resumed in accordance with General Grant's request and bombarded the fort until dusk, when nearly all our 10-inch and 15-inch shells were expended. The firing from the shore having ceased, we retired.

At 11:30 p.m. on the night of the 13th, Flag Officer Foote arrived below Fort Donelson with the ironclads *St. Louis*, *Louisville*, and *Pittsburgh*, and the wooden gunboats *Tyler* and *Conestoga*.

On the 14th, all the hard materials in the vessels, such as chains, lumber, and bags of coal, were laid on the upper decks to protect them from the plunging shots of the enemy. At 3:00 in the afternoon, our fleet advanced to attack the fort, the *Louisville* being on the west side of the river, the *St. Louis* (flag steamer) next, then the *Pittsburgh* and *Carondelet* on the east side of the river. The wooden gunboats were about a thousand yards in the rear. When we started in line abreast at a moderate speed, the *Louisville* and *Pittsburgh*, not keeping up to their positions were hailed from the flag steamer to "steam up."

At 3:30 p.m., when about a mile and a half from the fort, two shots were fired at us, both falling short. When within a mile of the fort the *St. Louis* opened fire, and the other ironclads followed, slowly and deliberately at first, but more rapidly as the fleet advanced. The flag officer hailed the *Carondelet* and ordered us not to fire so fast. Some of our shells went over the fort, and almost into our camp beyond.

As we drew nearer, the enemy's fire greatly increased in force and effect. But, the officers and crew of the *Carondelet* having recently been long under fire and having become practiced in fighting, her gunners were as cool and composed as old veterans. We heard the deafening crack of the bursting shells, the crash of the solid shot, and the whizzing of fragments of shell and wood as they sped through the vessel. Soon a 128-pounder struck our anchor, smashed it into flying bolts, and bounded over the vessel, taking away a part of our smokestack; then another cut away the iron boat davits as if they were pipe stems, whereupon the boat dropped into the water. Another ripped up the iron plating and glanced over. Another went through the plating and lodged in the heavy casemate. Another struck the pilot house, knocked the plating to pieces, and sent fragments of iron and splinters into the pilots, one of whom fell mortally wounded, and was taken below; another shot took away the remaining boat davits and the boat with them; and still they came, harder and faster, taking flagstaffs and smokestacks, and tearing off the side armor as lightning tears the bark from a tree.

Our men fought desperately, but, under the excitement of the occasion, loaded too hastily and the port rifled gun exploded. One of the crew, in his account of the explosion soon after it occurred, said: "I was serving the gun with shell. When it exploded it knocked us all down, killing none, but wounding over a dozen men and spreading dismay and confusion among us. For about two minutes I was stunned, and at least five minutes elapsed before I could tell what was the matter. When I found out that I was more scared than hurt, although suffering from the gunpowder which I had inhaled, I looked forward and saw our gun lying on the deck, split in three pieces. Then the cry ran through the boat that we were on fire, and my duty as pump man called me to the pumps. While I was there, two shots enter our bow ports and killed four men and wounded several others. They were borne past me, three with their heads off. The sight almost sickened me, and I turned my head away. Our master's mate came soon after and ordered us to our quarters at the gun. I told him the gun had burst, and that we had caught fire on the upper deck from the enemy's shell. He then said: "'Never mind the fire; go to your quarters.'" Then I took a station at the starboard tackle of another rifled bow-gun and remained there until the close of the fight." The carpenter and his men extinguished the flames.

When within 400 yards of the fort, and while the Confederates were running from their lower battery, our pilot house was struck again and another pilot wounded, our wheel was broken, and shells from the rear boats were bursting over us. All four of our boats were shot away and dragging in the water.

On looking out to bring our broadside guns to bear, we saw that the other gunboats were rapidly falling back out of line. The *Pittsburgh* in her haste to turn struck the stern of the *Carondelet*, and broke our starboard rudder, so that we were obliged to go ahead to clear the *Pittsburgh* and the point of rocks below. The pilot of the *St. Louis* was killed and the pilot of the *Louisville* was wounded. Both vessels had their wheel ropes shot away, and the men were prevented from steering the *Louisville* with the tiller ropes at the stern by the shells from the rear boats bursting over them. The *St. Louis* and *Louisville*, becoming unmanageable, were compelled to drop out of battle, and the *Pittsburgh* followed. All had suffered severely from the enemy's fire. Flag Officer Foote was wounded while standing by the pilot of the *St. Louis* when he was killed. We were then about 350 yards from the fort.

There was no alternative for the *Carondelet* in that narrow steam but to keep her head to the enemy and fire into the fort with her two bow guns, to prevent it, if possible, from returning her fire effectively. The enemy saw that she was in a manner left to his mercy, and concentrated the fire of all his batteries upon her. In return, the *Carondelet's* guns were well served to the last shot.

Our new acting gunner, John Hall, was just the man for the occasion. He came forward, offered his services, and with my sanction, took charge of the starboard bow rifled gun. He instructed the men to obey his warnings and follow his motions, and he told them that when he saw a shot coming he would call out "down" and stoop behind the breech of the gun as he did so. At the same instant the men were to stand away from the bow ports.

Nearly every shot from the fort struck the bows of the *Carondelet*. Most of them were fired on the ricochet level and could be plainly seen skipping on the water before they struck. The enemy's object was to sink the gunboat by striking her just below the waterline. They soon succeeded in planting two 32-pound shots in her bow, between wind and water, which made her leak badly, but her compartments kept her from sinking until we could plug up the shotholes.

Three shots struck the starboard case mating; four struck the port case mating forward of the rifle-gun; one struck on the starboard side, between the waterline and planksheer, cutting through the planking; six shots struck the pilot house, shattering one section into pieces and cutting through the iron casing. The smoke stacks were riddled.

Our gunners kept up a constant firing while we were falling back, and the warning words "Look out! Down!" were often heard and heeded by nearly all the gun crews. On one occasion, while the men were at the muzzle of the middle bow gun, loading it, the warning came just in time for them to jump aside as a 32-pounder struck the lower sill and, glancing up, struck the upper sill, then, falling on the inner edge of the lower sill, bounded on deck and spun around like a top, but hurt no one.

It was very evident that if the men who were loading had not obeyed the order to drop, several of them would have been killed. So I repeated the instructions and warned the men at the guns and the crew generally to bow or stand off from the ports when a shot was seen coming. But some of the young men, from a spirit of bravado or from a belief in the doctrine of fatalism, disregarded the instructions, saying it was useless to attempt to dodge a cannon ball, and they would trust to luck. The warning words "Look out! Down!" were again soon heard; down went the gunner and his men, as the whizzing shot glanced on the gun, taking off the gunner's cap and the heads of two of the young men who trusted to luck and in defiance of the order were standing up or passing behind him.

This shot killed another man also, who was at the last gun of the starboard side, and disabled the gun. It came in with a hissing sound; three sharp spats and a heavy bang told the sad fate of three brave comrades. Before the decks were well sanded, but there was so much blood on them that our men could not work the guns without slipping.

We kept firing at the enemy so long as he was within range, to prevent him from seeing us through the smoke. The *Carondelet* was the first in and the last out of the fight and was more damaged than any of the other gunboats, as the boat carpenters who repaired them subsequently informed me. She was much longer under fire than any other vessel of the flotilla; and, according to the report of the Secretary of the Navy, her loss in killed and wounded was nearly twice as great as that of all the other gunboats together. She fired more shot and shell into Fort Donelson than any other gunboat and was struck fifty-four times. These facts are given because a disposition was shown by correspondents and naval historians to ignore the services of the *Carondelet* on this and other occasions.

In the action of the 14th, all of the armored vessels were fought with the greatest energy, skill, and courage, until disabled by the enemy's heavy shot. In his official report of the battle the flag-officer said: "The officers and men in this hotly contested but unequal fight behaved with the greatest gallantry and determination."

Although the gunboats were repulsed in this action, the demoralizing effect of their cannonade and of the heavy and well-sustained fire of the *Carondelet* on the day before must have been very great and contributed in no small degree to the successful operations of the army on the following day.

After the battle I called upon the flag officer and found him suffering from his wounds. He asked me if I could have run past the fort, something I should not have ventured upon without permission.

The 15th was employed in the burial of our slain comrades. I read the Episcopal service on board the *Carondelet* under our flag at half-mast, and the sailors bore their late companions to a lonely field within the shadows of the hills. When they were about to lower the first coffin, a Roman Catholic priest appeared, and his services being accepted, he read the prayers for the dead. As the last service was ended, the sound of the battle being waged by General Grant, like the rumbling of distant thunder, was the only requiem for our departed shipmates.

On Sunday the 16th, at dawn, Fort Donelson surrendered and the gunboats steamed up to Dover. After religious services, the *Carondelet* proceeded back to Cairo, and arrived there on the morning of the 17th in such a dense fog that she passed below the town unnoticed and had great difficulty in finding the landing. There had been a report that the enemy was coming from Columbus to attack Cairo during the absence of its defenders, and while the *Carondelet* was cautiously feeling her way back and blowing her whistle, some people imagined she was a Confederate gunboat about to land, and made hasty preparations to leave the place. Our announcement of the victory at Fort Donelson changed their dejection into joy and exultation. On the following morning an order congratulating the officers and men of the *Carondelet* was received from Flag Officer Foote.

Needed Repairs

A few days later the *Carondelet* was taken up on the ways at Mound City, Illinois, six or seven miles above Cairo on the Ohio River, for repairs. A crowd of carpenters worked on her night and day. After the repairs were completed, she was ordered to make the experiment of backing upstream, which proved a laughable failure. She would sheer from one side of the river to the other, and with two anchors astern, she could not be held steady enough to fight her bow-guns downstream. She dragged both anchors alternately, until they came together, and the experiment failed completely.

On the morning of the 23d, the flag officer made a reconnaissance to Columbus, Kentucky, with four gunboats and two mortar-boats, accompanied by the wooden gunboat *Conestoga*, convoying five transports. The fortifications looked more formidable than ever. The enemy fired two guns, and sent up a transport with the pretext, it was said, of effecting an exchange of prisoners.

But at that time, as we learned afterward from a credible source, the evacuation of the fort (which General Grant's successes at Forts Henry and Donelson had made necessary) was going on, and the last raft and barge loads of all the movable munitions of war were descending the river, which, with a large quantity previously taken away, could and would have been captured by our fleet if we had received this information in time.

On the 4th of March, another reconnaissance in force was made with all the gunboats and four mortar boats, and the fortress had still a formidable, life-like appearance, though it had been evacuated two days before.

Duel with the *Grampus*

On the 5th of March, while we were descending the Mississippi in a dense fog, the flag steamer leading, the Confederate gunboat *Grampus*, or *Daredevil Jack*, the sauciest little vessel on the river,

> *She ran before us yawing and flirting about and blowing her alarm-whistle so as to announce our approach to the enemy who had now retired to Island Number Ten, a strong position sixty miles below Columbus.*

suddenly appeared across our track and "close aboard." She stopped her engines and struck her colors, and we all thought she was ours at last. But when the captain of the *Grampus* saw how slowly we moved, and as no gun was fired to bring him to, he started off with astonishing speed and was out of danger before the flag steamer could fire a gun. She ran before us yawing a flirting about and blowing her alarm-whistle so as to announce our approach to the enemy who had now retired to Island Number Ten, a strong position sixty miles below Columbus (and of the latitude of Forts Henry and Donelson), where General Beauregard, who was now in general command of our opponents, had determined to contest the possession of the river.

On March 15, the flotilla and transports continued on their way to Island Number Ten, arriving in its vicinity about nine in the morning. The strong and muddy current of the river had overflowed its banks and carried away every movable thing. Houses, trees, fences, and wrecks of all kinds were being swept rapidly downstream. The twists and turns of the river near Island Number Ten are certainly remarkable. Within a radius of eight miles from the island it crosses the boundary line of Kentucky and Tennessee three times, running on almost every point of the compass.

Island Number Ten

We were greatly surprised when we arrived above Island Number Ten and saw on the bluffs a chain of forts extending for four miles along the crescent-formed shore, with the white tents of the enemy in the rear. And there lay the island in the lower corner of the crescent, with the side fronting the Missouri shore lined with heavy ordnance, so trained that with the artillery on the opposite shore almost every point on the river between the island and the Missouri bank could be reached at once by all the enemy's batteries.

On the 17th, an attack was made on the upper battery by all the ironclads and mortar boats. The *Benton* (flag-steamer), lashed between the *Cincinnati* and *St. Louis*, was on the east side of the river; the *Mound City*, *Carondelet*, and *Pittsburgh* were on the west side; the last, however, changed her position to the east side of the river before the firing began. We opened fire on the upper fort at 1:20 p.m., and by order of the flag officer fired one gun a minute.

The enemy replied promptly, and some of his shot struck the *Benton*, but, owing to the distance from which they were fired, did but little damage. We silenced all the guns in the upper fort except one. During the action one of the rifled guns of the *St. Louis* exploded, killing, and wounding several of the gunners, another proof of the truth of the saying that the guns furnished the Western flotilla were less destructive to the enemy than to ourselves.

From March 17th to April 4th but little progress was made in the reduction of the Confederate works, the gunboats firing a few shot now and then at long range, but doing little damage. The mortar boats, however, were daily throwing 13 inch bombs, and so effectively at times that the Confederates were driven from their batteries and compelled to seek refuge in caves and other places of safety. But it was very evident that the great object of the expedition – the reduction of the works and the capture of the Confederate forces – could not be effected by the gunboats alone, owing to their mode of structure and to the disadvantage under which they were fought in the strong and rapid current of the Mississippi. This was the opinion not only of naval officers, but also of General Pope and other army officers.

On the 23rd of March, the monotony of the long and tedious investment was unfortunately varied in a very singular manner. The *Carondelet* being moored nearest the enemy's upper fort, under several large cottonwood trees, in order to protect the mortar-boats, suddenly, and without warning, two of the largest of the trees fell across her deck, mortally wounding one of the crew, severely wounding another, and doing great damage to the vessel. This was twelve days before I ran the gauntlet at Island Number Ten with the *Carondelet*.

To understand fully the importance of that adventure, some explanation of the military situation at and below Island Number Ten seems necessary. After the evacuation of New Madrid, which General Pope had forced by blockading the river twelve miles below at Point Pleasant, the Confederate forces occupied their fortified positions on Island Number Ten and the eastern shore of the Mississippi, where they were cut off by impassable swamps on the land side.

> *With our bow pointing to the island, we passed the lowest point of land without being observed it appears by the enemy.*

They were in a cul-de-sac, and the only way open for them to obtain supplies or to effect a retreat was by the river south of Island Number Ten.

General Pope with an army of 20,000 men was on the western side of the river below the island. Perceiving the defect in the enemy's position, he proceeded with great promptness and ability to take advantage of it. It was his intention to cross the river and attack the enemy from below, but he could not do this without the aid of a gunboat to silence the enemy's batteries opposite Point Pleasant and protect his army in crossing.

He wrote repeatedly to Flag Officer Foote, urging him to send down a gunboat past the enemy's batteries on Island Number Ten, and in one of his letters expressed the belief that a boat could pass down at night under cover of the darkness. But the flag officer invariably declined, saying in one of his letters to General Pope that the attempt "would result in the sacrifice of the boat, her officers and men, which sacrifice I would not be justified in making."

During this correspondence the bombardment still went on, but was attended with such poor results that it become a subject of ridicule among the officers of Pope's army, one of whom (Colonel Gilmore, of Chillicothe, Ohio) is reported to have said that often when they met, and inquiry was made respecting the operations of the flotilla, the answer would generally be: "Oh! It is still bombarding the State of Tennessee at long range." And a Confederate officer said that no casualties resulted and no damage was sustained at Island Number Ten from the fire of the gunboats.

On March 20th, Flag Officer Foote consulted his commanding officers, through Commander Stembel, as to the practicability of taking a gunboat past the enemy's forts to New Madrid, and all except myself were opposed to the enterprise, believing with Foote that the attempt to pass the batteries would result in the almost certain destruction of the boat. I did not think so, but believed with General Pope that under the cover of darkness and other favorable circumstances a gunboat might be run past the enemy's batteries, formidable as they were with nearly fifty guns. And although fully aware of the hazardous nature of the enterprise, I knew that the aid of a gunboat was absolutely necessary to enable General Pope to succeed in his operations against the enemy, and thought the importance of this success would justify the risk of running the gauntlet of the batteries on Island Number Ten and on the left bank.

The army officers were becoming impatient, and it was well known that the Confederates had a number of small gunboats below and were engaged in building several large and powerful vessels of which the renowned *Arkansas* was one. And there was good reason to apprehend that these gunboats would ascend the river and pass or silence Pope's batteries and relieve the Confederate forces on Island Number Ten and the eastern shore of the Mississippi. That Pope and Foote apprehended this appears from the correspondence between them.

The flag officer now called a formal council of war of all his commanding officers. It was held on board the flag steamer on the 28th or 29th of March and all except myself concurred in the opinion formerly expressed that the attempt to pass the batteries was too hazardous and ought not to be made. When I was asked to give my views, I favored the undertaking, and advised compliance with the requests of General Pope. When asked if I was willing to make the attempt with the *Carondelet,* I replied in the affirmative. Foote accepted my advice and expressed himself as greatly relieved from a heavy responsibility, as he had determined to send none but volunteers on an expedition which he regarded as perilous and of very doubtful success.

Having received written orders from the flag officer, under date of March 30th, I at once began to prepare the *Carondelet* for the ordeal. All the loose material at hand was collected, and on the 4th of April, the decks were covered with it to protect them against plunging shot. Hawsers and chain cables were placed around the pilothouse and other vulnerable parts of the vessel, and every precaution was adopted to prevent disaster. A coal barge laden with hay and coal was lashed to the part of the port side on which there was no iron plating to protect the magazine. It was truly said that the *Carondelet* at that time resembled a farmer's wagon prepared for market. The engineers led the escape steam through the pipes aft into the wheelhouse to avoid to puffing sound it made when blown through the smokestacks.

All the necessary preparations having been made, I informed the flag officer of my intention to run the gauntlet that night, and received his approval. Colonel N. B. Buford, who commanded the land forces temporarily with the flotilla, assisted me in preparing for the trip. The night of the 4th brought on board Captain Hottenstein of the 42nd Illinois and 23 sharp-shooters of his command, who volunteered their services, which were gratefully accepted. Colonel Buford remained on board until the last moment, to encourage us. I informed the officers and crew of the character of the undertaking, and all expressed a readiness to make the venture. In order to resist boarding parties, in case of being disabled, the sailors were well armed and pistols, cutlasses, muskets, boarding-pikes, and hand-grenades were within reach. Hose was attached to the boilers for throwing scalding water over any who might attempt to board. If it should be found impossible to save the vessel, it was designed to sink rather than burn her.

During the afternoon there was a promise of a clear, moonlight night, and it was determined to wait until the moon was down and then to make the attempt whatever the chances. Having gone so far, we could not abandon the project without an effect on the men almost as bad as failure.

At 10:00, the moon had gone down, and the sky, the earth, and the river were alike hidden in the black shadow of a thunder storm, which had now spread itself over all the heavens. As the time seemed favorable, I ordered the first master to cast off. Dark clouds now rose rapidly over us and enveloped us in almost total darkness, except when the sky was lighted up by the welcome flashes of vivid lightning to show us the perilous way we were to take. Now and then the dim outline of the landscape could be seen, and the forest bending under the roaring storm that came rushing up the river.

With our bow pointing to the island, we passed the lowest point of land without being observed it appears by the enemy. All speed was given to the vessel to drive her through the tempest. The flashes of lightning continued with frightful brilliancy, and "almost every second," wrote a correspondent, "every brace, post, and outline could be seen with startling distinctness, enshrouded by a bluish white glare of light, and then her form for the next minute would become merged in the intense darkness."

When opposite Battery Number 2 on the mainland, the smokestacks blazed up, but the fire was soon subdued. It was caused by the shoot becoming dry, as the escape-steam, which usually kept the stacks wet, had been sent into the wheelhouse, as already mentioned, to prevent noise. With such vivid lightning as prevailed during the whole passage, there was no prospect of escaping the vigilance of the enemy, but there was good reason to hope that he would be unable to point his guns accurately.

Again the smokestacks took fire and were soon put out. Then the roar of the enemy's guns began, and from Batteries Numbers 2, 3, and 4 on the mainland came to continued crack and scream of their rifle shells, which seemed to unite with the electric batteries of the clouds to annihilate us.

While nearing the island or some shoal point, during a few minutes of total darkness, we were startled by the order, "Hard a-port!" from our brave and skillful pilot, First Master William R. Hoel. We almost grazed the island, and it appears were not observed through the storm until we were close in, and the enemy having no time to point his guns fired at random. In fact, we ran so near that the enemy did not, probably could not, depress his guns sufficiently.

While close under the lee of the island and during a lull in the storm and in the firing, one of our pilots heard a Confederate officer shout, "Elevate your guns!" It is probable that the muzzles of those guns had been depressed to keep the rain out, and that the officers ordered the guns elevated just in time to save us from the direct fire of the enemy's heaviest fort; and this, no doubt, was the cause of our remarkable escape.

Having passed the principal batteries, we were greatly relieved from suspense, patiently endured however, by the officers and crew. But there was another formidable obstacle in the way – a floating battery, which was the great "war elephant" of the Confederates, built to blockade the Mississippi permanently. As we passed her, she fired six or eight shots at us, but without effect. One ball struck the coal barge, and one was found in a bale of hay. We found also one or two musket bullets.

We arrived at New Madrid about midnight with no one hurt, and were most joyfully received by our army. At the suggestion of Paymaster Nixon, all hands "spliced the main brace."

On Sunday the 6th, after prayers and thanksgiving, the *Carondelet* with General Gordon Granger, Colonel J.L. Kirby Smith of the 43d Ohio, and Captain Louis H. Marshall of General Pope's staff on board made a reconnaissance twenty miles down nearly to Tiptonville, the enemy's forts firing on her all the way down. We returned their fire and dropped a few shells into their camps beyond.

On the way back, we captured and spiked the guns of a battery of one 32 pounder and one 24 pounder in about 25 minutes opposite Point Pleasant. Before we landed to spike the guns, a tall Confederate soldier with cool and deliberate courage posted himself behind a large cottonwood tree and repeatedly fired upon us until our Illinois sharpshooters got to work on him from behind the hammock nettings. He had two rifles, which he soon dropped, fleeing into the woods with his head down. The next day he was captured and brought into camp at Tiptonville, with the tip of his nose shot off.

After the capture of this battery, the enemy prepared to evacuate his positions on Island Number Ten and the adjacent shores, and thus, as one of the historians of the civil war says, the *Carondelet* struck the blow that secured that victory.

Attack on New Madrid

Returning to New Madrid, we were instructed by General Pope to attack the enemy's batteries of six 64-pounders which protected his rear; and besides, another gunboat was expected. The *Pittsburgh* (Lieutenant Commander Thompson) ran the gauntlet without injury, during a thunder storm, at 2:00 in the morning of April 7th, and arrived at 5:00 o'clock; but she was not ready for service. The *Carondelet* attack the principal batteries at Watson's Landing alone and had nearly silenced them when the *Pittsburgh* came up astern and fired nearly over the *Carondelet's* upper deck, after she and the Confederates had ceased firing.

I reported to General Pope that we had cleared the opposite shores of the enemy and were ready to cover the crossing of the river and the landing of the army. Seeing themselves cut off, the garrison at Island Number Ten surrendered to Foote on the 7th of April, the day of the Confederate repulse at Shiloh. The other Confederates retreating before Pope's advance were nearly all overtaken and captured at 4:00 o'clock on the morning of the 8th; and about the same time, the cavalry under Colonel W.L. Elliott took possession of the enemy's deserted works on the Tennessee shore.

The result of General Pope's operations in connection with the services of the *Carondelet* below Island Number Ten was the capture of three generals (including General W.W. Mackall, who ten days before the surrender had succeeded General John P. McCown in the command at Madrid Bend), over 5,000 men, twenty pieces of heavy artillery, 7,000 stand of arms, and a large quantity of ammunition and provisions, without the loss of a man on our side.

On the 12th, the *Benton* (flag steamer) with the *Cincinnati*, *Mound City*, *Cairo* at St. Louis, passed Tiptonville and signaled the *Carondelet* and *Pittsburgh* to follow.

> *The Confederate fleet called the "River Defense" having been reinforced, they determined upon capturing the mortar boats or giving us battle.*

Battle on the Mississippi

Five Confederate gunboats came up the next day and offered battle; but after the exchange of a few shots at long range, they retired down the river. We followed them all the way to Craighead's Point, where they were under cover of their fortifications at Fort Pillow. I was not aware at the time that we were chasing the squadron of my esteemed shipmate of the U. S. Frigates *Cumberland* and *Merrimac*, Colonel John W. Dunnington, who afterward fought so bravely at Arkansas Post.

On the 14th, General Pope's army landed about six miles above Craighead's Point near Osceola under the protection of the gunboats. While he was preparing to attack Fort Pillow, Foote sent his executive officer twice to me on the *Carondelet* to inquire whether I would undertake, with my vessel and two or three other gunboats, to pass below the fort to cooperate with General Pope, to which inquiries I replied that I was ready at any time to make the attempt. But Pope and his army (with the exception of 1,500 men) were ordered away and the expedition against Fort Pillow was abandoned.

Between the 14th of April and the 10th of May, two or three of the mortar boats were towed down the river and moored near Craighead's Point, with a gunboat to protect them. They were employed in throwing 13-inch bombs across the point into Fort Pillow, two miles distant. The enemy returned our bombardment with vigor, but not with much accuracy or effect. Several of their bombs fell near the gunboats when we were three miles from the fort.

The Confederate fleet called the "River Defense" having been reinforced, they determined upon capturing the mortar boats or giving us battle. On the 8th, three of their vessels came to the point from which the mortar boats had thrown their bombs, but finding none returned.

Foote had given special orders to keep up steam and be ready for battle any moment, day or night. There was so much illness at that time in the flotilla that about a third of the officers and men were under medical treatment, and a great many were unfit for duty. On the 9th of May, at his own request, our distinguished commander-in-chief, Foote, was relieved from his arduous duties. He had become very much enfeebled from the wounds received at Fort Donelson and from illness. He carried with him the sympathy and regrets of all his command. He was succeeded by Flag Officer Charles Henry Davis, a most excellent officer.

This paper would not be complete without some account of the naval battles fought by the flotilla immediately after the retirement of Flag Officer Foote, under whose supervision and amid the greatest embarrassments it had been built, organized, and equipped. On the morning of the 10th of May, a mortar boat was towed down the river, as usual, at 5:00 a.m., to bombard Fort Pillow. The *Cincinnati* soon followed to protect her. At 6:35 eight Confederate rams came up the river at full speed. The *Carondelet* at once prepared for action and slipped her hawser to the "bare end," ready for orders to "go ahead."

No officer was on the deck of the *Benton* (flag steamer) except the pilot, Mr. Birch, who informed the flag officer of the situation and passed the order to the *Carondelet* and *Pittsburgh* to proceed without waiting for the flag steamer. General signal was also made to the fleet to get under way, but it was not visible on account of the light fog.

The *Carondelet* started immediately after the first verbal order; the others, for want of steam or some other cause, were not ready except the *Mound City*, which put off soon after we were fairly on our way to the rescue of the *Cincinnati*.

We had proceeded about a mile before our other gunboats left their moorings. The rams were advancing rapidly, and we steered for the leading vessel, *General Bragg*, a brig-rigged, sidewheel steam ram, far in advance of the others and apparently intent on striking the *Cincinnati*.

When about three quarters of a mile from the *General Bragg*, the *Carondelet* and *Mound City* fired on her with their bow guns, until she struck the *Cincinnati* on the starboard quarter, making a great hole in the shell room, through which the water poured with resistless force. The *Cincinnati* then retreated up the river and the *General Bragg* drifted down, evidently disabled.

The *General Price*, following the example of her consort, also rammed the *Cincinnati*. We fired our bow guns into the *General Price*, and she backed off disabled also. The *Cincinnati* was again struck by one of the enemy's rams, the *General Sumter*.

Having pushed on with all speed to the rescue of the *Cincinnati*, the *Carondelet* passed her in a sinking condition, and rounding to, we fired our bow and starboard broadside guns into the retreating *General Bragg* and the advancing rams, *General Jeff. Thompson*, *General Beauregard*, and *General Lovell*. Heading upstream, close to a shoal, the *Carondelet* brought her port broadside guns to bear on the *Sumter* and *Price*, which were dropping downstream.

At this crisis the *Van Dorn* and *Little Rebel* had run above the *Carondelet*; the *Bragg*, *Jeff. Thompson*, *Beauregard*, and *Lovell* were below her. The last three, coming up, fired into the *Carondelet*. She returned their fire with her stern guns, and, while in this position, I ordered the port rifled 50-pounder Dahlgren gun to be leveled and fired at the center of the *Sumter*. The shot struck the vessel just forward of her wheelhouse, and the steam instantly poured out from her ports and all parts of her casemates. We saw her men running out of them and falling or lying down on her deck.

None of our gunboats had yet come to the assistance of the *Carondelet*. The *Benton* and *Pittsburgh* had probably gone to aid the *Cincinnati*, and the *St. Louis* to relieve the *Mound City*, which had been badly rammed by the *Van Dorn*. The smoke at this time was so dense that we could hardly distinguish the gunboats above us. The upper deck of the *Carondelet* was swept with grapeshot and fragments of broken shell.

Some of the latter were picked up by one of the sharpshooters, who told me they were obliged to lie down under shelter to save themselves from the grape and other shot of the *Pittsburgh* above us and from the shot and broken shell of the enemy below us. Why some of our gunboats did not fire into the *Van Dorn* and *Little Rebel* while they were above the *Carondelet*, and prevent their escape, if possible, I never could make out.

As the smoke rose we saw that the enemy was retreating rapidly and in great confusion. The *Carondelet* dropped down to within half a mile above Craighead's Point and kept up a continual fire upon their vessels, which were very much huddled together. When they were nearly, if not quite beyond gunshot, the *Benton* having raised sufficient steam came down and passed the *Carondelet*, but the Confederates were under the protection of Fort Pillow before the *Benton* could reach them.

Our fleet returned to Plum Point, except the *Carondelet*, which dropped her anchor on the battlefield, two miles or more below the point and remained there two days on voluntary guard duty.

This engagement was sharp, but not decisive. From the first to the last shot fired by the *Carondelet* one hour and ten minutes elapsed. After the battle, long-range firing was kept up until the evacuation of Fort Pillow.

Fort Pillow Surrenders

On the 25th, seven of Colonel Ellet's rams arrived, a useful acquisition to our fleet. During the afternoon of June 4th, heavy clouds of smoke were observed rising from Fort Pillow, followed by explosions, which continued through the night. The last of which, much greater than the others, lit up the heavens and the Chickasaw bluffs with a brilliant light and convinced us that this was the parting salute of the Confederates before leaving for the lower Mississippi.

At dawn the next morning, the fleet was all astir to take possession of Fort Pillow, the flag steamer leading. We found the casemates, magazines, and breastworks blown to atoms.

On our way to Memphis, the enemy's steamer *Sovereign* was intercepted by one of our tugs. She was run ashore by her crew, who attempted to blow her up, but were foiled in their purpose by a boy of sixteen whom the enemy had pressed into service, who, after the abandonment of the vessel took the extra weights from the safety valves, opened the fire-doors and flue-caps, and put water on the fires, and, having procured a sheet, signaled the tug which came up and took possession. It may be proper to say that on our way down the river we respected private property and did not assail or molest any except those who were in arms against us.

Battle of Memphis

The morning of the 6th of June, we fought the battle of Memphis, which lasted one hour and ten minutes. It was begun by an attack upon our fleet by the enemy, whose vessels were in double line of battle opposite the city. We were then at a distance of a mile and a half or two miles above the city. Their fire continued for a quarter of an hour, when the attack was promptly met by two of our ram squadron, the *Queen of the West* (Colonel Charles Ellet) leading and the *Monarch* (Lieutenant Colonel A. W. Ellet, younger brother of the leader).

These vessels fearlessly dashed ahead of our gunboats, ran for the enemy's fleet, and at the first plunge succeeded in sinking one vessel and disabling another. The astonished Confederates received them gallantly and effectively. The *Queen of the West* and *Monarch* were followed in line of battle by the gunboats, under the lead of Flag Officer Davis, and all of them opened fire, which was continued from the time we got within good range until the end of the battle – two or three tugs keeping all the while a safe distance astern.

The *Queen of the West* was a quarter of a mile in advance of the *Monarch* and after having rammed one of the enemy's fleet, she was badly rammed by the *Beauregard*, which then, in company with the *General Price*, made a dash at the *Monarch* as she approached them. The *Beauregard*, however, missed the *Monarch* and struck the *General Price* instead on her port side, cutting her down to the waterline, tearing off her wheel instantly, and placing her *hors de combat*.

The *Monarch* then rammed the *Beauregard*, which had been several times raked fore and aft by the shot and shell of our ironclads, and she quickly sank in the river opposite Memphis. The *General Lovell*, after having been badly rammed by the *Queen of the West*, was struck by our shot and shell, and at about the same time and place as the *Beauregard*, sank to the bottom so suddenly as to take a considerable number of the officers and crew down with her, the others being saved by small boats and our tugs.

The *Price, Little Rebel* (with a shot-hole through her stream-chest), and our *Queen of the West*, all disabled, were run on the Arkansas shore opposite Memphis. The *Monarch* afterward ran into the *Little Rebel* just as our fleet was passing her in pursuit of the remainder of the enemy's fleet, then retreating rapidly down the river.

The *Jeff. Thompson*, below the point and opposite President's Island, was the next boat disabled by our shot. She was run ashore, burned, and blown up. The Confederate ram *Sumter* was also disabled by our shell and captured. The *Bragg* soon after shared the same fate and was run ashore, where her officers abandoned her and disappeared in the forests of Arkansas. All the Confederate rams which had been run on the Arkansas shore were captured. The *Van Dorn*, having a start, alone escaped down the river. The rams *Monarch* and *Switzerland* were dispatched in pursuit of her and a few transports, but returned without overtaking them, although they captured another steamer.

The scene at this battle was rendered most sublime by the desperate nature of the engagement and the momentous consequences that followed very speedily after the first attack. Thousands of people crowded the high bluffs overlooking the river. The roar of the cannon and shell shook the houses on shore on either side for many miles. First wild yells, shrieks, and clamors, then loud, despairing murmurs, filled and affrighted city. The screaming, plunging shell crashed into the boats, blowing some of them and their crews into fragments, and the rams rushed upon each other like wild beasts in deadly conflict. Blinding smoke hovered about the scene of all this confusion and horror; and, as the battle progressed and the Confederate fleet was destroyed, all the cheering voices on shore were silenced. When the last hope of the Confederate gave way, the lamentations which went up from the spectators were like cries of anguish.

Boats were put off from our vessels to save as many lives as possible. No serious injury was received by any one on board the United States fleet. Colonel Ellet received a pistol shot in the leg. A shot struck the *Carondelet* in the bow, broke up her anchor and anchor stock, and fragments were scattered over her deck among her officers and crew, wounding slightly Acting Master Gibson and two or three others who were standing at the time on the forward deck with me.

The heavy timber which was suspended at the waterline, to protect the boats from the Confederate rams, greatly impeded our progress, and it was therefore cut adrift from the *Carondelet* when that vessel was in chase of the *Bragg* and *Sumter*. The latter had just landed a number of her officers and crew, some of whom were emerging from the bushes along the bank of the river, unaware of the *Carondelet's* proximity, when I hailed them through a trumpet, and ordered them to stop or be shot. They obeyed immediately, and by my orders were taken on board a tug and delivered on the *Benton*.

General Jeff. Thompson, noted in partisan or border warfare, having signally failed with those rams at Fort Pillow, now resigned them to their fate. It was said that he stood by his horse watching the struggle and seeing at last his rams all gone, captured, sunk, or burned, he exclaimed, philosophically, "They are gone, and I am going," mounted his horse, and disappeared.

An enormous amount of property was captured by our squadron. In addition to the Confederate fleet, we captured at Memphis six large Mississippi steamers, each marked "C.S.A." We also seized a large quantity of cotton in steamers and on shore, and the property at the Confederate Navy Yard, and caused the destruction of the *Tennessee*, a large steam ram, on the stocks, which was to have been a sister ship to the renowned *Arkansas*. About one hundred Confederates were killed and wounded and one hundred and fifty captured. Chief of all results of the work of the flotilla was the opening of the Mississippi River once for all from Cairo to Memphis, and the complete possession of Western Tennessee by the Union forces.

Hi-Lights of a Hero's Life
Nathan Bedford Forrest

★ 1821 – (July 13) – Born in Bedford County, Tennessee.

★ 1841 – Joined a company of soldiers to protect Texas from a rumored invasion.

★ 1845 – (March 20) Shot and killed two men and wounded two others to avenge his uncle's death.

★ 1845 – (September 25) Married May Ann Montgomery.

★ 1846 – Son William was born.

★ 1836 – Daughter Fanny was born. She died in 1853.

★ 1851 – Moved to Memphis. He dealt in cotton, plantations, livestock, and slaves.

★ 1857 – Risked his life twice to save a man he did not know from being lynched by a mob.

★ 1858 – Was elected as a Memphis city alderman.

★ 1859 – Re-elected but resigned his seat to work a 2,000 acre plantation in Mississippi.

★ 1861 – Enlisted as a private in the Confederate army. Received instructions at Fort Wright.

★ 1861 – Joined and outfitted Company E Tennessee Mounted Rifle.

★ 1861 – (July 10) Promoted to lieutenant colonel and recruited the 7th Tennessee.

★ 1862 – (February 6) Battle of Fort Donelson.

★ 1862 – Promoted to colonel.

- ★ 1862 – (April 6-7) Wounded during the Battle of Shiloh.

- ★ 1862 – (July 13) Fought at the First Battle of Murfreesboro.

- ★ 1862 – (July 21) Promoted to brigadier general and took command of the 3rd Tennessee.

- ★ 1862 – (Fall to Spring 1863) Conducted raids in Tennessee, Kentucky, and Mississippi.

- ★ 1863 – (September 19-20) Fought bravely at the Battle of Chickamauga.

- ★ 1863 – (December 4) Promoted to major general.

- ★ 1864 – (March 25) Participated in the Battle of Paducah.

- ★ 1864 – (April 12) Captured Fort Pillow.

- ★ 1864 – (June 10) Fought brilliantly at Brice's Crossroad.

- ★ 1864 – (July 14-15) Lost at Tupelo.

- ★ 1864 – (August-October) Conducted raids in Tennessee.

- ★ 1864 – (November 29) Participated at the Battle of Spring Hill.

- ★ 1864 – (November 30) Survived the Battle of Franklin.

- ★ 1864 – (December 5-7) Participated in the Third Battle of Murfreesboro.

- ★ 1864 – (December 15-16) Fought at the Battle of Nashville.

- ★ 1865 – (February 28) Promoted to lieutenant general.

- ★ 1865 – (May 9) Surrendered his troops.

- ★ 1865 – Employed by the Marion and Memphis Railroad. Became company president.

- ★ 1867 – Joined the Ku Klux Klan in Tennessee and served as the Klan's Grand Wizard.

- ★ 1869 – Dissolved the Klan after Governor Brownlow was defeated.

- ★ 1873 – Financially ruined during the Panic of 1873.

- ★ 1875 – Leased land on President Island and worked convicts as farmhands.

- ★ 1877 – (October 29) Died and was buried at Elmwood Cemetery.

- ★ 1904 – His remains were disinterred and moved to Forrest Park in Memphis.

Fight Between the Batteries and Gunboats at Fort Donelson

By H.L. Bedford. *Southern Historical Society Papers, Volume 12, pages 166 – 172.*

The reports of Colonel James E. Bailey, commander of the garrison proper, and of Captain Jacob Culbertson, commander of the water batteries, are correct, and, as official documents, I suppose are complete; but they do not convey to the reader the disadvantages under which the batteries labored in this contest. The operations of the army at this place having proved disastrous to the Confederate cause, it has been condemned as a strategic point, and no one seems particularly anxious to acknowledge the responsibility of its selection.

It was the general impression at the Fort that its location had been ordered by the Tennessee authorities as being the most eligible point on the Cumberland River, in close proximity to Fort Henry, on the Tennessee. The original intention evidently was the obstruction of the Cumberland. The engineer in charge, Lieutenant Dixon, while tracing the outlines of the earthworks, never dreamed that a persistent stand against an invading army would ever be attempted, and I feel warranted in suggesting that General Albert Sidney Johnston regarded it simply as a protection to his rear.

When I received orders in October 1861 to report there as Instructor of Artillery, Colonel E.W. Munford, aide to General Johnston, informed me that he was instructed by his chief to impress upon me that the Cumberland river cut his rear, and the occupation of Bowling Green was dependent upon the proper guarding of that stream. If, then, Fort Donelson was intended to prevent the passage of gunboats, its location was an admirable one; it accomplished its mission, and its founder need feel no hesitation in claiming its paternity.

Nor does the final result of the operations of the land forces necessarily convict General Johnston of a mistake in the reinforcement of Donelson. At that time he was believed to possess that ability as a general which events soon verified, and his condemnation will have to rest on surer proofs than the charges of flippant writers. To the average mind the whole matter resolves itself into the simple question: Whether General Johnston sufficiently reinforced Fort Donelson to successfully resist the forces that invaded the State of Tennessee under General Grant by way of Fort Henry; and, if so, is he fairly chargeable with the blunders of his generals, in allowing themselves to be cooped in temporary trenches until reinforcements to the enemy could come up the Cumberland?

Any close student of the *Operations at Fort Donelson*, embraced in series Number 1, Volume 7, of the *Records of the Rebellion*, will probably detect by whom the mistakes were made. It is doubtless there recorded when and where the opportunity of withdrawing the Confederate forces was disregarded; that General Johnston was unfortunate in the selection, or rather the grouping of his lieutenants on this occasion is beyond controversy. His army consisted of raw recruits; his generals were ready made for him; their commissions were presumptions of merit; there had been no opportunity for development, and he had no alternative but to accept the patents of ability issued to them by the War Department.

The senior general arrived at the eleventh hour and seems to have been lacking in disposition or in power to hold his second in due subjection. The latter had been on the ground for about a week. He was full of energy and physical activity and possessed rare executive ability. He was restless under restraint, probably prone to insubordination, and it was almost impossible for him to yield his scepter to a new comer. He gave orders affecting the whole army without any known rebuke or remonstrance from his chief.

The performances of these two chieftains afford an apt illustration of a very homely old saying that will readily recur to most of you. This rule of duality of commanders, according to some of the official reports, seems to have obtained in the heavy batteries, but as it was not then known or recognized, it did not create any confusion. When I reported there for duty very little in the way of defense had been accomplished. Two 32-pounders carronades had been mounted on the river, and three 32-pounders were temporarily mounted on the crest of the bluff. The carronades were utterly useless, except against wooden boats at close quarters, while the three guns on the hill, on account of position, could not be made effectual against ironclads. The garrison, in command of Lieutenant Colonel Randle McGavock consisted of a part of Colonel Heiman's 10th Tennessee regiment, the nucleus of Colonel Sugg's 50th Tennessee (then called Stacker's regiment), and Captain Frank Maney's light battery.

As there were no heavy artillerists, Captain Beaumont's company of 50th Tennessee had been detailed for that duty. At the time of my arrival, there was considerable excitement at the Fort. Smoke was seen rising a few miles down the river, the long roll was being beat, and there was hurrying to and fro. Companies were getting under arms and into line with the rapidity of zealots, though wanting in the precision of veterans. The excitement subsided as the smoke disappeared. In a short while the companies were dismissed, and the men resumed their wonted avocations.

The local engineer was also in charge of the works at Fort Henry and was necessarily often absent. His duties were onerous and manifold. I, therefore, volunteered to remount the three 32-pounders and place them in the permanent battery; and as the completion of the defense was considered of more importance than the drilling of artillery, I was kept constantly on engineering duty until after the investment. General Tilghman arrived about the middle of December and took command. He manifested a good deal of interest in forwarding the work. The 50th Tennessee regiment (Colonel Suggs) was organized. The 30th Tennessee (Colonel Head) and the 49th Tennessee (Colonel Bailey) reported, and these, with Maney's light battery, constituted the garrison, Lieutenant Colonel McGavock having rejoined Colonel Heiman at Fort Henry.

    The work for the completion of the defenses and for the comfort of the soldiers was pushed on as rapidly as the means at hand would permit. There was no lagging, nor lukewarmness, nor shirking of duty. As one of the many evidences of the zeal manifested by the garrison, I would state that whenever a detail for work of any magnitude was made from any of the regiments, a field officer usually accompanied it, in order to secure promptness and concert of action. This, I believe, was the invariable rule with the 49th Tennessee.

    At the time of the arrival of reinforcements, the water batteries were not in that state of incompleteness and disorder which the report of a general officer charges, nor was there any gloom or despondency hanging over the garrison. It is true there was some delay in getting the 10-inch Columbiad in working condition, but no one connected with the Fort was responsible for it. The gun was mounted in ample time, but upon being tested, it came very nearly being dismounted by the running back of the carriage against the hurters.

    It was necessary to increase the inclination of the chassis, which was accomplished by obtaining larger rear traverse wheels from the iron works just above Dover. It was still found, even with a reduced charge of powder, that the recoil of the carriage against the counter hurters was of sufficient force to cut the ropes tied there as bumpers. There was no alternative but to dismount the piece and lower the front half of the traverse circle. By this means the inclination of the chassis was made so steep that the piece was in danger of getting away from the gunners when being run into battery and of toppling off in front.

    Any paper upon the subject of Fort Donelson would be incomplete without the mention of Lieutenant Colonel Wilton A. Haynes of the Tennessee artillery. He was, in the nomenclature of the volunteers, a "West Pointer," and was an accomplished artillerist. He came to Fort Donelson about the middle of January and found the Instructor of Artillery engaged in engineering duty and nothing being done in familiarizing the companies detailed for artillery service with their pieces. He organized an artillery battalion and made a requisition on General Polk at Columbus for two drill officers, and whatever of proficiency these companies attained as artillerists is due to him. He was physically unable to participate in the engagements and this may account for the failure of recognition in the official reports.

    The artillery battalion as organized by Colonel Haynes was fully competent to serve the guns with success, but General Pillow deemed otherwise and proceeded to the mistake of assigning Lieutenant Dixon to the command of the heavy batteries instead of attaching him to his personal staff, and availing himself of that officer's familiarity as an engineer with the topography of the battleground and of the surrounding country. The assignment was particularly unfortunate inasmuch as Dixon was killed before the main fight and the batteries were not only deprived of his services for that occasion, but the Confederate army lost an able engineer.

It must be remembered, however, that the great fear was of the gunboats. It was apprehended that their recent achievements at Fort Henry would be repeated at Donelson, and it was natural that the commanding general should make every other interest subservient to the efficiency of the heavy batteries.

The river defenses consisted of two batteries. The upper one was on the river bank immediately abreast of the earthworks. It was crescent shaped and contained one 32-pound caliber rifle gun and two 32-pounder carronades. The other battery was some 150 yards lower down and consisted of eight 32-pounders and one 10-inch Columbiad.

This lower battery, although essentially a straight line, ran *en echelon* to the left over the point of a hill that made down obliquely from the earthworks to the river, with the right piece resting on the brink of the river bank, and the Columbiad over in the valley of a stream, emptying into the river, some 150 yards lower down.

The backwater in this stream protected the batteries from a direct assault. About 900 yards below the lower battery, a floating abatis was placed in the river for the purpose of preventing the passage of boats. This was done by anchoring full length trees by the roots and allowing the tops to float. In ordinary stages of water this might have offered some impediment, but at the time of the attack the river was very high and the boats passed over without the least halt or break in their line of approach.

> *On the morning of the 14th, dense volumes of smoke were seen rising from down the river. It was evident that transports were landing troops.*

In all the accounts that I have seen from the Federal side, the armament of the water batteries is overestimated. Flag Officer Foot reports that there must have been about twenty heavy guns and General Lew Wallace places it at seventeen. Admiral Walke, while correctly stating the number in the lower battery, is in error in claiming that the upper was about the same in strength.

On the morning of the 12th of February, the finishing touches were put to the Columbiad, and the batteries were pronounced ready for gunboats, whereupon Lieutenant Dixon proceeded to the assignment of the guns. Captain R.R. Ross of the Maury Company Light Artillery whose company had been ordered to heavy batteries by General Pillow was placed in command of the rifle gun and the two carronades. Captain Beaumont's 50th Tennessee, Company A, and Captain Bidwell's company, 30th Tennessee, worked the 32-pounders, and the Columbiad was turned over to my command with a detachment of twenty men under Lieutenant Sparkman from Captain Ross' company to work it.

I received private instructions to continue the firing with blank cartridges, in the event the gun should dismount itself in action. The drill officers, Lieutenants McDaniel and Martin, were assigned to the 32-pounders, while Captains Culbertson and Shaster had special assignments or instructions, the nature of which I never knew.

As the artillerists, who were to serve the rifle and Columbiad, had no experience with heavy guns, most of them probably never having seen a heavy battery until that morning, it was important that they should be instructed in the manual of their pieces. Drilling, therefore, began immediately but had continued for a short time only when it was most effectually interrupted by the appearance of a gunboat down the river, which subsequently was ascertained to be the *Carondelet*. She fired about a dozen shots with remarkable precision and retired without any response from the batteries.

On the morning of the 13th drilling was again interrupted by the firing of this boat, and the same thing happened in the afternoon. It really appeared as if the boat was diabolically inspired, and knew the most opportune times to annoy us. Sometime during the day, probably about noon, she delivered her fire with such accuracy that forbearance was no longer endurable, and Lieutenant Dixon ordered the Columbiad and rifle to respond.

The first shot from the Columbiad passed immediately over the boat, the second fell short, but the third was distinctly heard to strike. A cheer of course followed, and Lieutenant Dixon, in the enthusiasm of the moment, ordered the 32-pounders to open fire, although the enemy was clearly beyond their range. The *Carondelet*, nothing daunted, continued the action, and soon one of her shells cut away the right cheek of one of Captain Bidwell's guns, and a flying nut passed through Lieutenant Dixon's head killing him instantly.

> *Several well directed shots raked the side and tore away her armor, according to the report of Lieutenant Sparkman, who was on the lookout.*

In this engagement, the flange of one of the front traverse wheels of the Columbiad was crushed and a segment of the front half of the traverse circle was cupped, both of which proved serious embarrassments in the action next day.

On the morning of the 14th, dense volumes of smoke were seen rising from down the river. It was evident that transports were landing troops. Captain Ross became impatient to annoy them, but having no fuse shells to his guns, he came over to the Columbiad and advised the throwing of shells down the river. The commander declined to do so without orders, whereupon Captain Culbertson, who had succeeded Lieutenant Dixon in the command of the batteries, was looked up, but he refused to give the order upon the ground that it would accomplish no good and that he did not believe in the useless shedding of blood.

Captain Ross, not to be outdone, set himself to the task of procuring the necessary order and returned to the Columbiad about 3:00 p.m. with a verbal order from General Floyd to harass the transports. In obedience to this order, we prepared to shell the smoke. A shell was inserted, the gun was given the proper elevation, the lanyard was pulled, and the missile went hissing over the bend of the river, plunged into a bank of smoke, and was lost to view. This was called by an army correspondent, claiming to have been on one of the gunboats, a shot of defiance. Before the piece could be reloaded, the prow of a gunboat made its appearance around the bend, quickly followed by three others, and arranging themselves in line of battle, steamed up to the attack.

When they had arrived within a mile and a half of the batteries, a solid shot having been substituted for a shell, the Columbiad began the engagement with a ricochet shot, the rifle gun a ready second. The gunboats returned the fire, right center boat opening, the others following in quick succession. After the third discharge the rifle remained silent on account of becoming accidentally spiked. This had a bad effect on the men at the Columbiad, causing them considerable uneasiness for their comrades at the upper battery.

The Columbiad continued the action unsupported until the boats came within the range of the 32-pounders, when the engagement became general, with ten guns of the batteries opposed to the twelve bow guns of the ironclads supplemented by those of the two wooden boats that remained in the rear throwing curvated shells.

As the boats drew nearer, the firing on both sides became faster until it appeared as if the battle had dwindled into a contest of speed in firing. When they arrived within 300 yards of the lower battery they came to a stand, and then it was that the bombardment was truly terrific. The roar of cannons was continuous and deafening, and commands, if necessary, had to be given by signs.

Pandemonium itself would hardly have been more appalling, but neither chaos nor cowardice obtruded themselves, and I must insist that General Wallace and Admiral Walke are mistaken in their assertions that the gunners were seen running from their guns. It is true there was some passing from the batteries to the Fort, but not by the artillerists in action, and as the passage was over an exposed place, in fact across the field of fire of the gunboats, it is a fair presumption that the transit was made as swiftly as possible.

Of one thing I am certain, there was no fleeing from the Columbiad, and although her discharges were necessarily very slow, I think every one in hearing that day will testify that her boom was almost as regular as the swinging of a pendulum. If these two Federal officers saw her condition when surrendered, they will admit that if was not likely that panic stricken cannoniers could have carried her safely through such a furious bombardment, especially to have done the execution with which she is accredited.

In his contribution to the *Century* of December 1884, doubtless by the cursory reading of Captain Bidwell's report, General Wallace is lead into the mistake of saying that each gunner selected his boat and stuck to her during the engagement. I am satisfied that the experienced officers who acted as gunners did not observe this rule. The Columbiad was rigidly impartial and fired on the boats as chance or circumstances dictated with the exception of the last few shots which were directed at the *Carondelet*.

This boat was hugging the eastern shore and was a little in advance of the others. She offered her side to the Columbiad, which was on the left and the most advanced gun of the batteries. Several well directed shots raked the side and tore away her armor, according to the report of Lieutenant Sparkman, who was on the lookout. Just as the other boats began to drift back, the *Carondelet* forged ahead for about a half length, as though she intended making the attempt to pass the battery, and it is presumable that she then received the combined fire of all the guns.

It is claimed that if Hannibal had marched on Rome immediately after the battle of Cannae, he could have taken the city, and by the same retrospective reasoning, it is probable that if Admiral Foote had stood beyond the range of 32-pounders he could have concentrated his fire on two guns. If his boats had fired with the deliberation and accuracy of the *Carondelet* on the previous day, he could have dismounted those guns, demolished the 32-pounders at his leisure, and shelled the Fort to his heart's content. But flushed with his victory at Fort Henry, his success there paved the way for his defeat at Donelson, a defeat that might have proved more disastrous could the Columbiad have used a full charge of powder and the rifle gun participated in the fight.

After the battle three of the gunboats were seen drifting helplessly down the stream, and a shout of exultation leaped from the lips of every soldier in the fort. It was taken up by the men in the trenches, and for awhile a shout of victory, the sweetest strain to the ears of those who win, reverberated over the hills and hollows around the little village of Dover.

While the cannoniers were yet panting from their exertion, Lieutenant Colonel Robb of the 49th Tennessee, who fell mortally wounded the next day, ever mindful of the comfort of those around him sent a grateful stimulant along the line of guns. Congratulations were the order of the hour. Generals Floyd and Pillow personally complimented the artillerists. They came to the Columbiad, called for the commander, and after congratulating him upon the performances of that day, promised that if the batteries would continue to keep back the gunboats, the infantry of their command would keep the land forces at a safe distance. That officer, who had been watching the smoke of the transports landing reinforcements, as he stood there before these generals, just thirty-six hours before surrender, receiving their assurances of protection, wondered if they were able to fulfill the promise, or if they were merely indulging an idle habit of braggadocio.

WHY WE FOUGHT

By Colonel Edward McCrady, Jr. *Southern Historical Society Papers, Volume 16*, pages 246 - 260.

It is with divided feelings, my comrades, that we meet upon this occasion. It is indeed doubtful which emotion is the stronger, that of pleasure in once more grasping the hands of those of us who survive, or of sadness in missing those who are not here to answer to our roll call. And so it must be with us on all such reunions as this.

Our bands are daily becoming smaller and smaller. No volunteers or recruits can now be enrolled in our ranks; nor any conscripts sent unwillingly to join us. In a few short years the coming generation will look with curiosity, at least, if we may not bespeak reverence upon any one who may live to say that he fought at Manassas or Gettysburg, who can tell how he marched with Jackson to victory, and perchance how at last he laid down his arms with Lee at Appomattox. Is it not natural then that we should draw closer together while we live, and that we should sometimes meet as we have done today to recall the times when together we offered our lives and shed our blood for our State, and suffered cold and hunger and thirst and sickness for the faith in which we were reared, and for the cause which we still maintain to have been righteous even though lost?

For what then did we fight? It is well, my comrades, that we who survive should take such occasions as this to tell to those who are growing up around us what were the great causes which impelled the young and the old of that time, the rich and the poor, the learned and ignorant, to take up arms and risk their lives in battle.

It has been said by a great historian that a man who risked and lost his life for a cause he believed a just one, though he was mistaken in so believing, is not among those whose fate deserves the most compassion, or whose career is least to be envied. But we were not mistaken in the cause for which we fought. We did not fight for slavery. Slavery was a burden imposed upon us by former generations of the world. A burden increased upon us by the falsely pretended philanthropic legislation of Northern States, which legislation did not

> **We did not fight for slavery. Slavery was a burden imposed upon us by former generations of the world. A burden increased upon us by the falsely pretended philanthropic legislation of Northern States, which legislation did not emancipate their slaves but forced them to be sent to the South and sold here.**

emancipate their slaves, but forced them to be sent to the South and sold here. Slavery was not the cause of the war, but the incidents upon which the differences between the North and the South and from which differences the war was inevitable from the foundation of our government were the cause.

That it was not slavery in itself for which we fought is shown by the thousands and thousands of volunteers who owned no slaves, and yet who were the first to hasten into our ranks. Take the instance of our own State.

The census of 1860 shows that there were but 26,701 slaveholders in South Carolina, and yet she gave 44,000 volunteers during the first eighteen months of the war. Supposing then that every slaveholder went into the service, we would have over 17,000 volunteers from the State who owned no slaves. But as you and I, my comrades, well know, the slaveholders, as a class, were by no means more prompt in offering their services than those who did not own slaves. You recollect that there was a provision in the Conscript Act actually exempting from service those who owned and worked a certain number of slaves. I think we may safely assume that two thirds of the volunteers owned no slaves.

I say it was not for slavery for which we fought, but that it was for the sovereignty of our State and for the supremacy of our race. The instinct of our people felt that the one was involved in the other.

An Understanding of Slavery's History

We fought for State's rights and State's sovereignty as a political principle. We fought for the State of South Carolina, with a loyal love that no personal sovereign has ever aroused. But more, you and I, my comrades, whether owning slaves or not, could not but foresee with the conviction of certainty, the calamities that would that must follow that have followed the emancipation of the Negro by the fanatical party which, by a mere minority of votes, obtained possession of the government in 1860.

We of this generation had no part in the establishment of slavery in this country – as early as 1741, South Carolina unsuccessfully endeavored to check the importation of slaves with which the mother country was crowding the province. But we were born to the question: what was to be done with an institution which we had inherited from England, which had been augmented by the casting off the slaves of the North upon the South? Northern philanthropists who had sent and sold their slaves to the South might safely, if not honestly, advocate their emancipation. But with us the question was not only as to the positive good or evil of the institution, but what would the Negro be, and what would we do with him, and what would he do to us if freed?

Had slavery never existed, I believe the war between the two sections of this country was inevitable, and, as we know, had all but commenced in 1832 while on the other hand, its existence rendered the political principle of State sovereignty more than a sentiment and a theory, and made it a practical question essential to the South in dealing with that institution.

We were so unfortunate as to permit the great underlying question at issue between the North and the South to turn, apparently, solely upon a matter on which the fanaticism of the world had been aroused. But I maintain with Alexander Stephens that while "slavery, so called, that legal subordination of the black race to the white, which existed in all but one of the States when the Union was formed, and in fifteen of them when the war began, was unquestionably the occasion of the war – the main exciting proximate cause on both sides – on the one as well as on the other, it was not the real, ultimate cause, the casa causans of it." Further, I believe and maintain that from the origin of our government, the war was inevitable had slavery never existed.

The war was not commenced in December 1860, when this State seceded or in April 1861, when we fired into Fort Sumter. Its seeds were in the Constitution, and it was declared in the Kentucky and Virginia Resolutions in 1798. The Convention which framed the Constitution was itself divided into the two parties which, after seventy years of discussion in the Senate chamber, adjourned the debate to the battlefields of our late war.

The one as the National party under the leadership of General Hamilton and the elder Adams and the other as the Federal party under Jefferson, at that early day organized the forces for strife and warred over the Virginia and Kentucky Resolutions and the Alien and Sedition Laws with a bitterness not exceeded in 1860.

As it is so often said that whatever may have been the nice theoretical distinctions as to the forms of government, the North became in favor of a strong consolidated central government, because its interests were in manufactures and protection, while the South was State's Rights in the defense of slavery and that thus the real cause of the war was the antagonism between free labor and slave labor. I would call attention to the fact that as early as 1796, a year before the first slave had been freed in the United States, when slavery still existed in every State in the Union, North as well as South, even then the different political theories of the government had already found for themselves more decidedly local habitations than names.

Washington, in his farewell address, observes: "In contemplating the causes which may disturb our Union, it occurs as a matter of serious concern that any ground should have been furnished for characterizing parties by geographical distinctions – Northern and Southern, Atlantic and Western – whence designing men may endeavor to excite a belief that there is a real difference of local interests and views."

Curiously enough, and may we not add pitifully, too, we read in his original drafts of this address a passage stricken out and on the margin, opposite the words "not important enough," which, when we come to examine, we find still more strongly indicated his apprehension for the Union from this very cause, i.e., the geographical location of parties.

It is well known that Mr. Jefferson, the author of the Kentucky Resolutions, was opposed to slavery, while on the other hand, the only vote in the First

Congress against the exclusion of slavery in the great Northwestern Territory – the munificent, or rather we should say under all the circumstances, looking now at it in the light of subsequent history, the prodigal and extravagant contribution of Virginia to the Union – came from the State of New York.

Mr. Davis, in his work on *The Rise and Fall of the Confederate Government* observes, "it was for climatic, industrial, and economical, not moral or sentimental reasons that slavery was abolished in the North, while it continued to exist in the Southern States." It was the climate and the soil that forbade African slavery there and not philanthropy.

History of Slavery in the North

Let us look at the facts. Vermont claims the honor of having first proposed to exclude slavery by her Bill of Rights in 1777, in anticipation of her separation from New York, but the census of 1790, the year before the separation took effect, shows that her frosts and snows had effectually done the work before, as there were in fact but seventeen slaves in the State to be emancipated.

Slavery was introduced into Massachusetts soon after its first settlement, and was so "tolerated" there that as late as 1833. Her Supreme Court could not say by what act, particularly, her institution was abolished. (Winchendon v. Hatfield. 4 Mass. 123. Commonwealth v. Aves, 18 Pick. 209.)

New Hampshire did not think it worth her while to pass an act to free the 158 slaves which only remained in that State in 1790 and so one of them lived a slave in that free State as late as 1840.

In the plantations of Rhode Island, slaves were more numerous than in the other New England States, as, indeed they well might be, when the merchants and sailors of this little State were the greatest traffickers in the slave trade. But as the Negro could not live in her latitude, the Rhode Islanders – the great Negro traders – provided a scheme of emancipation, which took a lifetime to work out, leaving in 1840 five slaves still in that State.

Connecticut was too much interested to indulge her philanthropy at the expense of a sudden emancipation. In 1790 there were 2,750 slaves, and so like Rhode Island, she adopted a gradual plan of emancipation, by the slow and prudent workings of which only 17 of her slaves remained as such in 1840.

Pennsylvania was in the same situation, having 3,737 slaves in 1790, and she, too, provided for gradual emancipation. The census of 1840 showed sixty-five Negroes still in slavery. In this

> ***Mark you, it was the Negroes their slave traders had landed upon their shores they wished to get rid of – not slavery. A provision of the law, then, that at a given day in the future all slaves would be free would accomplish the purpose because under such a law, the owners of slaves did not lose the value of their slaves, but were only required by a given time to send them to the South and to sell them there. This was the result of all the emancipation acts of the Northern States.***

State of Brotherly Love, as late as 1823, a Negro woman was sold by the sheriff to pay the debts of her master.

In New York, in which in 1790 there were 21,324 slaves, a similar act of gradual emancipation was passed (1799) by the operations of which in 1840, all but four slaves had been gotten rid of whether by emancipation, death, or shipment for sale at the South can only be conjectured.

New Jersey, though adopting the same scheme, was slower in getting rid of her slaves, 674 still remaining in 1840.

Now, my comrades, what did this scheme of gradual or future emancipation mean? You will at once see that if our Northern brethren had been earnest in freeing these people, in accordance with their righteous abhorrence of the institution of slavery and with their zealous love of universal freedom, they would all have been as philanthropic and disinterested as Vermont with her seventeen slaves and would have emancipated their Negroes as suddenly and more immediately than Mr. Lincoln did ours by his famous proclamation.

But such a course would have cost their citizens just the market value of their slaves. What, then could they do with these Negroes? The Negroes came from a warmer climate and could not live and thrive and be profitable with them. It was expedient, therefore, as an economical measure to get rid of the burden of their support and the plan of emancipation, at a given time in the future, would accomplish the purpose.

How? Mark you, it was the Negroes their slave traders had landed upon their shores they wished to get rid of – not slavery. A provision of the law, then, that at a given day in the future all slaves would be free would accomplish the purpose because under such a law, the owners of slaves did not lose the value of their slaves, but were only required by a given time to send them to the South and to sell them there. This was the result of all the emancipation acts of the Northern States.

The Northern people, as usual, beat us in the bargain. They sold their slaves to us, took our money for them, freed them without paying for them and then took credit for their philanthropy in freeing the Negroes they had sold to us.

> *There is an old proverb that the receiver is as bad as the thief. Unless history very much belies them, the righteous New Englanders notwithstanding their pious abhorrence of slavery have given a new reading to this old saw: that the receiver is worse than the thief.*

The Receiver is Worse than the Thief

Let us look at the conduct of our Northern brethren in another connection, and that in the worst feature with regard to slavery, and in doing so let us bear in mind that the superior morality and love of freedom in the North is supposed to have been peculiarly evinced in the suppression of this institution. If the Northern people were so zealous in freeing the Negroes from slavery, had they not been as active in putting them into slavery?

There is an old proverb that the receiver is as bad as the thief. Unless history very much belies them, the righteous New Englanders notwithstanding their pious abhorrence of slavery have given a new reading to this old saw: that the receiver is worse than the thief. They thought it no sin to fit out ships to steal Negroes to sell to Southerners, but their righteous souls were vexed at the idea that we should keep them in slavery after purchasing them.

During the four years that the ports of this State were opened for the slave trade (1804-1807), of the 202 vessels that arrived in Charleston harbor with slaves, sixty-one claimed to belong to Charleston and exactly the same number avowedly belonged to New England (Rhode Island fifty-nine, Boston one, Connecticut one) while seventy belonged to Britain. Of the other ten, three belonged to Baltimore, four to Norfolk, two to Sweden, and one to France.

I say the same number (61) claimed to belong to Charleston as avowedly belonged to New England, and, in using this expression, I, of course, mean to express my doubt if they did. I mean to say that a great number of these vessels which were claimed to belong to Charleston did not belong to Charleston, but were in fact owned by New Englanders or Old Englanders. If we look at the list of consignees, we will see that I am not probably mistaken in this supposition. Of the 202 vessels which brought in slaves, thirteen consignees were natives of Charleston, while eighty-one were natives of Rhode Island, ninety-one from Boston, and ten from France.

We may be very sure that every vessel really owned in Charleston was consigned to a Charlestonian, and we will not be very far wrong if we assume that all the eighty-eight vessels bringing slaves to Charleston, consigned to natives of Rhode Island, in fact belonged to Rhode Islanders, or at least to New Englanders. But there is further evidence that I am not mistaken in charging that Rhode Island had much more to do with this Negro importation than the people of this State, for it appears that but 2,006 of 39,075 slaves brought into Charleston were imported by our merchants and planters, while Rhode Islanders imported for us 8,338. (See Judge Smith's Statistics – Year Book City of Charleston, 1880.)

Again. More than fifty years after this, in 1858, the *London Times* charged that New York had become "the greatest slave trading mart in the world;" and Vice President Wilson, in his work upon the Rise and Fall of the Slave Power in America, quotes from the New York daily papers that there were "eighty-five vessels fitted out from New York from February 1859, to July 1860," for the slave trade. That "an average of two vessels each week clear out of our harbor, bound for Africa and a human cargo;" "that from 30,000 to 60,000 (Negroes) a year are taken from Africa to Cuba by vessels from the single port of New York." (Rise and Fall of Slave Trade in America, Volume II, page 618.)

Is it not absurd, with these historical facts upon record, for the Northern people, especially the New Englanders, to charge us with the moral offence of slavery?

The Differences Settle on Slavery

Slavery as an institution was doubtless the incident upon which the differences between the people of the North and the South settled and concentrated, but the moral offense of it that so aroused the fanaticism of the world was not the cause of the war. When slavery was prohibited in the Northwestern Territory in 1787, with the unanimous consent of the Southern delegates in Congress, but three of the Northern States had determined to put an end to slavery within their own borders, and of these three, Rhode Island and Pennsylvania freed no slaves then living, but only provided that those born after a certain time should be free. Vermont alone emancipated her seventeen slaves.

Franklin, it is true, had organized an Abolition Society in 1787. But for many years, during which the Federal and National parties continued their controversies as to the form of government, it was only proposed to bring to bear upon the institution of slavery the sentiment of the people of the States. The power of the Federal government to interfere in the matter was not even thought of.

The admission of Missouri, in 1820, no doubt was strenuously resisted because her Constitution permitted slavery and was only passed by Congress upon the compromise that slavery should not be introduced in the territories belonging then to the United States lying north of 36° 30'. But a moment's reflection will show that the moral offence of slavery could not have entered into the consideration of this compromise. For if slavery was wrong north of 36° 30', was it not wrong also south of it?

> *For if slavery was wrong north of 36° 30', was it not wrong also south of it?*

The opposition to the admission of more slave states arose from the fact that such states, by the Constitution, had representatives in Congress and in the Electoral College, not only for the white freemen but for three-fifths of their slaves also, which greatly added to their representation and power. That compromise was nothing more than the adjustment of the balance of political power between the states.

The admission of new states upon one condition or another, however affecting the interests of the slave states, was a fair subject of discussion. There was nothing in principle why a strict State's Rights Federalist might not have resisted the admission of another slave state, nor that one of the National party should not have advocated it. From other considerations, the Northern people were for the most part Consolidationists and Nationalists, while the Southern people were strict constructionists of the State's Rights school and upheld slavery. This was a coincidence of momentous consequence, but philosophically speaking, as regards slavery, it was nothing more.

And so it happened that for fifty years after the adoption of the Constitution, while the National party and the Whig party on the one hand, and the Federal party and the Democratic party on the other warred over the principles of the government, the opponents of the institution of slavery increased in numbers and energy, but without connection with the politics of the country. But during this time, this party in favor of a strong centralized National government had under one name or another gathered much strength.

As early as 1789, it had procured the passage of the famous 25th Section of the Judiciary Act, which allows an appeal from the final judgment of a state court to the Supreme Court of the United States in cases involving the construction of a law or treaty of the United States, thus asserting for the Federal Government the judicial construction of its measures as against the judicial views of the state. At the same session, another point was gained by the National party. Under the provision of the Constitution that makes it the duty of the President "to take care that the laws be carefully executed," the National party carried the point that the President, without the sanction of Congress, had the power to remove an officer of the government, the tenure of whose office was not fixed by the Constitution, at about the same time General Hamilton opened the question of the right of Congress to impose duties to encourage manufactures. Here, then, were three distinct issues – the real grounds of difference which culminated in our war.

Next followed the contest over the Virginia and Kentucky Resolutions and the Alien and Sedition Laws, which resulted in the election of Mr. Jefferson over Mr. Adams as President, and the temporary check to the rapid strides of the government to consolidation. But it was only a check – Mr. Jefferson could recover no lost ground for the State's Rights party.

Then, unfortunately, came the war of 1812 with Great Britain, absorbing the attention of his successor Mr. Madison and arresting all efforts to carry out the doctrines and policy which had brought the party into power and giving a strong impulse to centralization.

It is difficult to keep up with all the changes of names and organization of the parties during the fifteen years succeeding the war of 1812, but a study will show that under whatever name or disguise assumed, the great struggle still was between the State's Rights or local government and National or centralized government. The first measure of the old National party, then calling themselves The National Republican Party, in 1828, was the act known at the time as the Bill of Abominations, which, throwing aside the pretense of revenue, openly imposed a tax for protection – a measure which forms a prominent chapter in the history of this State.

As you all know, upon the passage of this act, Mr. Calhoun counseled resistance. Whether our great statesman contemplated, by the resistance he advised, a forcible resistance or a resistance through the courts, it is useless now to discuss. Its discussion would only revive the domestic dissension of the Nullification and Union parties of 1832.

It is enough that a large party in South Carolina understood his advice to be resistance by force, and acted upon it; and that the state took measures to maintain by arms its denial of the right of Congress to impose upon it duties not authorized by its construction of the Constitution, and that in doing so, it had the support of many of the ablest statesmen of the country and the volunteered aid from the people of other states. While on the other hand, President Jackson openly marshaled the forces of the Union to war upon South Carolina and upon those who upheld her.

The issue, which was so imminent, was avoided by mutual concessions of the United States and state governments. But I desire to call your attention, my comrades, to the fact that the late war in which we took part had all but commenced in 1832, and that the real question then was the same, the incident only different. The question in 1832 and in 1860 was as to the sovereignty of the state. The incident in 1832 was the tariff; the incident in 1860 was slavery. Well would it have been for us had the question in 1860 turned upon the same incident as that in 1832. Would that we might have fought and shed our blood upon the dry question of the tariff and taxation, instead of one upon which the world had gone mad.

I cannot but think that our Convention of 1860 made a great mistake in the declaration of the causes which induced the secession of the State, in resting our justification alone upon the conduct of the Northern people in regard to slavery, however gross a violation of the Constitution such conduct was; and it is a matter of satisfaction to us, my comrades, that our first and beloved commander, General Gregg, as a member of that Convention, opposed the adoption of the declaration on this very ground. I cannot but agree with him and think that the justification of the secession of the State was much more satisfactorily set out and rested upon much better grounds in the address to the people of the other Southern States, in which was so ably and well shown that the issue was the same as that in the Revolution of 1776 and like that turned upon the one great principle: self-government, self-taxation, the criterion of self-government.

> *If, then, my comrades, our cause was just, as just as that of our forefathers in 1776, and one for which we might well indeed have endured and risked our lives and shed our blood, need we be ashamed of the fight we made for it?*

This latter address went on to show that the Southern States stood exactly in the same position toward the Northern States that the Colonies did towards Great Britain. The Northern States having the majority in Congress claimed the same power of omnipotence in legislation as the British Parliament. That the general welfare was the only limitation of either, and the majority in Congress, as in the British Parliament, was the sole judges of the expediency of the legislation this general welfare required. That thus the government of the United States had become a consolidated government, and the people of the Southern States were compelled to meet the very despotism their fathers threw off in 1776.

If, then, my comrades, our cause was just, as just as that of our forefathers in 1776, and one for which we might well indeed have endured hardship and risked our lives and shed our blood, need we be ashamed of the fight we made for it?

It is said that when the war commenced, we vaunted that a single Southern soldier could whip three Yankees. Well, it was a very foolish boast, if made; as foolish as that of General Grant, about which I shall speak, and one which you, my comrades, will agree with me, was not heard among the men who had the whipping to do. We, who did meet the three Yankees, know well that we met men as brave as ourselves, if differing with us in temperament and in the manner of their warfare. But we did meet the three Yankees, and it did take, if not three, at least two and a half to one to destroy our armies at last. The total number of men called under arms by the Government of the United States, between April, 1861, and April, 1865, amounted to 2,759,049, of whom 2,656,053, were actually embodied in the Federal armies. Foreign military authorities have put down the number of men embodied in the Confederate armies as 1,100,000. But this we know to be a great exaggeration, taken from Northern sources; for even robbing the cradle and the grave, there was scarcely a million of men able to bear arms in the Confederate States, nor did we have arms to put in their hands had we so many.

Let me give you here, my comrades, my version of General Grant's famous unfulfilled boast, that "he would fight it out on this line if it took all the summer." I refer to this often quoted saying as a boast because it has been generally so understood; but I have always rather regarded it as a pledge or promise demanded of him alike by the manhood of the North as by the timidity of the officials at Washington.

Politics Devolve Into War

When the Confederate Government determined to subordinate military considerations to political, it required no greater strategical skill than was possessed by us of the line to perceive that we had offered to our enemy a most vulnerable point, which, unlike that of Achilles, was not only the most vulnerable, but the most vital point of the Confederacy, that its throat all through the war was bared to the knife whenever the Federal generals should be allowed to destroy rather than attempt to whip us; that the James River was the sure, if not easy road to the Confederate capital.

McClellan was too professional a soldier to be willing to strike anywhere else while that was open to him. So in the spring of 1862, he essayed the task with a force of 153,000 men, against which General Johnston had present for duty but 53,688 – just about one to three. After a month's resistance, McClellan approached Richmond on June 20, 1862, with a force of 115,102, against which General Lee, in the Seven Days battle, had but 80,762, scarcely more than one to two. Yet, with this force, McClellan was driven back to his gunboats. But, notwithstanding this reverse, the manhood of the North demanded again a fair fight on an open field and an answer to this boast that we would fight three to one.

No victory by mere strategical skill, aided by gunboats, would appease the Northern desire that the Army of Northern Virginia should be whipped on a fair field. So Pope was tried and, you recollect, my comrades, that after a march of sixty miles in two days, on three ears of green corn apiece for rations, we broke our fast on Westphalia hams, Mocha coffee, and sherry wine out of his stores and sent him back to Washington to tell that he was mistaken in telegraphing that he had captured Jackson and his corps. During those two terrible days (August 28-29), before Longstreet came up, our corps of 17,309 men withstood Pope's army of 74,578 – you recollect with what terrible sacrifice to our brigade. In the great battle of the 30th, after Longstreet had joined us, we had but 49,077 of all arms, and yet we gained a second victory on Manassas plains.

At Sharpsburg you fought 35,255 under Lee against 87,164, which McClellan states in his official report that he had in action. At Fredericksburg, in which our brigade again suffered so severely and where we lost our beloved leader, General Gregg, we fought 78,000 under Lee against 100,000 under Burnside, and at Chancellorsville 57,000 under Lee and Jackson defeated 132,000 under Hooker. At Gettysburg 62,000 under Lee made a drawn battle against 105,000 under Meade.

Upon This Line

When, then, Grant came, he found himself required to promise that he would not repeat the Vicksburg strategy, but would march straight to meet us in the open field. He might have all the men he wanted, provided only he would undertake to move straight on and crush us without the adventitious aid of the naval forces striking us where we were unable to resist. Such, I suppose, was somewhat the occasion of his promise to "fight it out on this line if it took all the summer." Did he fulfill his promise?

On the 1st of May, 1864, General Grant had 120,380 men of all arms, to which was added before he commenced active operations, 20,780, giving him a total of 141,160 men at the opening of the campaign, against which Lee had present for duty but 63,984. With these enormous odds in his favor he "fought it out" but a single month, during which time – to quote from our old friend, the Adjutant General of the Army of Northern Virginia, Colonel Taylor, from whom I have taken most of these figures – there had been an almost daily encounter of hostile arms, and the Army of Northern Virginia had placed hors de combat of the number under General Grant a number

> *So Pope was tried and, you recollect, my comrades, that after a march of sixty miles in two days, on three ears of green corn apiece for rations, we broke our fast on Westphalia hams, Mocha coffee, and sherry wine out of his stores and sent him back to Washington to tell that he was mistaken in telegraphing that he had captured Jackson and his corps.*

equal to its entire numerical strength at the commencement of the campaign; and notwithstanding its own heavy losses and the reinforcements received by the enemy, still presented an impregnable front to its opponent, and constituted an insuperable barrier to General Grant's "On to Richmond."

Let me use the language of a foreign writer to describe the scenes of the second great battle of Cold Harbor, which brought to an end Grant's promise to fight it out on that line: But the June of 1864, says Colonel Chesney, "found Grant almost in sight of the city, upon the very ground which McClellan had held on the banks of the Chickahominy two years before. Four times he had changed the line of operation chosen in obedience to Lincoln's strong desire, on which he had declared his intention to 'fight it out all the summer.'" Four times he had recoiled from the attempt to force his way direct to the rebel capital, for his indomitable and watchful adversary ever barred the way.

Once more, on the morning of June 3rd, he flung his masses fiercely against the line held by Lee, which ran across the very field of battle where that General had won his first triumph over McClellan. The result was so fearful and useless a slaughter that, according to the chief Union historian, when later in the day orders were issued to renew the assault, the whole army correctly appreciating what the inevitable result must be, silently disobeyed."

Again, the same writer says: "The most eulogistic biographer of the great Federal general speaks as though it were under his breath when he tells the story of the battle of Cold Harbor. 'There was a rush,' says such a one; 'a bitter struggle, a rapid interchange of deadly fire and the (Federal) army became conscious that the task was more than it could do.'"

The testimony of Swinton, himself an eyewitness, is more emphatic and complete: "It took hardly more than ten minutes to decide the battle. There was along the whole line a rush – the spectacle of impregnable works, a bloody loss, a sullen falling back, and the action was decided."

What an ignominious end to a boast, or what a failure in the fulfillment of a promise that he would fight his way to Richmond over the land route if it took him all the summer! By the first of June, Grant had not only failed in this boastful promise, but he had so lost the confidence and command of his grand army that it absolutely refused his order to advance again.

The summer had thus scarcely begun when Grant was obliged to abandon the idea of fighting it out on the line he had been so ready to undertake. But abandon it he must, for he had learnt by bitter experience, as Colonel Chesney observes, that the "continuous hammering" in which he had trusted, might break the instrument while its work was yet unfinished. Not even the vast resources on which he had power to draw could long spare 20,000 men a week for the continuance of the experiment. He had lost in the first three weeks of battle with Lee 60,000 men; and as Lee had only commenced the campaign with 63,000, Grant could not but reflect that had their armies been equal Lee would not have left him a vestige of his with which to retreat.

But with the abandonment of his boast or promise came the beginning of the end to us. From this time forth, Grant contented himself with resuming the work from which McClellan had been called in disgrace, but unlike McClellan, he was furnished with all the men and material a siege required. Butler had joined him, and he now had 150,000 men with which to commence the slow but sure if not glorious work of wearing out the remnant of the Army of Northern Virginia.

In speaking of that last terrible struggle of nearly a year, let me use the language of the distinguished English soldier and essayist rather than my own: "Not in the first flush of triumph when his army cheered his victory over McClellan," writes Colonel Chesney, "not when hurling back Federal masses three times the weight of his own on the banks of the Rappahannock, nor even when advancing the commander of victorious legions to carry the war away from his loved Virginia into the North had Lee seemed so great, or won the love of his soldiers so closely as through the dark winter that followed. Overworked his men were sadly, with forty miles of entrenchments for that weakened army to guard. Their prospects were increasingly gloomy as month passed by after month bringing them no reinforcements, while their enemy became visibly stronger. Their rations grew scantier and poorer, while the jocund merriment of the investing lines told of abundance, often raised to luxury by voluntary tribute from the wealth of the North." "But the confidence of the men in their beloved chief," says Colonel Chesney, "never faltered, their sufferings were never laid on Uncle Robert. The simple piety which all knew to be the rule of his life acted upon thousands of those under him with a power which those can hardly understand who know not how community of hope, suffering, and danger fairly shared amid the vicissitudes of war quickens the sympathies of the roughest and lowest, as well as those above them."

In Lee's own language the line of defense "stretched so long as to break," at last gave way and the end came. A single Southern soldier had not in the long run been able to whip three Yankees, however gloriously he had fought. Numbers, material, and discipline at last triumphed over individual heroism.

I need not recall the agony of those last days. Let me rather quote again from Colonel Chesney's memoir of General Lee. He says: "The day will come when the evil passions of the great civil strife will sleep in oblivion, and North and South do justice to each other's motives and forget each other's wrongs. Then history will speak with clear voice of the deeds done on either side, and the citizens of the whole Union do justice to the memories of the dead, and place above all others the name of the great chief of whom we have written. In strategy mighty, in battle terrible, in adversity as in prosperity a hero indeed. With the simple devotion to duty and the rare purity of the ideal Christian knight he joined all the kingly qualities of a leader of men."

It was in one of these last terrible days, my comrades, that your first captain and your last colonel fell mortally wounded. In the fight at Hatcher's Run on the 30th March, 1864, Colonel C.W.

McCreary was shot through the lungs and died as he was carried to the breastworks. I need not remind you how admirable a soldier he was, how brave in battle, how skillfully he could handle a regiment in action, and how gentle he was to all around him. Educated in the State Military Academy, he was fully prepared for the command of the regiment to which he succeeded and which he led so gallantly and successfully in many engagements. I can still hear his voice ringing through the din of battle as he aligned the regiment for some desperate work. You had reason to be proud of him while he lived, to mourn him when he died, and should now revere his memory.

It is, my comrades, one of the greatest misfortunes of our defeat that not even the names of those who fought and bled and died in our glorious struggle have been preserved, and that unless collected and enrolled by us now will soon be forgotten and their memory lost with the cause for which they warred.

To subscribe, send an email to: thestainlessbanner@gmail.com or visit our website at www.thestainlessbanner.com

Subscription is free.

The Stainless Banner

An e-zine dedicated to the armies of the Confederacy

Volume 3, Issue 3
March 2012

THE THIRD DOMINO FALLS!
THE LOSS OF MISSOURI

Nine days ago, Earl Van Dorn had been in Virginia, commanding the cavalry in the Army of Northern Virginia. Now, he was in Pocahontas, Arkansas, in command of Trans-Mississippi Department Two and making plans to drive the enemy back up the Mississippi River and capture St. Louis. His first order would consolidate the three forces in his district.

Two of the armies were across the state, encamped in the Boston Mountains near the city of Fayetteville. General Benjamin McCulloch's army, 8,000 strong, was still basking in their victory won along Wilson Creek. After suffering a series of losses at the hands of a larger Union force, Sterling Price's 7,000 Missourians had retreated across the Missouri border to regroup. Bad blood existed between McCulloch and Price, and, now that they were in close proximity to each other, they picked up their argument where they had left off. Dispatches were sent urging Van Dorn to settle their long standing differences. Van Dorn would do them one better. He would come in person and lead his troops "to glory and immortal renown."[1] As he rode into camp, the Confederates gave him a salute of forty guns, befitting his rank of major general.

Albert Pike commanded the third army in Van Dorn's department. Pike moved quickly to put his 2,000 warriors from the Five Nations on the road to

What's Inside:

Van Dorn's Official Report	6
Price's Official Report	10
Commendation	13
Pike's Official Report	14
Battle of Elkhorn Tavern	21
Earl Van Dorn	30
Cleburne's Slave Proposal	31
Letters	38
Slaves in our Army	42

---

[1] Shelby Foot. The Civil War: A Narrative, Volume 2. (Alexandria: Time-Life Books), page 127.

Arkansas when his plans hit an unexpected glitch. The tribes had signed treaties that gave them certain protections. One of those protections was that they could not be forced from their homes without their consent. Which they would gladly give...for gold. Price had no choice but to pay up. Three days later, the warriors were on their way to the Boston Mountains.

Once his armies were together, Van Dorn gave the order. Cook three days' rations and prepare for a forced march, which would end in battle, and, if the gods of war smiled favorably on their endeavor, victory. Van Dorn's plan was very simple. The Union army in pursuit of Price was spread throughout the countryside. He would run them to ground and destroy them in detail.

The next morning, the Confederates set out: 17,000 men and sixty guns. Snow fell, blanketing the ground and a wicked wind whipped through the trees. Price's Missourians led the way, happy to be on the offensive again and sure that they would be victorious once more within the borders of their home state. McCulloch's Texans were next. On the flank, in a long file, marched the warriors from the Five Nations.

Van Dorn bounced along side in an ambulance. Before leaving Virginia, he had suffered a horrendous fall jumping a ditch and was still feeling the effects. He was also suffering chills and a fever, the results of swimming across an icy river hurrying to Fayetteville.

The Meaning of a Forty Gun Salute

Union General Samuel Curtis sat in his tent near Cross Hollows, Arkansas, and tried to ignore the cold by writing a letter home. The faint rumble of cannon echoed across the countryside. Forty guns! The salute reserved for a commanding general. Had a new chieftain come to fight? He did not have to wait long for an answer. Intelligence came in that night. His scouts, including the soon to be famous Wild Bill Hickok, rode into camp and informed him that the Confederates were marching north looking for a fight.

Curtis sent word to General Franz Sigel to fall back to Sugar Creek, near the Missouri border, where Curtis would be waiting. Sigel fell back, skirmishing as he did. He arrived on March 6th.

Curtis set his line along the creek bank. Behind his defenses, about a mile away, was the small village of Leetown, which was nothing more than a collection of about a dozen cabins scattered around a store and a blacksmith. Approximately a mile northwest of Leetown was Pea Ridge. The Springfield-Fayetteville road crossed the creek and ran down a small

> *Osterhaus was falling back before a horde of feather wearing, hatching bearing Indians. Confederate soldiers were one thing... Indians another matter altogether. Osterhaus' Germans broke and ran as fast they could from field.*

valley. This road was known to the locals as the Telegraph road. The poles and wires ran along side the road and came to a sudden end at a small tavern on the eastern edge of Pea Ridge. Elkhorn Tavern received its name from the large elk skull, complete with antlers, nailed to the beamed ceiling.

Winter winds swept in, bringing with it the snow that had accompanied Van Dorn on his march. The Federals stroked their campfires higher and bedded down to wait.

Flanking the Enemy

Through the cold and snow, Van Dorn's army had marched fifty miles in three days. The men huddling on the opposite side of Sugar Creek were cold, tired, and hungry, having consumed their rations days ago. But Van Dorn had no intention of letting his men rest for the night or plunging straight ahead at entrenched breastworks the next morning. There was still marching to be done. Ordering the men to leave their campfires burning, he sent Price down the Bentonville and Keetsville Road, behind Pea Ridge, to Frost Hill before ordering them to double back toward Elkhorn Tavern and attack the Union left rear at dawn. McCulloch and Pike drew the shorter march. Their path would take them through Leetown to strike Curtis' right flank. Van Dorn would go into battle outnumbering his opponent.

Price's march was delayed by felled trees and other obstructions the Federals had strewn across the Bentonville Road. By time, he made the turn south toward Elkhorn Tavern, the sun was up and Curtis was aware that the Confederates were about to sweep down on his rear.

Curtis Makes His Decision

Curtis' options were limited. He could beat a hasty retreat back toward Springfield. But he quickly ruled it out. If he marched down the Telegraph road, the Confederates would overrun his flank. If he fled southward, then the Confederates would be across his supply line and communications. The other option was to wheel around and fight with his back to Sugar Creek and his own entrenchments.

He chose to fight. His men were rousted from their positions and quickly about-faced. Colonel Eugene Carr's division was sent to meet the threat from Elkhorn Tavern. Colonel Peter Osterhaus' division was sent running to Leetown to protect the flank and a colonel with the ironic name of Jefferson Davis was sent with his division to reinforce Osterhaus. General Alexander Asboth's division was held in reserve. Before Price could unleash his attack, Curtis had successfully turned his rear into his front.

Carr threw up a strong defensive position with batteries staggered along the Telegraph road and a line of infantry thrown out in advance with three other lines prepared to fire over their heads into the Confederates.

The Confederates Surge

Price's men screamed down the small valley, firing as they came. Southern cannon quickly wrecked Carr's defensive line, destroying three out of four guns in one battery, blowing up the ammunition caissons and killing all the cannoneers.

Carr's first line retreated to just north of the tavern, where they barely managed to hold off Price's men. Carr sent word to Curtis. The Confederates were reforming and when they came, they would overwhelm his defensives. Send reinforcements on the double-quick.

Curtis received Carr's message, but before he could respond, an awful racket rose from Leetown. High-pitched screaming, unlike anything Curtis had heard before. This was not the usual Rebel yell but an unearthly sound of death. A rider galloped from the direction of the sound. Osterhaus was falling back before a horde of feather wearing, hatching bearing Indians. Confederate soldiers were one thing… Indians another matter altogether. Osterhaus' Germans broke (who could blame them) and ran as fast they could from the field. Davis was holding, if one could count falling back, as holding. Reinforcements were needed immediately!

Who would get Curtis' lone division of reserves? It was a question Curtis could not answer – not yet anyway. He sent word for both lines to hold. The battle would determine who would be reinforced.

Trouble on the Front

Pike's Indians chased the Federals until they reached the abandoned cannon, where they stopped to celebrate their victory. Wearing the harnesses from the dead horses, they danced and chanted, ignoring Pike's frantic calls to get back in line. McCulloch had run into stiff opposition on the left, and Pike desired to go to his aid.

When Pike insisted that his warriors continue their pursuit, the Indians rebelled. They did not like fighting white man style, where cannonballs crashed through their midst, chewing up large numbers of casualties. They had never fought like that before and they would never do so again. From now on, they would do their fighting from behind trees or rocks or any other shield they could find. Nothing Pike could say or threaten could get the warriors back in line. Only a cavalry battalion under the leadership of Colonel Stand Watie obeyed Pike's orders.

More trouble developed when McCulloch was killed leading a skirmish line forward. The effect of his death was as debilitating to his assault as the Indians' refusal to get back in line was to Pike's. The Texans were at a loss as to what to do now that their leader was gone. Their advance slowed then stopped; the men throwing down their guns and wandering off the field to follow their fallen leader to the rear. On the other side of the field, Davis did not question it. His men were exhausted, and he was more than content to let the battle whimper to a halt.

Pike gathered all the men he could and hastened toward Elkhorn Tavern to aid Price whose men were still fighting.

The Battle Grinds to a Halt

The sun was setting as the Missourians plunged forward again. The sight of the surging Confederates was too much to bear. As the Federals abandoned their cannon and retreated behind the infantry, a soldier from Iowa threw a smoldering quilt across a caisson. The

Iowan had barely joined his comrades when the caisson exploded. A plume of smoke and blown-off body parts rose into the sky.

The resulting explosion was felt in Leetown, solving Curtis' dilemma as to where to send his reinforcements – he sent them to Elkhorn Tavern. It was too little too late. By time Asboth's division arrived on the field, the fighting was over. Carr had held. His exhausted men threw themselves down on the ground to sleep. A half a mile away, Price's men did the same.

March 8

Dawn broke to reveal a battlefield wreathed in gun smoke. The sun rose as red as it had been when it had set. After a long night's march over Pea Ridge, Pike and the remnants of both his command and McCulloch's Texans arrived at Elkhorn Tavern. With little to no ammunition (the wagon trains had gone south to safety) and his advantage in numbers gone, Van Dorn's only alternative was to dig in and see if he could provoke the Federals into attacking.

When Confederate cannon opened fire, Curtis recognized the weak display as a ruse to run his men from the field. Instead, he ordered Sigel to get his men up and ready to attack. The German rolled his batteries forward and destroyed battery after battery of Confederate guns arrayed along the ridge in a spectacular display of accuracy. Then Sigel gave the signal and his men moved forward.

Carr and Davis faced a mutiny in their ranks. It was not fair that Sigel's men were going to get the spoils, their men complained bitterly, not after they had carried the battles yesterday. To mollify their men, Carr and Davis ordered their lines forward.

Overwhelmed and out of ammunition, the Confederates broke and ran from the menacing blue lines…past Elkhorn Tavern and through the valley leading to Pea Ridge. The battle was over.

Van Dorn's shattered army returned to their camps in the Boston Mountains. Van Dorn wrote Albert Sidney Johnston to assure him that he had not been defeated but only thwarted in his objective to capture St. Louis. Once his men were rested, he would try again.

Johnston was not convinced. He ordered Van Dorn to cross the Mississippi and join him at Corinth, for all intents and purposes, ending the fight for Missouri.

In less than a month, the heart of the South had been exposed. Kentucky, Missouri, the western two-thirds of Tennessee, including Nashville and Memphis, were in Union hands. Control of the Mississippi, Tennessee and Cumberland rivers gave the Federals an easy route into Louisiana, Mississippi and Alabama.

The desperate Confederates had been pushed back as far as they could go. It would be up to Johnston to turn the tide. He would get his chance at a small steamboat stop on the Tennessee River known to the locals as Pittsburg Landing.

Van Dorn's Report on the Battle of Elkhorn Tavern

HEADQUARTERS TRANS-MISSISSIPPI DISTRICT,
Jacksonport, Arkansas, March 27, 1862

General Beauregard, Commanding, &c.

GENERAL: I have the honor to report that while at Pocahontas I received dispatches on February 22, informing me that General Price had rapidly fallen back from Springfield before a superior force of the enemy and was endeavoring to form a junction with the division of General McCulloch in Boston Mountains. For reasons which seemed to me imperative, I resolved to go in person and take command of the combined forces of Price and McCulloch.

I reached their headquarters March 3rd, and being satisfied that the enemy, who had halted at Sugar Creek, fifty-five miles distant, was only waiting large reinforcements before he would advance, I resolved to attack him at once. Accordingly, I sent for General Pike to join me near Elm Springs with the forces under his command, and, on the morning of March 4th, moved with the divisions of Price and McCulloch by way of Fayetteville and Bentonville to attack the enemy's main camp on Sugar Creek. The whole force under my command was about 16,000 men.

On the 6th we left Elm Springs for Bentonville, and from prisoners captured by our scouting parties on the 5th, I became convinced that up to that time no suspicion was entertained of our advance, and that there were strong hopes of our effecting a complete surprise and attacking the enemy before the large detachments encamped at various points in the surrounding country could rejoin the main body. I therefore endeavored to reach Bentonville, eleven miles distant, by rapid march, but the troops moved so very slowly that it was 11:00 a.m. before the head of the leading division (Price's) reached the village, and we had the mortification to see Sigel's division, 7,000 strong, leaving it as we entered. Had we been one hour sooner, we should have cut him off with his whole force and certainly have beaten the enemy the next day.

We followed him, our advance skirmishing with his rear guard, which was admirably handled, until we had gained a point on Sugar Creek about seven miles beyond Bentonville and within one or two miles of the strongly entrenched camp of the enemy.

In conference with Generals McCulloch and McIntosh, who had an accurate knowledge of this locality, I had ascertained that by making a detour of

> *So long as brave deeds are admired by our people, the names of McCulloch and McIntosh will be remembered and loved.*

eight miles, I could reach the Telegraph road leading from Springfield to Fayetteville and be immediately in rear of the enemy and his entrenchments. I had resolved to adopt this route and, therefore, halted the head of my column near the point where the road by which I proposed to move diverges, threw out my pickets, and bivouacked as if for the night. But soon after dark, I marched again moving with Price's division in advance, and taking the road by which I hoped before daylight to reach the rear of the enemy. Some obstructions, which the enemy had hastily thrown in the way, so impeded our march that we did not gain the Telegraph road until near 10:00 a.m. of the 7th.

By prisoners, with forage wagons, whom our cavalry pickets brought in, we were assured that we were not expected in that quarter, and that the promise was fair for a complete surprise.

I at once made dispositions for attack, and directing General Price to move forward cautiously, soon drew the fire of a few skirmishers, who were rapidly reinforced, so that before 11:00, we were fairly engaged, the enemy holding very good positions and maintaining a heavy fire of artillery and small arms upon the constantly advancing columns which were being pressed upon him.

I had directed General McCulloch to attack with his forces the enemy's left, and before 2:00, it was evident that if his division could advance or even maintain its ground, I could at once throw forward Price's left, advance his whole line and end the battle. I sent him a dispatch to this effect, but it was never received by him. Before it was penned his brave spirit had winged its flight, and one of the most gallant leaders of the Confederacy had fought his last battle.

About 3:00 p.m. I received, by aides-de-camp, the information that Generals McCulloch and McIntosh and Colonel Hebert were killed, and that the division was without any head. I nevertheless pressed forward with the attack, and at sunset, the enemy was fleeing before our victorious troops at every point in our front, and when night fell, we had driven him entirely from the field of battle.

Our troops slept upon their arms nearly a mile beyond the point at which he made his last stand, and my headquarters for the night were at the Elkhorn Tavern. We had taken during the day seven cannon and about 200 prisoners.

In the course of the night, I ascertained that the ammunition was almost exhausted, and that the officer in charge of the ordnance supplies could not find his wagons, which with the subsistence train had been sent to Bentonville. Most of the troops had been without any food since the morning of the 6th, and the artillery horses were beaten out. It was therefore with no little anxiety that I awaited the dawn of day. When it came, it revealed to me the enemy in a new and strong position, offering battle. I made my dispositions at once to accept, and by 7:00, the cannonading was as heavy as that of the previous day.

On the side of the enemy, the fire was much better sustained, for being freed from the attack of my right wing; he could now concentrate his whole artillery force. Finding that my right wing was much disorganized, and that the batteries were one after the other retiring from the field with every shot expended, I resolved to withdraw the army, and at once placed the ambulances, with all the wounded they could bear, upon the Huntsville road, and a portion of McCulloch's division, which had joined me during the night, in position to follow, while I so disposed of my remaining forces as best to deceive the enemy as to my intention and to hold him in check while executing it.

About 10:00, I gave the order for the column to march and soon afterwards for the troops engaged to fall back and cover the rear of the army. This was done very steadily; no attempt was made by the enemy to follow us, and we encamped about 3:00 p.m. about ten miles from the field of battle. Some demonstrations were made by his cavalry upon my baggage train and the batteries of artillery, which returned by different routes from that taken by the army, but they were instantly checked, and thanks to the skill and courage of Colonel Stone and Major Wade, all of the baggage and artillery joined the army in safety.

So far as I can ascertain, our loss amounts to about 600 killed and wounded and 200 prisoners: and one cannon, which, having become disabled, I ordered to be thrown into a ravine.

The best information I can procure of the enemy's loss places his killed at more than 700, with at least an equal number of wounded. We captured about 300 prisoners, making his total loss about 2,000. We brought away four cannon and ten baggage wagons, and we burned upon the field three cannon taken by McIntosh in his brilliant charge. The horses having been killed, these guns could not be brought away.

The force with which I went into action was less than 14,000 men. That of the enemy is variously estimated at from 17,000 to 24,000.

During the whole of this engagement, I was with the Missouri division, under Price, and I have never seen better fighters than these Missouri troops and more gallant leaders than General Price and his officers. From the first to the last shot, they continually pushed on and never yielded an inch they had won, and when at last they received the order to fall back, they retired steadily and with cheers. General Price received a severe wound early in the action but would neither retire from the field nor cease to expose himself to danger.

No successes can repair the loss of the gallant dead who fell on this well fought field. McCulloch was the first to fall. I had found him, in the frequent conferences I had with him, a sagacious, prudent counselor, and a bolder soldier never died for his country.

McIntosh had been very much distinguished all through the operations which have taken place in this region; and during my advance from Boston Mountains, I placed him in command of the cavalry brigade and in charge of the pickets. He was alert, daring, and devoted to his duty. His kindness of disposition, with his reckless bravery, had attached the troops strongly to him, so

> *Our troops slept upon their arms nearly a mile beyond the point at which he made his last stand, and my headquarters for the night were at the Elkhorn Tavern.*

that after McCulloch fell, had he remained to lead them, all would have been well with my right wing. But after leading a brilliant charge of cavalry and carrying the enemy's battery, he rushed into the thickest of the fight again at the head of his old regiment and was shot through the heart. The value of these two officers was best proven by the effect of their fall upon the troops. So long as brave deeds are admired by our people, the names of McCulloch and McIntosh will be remembered and loved.

General Slack, after gallantly maintaining a long continued and successful attack, was shot through the body; but I hope his distinguished services will be restored to his country.

A noble boy, [S.] Churchill Clark, commanded a battery of artillery, and during the fierce artillery actions of the 7th and 8th was conspicuous for the daring and skill which he exhibited. He fell at the very close of the action. Colonel Rives fell mortally wounded about the same time and was a great loss to us. On a field where many gallant gentlemen were, I remember him as one of the most energetic and devoted of them all.

To Colonel Henry Little my especial thanks are due for the coolness, skill, and devotion with which, for two days, he and his gallant brigade bore the brunt of the battle. Colonel Burbridge, Colonel Rosser, Colonel Gates, Major Lawther, Major Wade, Captain MacDonald, and Captain Schaumberg are some of those who attracted my special attention by their distinguished conduct.

In McCulloch's division, the Louisiana regiment, under Colonel Louis Hebert and the Arkansas regiment under Colonel McRae are especially mentioned for their good conduct. Major Montgomery, Captain Bradfute, Lieutenants Lomax, Kimmel, Dillon, and Frank Armstrong, assistant adjutant general, were ever active and soldierly. After their services were no longer required with their own division, they joined my staff, and I am much indebted to them for the efficient aid they gave me during the engagement of the 8th. They are meritorious officers, whose value is lost to the service by their not receiving rank more accordant with their merit and experience than that they now hold.

Being without my proper staff, I was much gratified by the offer of Colonel Shands and Captain Barrett of the Missouri Army of their services as aides. They were of great assistance to me by the courage and intelligence with which they bore my orders; also Colonel Lewis, of Missouri.

None of the gentlemen of my personal staff, with the exception of Colonel Maury, assistant adjutant-general, and Lieutenant C. Sulivane, my aide-de-camp, accompanied me from Jacksonport, the others having left on special duty. Colonel Maury was of invaluable service to me both in preparing for and during the battle. Here, as on other battlefields where I have served with him, he proved to be a zealous patriot and true soldier; cool and calm under all circumstances, he was always ready, either with his sword or his pen. His services and Lieutenant Sulivane's were distinguished. The latter had his horse killed under him while leading a charge, the order for which he had delivered.

You will perceive from this report, general, that, although I did not, as I hoped, capture or destroy the enemy's army in Western Arkansas, I have inflicted upon it a heavy blow and compelled him to fall back into Missouri. This he did on the 16th instant.

For further details concerning the action and for more particular notices of the troops engaged I respectfully refer you to the reports of the subordinate officers, which accompany this report.

Very respectfully, sir, your obedient servant,
Earl Van Dorn,
Major-General

Price's Official Report on Elkhorn Tavern

HEADQUARTERS MISSOURI STATE GUARD,
Camp Ben. McCulloch, Missouri, March 22, 1862.

Col. D. H. MAUEY,
Assistant Adjutant General

COLONEL: I have the honor to submit to Major General Van Dorn the following report of the part taken by the Missouri troops in the action of the 6th, 7th, and 8th instant:

That officer having arrived at Cove Creek and assumed command of the Confederate forces in Western Arkansas, I gladly placed myself and my army under his orders, and in obedience to these, took up the line of march in the direction of Bentonville on the morning of March 4th, provided with three days' cooked rations, and leaving my baggage and supply trains to follow slowly in the rear.

My forces consisted of the First Brigade Missouri Volunteers, Col. Henry Little commanding; the Second Brigade, Brigadier-General Slack commanding; a battalion of cavalry, under command of Lieutenant Colonel Cearnal, and the State troops, under the command of Brigadier Generals Rams, Green, and Frost, Colonels John B. Clark, Jr., and James P. Saunders, and Major Lindsay, numbering in all 6,818 men with eight batteries of light artillery.

With these I reached Elm Springs on the evening of the 5th, and on the morning of the 6th, advanced to Bentonville where burning houses indicated the presence of the enemy. Colonel Gates' regiment of cavalry, Lieutenant Colonel Cearnal's battalion, and the mounted men of General Rains' command were rapidly pushed forward to the east of the town and soon became briskly engaged with what proved to be the rear guard of General Sigel's forces, the main body of which had passed through Bentonville that morning in the direction of Elkhorn Tavern, near which the enemy were encamped in force and strongly entrenched.

Skirmishing between our advance and this rear guard was kept up throughout the day and resulted in the capture by us of quite a number of prisoners, from whom we gained much useful information.

Towards evening we bivouacked as if for the night within five or six miles of the enemy, but resumed the line of march at 8:00 p.m., and, in spite of the impediments with which the enemy had sought to obstruct our way, reached a point on the Telegraph road to the north and in the rear of the enemy's position. A march of about two miles along the deep valley through which the road leads brought us within view of the plateau upon which the enemy were posted, and which lay to the north of the Elkhorn Tavern.

Our advance had already begun to skirmish with the vedettes of the enemy, when I discovered that they were about to place a battery in position to command the road. I at once deployed the brigades of General Slack and Colonel Little to the right and the rest of my forces to the left and took possession of the heights on either hand. This movement gave my artillery on the left a very commanding position, from which they were enabled not only to check the enemy's advance upon our left, but also to support our right in its advance upon the enemy.

The brunt of the action fell during the early part of the day upon my right wing, consisting of General Slack's and Colonel Little's brigades. They pushed forward gallantly against heavy odds and the most stubborn resistance, and were victorious everywhere.

At this time and here fell two of my best and bravest officers, Brigadier General William Y. Slack and Lieutenant Colonel Cearnal, the former mortally and the latter severely wounded.

I now advanced my whole line, which gradually closed upon the enemy and drove them from one position to another, until we found them towards evening in great force on the south and west of an open field, supported by masked batteries.

The artillery and infantry of my left wing were brought up to attack them, and they did so with a spirit and determination worthy of all praise. The fiercest struggle of the day now ensued; but the impetuosity of my troops was irresistible, and the enemy was driven back and completely routed. My right had engaged the enemy's center at the same time with equal daring and equal success, and had already driven them from their position at Elkhorn Tavern.

Night alone prevented us from achieving a complete victory, of which we had already gathered some of the fruits, having taken two pieces of artillery and a quantity of stores. My troops bivouacked upon the ground which they had so nobly won almost exhausted and without food, but fearlessly and anxiously awaiting the renewal of the battle in the morning.

> **My troops bivouacked upon the ground which they had so nobly won almost exhausted and without food, but fearlessly and anxiously awaiting the renewal of the battle in the morning.**

The morning disclosed the enemy strengthened in position and numbers and encouraged by the reverses which had unhappily befallen the other wing of the army, when the brave Texan chieftain, Benjamin McCulloch, and his gallant comrade, General McIntosh, had fallen fearlessly and triumphantly leading their devoted soldiers against the invaders of their native land. They knew, too, that Hebert – the accomplished leader of that veteran regiment the 3rd Louisiana, which won so many laurels on the bloody field of the Oak Hills, and which then, as well as now, sustained the proud reputation of Louisiana – was a prisoner in their hands.

The enemy was not slow to renew the attack. They opened upon us vigorously, but my trusty men faltered not. They held their position unmoved until (after several of the batteries not under my command had left the field) they were ordered to retire. My troops obeyed it unwillingly, with faces turned defiantly against the foe.

It was then that I lost two officers of whom any nation might be proud. The one, Colonel Benjamin A. Rives, fell in the prime of his manhood, at the zenith of his usefulness. No braver or more gallant officer, no more accomplished gentleman, no more unselfish patriot ever led a regiment or died for his country's honor. The other, (S.) Churchill Clark was, as Colonel Little justly observes in his report, "a child in simplicity and piety of character, a boy in years, but a soldier in spirit and a hero in action. They fell at the very close of the hard fought battle, well deserving the glowing praises which their immediate commander bestows upon them.

My forces were withdrawn in perfect order without the loss of a gun. For the details of all this, I beg leave to make reference to the accompanying reports of my subordinate officers.

The conduct of nearly every officer and soldier under my command was such as to win my admiration. It is the less necessary that I should commend any one particularly to the notice of the major general commanding, as the operations of my arms were conducted under his eye, while his presence and gallant bearing, as well as his skill, contributed immeasurably to the success of our cause.

I am, very respectfully, your obedient servant,
STERLING PRICE,
Major General, Commanding Missouri State Guard

COMMENDATION FOR THE FIGHT AT ELKHORN TAVERN
GENERAL ORDER 7
FROM GENERAL VAN DORN'S HEADQUARTERS

HDQRS. TRANS-MISSISSIPPI DISTRICT,
Van Buren, Arkansas, March 16, 1862

The major general commanding this district desires to express to the troops his admiration for their conduct during the recent expedition against the enemy. Since leaving camp in Boston Mountains they have been incessantly exposed to the hardships of a winter campaign and have endured such privations as troops rarely encounter.

In the engagements of the 6th, 7th, and 8th instant, it was the fortune of the major-general commanding to be immediately with the Missouri division, and he can therefore bear personal testimony to their gallant bearing.

From the noble veteran who has led them so long to the gallant S. Churchill Clark, who fell while meeting the enemy's last charge, the Missourians proved themselves devoted patriots and staunch soldiers. He met the enemy on his chosen positions and took them from him. They captured four of his cannon and many prisoners. They drove him from his field of battle and slept upon it.

The victorious advance of McCulloch's division upon the strong position of the enemy's front was inevitably checked by the misfortunes which now sadden the hearts of our countrymen throughout the Confederacy. McCulloch and McIntosh fell in the very front of the battle and in the full tide of success. With them went down the confidence and hope of their troops. No success can repair the loss of such leaders. It is only left to us to mourn their untimely fall, emulate their heroic courage and avenge their death.

You have inflicted upon the enemy a heavy blow, but we must prepare at once to march against him again. All officers and men must be diligent in perfecting themselves in knowledge of tactics and of camp discipline. The regulations of the army upon this subject must be rigidly enforced.

By order of Major General Earl Van Dorn:
DABNEY H. MAURY,
Assistant Adjutant-General.

Brigadier General Albert Pike's Report on the Elkhorn Tavern

DWIGHT MISSION, CHEROKEE NATION, INDIAN TERRITORY,
March 14, 1862

Capt. D.H. MAURY,
Assistant Adjutant-General

SIR: On February 25, I reached Cantonment Davis, near Fort Gibson, with Colonel Cooper's Choctaw and Chickasaw battalion, which had been encamped near the mouth of the Canadian. The same evening Colonel D. N. McIntosh's regiment of Creeks arrived at the same point. I had, in charge, a large amount of coin and other moneys for the different Indian tribes and found delegations of the Osages, Comanches, and Reserve Indians awaiting me, and the disposition of the moneys left unexpectedly in my hands, together with the dealings with the Indian tribes, detained me there three days.

The Choctaws, Chickasaws, and Creeks refused to march until they were paid off, and, as by their treaties with us they could not be taken out of the Indian country without their consent, I had no alternative but to submit. The payment of the Choctaws and Chickasaws occupied three days.

On the morning of the third day, I left them behind at Fort Gibson, except O. G. Welch's squadron of Texans, part of the First Choctaw and Chickasaw Regiment, and with the Creek regiment whom I persuaded to move by the promise that they should be paid at the Illinois River, I marched to Park Hill, near that river, remained there one day, and not being overtaken, as I expected to be, by the Choctaw and Chickasaw troops, moved the next day, Monday, March 3, towards Evansville, and the next day to Cincinnati, on the Cherokee line, where I overtook Colonel Stand Watie's regiment of Cherokees.

The next day, Wednesday, with Colonel Watie's regiment and Captain Welch's squadron, I reached Freschlag's Mill, and on Thursday overtook Colonel Drew's regiment of Cherokees at Smith's Mill, and came up with the rear of General McCulloch's division late in the afternoon.

That night I encamped within two miles of Camp Stephens. At 9:30, I received General Van Dorn's order, to the effect that the army would move at 8:00, and that I would follow General McCulloch's division. I sent to General McCulloch to ascertain at what hour the road would be clear for me to move, and received his reply that it would be clear at 12:00 and that his train would not move until daylight. At 12:00, I marched with my command, overtook and passed General McCulloch's train, which was in motion, and had to wait until sunrise a little south of Sugar Creek until his infantry had passed it on a little bridge of rails.

We followed closely in his rear until the head of my command had passed the houses on what is called Pea Vine Ridge, where we were halted, and Colonel Sims' Texas regiment, countermarching, passed us to the rear, an officer informing me that I was to countermarch and follow the other troops. I did so, and we were then marched off the Bentonville road to the south through the woods.

Soon after Captain Lomax, of General McCulloch's staff, informed me that the enemy had fortified a little place called Leetown, about four miles to the south, which we were marching to attack, and that General McCulloch's orders were that my command, on reaching the spot, should form in line in rear of General McIntosh's brigade, which would itself be in rear of a line of infantry, and that when the firing should begin all were to dismount and charge together.

We had marched from the road in a southeasterly direction about a mile from the point where we left it, and were passing along a narrow road, between a piece of woods on our left and a fenced field on our right, when we discovered in front of us, at the distance of about 300 yards, a battery of three guns, protected by five companies of regular cavalry. A fence ran from east to west through the woods, and behind this we formed in line, with Colonel Sims' regiment on the right, the squadron of Captain Welch next to him, and the regiments of Colonels Watie and Drew in continuation of the line on the left.

The enemy was in a small prairie, about 250 yards across, on the right of which was the fenced field, and on our left it extended to a large prairie field, bounded on the east by a ridge. In rear of the battery was a thicket of underbrush, and on its right, a little to the rear, a body of timber.

General McIntosh's cavalry had passed on into the large prairie field to our left and the infantry were quite across it, close to the ridge, about 600 yards from us. My whole command consisted of about 1,000 men, all Indians, except one squadron. The enemy opened fire into the woods where we were, the fence in front of us was thrown down, and the Indians (Watie's regiment on foot and Drew's on horseback), with part of Sims' regiment, gallantly led by Lieutenant Colonel Quayle, charged full in front through the woods and into the open ground with loud yells, routed the cavalry, took the battery, fired upon and pursued the enemy, retreating through the fenced field on our right, and held the battery, which I afterwards had drawn by the Cherokees into the woods. Four of the horses of the battery alone remained on the ground, the others running off with the caissons, and for want of horses and harness we were unable to send the guns to the rear.

The officers of my staff, Captains Schwarzman and Hewitt and Lieutenant Pike, with Captain Lee, of Acting Brigadier General Cooper's staff, rode with us in the charge. Our loss was two of Colonel Drew's men killed and one wounded. Colonel Sims had one man killed and one wounded. Of the enemy, between thirty and forty were killed in the field and around the guns. The charge was made just at noon.

We remained at the battery for some twenty minutes, when Colonel Watie informed me that another battery was in our front, beyond the skirt of underbrush, protected by a heavy force of infantry. General McIntosh's force was not near us, nor do I know where it then was. The infantry were still in their position near the ridge, across the large field on the left, and did not approach us; indeed, at one time moved farther off along the ridge. Colonel Drew's regiment was in the field on our right, and around the taken battery was a mass of Indians and others in the utmost confusion, all talking, riding this way and that, and listening to no orders from any one. I directed Captain Roswell W. Lee, of Acting Brigadier General Cooper's staff, always conspicuous for gallantry and coolness, to have the guns which had been taken faced to our front, that they might be used against the battery just discovered; but he could not induce a single man to assist in doing so.

At this moment the enemy sent two shells into the field, and the Indians retreated hurriedly into the woods out of which they had made the charge. Well aware that they would not face shells in the open ground, I directed them to dismount, take their horses to the rear, and each take to a tree, and this was done by both regiments, the men thus awaiting patiently and coolly the expected advance of the enemy, who now and for two hours and a half afterwards, until perhaps twenty minutes before the action ended, continued to fire shot and shell into the woods where the Indians were from their battery in front, but never advanced. This battery also was thus, with its supporting force, by the presence of the Indians, rendered useless to the enemy during the action.

In the meantime, our artillery had come into action some distance to our left and front, beyond a large field, extending from the woods in which we were to a line of woods beyond it, which hid the conflict from our view. Leaving the Indians in the woods, I passed beyond them to the left into the open ground nearer the conflict, and remained some time.

About 1:30, there was a very heavy fire of musketry for about ten minutes, and soon after, about two regiments of our cavalry came into the field on our left front and formed in line facing the woods on that side. Colonel Drew then came to me with his regiment, about 500 strong, and I sent him across the field, directing him to form in rear of the line of cavalry, and if they advanced through the woods to follow them, dismount his men near the other edge, and let them join in the fight in their own fashion. They crossed the field and took the position indicated.

It was just after this that I directed Sergeant Major West, of Colonel Watie's regiment, to take some of the Cherokees and drag the captured guns into the woods, which was done, the enemy still firing over them into the woods, where he placed a guard of Cherokees over the cannon.

Soon after the cavalry force crossed to our side of the field and formed in line in front of the woods in which the Indians were and remained there until the enemy threw a shot in that direction, when they also took shelter in the woods. During all this time I received no orders whatever nor any message from any one.

About 3:00 I rode towards the fenced field. I saw nothing of our cavalry, but found a body of our infantry halted on the road running along the fence by which we had originally come. It consisted of the regiments of Colonels Churchill, Hill, and Rector, and Major Whitfield's battalion. Major Whitfield informed me that Generals McCulloch and McIntosh were both killed, and that 7,000 of the enemy's infantry were marching to gain our left, one body of which, at least 3,000 strong, he had himself seen. Totally ignorant of the country and the roads, not knowing the number of the enemy, nor whether the whole or what portion of General McCulloch's command had been detached from the main body for this action, I assumed command and prepared to repel the supposed movement of the enemy.

To our left, beyond the field where our infantry had first been seen by me in the forenoon, was a wooded ridge of no great height, with a fence running along the foot of it on the west and northwest; between it and the Bentonville road was open and level ground. I marched the infantry, Welch's squadron and Watie's regiment, across the field, dismounted the horsemen, directed all to be posted behind the fences, and sent Major Boudinot, of Watie's regiment, to inform General Van Dorn that I would try to hold the position; but upon riding up and along the ridge to the rear I found the position not tenable, as the enemy could cross it and descend upon our rear by an open road that ran over it.

> *Totally ignorant of the country and the roads, not knowing the number of the enemy, nor whether the whole or what portion of General McCulloch's command had been detached from the main body for this action, I assumed command and prepared to repel the supposed movement of the enemy.*

At this time the firing on the field had ceased, and I saw coming into the road at the farm house a large body of cavalry and Good's battery. It was evident enough that the field was left to the enemy, and as we were not in sufficient numbers to resist them and the ground afforded no defensive position, I determined to withdraw the troops and lead them to General Van Dorn. Indeed, the officers assured me that the men were in such condition that it would be worse than useless to bring them into action again that day

I, accordingly, sent orders to the artillery and cavalry to join me. What had become of the other troops engaged no one could inform me. I concluded they had retreated towards Camp Stephens, gaining the road by which we had come in the morning. Colonel Stone and Captain Good came to me, and I informed them of my purpose. Placing the squadron of Captain Welch in front, the infantry marching next, followed by Good's battery, with the Cherokees on the flanks, and, as I supposed, Colonel Stone's regiment in the rear, we gained the Bentonville road, and marched on it in perfect order to the Telegraph road.

The order sent to the Cherokees to join us had not, by some accident, reached Colonel Drew, and his regiment remained in the woods, and after a time retreated towards Camp Stephens, where, he informed me, he found Colonel Stone's regiment had arrived before him. This regiment understanding, I learned that part of the enemy's force was marching to attack the trains, took that direction.

The infantry had in three days marched sixty miles, had been on foot all the preceding night, and fought that day without water, and Colonel Churchill begged me to leave them where they could procure it. When we reached the Telegraph road I was about to conduct them to headquarters; but unable to learn the position of the two armies or how the road came upon the field, and learning that where our forces were there was no water and that there was a running stream on the Pineville road about a mile and a half from the point where the Bentonville road descends into the valley, I led them to and on the Pineville road, intending to halt at the water, and letting the men have that at least:, as they had nothing to eat, to join the main army early in the morning. Orders from General Van Dorn caused us to retrace our steps and march to his headquarters, which we reached long after dark.

On Saturday morning I was directed by General Van Dorn to post part of Colonel Watie's men, who were my whole command, except Captain Welch's squadron, on time high ridge to our right and the residue on another ridge on the left, to observe the enemy and give him information if any attempt was made by them in force to turn his left flank. I accompanied those sent on the ridge to the right, and sent Captain Fayette Hewitt, of my staff, to post the others. To Captain Welch I gave permission to join any Texan regiment he chose; and he joined that of Colonel Greer and remained with it until the action ended.

After remaining for some two hours near the foot of the ridge, on the south side observing the enemy's infantry, heavy columns of which were in the fields beyond and the fire of their batteries in full view of me and seeing no movement of the infantry to the left, I recrossed the ridge, descended it, and went towards General Van Dorn's headquarters. Being told that he and General Price were in the field to the left of his headquarters, I took the road that led there and halted on the first hill below headquarters, where a battery was posted, facing the Telegraph road, and which I was told had been sent to the rear for ammunition. Here I heard that orders had been given for the army to fall back and take a new position. Another battery came up and the captain asked me for orders. I told him he had better place his battery in position in line with the others to play upon the road, and then send to General Van Dorn for orders. In the meantime, I sent two officers to the general to deliver him the message, and I remained with the batteries.

We now heard long continued cheering in front. Bodies of our troops had come across the ridge on the right and down the Hospital Hollow, in good order apparently, and I suppose they were marching to the left to repel perhaps the attempt upon our left flank, apprehended by General Van Dorn in the morning. Seeing no fugitives on the Telegraph road we supposed the cheering to proceed from our own troops and that the day was ours, when an officer rode down and informed me that the field was occupied by Federal troops.

Soon after, another came and told me that no one had seen either General Van Dorn or General Price for some time and it was supposed they were captured, as the field where they were last seen was full of Federals. He remarked to me, "You are not safe here, for the enemy's cavalry are within 150 yards of you."

The troops that had come across the ridge and down the Hospital Hollow were now below us on the Telegraph road. Colonel Watie had sent to me for orders. I had sent to him to bring his men from the ridge down into the valley and there halt for orders, and I supposed he had done so; but he did not receive the order and remained on the mountain, from which he went direct to Camp Stephens.

Just at this moment, the two batteries close to me commenced to wheel and hurried down the hill into the road. I do not know that any one gave them any order to fall back. The captain of one battery said that someone ordered it, but I think that the information of the capture of our generals was overheard and that no order was given. No one was there to give an order. The batteries rattled down the steep hill and along the Telegraph road, and as I rode by the side of them I heard an officer cry out, "Close up, close up, or you will all be cut to pieces."

On reaching the road, I rode past the batteries to reach a point at which to make a stand, for, having passed the road but once and then in the night, it was all an unknown land to me. When we reached the first open level ground I halted the leading gun, directed the captain of the company in front to come into battery, facing to the rear on the right of the plain going northward. The battery in the rear I knew had no ammunition. Saw the first gun so placed in position, I rode back to the second battery and directed the only officer I could find to do the same on the left of the plain. When I turned around to go to the front, I found that the gun faced to the rear had been again turned into the road, and that the whole concern was again going up the road northward.

I rode again to the front and halted the leading battery at the foot of the next level, ordered it into line, facing to the rear, gave the necessary commands myself, and had three guns brought into position. Two regiments of infantry were standing there in lines ranging up and down the valley, the flank of each to the enemy. I directed them to form in the rear of the batteries. At this moment a shell was sent by the enemy up the road from the point of the hill around which we had just passed. The cry of "the cavalry are coming" was raised and everything became confusion.

It was impossible to bring the other guns into battery. Those already faced turned again into the road; and supposing that, of course, they would take the Bentonville road, which, at leaving the other, ascended a steep hill, and thinking I could certainly halt them after a slow ascent on its summit, I galloped through the bottom and up the ravine on the left of the hill, dismounted and climbed the hill on foot, remounted at the summit, rode to the brow of the hill, looked down into the road, and found that our retreating troops, batteries and all, had passed by on the Telegraph road, the enemy's cavalry pursuing, en route for Springfield, Missouri.

Captain Hewitt and my aide-de-camp, Lieutenant W.L. Pike, had followed me and, except half a dozen stragglers, we were alone. We waited a few moments on the brow of the hill uncertain of what course to pursue, when, on our right, as we faced the valley, and at a distance of about 100 yards, a gun of the enemy sent a shot into the valley, and another on the other side, farther off, replied with another.

We then turned and rode up the road towards Bentonville, and after riding about a mile found that the enemy's cavalry were pursuing at full speed. Leaving them in the rear by rapid riding, we turned into the woods on the right, passed around the farm house on the Pea Vine Ridge, and road westward between the Pineville and Bentonville roads.

We had been informed by my brigade commissary, who had come up from Camp Stephens about 10:00, that our whole train had been turned back and was encamped at Pea Vine Ridge.

Three miles from the Telegraph road, we saw a small body of our retreating horsemen fired upon by the enemy's infantry, and concluded, as they had evidently anticipated our retreat and made every arrangement necessary in view of it to destroy our retreating forces, that General Sigel returning by the route up Sugar Creek by which he had retreated was in front of our train and it was lost.

> *Owing to the circuit which we were constrained to make and to the fatigued condition of our starved horses, we were unable to gain the front of our retreating forces until after they had left Elm Springs...*

Owing to the circuit which we were constrained to make and to the fatigued condition of our starved horses, we were unable to gain the front of our retreating forces until after they had left Elm Springs; and learning that the Indian troops had marched from that point to Cincinnati, we joined them at that place.

The enemy, I learn, had been encamped at Pea Vine Ridge for three weeks, and Sigel's advance was but a ruse to induce our forces to march northward and give them battle in positions selected by themselves.

I may add that in their pursuit of our retreating train they followed no farther than Bentonville and returned from that point. I was within five miles of that place on Monday morning and was misled by information that they had taken it that morning; but they did not enter it until the afternoon.

I did not know until I reached Cincinnati what had become of the main body of our threes. I there met Captain Schwarzlman and Major Lanigan, who informed me of their retreat, and that Generals Van Dorn and Price were marching from Huntsville to Van Buren, and also heard of the order to burn all the wagons on the Cove Creek road that could not cross Boston Mountains.

Just before night, Saturday afternoon, I had met Colonel Rector in the hills, who told me he had about 500 men with him; that they were in such condition that they could not go more than six or eight miles a day, and that he thought he would take them into the mountains, hide their arms in a secure place, and, as he could not keep them together and feed them, let them disperse. He asked my opinion as to this, and I told him that no one knew where the rest of the army was; that Generals Van Dorn and Price were supposed to be captured and the train taken; that if his men dispersed with their arms they would throw them away, and that I thought the course he proposed was the wisest one under the circumstance The enemy was pursuing on all the roads, and as it was almost impossible for even a dozen men in a body to procure food, I still do not see what better he could have done.

General Cooper, with his regiment and battalion of Choctaws and Chickasaws, and Colonel McIntosh, with 200 men of his regiment of Creeks, came up with our retreating train at Camp Stephens, where they found Colonel Drew's regiment, and remained with General Green, protecting the train until it reached Elm Springs, where they were all ordered to march with their own train to Cincinnati.

I am, captain, very respectfully, yours,
ALBERT PIKE,
Brigadier General, Commanding
Department of Indian Territory

THE BATTLE OF ELKHORN TAVERN

Taken from the *Confederate Military History, Volume 10, Chapter IV*. (Edited by C.L. Gray)

On January 29, 1862, with headquarters at Little Rock, General Van Dorn assumed command of the district, which comprised Missouri, Louisiana north of Red River, Arkansas west of the St. Francis River and Indian Territory. Headquarters were established at Pocahontas, Arkansas.

On February 6th, General McCulloch was commanded by Van Dorn to order two regiments of infantry, two of cavalry, and one battery of artillery to proceed at once to Pocahontas, where they would be stationed for the time being.

The appointment of Major General Van Dorn to the command of the Trans-Mississippi district was no doubt made in order to bring about harmony of action between the Missouri and Arkansas troops, or, rather, between the commanders of the respective forces, the soldiers being on the best of terms and their sympathies alike in many respects.

The Arkansans were eager to advance against the enemy wherever they could find him, and were equally indignant at the cruelties of war inflicted upon the once prosperous and happy districts of Missouri, which the enemy had invaded and ravaged. They were sorry they could not have a chance at Fremont, who had induced the large enlistment of Germans in the Federal army – "Dutch," as they were called all alike – immigrants lately from a strange land, but eager to precipitate themselves into a conflict growing out of questions that were supposed to be settled and compromised in the formation of the government that offered them an asylum. They were principally from the servile grades of their own land; ignorant, brutal, and needing to be instructed in matters of government and conduct of civilized warfare more than the Negroes. The Confederates wished to have vengeance especially upon these intruders, who insulted women, burned homes of noncombatants, and murdered prisoners of war.

The difficulties between the Texan commander of the Arkansas troops and General Price requiring settlement were:

> *The appointment of Major General Van Dorn to the command of the Trans-Mississippi district was no doubt made in order to bring about harmony of action between the Missouri and Arkansas troops, or, rather, between the commanders of the respective forces...*

1) rank and precedence;
2) the proper field of action;
3) widely divergent views of military strategy.

General Price, holding the higher rank, had yielded the command of the combined forces on a former occasion. It could not be expected of him that he should do so continuously, especially since he had shown, by practical successes, that he could cope with the enemy and attract thousands to his standard, unaided, and on his own motion had displayed an energy and enterprise in military campaign that has rarely been equaled.

General McCulloch had an unconquerable distrust of the military judgment and capacity of General Price, notwithstanding his achievements, and of the stability and subordination of the recruits he had drawn to his standard. He avoided the association with earnestness, claiming that he was assigned to the Indian Territory, and was not authorized to march his command into Missouri. He was as much bent upon retaining his Indian command as General Price was anxious for the occupation and redemption of Missouri. If there had been forces adequate, it might have been well enough to keep the Indian country under military control; but it was of secondary importance in comparison with other fields.

There was, however, reason for believing General Price's designs in Missouri could not be carried out. Its strategical effect in preventing the reinforcement of Grant was its chief importance. The eastern boundary of Missouri was occupied by large bodies of the enemy, and other forces could be sent out from the Ohio River on short notice. Kansas, to the west, swarmed with the enemies of the South. Were there available forces of the Confederates sufficient to hold Missouri, should they succeed in occupying it? Yet it was strategy to make war in Missouri.

In fact, the soldiers of both commands, Arkansans and Missourians, were otherwise likely to have to go to the assistance of Polk or of Johnston and Beauregard east of the Mississippi River, where the great wager of battle was being listed, not for a district, but for the entire country. A vigorous movement into Missouri might have rendered such transfer unnecessary.

Very openly it was said by some that the object of Van Dorn's assignment was to accomplish this transfer. The circumstance of his prompt establishment of headquarters at Pocahontas, in striking distance of Point Pleasant on the Mississippi, the route by which Hardee's command had been transferred, confirmed this opinion in many minds.

Halleck's strategy was to prevent this. General John Pope, who had been in command of the enemy's forces in Missouri between the Missouri and Osage rivers, had sent "Merrill's Horse" through Saline county, where they were bombarded with mortars loaded with mud by Jo Shelby and his men, near Waverly. They stripped farms, impressed stock from women, and on February 19th, captured several companies of Confederate recruits at Blackwater Creek, near Knobnoster, under Colonels Robinson, Alexander and McOiffin, of which achievement Generals Pope and Halleck made much boast to Washington.

On December 23, 1861, Brigadier General S.R. Curtis was assigned to the command of the Federal forces of the southwestern district of Missouri. On December 2nd, martial law had been declared in Missouri by Mr. Lincoln, and Curtis was without restraint. On January 3, 1862, the men under him burned the towns of Dayton and Columbus and with a largely superior force proceeded southward, confronted by Price's men.

Taking Springfield, after a skirmish on February 12th, and fighting at Crane Creek on the 14th, and near Flat Creek on the 15th, Curtis met a more stubborn resistance by Price's men at Sugar Creek, Arkansas on the 17th. Sustaining considerable loss, he encamped on the battleground, waiting for Sigel, who was a few miles behind, to reinforce him.

While the Confederates under Price were camped at Cross Hollows, a cavalry force of Federals under General Asboth, on the 18th, took Bentonville, Arkansas, which the Confederates had evacuated. The same officer, on the 23rd, marched into Fayetteville, occupied only by a Confederate picket of Colonel W. H. Brooks' battalion.

Fayetteville is the principal town of northwest Arkansas, north of the Boston Mountains, the center of a fine region of rolling black lands, where grow the famous big, red apples. Its permanent occupation would signify the subjugation of a populous section of the State, most of whose men were in the Confederate army, and was a menace to Van Buren and Fort Smith.

General Van Dorn was at Pocahontas when, on February 23, 1862, he received dispatches informing him of the retreat of Price, followed by Curtis and Sigel, and the battle of Sugar Creek. Van Dorn immediately sent McCulloch orders to form a junction with Price without loss of time, to which McCulloch replied that he had ordered the command to march, as soon as the commanding general should arrive, with six days' cooked rations.

McCulloch's command marched the next day across Boston Mountains to Elm Springs, Arkansas, where it would be joined by General Van Dorn and the Indian forces of General Albert Pike, who had been given command of the department of the Indian Territory on November 22, 1861. The main body of Price's Missouri State Guard was camped near Elm Springs. The march of the division over the Boston Mountains was toilsome and slow. It reached the place of rendezvous on the 3rd, where the commanding general had arrived.

On March 4th, without waiting for General Pike, Van Dorn moved out for Bentonville, where Sigel, with his Germans, had arrived and taken possession. Two bodies of cavalry, one under McIntosh and one under Gates, were pushed forward, the former to go around the town on the west, the latter on the east, in an effort to cut off Sigel from the main body of the enemy at Sugar Creek. But McIntosh found the country north of Bentonville so rough with rocks, ravines, and mountains, guarded by a natural cheval-de-frise of small oaks and blackjacks, that he could not hope to form a junction with Gates.

Coming upon the Federals in force on these heights, and being fired upon from an ambuscade, he made an effort to charge the enemy in position, but the ground was impracticable for cavalry, and he drew back to Bentonville, which by that time had been evacuated by Sigel. Sigel left the north side of the town as Price's division entered on the south; his departure marked by burning depots and forage piles.

Van Dorn says in his report: "Owing to bad roads and delay, though the distance from Bentonville to Elm Springs is only eleven miles, it was 11:00 before the leading division (Price's) reached the village. If we had arrived an hour sooner, we could have cut off Sigel and beaten the enemy easily the next day."

Colonel Gates pressed upon the retreating Germans and charged their rear guard on the road to Springfield, killing and wounding several of the guard, and capturing a baggage-wagon laden with arms and ammunition. He accelerated Sigel's march by continuing the pursuit and attack until the enemy disappeared in the uncertain light of the winter night. Sigel continued his march in the darkness until he joined the main body in its stronghold, on the heights commanding the valley of Sugar Creek.

Snow fell during the night and clothed both hill and valley in a mantle of white. The hills are high on both sides; the valley deep, about half a mile in width. The main road from Fayetteville to Springfield, via Cross Hollows, crosses the valley at right angles, and the road from Fayetteville leading to Keetsville, Missouri, after making a circuit through the hills, also passes through this valley. Going north, a road takes off to the left nearly parallel with it, some three or four miles distant, returning to the Telegraph road on the "divide," called Pea Ridge, or Peavine Ridge. These roads Curtis had blockaded with trees felled across them. He had erected formidable breastworks on the headlands, and the approach by the main road from Bentonville he had "completely shielded by earthworks."

As Van Dorn well knew, to attack the enemy's line from the south, with his infantry and artillery in chosen positions, would be storming a stronghold. He resolved to make a formidable demonstration in front, while he should lead his main attack against the enemy's left (northeast) flank, marching around on the north of the Federal line. Camping with his whole force within a mile of the enemy's front, he lighted the snow clad hills with the fires of an army, as if in position to give battle next day from the alignment then occupied.

After the men had eaten supper, Van Dorn and Price, with the Missouri division, leaving their campfires burning, resumed the march in the night, moving on the parallel road which would lead them into the Telegraph road, by a long and toilsome circuit, it is true, but well in the enemy's rear, and in an equal position on Pea Ridge near Elkhorn Tavern, to the north of the enemy. The large trees felled across the roads by Curtis, to block the approaches on his left and rear, proved formidable obstacles to cut away for the passage of the Confederate artillery and ordnance wagons, and the flanking column did not reach the ridge in the enemy's rear until 10:00 a.m. of the 7th. Its march had not been molested and it took the desired position unopposed.

The roar of artillery and rattle of small arms came from the distant front and center as this line of attack was formed in the rear of the carefully-established lines of the enemy. Completely surprised, Curtis had necessarily to reverse his front at the place of attack, which was his extreme left, and now became his right, at the same time that his established right center was engaged from the front.

When Price's division ascended to the plateau of Pea Ridge, there ensued an artillery duel of more than an hour's duration, between the batteries of Captains Wade and Clark, and the enemy's batteries commanded by Colonel Carr. The guns of the enemy first ceased firing. Gates' Missouri cavalry charged the position occupied by the batteries, but

> *Meanwhile, on the field near Elkhorn Tavern, before 2:00, it was evident, Van Dorn reported afterward, that if McCulloch could advance or even maintain his ground, Price's left could be thrown forward, the whole line advanced, and victory won.*

was repulsed; then, dismounting, went into line under General Little.

The enemy charged Little's brigade twice and were repulsed. Having placed a battery in position which played upon the enemy's lines, the commands of Little and Slack charged the position and held it. A general advance was still deferred, waiting for McCulloch's demonstration against the enemy's front.

McCulloch was necessarily delayed in arraying the disorganized detachments which choked the narrow roads – General Pike with his Choctaws, Cherokees, and Creeks, Stand Watie's regiment on foot, D.N. McIntosh's Creeks on foot, Drew's Choctaws, pony-mounted, and a "squadron," as General Pike named it, of mounted whites – in all only 1,000 men. General Douglas Cooper's Indian command contained Chilly McIntosh, the Creek war chief, and John Jumper, Boudinot, and other celebrated Cherokees, all of whom had come up late on the 6th.

"It was about 10:30 a.m.," says Colonel Evander McNair of the Fourth Arkansas, on the extreme right of Hébert's (Second) brigade, "before that brigade, under the lead of McCulloch, was ordered into action." The brigade was composed of the Arkansas regiments of Colonel McIntosh, Colonel McNair and Colonel Mitchell, Hébert's Third Louisiana, and McRae's battalion.

There were nominally attached to the brigade, Brooks' Arkansas battalion, Good's, Hart's and Provence's Arkansas batteries, Gaines' Texas battery, the Third (Greer's) Texas cavalry, and Whitfield's battalion Texas cavalry. The other brigade, called the First brigade, sometimes led by McIntosh, was commanded by Colonel Elkanah Greer of the Third Texas and was composed of Churchill's Arkansas rifles, the Second Arkansas regiment, the South Kansas-Texas regiment, and three commands of Texas cavalry.

Colonel McIntosh usually left the command of his regiment to Lieutenant Colonel Embry, and forming a brigade of mounted men from the five regiments, led them as cavalry, which was the arm of the service preferred by that dashing soldier. The colonels of Arkansas regiments, in both of these brigades, had already greatly distinguished themselves.

General McCulloch, in person, directed the movement against the enemy's front and center, near Leetown, up the valley and along its sides. For this the enemy was prepared and resisted with a storm of shot and shell from his batteries in position, and with infantry behind his breastworks. There were vacant fields, separated by strips of timber and dense undergrowth in the valley, and fallen timber, which the Confederates had to pass. This they did with difficulty, but with undaunted resolution under a harassing crossfire from the enemy upon the heights. They ran upon ambuscades of the infantry in the underbrush, which they drove back, and when opposed by a new formation, repulsed that also, until, penetrating the cul-de-sac formed by the valley, they were met by large bodies of the enemy's infantry. The Confederates reformed their disordered lines and charged, driving back the enemy and capturing a battery which had been playing upon them at a distance of nearly 200 yards.

It was when the enemy had concentrated his forces to meet this charge that General McCulloch fell, shot from the brush, and Colonel Hébert, leading an advancing party of the brigade which became disconnected, was surrounded and captured. Four times the Confederates repulsed the enemy's lines in this advance up the valley, driving batteries and repulsing assaults by cavalry on their flanks, with great slaughter of men and horses. But finding the enemy strongly entrenched and increasing in numbers, beginning to enfilade their lines and threatening to surround them, being themselves unsupported by reinforcements from their own lines, and "not hoping to obtain any advantage by persistence in the attack, they fell back in good order, no one pursuing them," to a position which Colonel Greer, who now commanded the division, ordered to be occupied until further orders.

Colonel McIntosh had led a cavalry charge with five regiments across a field and, driving away the gunners, carried a battery of the enemy. With his usual fearless energy, he returned to the assault in a second charge, and was shot dead at the head of his men. The consequence of the loss of these leaders, to whom the entire command looked for direction in the disposition of their forces in the action, caused a paralysis of this wing of the army. Officers rode about trying to learn the position of commands, what movement next should be made, and who was to take the place of the dead commanders, while the men stood or rested in their lines, in a state of inaction, until after 2:00. Then, after correspondence with the commanding general, several miles distant, they were ordered to his assistance.

Meanwhile, on the field near Elkhorn Tavern, before 2:00, it was evident, Van Dorn reported afterward, that if McCulloch could advance or even maintain his ground, Price's left could be thrown forward, the whole line advanced, and victory won. A dispatch to this effect was sent to McCulloch, but was never received by him. "Before it was penned, his brave spirit had winged its flight, and one of the most gallant leaders of the Confederacy had fought his last battle."

It was getting late in the day, and General Price sent instructions to his subordinate commanders that they would press the enemy at once, and drive him from the field, or be driven, and to prepare for a general advance. The brunt of the action had fallen during the early part of the day on the brigades of Slack and Little, and they were everywhere victorious, though Slack fell mortally wounded.

Those that remained of McCulloch's wing, after the battle of the 7th, followed the route taken the previous night by Price; and marching all night, a little before daylight on the morning of the 8th reached Van Dorn, and were disposed to the right and left of the line at Elkhorn Tavern. Here, upon the renewal of the battle on the 8th, the greater part of the troops remained inactive, while the

> *Governor Harris had written Van Dorn from Clarksville, Tennessee, that General Beauregard desired Van Dorn to join his forces with those of Beauregard on the Mississippi River, if possible.*

cannonading on both sides continued, until ordered to fall back on Huntsville. Human endurance could stand no further tax. Some of the cavalry were dispatched to protect the flanks, or, as Colonel Greer expressed it, "to keep the cavalry out of the way of the infantry bringing up the rear of the retreating army."

Van Dorn and Price grandly carried out the plan of campaign on their part, but they were defeated in the end by a series of accidents, the like of which rarely occur, though similar ones caused disaster in other great battles with more fateful results.

On March 9, 1862, General Van Dorn requested of General Curtis that, according to the usages of war, his burial parties be permitted to collect and inter the bodies of officers and men who fell during the engagement of the 7th and 8th, to which the Federal commander replied that all possible facilities would be given, and that many of the dead had already been interred. He added that quite a number of Confederate surgeons had been captured (engaged in the hospitals during the battle) and permitted to act under parole, and further liberty would be allowed if such accommodations would be reciprocated.

Curtis regretted to state that many of the Federal dead had been tomahawked and scalped, and their bodies shamefully mangled, contrary to civilized warfare, and expressed a hope that this important struggle would not degenerate into a savage warfare. To this note Colonel D. H. Maury, Van Dorn's adjutant-general, made an immediate reply, as follows:

General: I am instructed by Major-General Van Dorn, commanding this district, to express to you his thanks and gratification on account of the courtesy extended by yourself and the officers under your command to the burial party sent by him to your camp on the 9th instant. He is pained to learn, by your letter brought to him by the commanding officer of the party, that the remains of some of your soldiers have been reported to you to have been scalped, tomahawked, and otherwise mutilated. He hopes you have been misinformed. The Indians who formed part of his forces have for many years been regarded as a civilized people. He will, however, most cordially unite with you in repressing the horrors of this unnatural war. That you may cooperate with him to this end more effectually, he desires me to inform you that many of our men who surrendered themselves prisoners of war were reported to him as having been murdered in cold blood by their captors, who were alleged to be Germans. The privileges which you extend to our medical officers will be reciprocated, and as soon as possible, means will be taken for an exchange of prisoners.

On the 18th of March, 1862, General Van Dorn reported that the entire army he had marched against the enemy some days since was in camp a few miles from Van Buren, and that he would march in a few days for Pocahontas to make a junction with whatever force might be assembled at that point. His intention was then to attack the enemy near New Madrid or Cape Girardeau, and, if practicable, march on St. Louis, and thus withdraw the forces threatening that part of Arkansas. A heavy blow had been struck the Federals. Van Dorn proposed to seek another field before they recovered. If he gave battle near New Madrid, he would relieve Beauregard, in command at Corinth. If that were not advisable, he would march boldly and rapidly toward St. Louis.

On March 7th, Governor Isham G. Harris had written Van Dorn from Clarksville, Tennessee that General Beauregard desired Van Dorn to join his forces with those of Beauregard on the Mississippi River, if possible. To this, General Van Dorn replied that he would unite all his troops at Pocahontas, about the 7th of April, and would have about 20,000 men, maybe more. He also wrote that the enemy in Arkansas had fallen back to Springfield.

On the 17th of March, he sent a message to General Albert Sidney Johnston that by the 22nd he would get off and reach Pocahontas on April 7th with 15,000 men. He received a letter from General R. E. Lee, dated March 19th, informing him that all the troops called from Arkansas and Texas and by Hébert from the coast were ordered to him (Johnston).

March 19th, General Van Dorn ordered Colonel T. J. Churchill, with his brigade, and Gates' battalion of cavalry, to make an expedition against Springfield, Missouri, and endeavor to capture and destroy the stores of the enemy there.

On the same day the First division, army of the West, under command of Major General Price, was ordered to be ready to march on the 25th instant.

General Pike was to continue in command of the troops in the Indian Territory, and Woodruff's battery, reorganized at Little Rock, was ordered to report to him at Van Buren. Major W.L. Cabell, at Pocahontas, that it had been decided to make Des Arc, Arkansas, the point of rendezvous and of deposit for supplies.

Brigadier General Albert Rust was ordered to assume command of the lower Arkansas from Clarksville to its mouth, and of White River from Des Arc to its mouth, and that all companies organized under the call of Governor Rector for the Confederate service should report to Colonel James P. Major at Des Arc.

On the 28th of March, General T.J. Churchill was urged to reach Des Arc by the earliest possible day. All these orders pointed to the transfer of the army of the West to the east side of the Mississippi, to reinforce Generals Johnston and Beauregard at Corinth, Mississippi.

General Price, for the Missourians, had acquiesced and relinquished his former rank in the State Guard for the same rank in the Confederate army. Special orders announced that the First brigade of Price's division would embark for Memphis on April 8th, and Colonel Little would take command. At Des Arc, General Price bade farewell to the soldiers of the State Guard in a touching and eloquent order.

HI-LIGHTS OF A HERO'S LIFE
EARL VAN DORN

- 1820 – (September 17) Born near Port Gibson in Claiborne County, Mississippi.
- 1842 – Graduated 52nd out of 56 cadets from West Point.
- 1842 – (July 1) Appointed brevet second lieutenant in the 7th U.S. Infantry.
- 1843 – Married Caroline Godbold.
- 1844 – (November 30) Promoted to second lieutenant.
- 1846 – (September 21-23) Saw action at the Battle of Monterrey.
- 1847 – (March 3) Promoted to first lieutenant.
- 1847 – (March 9-29) Participated in the Siege of Vera Cruz.
- 1847 – (April 18) Promoted to brevet captain at the Battle of Cerro Gordo.
- 1847 – (August 20) Promoted to brevet major at the Battle of Churubusco.
- 1850 – Saw action in Florida against the Seminoles.
- 1855 – (March 3) Promoted to captain in the 2nd Cavalry.
- 1858 – (October 1) Severely injured at the Battle of Rush Springs.
- 1860 – (June 28) Promoted to major.
- 1861 – (January 31) Resigned his army commission.
- 1861 – (March 16) Entered the Confederate Army as a colonel of infantry.
- 1861 – (April 11) Given command of the Confederate forces in Texas.
- 1861 – (April 25) Appointed colonel of the 1st C.A. Regular Cavalry in Virginia.
- 1861 – (June 5) Promoted to brigadier general.
- 1861 – (September 19) Promoted to major general.
- 1862 – (January 29) Assumed command of the Trans Mississippi District.
- 1862 – (March 6-8) Battle of Pea Ridge (Elkhorn Tavern).
- 1862 – (October 3-4) Second Battle of Corinth.
- 1862 – Due to his performance at Corinth, was relieved of his district command.
- 1862 – (December 20) Raided Union supply depot at Holly Springs.
- 1863 – (January 13) Appointed to command cavalry in the Department of Mississippi and East Louisiana.

- 1863 – (March 5) Battle of Thompson's Station.
- 1863 – (March 16) Given command of the Army of Tennessee's Cavalry Corps.
- 1863 – (May 7) Shot by Dr. James Bodie Peters.
- 1863 – Laid to rest in the Wintergreen Cemetery in Port Gibson.

Patrick Cleburne's Slave Enlistment Proposal

JANUARY 2, 1864

COMMANDING GENERAL, THE CORPS, DIVISION, BRIGADE, AND REGIMENTAL COMMANDERS OF THE ARMY OF TENNESSEE:

GENERAL: Moved by the exigency in which our country is now placed, we take the liberty of laying before you, unofficially, our views on the present state of affairs. The subject is so grave, and our views so new, we feel it a duty both to you and the cause that before going further we should submit them for your judgment and receive your suggestions in regard to them. We therefore respectfully ask you to give us an expression of your views in the premises.

We have now been fighting for nearly three years, have spilled much of our best blood, and lost, consumed, or thrown to the flames an amount of property equal in value to the specie currency of the world. Through some lack in our system the fruits of our struggles and sacrifices have invariably slipped away from us and left us nothing but long lists of dead and mangled. Instead of standing defiantly on the borders of our territory or harassing those of the enemy, we are hemmed in today into less than two-thirds of it, and still the enemy menacingly confronts us at every point with superior forces. Our soldiers can see no end to this state of affairs except in our own exhaustion; hence, instead of rising to the occasion, they are sinking into a fatal apathy, growing weary of hardships and slaughters which promise no results.

In this state of things, it is easy to understand why there is a growing belief that some black catastrophe is not far ahead of us, and that unless some extraordinary change is soon made in our condition we must overtake it. The consequences of this condition are showing themselves more plainly every day; restlessness of morals spreading everywhere, manifesting itself in the army in a growing disregard for private rights; desertion spreading to a class of soldiers it never dared to tamper with before; military commissions sinking in the estimation of the soldier; our supplies failing; our firesides in ruins. If this state continues much longer we must be subjugated.

Every man should endeavor to understand the meaning of subjugation before it is too late. We can give but a faint idea when we say it means the loss of all we now hold most sacred – slaves and all other personal property, lands, homesteads, liberty, justice, safety, pride, manhood. It means that the history of this heroic struggle will be written by the enemy. That our youth will be trained by Northern school teachers; will learn from Northern school books their version of the war; will be impressed by all the influences of history and education to regard our gallant dead as traitors, our maimed veterans as fit objects for derision.

It means the crushing of Southern manhood, the hatred of our former slaves, who will, on a spy system, be our secret police. The conqueror's policy is to divide the conquered into factions and stir up animosity among them, and in training an army of Negroes, the North no doubt holds this thought in perspective.

We can see three great causes operating to destroy us: First, the inferiority of our armies to those of the enemy in point of numbers; second, the poverty of our single source of supply in comparison with his several sources; third, the fact that slavery, from being one of our chief sources of strength at the commencement of the war, has now become, in a military point of view, one of our chief sources of weakness.

The enemy already opposes us at every point with superior numbers, and is endeavoring to make the preponderance irresistible. President Davis, in his recent message, says the enemy "has recently ordered a large conscription and made a subsequent call for volunteers, to be followed, if ineffectual, by a still further draft." In addition, the President of the United States announces that "he has already in training an army of 100,000 Negroes as good as any troops," and every fresh raid he makes and new slice of territory he wrests from us will add to this force.

Every soldier in our army already knows and feels our numerical inferiority to the enemy. Want of men in the field has prevented him from reaping the fruits of his victories and has prevented him from having the furlough he expected after the last reorganization, and when he turns from the wasting armies in the field to look at the source of supply, he finds nothing in the prospect to encourage him.

Our single source of supply is that portion of our white men fit for duty and not now in the ranks. The enemy has three sources of supply: First, his own motley population; secondly, our slaves; and thirdly, Europeans whose hearts are

> *In this state of things, it is easy to understand why there is a growing belief that some black catastrophe is not far ahead of us, and that unless some extraordinary change is made in our condition, we must overtake it.*

fired into a crusade against us by fictitious pictures of the atrocities of slavery and who meet no hindrance from their governments in such enterprise because these governments are equally antagonistic to the institution. In touching the third cause, the fact that slavery has become a military weakness, we may rouse prejudice and passion, but the time has come when it would be madness not to look at our danger from every point of view, and to probe it to the bottom.

Apart from the assistance that home and foreign prejudice against slavery has given to the North, slavery is a source of great strength to the enemy in a purely military point of view, by supplying him with an army from our granaries; but it is our most vulnerable point, a continued embarrassment and in some respects an insidious weakness.

Wherever slavery is once seriously disturbed, whether by the actual presence or the approach of the enemy, or even by a cavalry raid, the whites can no longer with safety to their property openly sympathize with our cause. The fear of their slaves is continually haunting them, and from silence and apprehension many of these soon learn to wish the war stopped on any terms.

The next stage is to take the oath to save property, and they become dead to us, if not open enemies. To prevent raids, we are forced to scatter our forces and are not free to move and strike like the enemy; his vulnerable points are carefully selected and fortified depots. Ours are found in every point where there is a slave to set free. All along the lines slavery is comparatively valueless to us for labor, but of great and increasing worth to the enemy for information.

It is an omnipresent spy system, pointing out our valuable men to the enemy, revealing our positions, purposes, and resources, and yet acting so safely and secretly that there is no means to guard against it. Even in the heart of our country, where our hold upon this secret espionage is firmest, it waits but the opening fire of the enemy's battle line to wake it, like a torpid serpent, into venomous activity.

> *As between the loss of independence and the loss of slavery, we assume that every patriot will freely give up the latter – give up the slave rather than be a slave himself.*

In view of the state of affairs what does our country propose to do? In the words of President Davis "no effort must be spared to add largely to our effective force as promptly as possible. The sources of supply are to be found in restoring to the army all who are improperly absent, putting an end to substitution, modifying the exemption law, restricting details, and placing in the ranks such of the able-bodied men now employed as wagoners, nurses, cooks, and other employees, as are doing service for which the Negroes may be found competent." Most of the men improperly absent, together with many of the exempts and men having substitutes, are now without the Confederate lines and cannot be calculated on.

If all the exempts capable of bearing arms were enrolled, it will give us the boys below eighteen, the men above forty-five, and those persons who are left at home to meet the wants of the country and the army, but this modification of the exemption law will remove from the fields and manufactories most of the skill that directed agricultural and mechanical labor, and, as stated by the President, "details will have to be made to meet the wants of the country," thus sending many of the men to be derived from this source back to their homes again.

Independently of this, experience proves that striplings and men above conscript age break down and swell the sick lists more than they do the ranks. The portion now in our lines of the class who have substitutes is not on the whole a hopeful element, for the motives that created it must have been stronger than patriotism, and these motives added to what many of them will call breach of faith, will cause some to be not forthcoming, and others to be unwilling and discontented soldiers.

The remaining sources mentioned by the President have been so closely pruned in the Army of Tennessee that they will be found not to yield largely. The supply from all these sources, together with what we now have in the field, will exhaust the white race, and though it should greatly exceed expectations and put us on an equality with the enemy, or even give us temporary advantages, still we have no reserve to meet unexpected disaster or to supply a protracted struggle.

Like past years, 1864 will diminish our ranks by the casualties of war, and what source of repair is there left us? We therefore see in the recommendations of the President only a temporary expedient, which at the best will leave us twelve months hence in the same predicament we are in now.

The President attempts to meet only one of the depressing causes mentioned; for the other two he has proposed no remedy. They remain to generate lack of confidence in our final success, and to keep us moving down hill as heretofore. Adequately to meet the causes which are now threatening ruin to our country, we propose, in addition to a modification of the President's plans, that we retain in service for the war all troops now in service, and that we immediately commence training a large reserve of the most courageous of our slaves, and further that we guarantee freedom within a reasonable time to every slave in the South who shall remain true to the Confederacy in this war. As between the loss of independence and the loss of slavery, we assume that every patriot will freely give up the latter – give up the Negro slave rather than be a slave himself. If we are correct in this assumption, it only remains to show how this great national sacrifice is, in all human probabilities, to change the current of success and sweep the invader from our country.

Our country has already some friends in England and France, and there are strong motives to induce these nations to recognize and assist us, but they cannot assist us without helping slavery, and to do this would be in conflict with their policy for the last quarter of a century. England has paid hundreds of millions to emancipate her West India slaves and break up the slave trade. Could she now consistently spend her treasure to reinstate slavery in this country? But this barrier once removed, the sympathy and the interests of these and other nations will accord with our own, and we may expect from them both moral support and material aid. One thing is certain, as soon as the great sacrifice to independence is made and known in foreign countries, there will be a complete change of front in our favor of the sympathies of the world. This measure will deprive the North of the moral and material aid which it now derives from the bitter prejudices with which foreigners view the institution, and its war, if continued, will henceforth be so despicable in their eyes that the source of recruiting will be dried up.

It will leave the enemy's Negro army no motive to fight for and will exhaust the source from which it has been recruited. The idea that it is their special mission to war against slavery has held growing sway over the Northern people for many years and has at length ripened into an armed and bloody crusade against it. This baleful superstition has so far supplied them with a courage and constancy not their own. It is the most powerful and honestly entertained plank in their war platform. Knock this away and what is left? A bloody ambition for more territory, a pretended veneration for the Union, which one of their own most distinguished orators (Doctor Beecher in his Liverpool speech) openly avowed was only used as a stimulus to stir up the anti-slavery crusade, and lastly the poisonous and selfish interests which are the fungus growth of the war itself.

Mankind may fancy it a great duty to destroy slavery, but what interest can mankind have in upholding this remainder of the Northern war platform? Their interests and feelings will be diametrically opposed to it. The measure we propose will strike dead all John Brown fanaticism and will compel the enemy to draw off altogether or in the eyes of the world to swallow the Declaration of Independence without the sauce and disguise of philanthropy. This delusion of fanaticism at an end, thousands of Northern people will have leisure to look at home and to see the gulf of despotism into which they themselves are rushing.

The measure will at one blow strip the enemy of foreign sympathy and assistance and transfer them to the South; it will dry up two of his three sources of recruiting; it will take from his Negro army the only motive it could have to fight against the South and will probably cause much of it to desert over to us; it will deprive his cause of the powerful stimulus of fanaticism and will enable him to see the rock on which his so called friends are now piloting him.

The immediate effect of the emancipation and enrollment of Negroes on the military strength of the South would be: To enable us to have armies numerically superior to those of the North, and a reserve of any size we might think necessary; to enable us to take the offensive, move forward and forage on the enemy. It would open to us in prospective another and almost untouched source of supply and furnish us with the means of preventing temporary disaster, and carrying on a protracted struggle.

It would instantly remove all the vulnerability, embarrassment, and inherent weakness which result from slavery. The approach of the enemy would no longer find every household surrounded by spies; the fear that sealed the master's lips and the avarice that has, in so many cases, tempted him practically to desert us would alike be removed. There would be no recruits awaiting the enemy with open arms, no complete history of every neighborhood with ready guides, no fear of insurrection in the rear, or anxieties for the fate of loved ones when our armies moved forward. The chronic irritation of hope deferred would be joyfully ended with the Negro, and the sympathies of his whole race would be due to his native South.

It would restore confidence in an early termination of the war with all its inspiring consequences, and even if contrary to all expectations the enemy should succeed in overrunning the South, instead of finding a cheap, ready-made means of holding it down, he would find a common hatred and thirst for vengeance, which would break into acts at every favorable opportunity, would prevent him from settling on our lands, and render the South a very unprofitable conquest. It would remove forever all selfish taint from our cause and place independence above every question of property.

The very magnitude of the sacrifice itself, such as no nation has ever voluntarily made before, would appall our enemies, destroy his spirit and his finances, and fill our hearts with a pride and singleness of purpose which would clothe us with new strength in battle.

> *In addition, the President of the United States announces that "he has already in training an army of 100,000 negroes as good as any troops," and every fresh raid he makes and new slice of territory he wrests from us will add to this force.*

Apart from all other aspects of the question, the necessity for more fighting men is upon us. We can only get a sufficiency by making the Negro share the danger and hardships of the war. If we arm and train him and make him fight for the country in her hour of dire distress, every consideration of principle and policy demand that we should set him and his whole race who side with us free. It is a first principle with mankind that he who offers his life in defense of the State should receive from her in return his freedom and his happiness, and we believe in acknowledgment of this principle.

The Constitution of the Southern States has reserved to their respective governments the power to free slaves for meritorious services to the State. It is politic besides. For many years, ever since the agitation of the subject of slavery commenced, the Negro has been dreaming of freedom, and his vivid imagination has surrounded that condition with so many gratifications that it has become the paradise of his hopes. To attain it, he will tempt dangers and difficulties not exceeded by the bravest soldier in the field. The hope of freedom is perhaps the only moral incentive that can be applied to him in his present condition. It would be preposterous then to expect him to fight against it with any degree of enthusiasm, therefore we must bind him to our cause by no doubtful bonds; we must leave no possible loophole for treachery to creep in. The slaves are dangerous now, but armed, trained, and collected in an army they would be a thousand fold more dangerous: therefore when we make soldiers of them we must make free men of them beyond all question, and thus enlist their sympathies also.

We can do this more effectually than the North can now do, for we can give the Negro not only his own freedom, but that of his wife and child, and can secure it to him in his old home. To do this, we must immediately make his marriage and parental relations sacred in the eyes of the law and forbid their sale. The past legislation of the South concedes that large free middle class of Negro blood, between the master and slave, must sooner or later destroy the institution. If, then, we touch the institution at all, we would do best to make the most of it, and by emancipating the whole race upon reasonable terms, and within such reasonable time as will prepare both races for the change, secure to ourselves all the advantages, and to our enemies all the disadvantages that can arise, both at home and abroad, from such a sacrifice.

Satisfy the Negro that if he faithfully adheres to our standard during the war he shall receive his freedom and that of his race. Give him as an earnest of our intentions such immediate immunities as will impress him with our sincerity and be in keeping with his new condition, enroll a portion of his class as soldiers of the Confederacy, and we change the race from a dreaded weakness to a position of strength.

Will the slaves fight? The helots of Sparta stood their masters good stead in battle. In the great sea fight of Lepanto where the Christians checked forever the spread of Mohammedanism over Europe, the galley slaves of portions of the fleet were promised freedom, and called on to fight at a critical moment of the battle. They fought well, and civilization owes much to those brave galley slaves. The Negro slaves of Saint Domingo, fighting for freedom, defeated their white masters and the French troops sent against them. The Negro slaves of Jamaica revolted, and under the name of Maroons held the mountains against their masters for 150 years; and the experience of this war has been so far that half-trained Negroes have fought as bravely as many other half-trained Yankees. If, contrary to the training of a lifetime, they can be made to face and fight bravely against their former masters, how much more probable is it that with the allurement of a higher reward, and led by those masters, they would submit to discipline and face dangers.

We will briefly notice a few arguments against this course. It is said Republicanism cannot exist without the institution. Even were this true, we prefer any form of government of which the Southern people may have the molding, to one forced upon us by a conqueror. It is said the white man cannot perform agricultural labor in the South. The experience of this army during the heat of summer from Bowling Green, Kentucky to Tupelo, Mississippi is that the white man is healthier when doing reasonable work in the open field than at any other time.

It is said an army of Negroes cannot be spared from the fields. A sufficient number of slaves is now administering to luxury alone to supply the place of all we need, and we believe it would be better to take half the able bodied men off a plantation than to take the one master mind that economically regulated its operations. Leave some of the skill at home and take some of the muscle to fight with.

It is said slaves will not work after they are freed. We think necessity and a wise legislation will compel them to labor for a living. It is said it will cause terrible excitement and some disaffection from our cause. Excitement is far preferable to the apathy which now exists, and disaffection will not be among the fighting men.

It is said slavery is all we are fighting for, and if we give it up, we give up all. Even if this were true, which we deny, slavery is not all our enemies are fighting for. It is merely the pretense to establish sectional superiority and a more centralized form of government, and to deprive us of our rights and liberties.

We have now briefly proposed a plan which we believe will save our country. It may be imperfect, but in all human probability it would give us our independence. No objection ought to outweigh it which is not weightier than independence. If it is worthy of being put in practice it ought to be mooted quickly before the people, and urged earnestly by every man who believes in its efficacy. Negroes will require much training; training will require time, and there is danger that this concession to common sense may come too late.

GENERAL WALKER'S LETTER TO DAVIS REGARDING CLEBURNE'S PROPOSAL

HEADQUARTERS DIVISION,
Near Dalton, January 12, 1864

His Excellency JEFFERSON DAVIS,
President of Confederate States:

I feel it my duty as an officer of the Army to lay before the Chief Magistrate of the Southern Confederacy the within document, which was read on the night of the 2nd of January, 1864, at a meeting which I attended in obedience to the following order:

HEADQUARTERS HARDEE'S CORPS,
Dalton, Georgia, January 2, 1864.

Major General WALKER,

GENERAL: Lieutenant General Hardee desires that you will meet him at General Johnston's headquarters this evening at 7:00.

>Very respectfully, your obedient servant,
>D.H. POOLE,
>*Assistant Adjutant General*

Having, after the meeting adjourned, expressed my determination to apply to General Cleburne for a copy of the document to forward to the War Department, some of the gentlemen who were present at that meeting insisted upon their sentiments on so grave a subject being known to the Executive. I informed them that I would address a letter to each of the gentlemen present at the meeting, which I did. I addressed a note to General Cleburne, asking him for a copy of the document, informing him that I felt it my duty to forward it to the War Department; that should he do so I would, of course, give him a copy of the endorsement I made on it. He furnished me with a copy, and avowed himself its author.

I applied to the commanding general for permission to send it to the War Department through the proper official channel, which, for reasons satisfactory to himself, he declined to do; hence the reason for it not reaching you through the official channel. The gravity of the subject, the magnitude of the issues involved, my strong convictions that the further agitation of such sentiments and propositions would ruin the efficacy of our Army and involve our cause in ruin and disgrace constitutes my reasons for bringing the document before the Executive.

>W.H.T. WALKER,
>*Major-General*

JEFFERSON DAVIS' LETTER TO GENERAL WALKER REGARDING CLEBURNE'S PROPOSAL

>RICHMOND, *VIRGINIA, January* 13, 1864

General W.H.T. WALKER,

GENERAL: I have received your letter, with its enclosure, informing me of the propositions submitted to a meeting of the general officers on the 2nd instant, and thank you for the information. Deeming it to be injurious to the public service that such a subject should be mooted, or even known to be entertained by persons possessed of the confidence and respect of the people, I have concluded that the best policy under the circumstances will be to avoid all publicity, and the Secretary of War has therefore written to General Johnston requesting him to convey to those concerned my desire that it should be kept private. If it be kept out of the public journals its ill effect will be much lessened.

>Very respectfully and truly, yours,
>JEFFERSON DAVIS

SECRETARY OF WAR SEDDON'S LETTER TO GENERAL JOHNSTON REGARDING CLEBURNE'S PROPOSAL

WAR DEPARTMENT, *C.S.A.*,
Richmond, Virginia, January 24, 1861.

General Joseph E. Johnston,
Dalton, Georgia:

GENERAL: Major General Walker has communicated directly to the President copies of a memorial prepared by Major General Cleburne, lately the subject of consultation among the generals of division in your command, as also of a letter subsequently addressed by himself to the Generals present, asking the avowal of the opinions entertained by them in relation to such memorial, with their replies. I am instructed by the President to communicate with you on the subject.

He is gratified to infer, from your declining to forward officially General Walker's communication of the memorial, that you neither approved the views advocated in it, nor deemed it expedient that, after meeting as they happily did the disapproval of the council, they should have further dissemination or publicity.

The motives of zeal and patriotism which have prompted General Walker's action are, however, fully appreciated, and that action is probably fortunate, as it affords an appropriate occasion to express the earnest conviction of the President that the dissemination or even promulgation of such opinions under the present circumstances of the Confederacy, whether in the Army or among the people, can be productive only of discouragement, distraction, and dissension. The agitation and controversy which must spring from the presentation of such views by officers high in public confidence are to be deeply deprecated, and while no doubt or mistrust is for a moment entertained of the patriotic intents of the gallant author of the memorial, and such of his brother officers as may have favored his opinions, it is requested that you will communicate to them, as well as all others present on the occasion, the opinions, as herein expressed, of the President, and urge on them the suppression, not only of the memorial itself, but likewise of all discussion and controversy respecting or growing out of it.

I would add that the measures advocated in the memorial are considered to be little appropriate for consideration in military circles, and indeed in their scope pass beyond the bounds of Confederate action and could, under our constitutional system, neither be recommended by the Executive to Congress nor be entertained by that body. Such views can only jeopardize among the States and people unity and harmony, when for successful co-operation and the achievement of independence both are essential.

With much respect, very truly, yours,
JAMES A. SEDDON,
Secretary of War

GENERAL JOHNSTON TO SECRETARY OF WAR SEDDON REGARDING CLEBURNE'S PROPOSAL

DALTON, *February* 2, 1864.

Honorable James A. Seddon,
Secretary of War:

SIR: I had the honor to receive the letter in which you express the views of the President in relation to the memorial of Major General Cleburne on the 31st ultimo, and immediately transmitted his instructions in your own language to the officers concerned. None of the officers to whom the memorial was read favored the scheme; and Major General Cleburne, as soon as that appeared, voluntarily announced that he would be governed by the opinion of those officers, and put away his paper. The manner of strengthening our armies by using Negroes was discussed, and no other thought practicable than that which I immediately promised to the President.

I regarded this discussion as confidential, and understood it to be so agreed before the party separated. This and General Cleburne's voluntary promise prevented any apprehension in my mind of the agitation of the subject of the memorial. I have had no reason since to suppose that it made any impression.

Most respectfully, your obedient servant,
J.E. JOHNSTON

✯ ✯ ✯

It is said slavery is all we are fighting for, and, if we give it up, we give up all. Even if this were true, which we deny, slavery is not all our enemies are fighting for. It is merely the pretense to establish sectional superiority and a more centralized form of government and to deprive us of our rights and liberties.

Patrick Cleburne

Slaves In Our Army

By Buck Irving, Assistant Adjunct General, Cleburne's Division. *Richmond Dispatch, August 5, 1904.*

In the spring of 1897, I had a letter from the War Department at Washington, asking me to authenticate a document in the files of the Confederate Record Office. Considering that paper of the first interest and value, I send, herewith, a copy, and will give your readers the circumstances surrounding it.

After the disgraceful defeat of the Confederate army at Missionary Ridge in front of Chattanooga on the 25th of November, 1864, the bulk of it retreated to Dalton, Georgia, Cleburne's Division, which was the rear guard. On the 27th, the division made a stand at Ringgold Gap, and without assistance, and single-handed, checked and defeated the attempt of the pursuing army under General Hooker to capture the wagon, artillery, and ordnance train of Bragg's army. Holding the position until the safety of these were assured, the division retired, under orders to Tunnel Hill, some ten miles north of Dalton, where it remained on outpost.

Cleburne Absorbed

In December following, I noticed that General Cleburne was for several days deeply preoccupied and engaged in writing. Finally he handed me his manuscript, which upon reading, I found to be an advocacy of freeing the Negroes and their enlistment in our military service. In reply to his question as to what I thought of it, I said while I fully concurred in his opinion as to the absolute necessity of some such step to recruit the army and recognized the force of his arguments, still I doubted the expediency, at that time, of his formulating these views. First, because the slave holders were very sensitive as to such property and were totally unprepared to consider such a radical measure, and many, not being in our service, could not properly appreciate that it had become a matter of self-preservation that our ranks should be filled to meet, in some degree, the numerical superiority of the enemy – consequently, it would raise a storm of indignation against him. And next, that one of the corps of our army was without

> *He answered that a crisis was upon the South, the danger of which he was convinced would most quickly be averted in the way outlined and feeling it was his duty to bring this before the authorities, he would try to do so, irrespective of any personal result.*

a lieutenant general, that he, General Cleburne, had already achieved, unaided, a signal success at Ringgold, for which he had received the thanks of Congress, and stood in reputation first among the major generals, and might justly expect to be advanced to this vacancy, and I felt assured the publicity of this paper would be used detrimentally to him, and his chances of promotion destroyed.

To that he answered that a crisis was upon the South, the danger of which he was convinced could most quickly be averted in the way outlined and, feeling it to be his duty to bring this before the authorities, he would try to do so, irrespective of any personal result. To my question as to whether or not the Negroes would make efficient soldiers, he said that with reasonable and careful drilling, he had no doubt they would and, as deep as was his attachment to his present command, he would cheerfully undertake that of a Negro division in this emergency.

Copies of the Plan

Under his instructions I made, from his notes, a plain copy of the document, which was read to, and free criticisms invited from members of his staff, one of whom, Major Calhoun Benham, strongly dissented, asked for a copy with the purpose of writing a reply in opposition.

The division brigadiers were then called together, and my recollection is, that their endorsement was unanimous – namely: Polk, Lowery, Govan, and Granberry. Later, a meeting of the general officers of the army, including its commander, General Joseph E. Johnston, was held at General Hardee's headquarters and the paper submitted. It was received with disapproval by several, and before this assemblage, Major Benham read his letter of protest. Not having been present, I am unable to state the individual sentiment of the higher officers, but my impression is, that Generals Hardee and Johnston were favorably disposed, though the latter declined to forward it to the War Department on the ground that it was more political than military in tenor.

> *The proposal was received with disapproval by several, and before this assemblage, Major Benham read his letter of protest.*

That was a sore disappointment to Cleburne, who supposed his opportunity of bringing the matter before the President was lost, as he was too good a soldier and strict a disciplinarian to think of sending it over the head of his superior.

Queer Outcome

The day following, Major General W.H.T. Walker addressed him a note, stating that this paper was of such a dangerous (I think he said incendiary) character, that he felt it his duty to report it to the President, and asking if General Cleburne would furnish him a copy and avow himself its author.

Both requests were promptly complied with, Cleburne remarking that General Walker had done him an unintentional service in accomplishing his desire that this matter be brought to the attention of the Confederate authorities. Communication with Richmond was then very slow and uncertain. General Cleburne, naturally, felt somewhat anxious as to the outcome of the affair, though manifesting no regrets, and in discussing the matter and possibilities, said the worst that could happen to him would be court martial and cashiering, if which occurred, he would immediately enlist in his old regiment, the 15th Arkansas, then in his division; that if not permitted to command, he could at least do his duty in the ranks.

After the lapse of some weeks the paper was returned endorsed by President Davis, substantially, if not verbatim, as follows:

"While recognizing the patriotic motives of its distinguished author, I deem it inexpedient, at this time, to give publicity to this paper, and request that it be suppressed. J. D."

Upon receipt of this, General Cleburne directed me to destroy all copies, except the one returned from Richmond. This was filed in my office desk, which was subsequently captured and burned with its contents by the Federal cavalry during the Atlanta campaign.

Comes to Light

After the war, I was several times solicited, from both Confederate and Federal sources, to furnish copies, which was impossible, as I felt sure the only one retained had been destroyed, as above stated, and that no other existed. A few years ago Major Benham died in California, and to my extreme surprise and delight, a copy, the one supplied him at Tunnel Hill, was found among his papers. This was forwarded to Lieutenant L.H. Mangum, Cleburne's former law partner and afterwards aide-de-camp, who sent it to me to identify, which I readily did. Mangum afterwards placed it in the hands of General Marcus J. Wright, agent of the War Department, for collection of Confederate records, and it was this paper I was called upon to authenticate, the reason for which being that as it is a copy and not an original, some such official certification was desirable.

His Policy Adopted

A short while before his death, on the fatal field of Franklin, Cleburne had the gratification of knowing that a bill, embodying exactly his proposition, was advocated upon the floor of the Confederate Congress. This was subsequently passed and became a law, by executive approval.

It is scarcely a matter of speculation to tell what the result of this measure would have been, had it gone promptly into effect early in the spring of 1864. General Hood, whose opinion is entitled to weight, probably states it correctly in his book, *Advance and Retreat*, when referring to Cleburne, says: "He was a man of equally quick perception and strong character, and was, especially in one respect, in advance of many of our people. He possessed the boldness and wisdom to earnestly advocate at an early period of the war the freedom of the Negro and enrollment of the young and able-bodied men of that race. This stroke of policy and additional source of strength to our armies would, in my opinion, have given us our independence."

To subscribe, send an email to:
thestainlessbanner@gmail.com
or visit our website at www.thestainlessbanner.com

Subscription is free.

The Stainless Banner

An e-zine dedicated to the armies of the Confederacy

Volume 3, Issue 4
April 2012

THE FINAL DOMINO FALLS!
THE DEATH OF ALBERT SIDNEY JOHNSTON

In the days following the fall of Fort Donelson, Albert Sidney Johnston labored to bring the remainder of his army over the Cumberland River to Nashville. But Nashville was not his final destination. For some inexplicable reason, the citizens of Nashville had not thought it necessary to build defenses to protect the city from a possible attack from the river. Johnston knew it was only a matter of time until Don Carlos Buell moving south from Kentucky and Union gunboats sailing down the Cumberland joined forces to seize this vital Confederate depot.

Not wanting to lose the supplies stored in the warehouses along the river, Johnston ordered General John Floyd to send the supplies south. Floyd turned the task over to Nathan Bedford Forrest and quickly left the city. Once the warehouses were empty, Forrest rode south to join the army, now assembling at Corinth, Mississippi.

With both the Tennessee and Cumberland rivers lost, the railroads became the army's lifeline. At Corinth, just over the Tennessee/Mississippi border, the Memphis & Charleston and the Mobile & Ohio intersected, bringing supplies and, more importantly, men from the east, west, and south. Already troops from the garrison at Pensacola, Florida, under the command of Braxton Bragg, were on their way, as were reinforcements from New Orleans and Mobile. Earl Van Dorn's army in Arkansas had been ordered to cross the Mississippi and join Beauregard, who had safely brought half of the army across the Tennessee River to Corinth after Fort Henry fell.

What's Inside:

Beauregard's Report on the Battle	5
Bragg's Report on the Battle	14
The Battle of Shiloh	22
Albert Sidney Johnston	32
Albert Sidney Johnston at Shiloh	33

Johnston realized his army was vulnerable to being attacked and defeated in detail by the two Union forces moving south. So, he bluffed once more to convince Grant he was moving east to re-establish his headquarters at Chattanooga. His ruse was successful. Grant was absolutely convinced that the Confederates were headed to Chattanooga.

Once the army reached Murfreesboro, Johnston turned south to Decatur, Alabama, crossed the river and headed due west toward Corinth, risking a possible attack from Grant who was traveling along the Tennessee River toward Corinth. A month after leaving Nashville, Johnston arrived at Corinth.

Beauregard

For Beauregard coming west was only a detour, albeit a detour destined to bring him glory and honor, before returning to Virginia to win the war. Observing the raw troops being hastily assembled at Corinth, he realized he could do neither. It was such a shock to his delicate nervous system that he took to his bed and refused to take official command. From his sickbed, he bombarded Richmond with hysterical telegrams begging them to send Braxton Bragg to Corinth, even offering to serve under Bragg, if that would help persuade Bragg into coming.

What began with Polk parading into Columbus…now ended in a secluded spot on a battlefield near a small log church in Tennessee.

Once Bragg arrived, he was able to talk Beauregard down…or up off his sick bed. A shaky Beauregard finally assumed command of the army, only to suffer another set back to his recovery. Leonidas Polk had arrived and seemed to be as rattled and flustered as Beauregard. "The two fed the fires of each other's worry whenever they were together, prompting Bragg to confide in a letter to his wife that 'every interview with General Polk turns (Beauregard's recovery) back a week.'"[1]

It was a very happy and relieved Beauregard who welcomed Johnston and the 17,000 men with him to Corinth.

Johnston Prepares for Battle

Forty thousand men were now assembled at Corinth. Green troops Johnston did not have time to properly train or equip. For Grant was now at Pittsburg Landing on the Tennessee River thirty miles from Corinth, and Buell was not far behind. Johnston had to defeat Grant before Buell came up. On April 2nd, Johnston issued orders for the army to advance.

The men should have made the thirty miles in one day, but Johnston had given Beauregard responsibility for the logistics of the march. Beauregard quickly turned the march into a fiasco by having the divisions with their trains marching

---

[1] Steven E. Woodworth. *Jefferson Davis and His Generals.* (Lawrence: University Press of Kansas, 1990), 95.

down different roads only to converge and pass at intersections within minutes of each other. The army was not up to such split second timing, which resulted in a terrific traffic jam. For three days, the army struggled to reach Corinth.

When report came that one of Bragg's divisions had simply disappeared into the Mississippi countryside, Johnston finally lost his cool. "This is perfectly puerile!" He declared. "This is not war. Let us have our horses."[2] With his staff trailing behind him, he set out to find the missing men, which he did, standing off to one side of the road, waiting, while Polk's wagon train passed by. Johnston ordered the wagons from the road, and the missing division was soon off to reunite with the rest of the corps.

On the evening of April 5th, Johnston's army was just south of a small log church at Shiloh, Tennessee. In front of the Confederates, Grant's army was spread out between Owl and Lick creeks with their backs to the Tennessee River.

Johnston's Plan of Attack

Johnston's strategy was not complicated. With raw troops, he could not afford to be. He planned to assault the Union line along its entirety with the main thrust targeting the left in order to rip the Federal flank away from Owl Creek, roll up the line and seize Pittsburg Landing. This would cut Grant off from his transports and force him to either surrender or be destroyed.

Johnston sat down with Beauregard and went over the plan. Polk would take command of the left, Hardee the center, and Bragg the right. Unfortunately, Beauregard either didn't understand or changed the plan. Instead of each corps assailing their section of the line, Beauregard strung out the corps in three parallel lines. One line attacking behind the other.

It was a disaster in the making. Commanders from corps level down to brigade lost control of their men. The lines bunched together and regiments became entangled, slowing and, at times, stalling the attack. By time Johnston realized what Beauregard had done, it was too late to change it.

When Johnston made his plans to attack Grant, a major component of his strategy was to surprise the Federals. But any hope of surprise had been lost in the three days it took the army to reach Shiloh. Once the men were in their camps, they began shooting off their guns to make sure they would fire after being drenched in the torrential rains that had fallen during their long trek.

With the element of surprise gone, Beauregard lobbied Johnston to call off the attack, return to Corinth and wait until a better opportunity presented itself. Beauregard gained an unexpected ally when Bragg rode to headquarters to express the same concerns.

What both men did not realize was that the Federals had no idea that the army was so near. Sherman, who was encamped near Shiloh Church, believed the gunfire he heard was a wayward Confederate cavalry unit or his own men shooting off their guns to make sure they would work. He relayed a message to Grant that there was not a Confederate in sight. Grant's telegram that night to

---

[2] Ibid., 97.

Henry Halleck confirmed that he did not expect an attack the next day.

As dawn broke, more officers, concerned about the lack of surprise, arrived at headquarters to talk Johnston out of attacking. The sound of gunfire broke out. Johnston brought the meeting to a quick end. "The battle has opened, gentlemen," he simply said. "It is too late to change our dispositions."[3] Then he rode toward the front.

Johnston on the Battlefield

In the opening hours, the Confederates fought like seasoned veterans. Johnston was everywhere, encouraging the men to keep moving forward, stopping the looting that broke out, and, when he came across a clutch of wounded Union soldiers, dispatching his physician to care for them.

At the Hornet's Nest, Braxton Bragg valiantly tried to dislodge a division of Federals, but once again Beauregard's botched disposition of the troops limited Bragg from hitting the Nest with any strength. He could only feed in troops piecemeal as they came to him.

In the early afternoon, Johnston arrived to lend a hand. Coming upon an exhausted Tennessee regiment, he rode to the front and exhorted the soldiers to hit the enemy with the bayonet. As he neared the Nest, a bullet cut into his calf.

When his aide, Tennessee Governor Isham Harris, asked if he was wounded, Johnston answered in the affirmative. The bullet had ripped open an artery. Harris quickly got Johnston to safety and helped him down from his horse. But there was little Harris could do. Within minutes, Johnston was dead.

The Loss of a Lion

What began with Polk parading into Columbus – shattering Kentucky's neutrality and opening up the Cumberland and Tennessee rivers to Union gunboats – now ended in a secluded spot on a battlefield near a small log church in Tennessee. Albert Sidney Johnston was dead, and with his passing went any hope of winning the war in the West. For his death unleashed a cancer within the soon to be named Army of Tennessee. It would consume and destroy its new commander: Braxton Bragg. That cancer was the unrestrained ambition of Leonidas Polk.

Is that a fair statement? Look at the evidence. When Beauregard was replaced by Bragg, Polk began a campaign to bring about Bragg's destruction so he could assume command of the army. In slandering Bragg to his fellow officers, Polk created a mutinous atmosphere where obeying Bragg's orders became optional within the army. Perryville, Stones River, Chickamauga Creek were all campaigns where defeat was snatched from the jaws of victory because Polk and his cadre of conspirators refused to follow orders on the battlefield. When those campaigns failed, as they must, Polk wrote long letters to Jefferson Davis blaming Bragg for the very failures Polk was responsible for.

Bragg worked diligently to rid the army of Polk, but Davis did not believe Bragg's many reports of Polk's behavior. He simply did not accept that his former school mate could behave in such a

---

[3] Ibid., 99.

conniving manner. Yet, Polk did, and Bragg lost battles that should have been won and lost territory that should have not been lost except for the blind ambition of Leonidas Polk.

But Albert Sidney Johnston was a great friend of Davis. Far more than Polk had been. Under Johnston, Polk may have never launched his subversive campaign to gain command, and the battles fought would have been toward victory and not to defeat so Polk could gain what his machinations had caused Bragg to lose – command.

Again, we shall never know. All we have left are historians' opinions on what might have been. What we do know for sure is that the war in the West suffered a fatal and irreparable blow when Albert Sidney Johnston died at Shiloh.

We may be annihilated, but we cannot be conquered.

Albert Sidney Johnston

BEAUREGARD'S REPORT ON THE BATTLE OF SHILOH

HEADQUARTERS ARMY OF THE MISSISSIPPI,
Corinth, Miss, April 11, 1862.

General S. COOPER,
Adjutant and Inspector General C. S. Army, Richmond, Virginia

GENERAL: On the 2nd ultimo, having ascertained conclusively, from the movements of the enemy on the Tennessee River and from reliable sources of information, that his aim would be to cut off my communications in West Tennessee with the Eastern and Southern States, by operating from the Tennessee River, between Crump's Landing and Eastport, as a base, I determined to foil his designs by concentrating all my available forces at and around Corinth.

Meanwhile, having called on the Governors of the States of Tennessee, Mississippi, Alabama, and Louisiana to furnish additional troops, some of them (chiefly regiments from Louisiana) soon reached this vicinity, and with two divisions of General Polk's command from Columbus, and a fine corps of troops from Mobile and Pensacola, under Major General Bragg, constituted the Army of the Mississippi. At the same time General Johnston, being at Murfreesboro, on the march to form a junction of his forces with mine, was called on to send at least a brigade by railroad, so that we might fall on and crush the enemy, should he attempt an advance from under his gunboats.

The call on General Johnston was promptly complied with. His entire force was also hastened in this direction, and by April 1, our united forces were concentrated along the Mobile and Ohio Railroad from Bethel to Corinth and on the Memphis and Charleston Railroad from Corinth to Iuka.

It was then determined to assume the offensive, and strike a sudden blow at the enemy, in position under General Grant on the west bank of the Tennessee, at Pittsburg, and in the direction of Savannah, before he was reinforced by the army under General Buell, then known to be advancing for that purpose by rapid marches from Nashville via Columbia. About the same time General Johnston was advised that such an operation conformed to the expectations of the President.

By a rapid and vigorous attack on General Grant, it was expected he would be beaten back into his transports and the river, or captured, in time to enable us to profit by the victory, and remove to the rear all the stores and munitions that would fall into our hands in such an event before the arrival of General Buell's army on the scene. It was never contemplated, however, to retain the position thus gained and abandon Corinth, the strategic point of the campaign:

Want of general officers needful for the proper organization of divisions and brigades of an army brought thus suddenly together and other difficulties in the way of an effective organization delayed the movement until the night of the 2nd instant, when it was heard, from a reliable quarter, that the junction of the enemy's armies was near at hand. It was then, at a late hour, determined that the attack should be attempted at once, incomplete and imperfect as were our preparations for such a grave and momentous adventure. Accordingly, that night at 1:00 a.m., the preliminary orders to the commanders of corps were issued for the movement.

> *By a rapid and vigorous attack on General Grant, it was expected he would be beaten back into his transports and the river, or captured, in time to enable us to profit by the victory...*

On the following morning the detailed orders of movement were issued, and the movement, after some delay, commenced, the troops being in admirable spirits. It was expected we should be able to reach the enemy's lines in time to attack him early on the 5th instant. The men, however, for the most part, were unused to marching, and the roads, narrow and traversing a densely wooded country, became almost impassable after a severe rainstorm on the night of the 4th, which drenched the troops in bivouac; hence our forces did not reach the intersection of the roads from Pittsburg and Hamburg, in the immediate vicinity of the enemy, until late Saturday afternoon.

It was then decided that the attack should be made on the next morning, at the earliest hour practicable, in accordance with the orders of movement; that is, in three lines of battle, the first and second extending from Owl Creek, on the left, to Lick Creek, on the right, a distance of about three miles, supported by the third and the reserve. The first line, under Major General Hardee, was constituted of his corps, augmented on his right by Gladden's brigade, of Major General Bragg's corps, deployed in line of battle, with their respective artillery following immediately by the main road to Pittsburg and the cavalry in rear of the wings. The second line, composed of the other troops of Bragg's corps, followed the first at a distance of 500 yards in the same order as the first. The army corps under General Polk followed the second line, at a distance of about 800 yards, in lines of brigades deployed, with their batteries in rear of each brigade, moving by the Pittsburg road, the left wing supported by cavalry. The reserve, under Brigadier General Breckinridge, followed closely the third line in the same order, its right wing supported by cavalry.

These two corps constituted the reserve, and were to support the front lines of battle, by being deployed, when required, on the right and left of the Pittsburg road, or otherwise act according to the exigencies of the battle.

At 5:00 a.m. on the 6th instant, a reconnoitering party of the enemy having become engaged with our advance pickets, the commander of the forces gave orders to begin the movement and attack as determined upon, except that Trabue's brigade, of Breckinridge's division, was detached and advanced to support the left of Bragg's corps and line of battle when menaced by the enemy, and the other two brigades were directed to advance by the road to Hamburg to support Bragg's right; and at the same time Maney's regiment, of Polk's corps, was advanced by the same road to reinforce the regiment of cavalry and battery of four pieces already thrown forward to watch and guard Greer's, Tanner's, and Borland's Fords, on Lick Creek.

At 5:30 a.m. our lines and columns were in motion, all animated, evidently, by a promising spirit. The front line was engaged at once but advanced steadily, followed in due order, with equal resolution and steadiness, by the other lines, which were brought successively into action with rare skill, judgment, and gallantry by the several corps commanders as the enemy made a stand, with his masses rallied for the struggle for his encampments.

Like an Alpine avalanche our troops moved forward, despite the determined resistance of the enemy, until after 6:00 p.m., when we were in possession of all his encampments between Owl and Lick Creeks but one; nearly all of his field artillery; about thirty flags, colors, and standards; over 3,000 prisoners, including a division commander (General Prentiss), and several brigade commanders; thousands of small arms; an immense supply of subsistence, forage, and munitions of war, and a large amount of means of transportation – all the substantial fruits of a complete victory, such, indeed, as rarely have followed the most successful battles; for never was an army so well provided as that of our enemy.

The remnant of his army had been driven in utter disorder to the immediate vicinity of Pittsburg, under the shelter of the heavy guns of his ironclad gunboats, and we remained undisputed masters of his well selected, admirably provided cantonments, after ever twelve hours of obstinate conflict with his forces, who had been beaten from them and the contiguous covert, but only by a sustained onset of all the men we could bring into action.

Our loss was heavy, as will appear from the accompanying return, marked B. Our commander-in-chief, General Johnston, fell mortally wounded, and died on the field at 2:30 p.m., after having shown the highest qualities of the commander and a personal intrepidity that inspired all around him and gave resistless impulsion to his columns at critical moments.

The chief command then devolved upon me, though at the time I was greatly prostrated and suffering from the prolonged sickness with which I had been afflicted since early in February. The responsibility was one which in my physical condition I would have gladly avoided, though cast upon me when our forces were successfully pushing the enemy back upon the Tennessee River, and though supported on the immediate field by such corps commanders as Major Generals Polk, Bragg, and Hardee, and Brigadier General Breckinridge, commanding the reserve.

It was after 6:00 p.m., as before said, when the enemy's last position was carried, and his forces finally broke and sought refuge behind a commanding eminence covering the Pittsburg Landing, not more than half a mile distant, and under the guns of the gunboats, which opened on our eager columns a fierce and annoying fire with shot and shell of the heaviest description.

Darkness was close at hand. Officers and men were exhausted by a combat of over twelve hours without food and jaded by the march of the preceding day through mud and water. It was, therefore, impossible to collect the rich and opportune spoils of war scattered broadcast on the field left in our possession, and impracticable to make any effective dispositions for their removal to the rear.

> *The remnant of his army had been driven in utter disorder to the immediate vicinity of Pittsburg, under the shelter of the heavy guns of his ironclad gunboats.*

I accordingly established my headquarters at the church of Shiloh, in the enemy's encampments, with Major General Bragg, and directed our troops to sleep on their arms in such positions in advance and rear as corps commanders should determine, hoping, from news received by a special dispatch, that delays had been encountered by General Buell in his march from Columbia, and that his main force, therefore, could not reach the field of battle in time to save General Grant's shattered fugitive forces from capture or destruction on the following day.

During the night the rain fell in torrents, adding to the discomforts and harassed condition of the men. The enemy, moreover, had broken their rest by a discharge at measured intervals of heavy shells thrown from the gunboats; therefore on the following morning the troops under my command were not in condition to cope with an equal force of fresh troops, armed and equipped like our adversary, in the immediate possession of his depots and sheltered by such an auxiliary as the enemy's gunboats.

About 6:00 a.m. on the morning of April 7, however, a hot fire of musketry and artillery, opened from the enemy's quarter on our advanced line, assured me of the junction of his forces, and soon the battle raged with a fury which satisfied me I was attacked by a largely superior force. But from the outset our troops, notwithstanding their fatigue and losses from the battle of the day before, exhibited the most cheering, veteran-like steadiness.

On the right and center the enemy was repulsed in every attempt he made with his heavy columns in that quarter of the field. On the left, however, and nearest to the point of arrival of his reinforcements, he drove forward line after line of his fresh troops, which were met with a resolution and courage of which our country may be proudly hopeful.

Again and again our troops were brought to the charge, invariably to win the position in issue, invariably to drive back their foe. But hour by hour, thus opposed to an enemy constantly reinforced, our ranks were perceptibly thinned under the unceasing, withering fire of the enemy, and by 12:00 p.m. eighteen hours of hard fighting had sensibly exhausted a large number.

My last reserves had necessarily been disposed of, and the enemy was evidently receiving fresh reinforcements after each repulse; accordingly about 1:00 p.m., I determined to withdraw from so unequal a conflict, securing such of the results of the victory of the day before as was then practicable.

Officers of my staff were immediately dispatched with the necessary orders to make the best dispositions for a deliberate, orderly withdrawal from the field, and to collect and post a reserve to meet the enemy, should he attempt to push after us.

In this connection I will mention particularly my adjutant general Colonel Jordan, who was of much assistance to me on this occasion, as he had already been on the field of battle on that and the preceding day.

About 2:00 p.m., the lines in advance, which had repulsed the enemy in their last fierce assault on our left and center, received the orders to retire. This was done with uncommon steadiness and the enemy made no attempt to follow.

The line of troops established to cover this movement had been disposed on a favorable ridge commanding the ground of Shiloh Church. From this position our artillery played upon the woods beyond for a while, but upon no visible enemy and without reply. Soon satisfied that no serious pursuit would be attempted this last line was withdrawn, and never did troops leave a battlefield in better order; even the stragglers fell into the ranks and marched off with those who had stood more steadily by their colors.

A second strong position was taken up about a mile in rear, where the approach of the enemy was awaited for nearly an hour, but no effort to follow was made, and only a small detachment of horsemen could be seen at a distance from this last position, warily observing our movements.

Arranging through my staff officers for the completion of the movements thus begun, Brigadier General Breckinridge was left with his command as a rear guard to hold the ground we had occupied the night preceding the first battle, just in front of the intersection of the Pittsburg and Hamburg roads, about four miles from the former place, while the rest of the army passed to the rear in excellent order.

On the following day General Breckinridge fell back about three miles, to Mickey's, which position we continued to hold, with our cavalry thrown considerably forward in immediate proximity to the battlefield.

Unfortunately, toward night of the 7th instant it began to rain heavily. This continued throughout the night; the roads became almost impassable in many places, and much hardship and suffering now ensued before all the regiments reached their encampments; but, despite the heavy casualties of the two eventful days of April 6 and 7, this army is more confident of ultimate success than before its encounter with the enemy.

To Major Generals Polk, Bragg, and Hardee, commanding corps, and to Brigadier General Breckinridge, commanding the reserve, the country is greatly indebted for the zeal, intelligence, and energy with which all orders were executed; for the foresight and military ability they displayed in the absence of instructions in the many exigencies of the battle on a field so densely wooded and broken, and for their fearless deportment as they repeatedly led their commands personally to the onset upon their powerful adversary. It was under these circumstances that General Bragg had two horses shot under him; that Major General Hardee was slightly wounded, his coat rent by balls, and his horse disabled, and that Brigadier General Breckinridge was twice struck by spent balls.

For the services of their gallant subordinate commanders and of other officers, as well as for the details of the battle-field, I must refer to the reports of corps, division, and brigade commanders, which shall be forwarded as soon as received.

To give more in detail the operations of the two battles resulting from the movement on Pittsburg than now attempted must have delayed this report for weeks and interfered materially with the important duties of my position. But I may be permitted to say that not only did the obstinate conflict for twelve hours on Sunday leave the Confederate Army masters of the battlefield and our adversary beaten, but we left that field on the next day only after eight hours' incessant battle with a superior army of fresh troops, whom we had repulsed in every attack on our lines – so repulsed and crippled, indeed, as to leave it unable to take the field for the campaign for which it was collected and equipped at such enormous expense and with such profusion of all the appliances of war.

> **The chief command then devolved upon me, though at the time I was greatly prostrated and suffering from the prolonged sickness with which I had been afflicted since early in February.**

These successful results were not achieved, however, as before said, without severe loss – a loss not to be measured by the number of the slain or wounded, but by the high social and personal worth of so large a number of those who were killed or disabled, including the commander of the forces, whose high qualities will be greatly missed in the momentous campaign impending.

I deeply regret to record also the death of the Honorable George W. Johnson, Provisional Governor of Kentucky, who went into action with the Kentucky troops, and continually inspired them by his words and example. Having his horse shot under him on Sunday, he entered the ranks of a Kentucky regiment on Monday, and fell mortally wounded toward the close of the day. Not his State alone, but the whole Confederacy, has sustained a great loss in the death of this brave, upright, and able man.

Another gallant and able soldier and captain was lost to the service of the country when Brigadier General Gladden, commanding the First Brigade, Withers' division, Second Army Corps, died from a severe wound received on the 6th instant, after having been conspicuous to his whole corps and the army for courage and capacity.

Major General Cheatham, commanding First Division, First Corps, was slightly wounded and had three horses shot under him.

Brigadier General Clark, commanding Second Division, of the First Corps, received a severe wound also on the first day, which will deprive the army of his valuable services for some time.

Brigadier General Hindman, engaged in the outset of the battle, was conspicuous for a cool courage, efficiently employed in leading his men ever in the thickest of the fray, until his horse was shot under him and he was unfortunately so severely injured by the fall that the army was deprived on the following day of his chivalrous example.

Brigadier Generals B.R. Johnson and Bowen, most meritorious officers, were also severely wounded in the first combat, but it is hoped will soon be able to return to duty with their brigades.

To mention the many field officers who died or were wounded while gallantly leading their commands into action and the many brilliant instances of individual courage displayed by officers and men in the twenty hours of battle is impossible at this time, but their names will be duly made known to their countrymen.

The immediate staff of the lamented commander-in-chief, who accompanied him to the field, rendered efficient service, and, either by his side or in carrying his orders, shared his exposure to the casualties of the well-contested battle-field. I beg to commend their names to the notice of the War Department, namely: Captains. H. P. Brewster and N. Wickliffe, of the adjutant and inspector general's department; Captain Theodore O'Hara, acting inspector-general; Lieutenants George Baylot and Thomas M. Jack, aides-de-camp. Volunteer aides-de-camp Colonel William Preston, Major D. M. Hayden, E. W. Munford, and Calhoun Benham. Major Albert J. Smith and Captain Wickham, of the quartermaster's department.

To these gentlemen was assigned the last sad duty of accompanying the remains of their lamented chief from the field, except Captains Brewster and Wickliffe, who remained and rendered valuable services as staff officers on April 7.

Governor Isham G. Harris, of Tennessee, went upon the field with General Johnston, was by his side when he was shot, aided him from his horse, and received him in his arms when he died. Subsequently the Governor joined my staff and remained with me throughout the next day, except when carrying orders or employed in encouraging the troops of his own State, to whom he gave a conspicuous example of coolness, zeal, and intrepidity.

I am also under many obligations to my own general, personal, and volunteer staff, many of whom have been so long associated with me. I append a list of those present on the field on both days and whose duties carried them constantly under fire, namely: Colonel Thomas Jordan, Captain Clifton II. Smith, and Lieutenant John M. Otey, adjutant general's department; Major George W. Brent, acting inspector-general; Colonel R.B. Lee, chief of subsistence, whose horse was wounded; Lieutenant Colonel S.W. Ferguson and Lieutenant A.R. Chisolm, aides-de-camp. Volunteer aides-de-camp Colonel Jacob Thompson, Majors Numa Augustin and H.E. Peyton, and Captains Albert Ferry and B.B. Waddell. Captain W.W. Porter, of Major General Crittenden's staff, also reported for duty, and shared the duties of my volunteer staff on Monday. Brigadier General Trudeau, of Louisiana Volunteers, also for a part of the first day's conflict was with me as a volunteer aide. Captain E.H. Cummins, signal officer, also was actively employed as staff officer on both days.

Nor must I fail to mention that Private W.E. Goolsby, Eleventh Regiment Virginia Volunteers, orderly to my headquarters since last June, repeatedly employed to carry my verbal orders to the field, discharged the duty with great zeal and intelligence.

Other members of my staff were necessarily absent from the immediate field of battle, entrusted with responsible duties at these headquarters, namely: Captain F.H. Jordan, assistant adjutant general, in charge of general headquarters; Major Eugene E. McLean, chief quartermaster, and Captain E. Deslonde, quartermaster's department. Lieutenant Colonel Ferguson, aide-de-camp, early on Monday was assigned to command and directed the movements of a brigade of the Second Corps.

Lieutenant Colonel Gilmer, chief engineer, after having performed the important and various duties of his place with distinction to himself and material benefit to the country, was wounded late on Monday. I trust, however, I shall not long be deprived of his essential services. Captain Lockett Engineer Corps, chief assistant to Colonel Gilmer, after having been employed in the duties of his corps on Sunday, was placed by me on Monday in command of a battalion without field officers. Captain Fremaux, Provisional Engineers, and Lieutenants Steel and Helm also rendered material and even dangerous service in the line of their duty. Major General (now General) Braxton Bragg, in addition to his duties of chief of staff, as has been before stated, commanded his corps – much the largest in the field – on both days with signal capacity and soldiership.

Surgeons Foard, medical director; R.L. Brodie and S. Choppin, medical inspectors, and D.W. Yahdell, medical director of the Western Department, with General Johnston, were present in the discharge of their arduous and high duties, which they performed with honor to their profession.

Captain Tom Saunders, Messrs. Scales and Metcalf, and Mr. Tully, of New Orleans, were of material aid on both days, ready to give news of the enemy's positions and movements regardless of exposure.

While thus partially making mention of some of those who rendered brilliant, gallant, or meritorious service on the field, I have aimed merely to notice those whose position would most probably exclude the record of their services from the reports of corps or subordinate commanders.

From this agreeable duty I turn to one in the highest degree unpleasant; one due, however, to the brave men under me as a contrast to the behavior of most of the army who fought so heroically. I allude to the fact that some officers, non commissioned officers, and men abandoned their colors early on the first day to pillage the captured encampments; others retired shamefully from the field on both days while the thunder of cannon and the roar and rattle of musketry told them that their brothers were being slaughtered by the fresh legions of the enemy. I have ordered the names of the most conspicuous on this roll of laggards and cowards to be published in orders.

It remains to state that our loss on the two days, in killed outright, was 1,728; wounded, 8,012, and missing, 959; making an aggregate of casualties, 10,699.

This sad list tells in simple language of the stout fight made by our countrymen in front of the rude log chapel of Shiloh, especially when it is known that on Monday, from exhaustion and other causes, not 20,000 men on our side could be brought into action.

Of the losses of the enemy I have no exact knowledge. Their newspapers report it as very heavy. Unquestionably it was greater even in proportion than our own on both days, for it was apparent to all that their dead left on the field outnumbered ours two to one. Their casualties, therefore, cannot have fallen many short of 20,000 in killed, wounded, prisoners, and missing.

Through information derived from many sources, including the newspapers of the enemy, we engaged on Sunday the divisions of Generals Prentiss, Sherman, Hurlbut, McClernand, and Smith, of 9,000 men each, or, at least, 45,000 men. This force was reinforced Sunday night by the divisions of Generals Nelson, McCook, Crittenden, and Thomas, of Major General Buell's army, some 25,000 strong, including all arms; also General L. Wallace's division, of General Grant's army, making at least 33,000 fresh troops, which, added to the remnant of General Grant's forces – on Monday morning amounting to over 20,000 – made an aggregate force of some 53,000 men, at least, arrayed against us on that day.

In connection with the results of the battle I should state that most of our men who had inferior arms exchanged them for the improved arms of the enemy; also that most of the property, public and personal, in the camps from which the enemy was driven on Sunday was rendered useless or greatly damaged, except some of the tents.

All of which is respectfully submitted through my volunteer aide-de-camp, Colonel Jacob Thompson, of Mississippi, Who has in charge the flags, standards, and colors captured from the enemy.

I have the honor to be, general, your obedient servant,
P.G.T. BEAUREGARD,
General, Commanding

Bragg's Report on the Battle of Shiloh

HDQRS. SECOND CORPS, ARMY OF THE MISSISSIPPI,
Corinth, Mississippi, April 30, 1862

Brigadier General THOMAS JORDAN,
Chief of Staff

General: In submitting a report of the operations of my command, the Second Army Corps, in the actions of Shiloh, on the 6th and 7th of April, it is proper that the narrative of events on the field be preceded by a sketch of the march from here.

But few regiments of my command had ever made a day's march. A very large proportion of the rank and file had never performed a day's labor. Our organization had been most hasty, with great deficiency in commanders, and was therefore very imperfect. The equipment was lamentably defective for field service, and our transportation, hastily impressed in the country, was deficient in quantity and very inferior in quality. With all these drawbacks the troops marched late in the afternoon of the 3rd, a day later than intended, in high spirits and eager for the contest.

The road to Monterey (11 miles) was found very bad, requiring us until 11:00 a.m. on the 4th to concentrate at that place, where one of my brigades joined the column. Moving from there, the command bivouacked for the night near the Mickey house immediately in rear of Major General Hardee's corps; Major General Polk's corps being just in our rear.

Our advanced cavalry had encountered the enemy during the day and captured several prisoners, being compelled, however, to retire. A reconnaissance in some force from the enemy made its appearance during the evening in front of General Hardee's corps and was promptly driven back.

The commanders of divisions and brigades were assembled at night, the order of battle was read to them, and the topography of the enemy's position was explained, as far as understood by us. Orders were then given for the troops to march at 3:00 a.m., so as to attack the enemy early on the 5th.

About 2:00 a.m. a drenching rainstorm commenced, to which the troops were exposed, without tents, and continued until daylight, rendering it so dark and filling the creeks and ravines to such an extent as to make it impracticable to move at night. Orders were immediately sent out to suspend the movement until the first dawn of day. Continued firing by volleys and single shots was kept up all night and until 7:00 a.m. next morning by the undisciplined troops of our front, in violation of positive orders. Under such circumstances little or no rest could be obtained by our men, and it was -- o'clock in the morning before the road was clear so as to put my command in motion, though it had been in ranks and ready from 3:00 a.m., in the wet and cold, and suffering from inaction.

At this juncture the commanding general arrived at our position. My column, at last fairly in motion, moved on without delay until arriving near where the Pittsburg road leaves the Bark road, when a message from Major General Hardee announced the enemy in his front and that he had developed his line. As promptly as my troops could be brought up in a narrow road, much encumbered with artillery and baggage wagons, they were formed, according to order of battle, about 800 yards in rear of Hardee's line, my center resting on the Pittsburg road, my right brigade, Gladden's, of Withers' division, thrown forward to the right of the first line, Major General Hardee's force not being sufficient for the ground to be covered.

In this position we remained, anxiously awaiting the approach of our reserves to advance upon the enemy, now but a short distance in our front. The condition of the roads and other untoward circumstances delayed them until late in the afternoon, rendering it necessary to defer the attack until next morning.

The night was occupied by myself and a portion of my staff in efforts to bring forward provisions for a portion of the troops then suffering from their improvidence. Having been ordered to march with five days' rations, they were found hungry and destitute at the end of three days. This is one of the evils of raw troops, imperfectly organized and badly commanded; a tribute, it seems, we must continue to pay to universal suffrage, the bane of our military organization. In this condition we passed the night, and at dawn of day prepared to move.

The enemy did not give us time to discuss the question of attack, for soon after dawn he commenced a rapid musketry fire on our pickets. The order was immediately given by the commanding general and our lines advanced. Such was the ardor of our troops that it was with great difficulty they could be restrained from closing up and mingling with the first line. Within less than a mile the enemy was encountered in force at the encampments of his advanced positions, but our first line brushed him away, leaving the rear nothing to do but to press on in pursuit. In about one mile more we encountered him in strong force among almost the entire line. His batteries were posted on eminences, with strong infantry supports.

> *"Forward! Let every order be forward."*

Finding the first line was now unequal to the work before it, being weakened by extension and necessarily broken by the nature of the ground, I ordered my whole force to move up steadily and promptly to its support. The order was hardly necessary, for subordinate commanders, far beyond the reach of my voice and eye in the broken country occupied by us, had promptly acted on the necessity as it arose, and by the time the order could be conveyed the whole line was developed and actively engaged.

From this time, about 7:30 a.m., until night the battle raged with little intermission. All parts of our line were not constantly engaged, but there was no time without heavy firing in some portion of it. My position for several hours was opposite my left center (Ruggles' division), immediately in rear of Hindman's brigade of Hardee's corps.

In moving over the difficult and broken ground the right brigade of Ruggles' division, Colonel Gibson commanding, bearing to the right, became separated from the two left brigades, leaving a broad interval.

Three regiments of Major General Polk's command opportunely came up and filled this interval. Finding no superior officer with them, I took the liberty of directing their movements in support of Hindman, then, as before, ardently pressing forward and engaging the enemy at every point.

On the ground which had come under my immediate observation we had already captured three large encampments and three batteries of artillery. It was now about 10.30 a.m.

Our right flank, according to the order of battle, had pressed forward ardently under the immediate direction of the commanding general and swept all before it: Batteries, encampments, storehouses, munitions in rich profusion, were ours, and the enemy, fighting hard and causing us to pay dearly for our successes, was falling back rapidly at every point. His left, however, opposite our right, was his strongest ground and position, and was disputed with obstinacy.

It was during this severe struggle that my command suffered an irreparable loss in the fall of Brigadier General Gladden, commanding First Brigade, Withers' division, mortally, and Colonel D.W. Adams, Louisiana Regular Infantry, his successor, severely, wounded. Nothing daunted, however, by these losses, this noble division, under its gallant leader, Withers, pressed on with the other troops in its vicinity and carried all before them. Their progress, however, under the obstinate resistance made was not so rapid as was desired in proportion to that of the left, where the enemy was less strong; so that, instead of driving him, as we intended, down the river leaving the left open for him to pass, we had really enveloped him on all sides and were pressing him back upon the landing at Pittsburg.

Meeting at about 10.30 a.m. upon the left center with Major General Polk, my senior, I promptly yielded to him the important command at that point, and moved toward the right, in the direction in which Brigadier General Hindman, of Hardee's line, had just led his division. Here we met the most obstinate resistance of the day, the enemy being strongly posted, with infantry and artillery, on an eminence immediately behind a dense thicket. Hindman's command was gallantly led to the attack, but recoiled under a murderous fire. The noble and gallant leader fell, severely wounded, and was borne from the field he had illustrated with a heroism rarely equaled.

The command soon returned to its work, but was unequal to the heavy task. Leaving them to hold their position, I moved farther to the right, and brought up the First Brigade (Gibson), of Ruggles' division, which was in rear of its true position, and threw them forward to attack this same point. A very heavy fire soon opened, and after a short conflict this command fell back in considerable disorder. Rallying the different regiments, by means of my staff officers and escort, they were twice more moved to the attack, only to be driven back by the enemy's sharpshooters occupying the thick cover. This result was due entirely to want of proper handling.

Finding that nothing could be done here, after hours of severe exertion and heavy losses, and learning of the fall of our commander, who was leading in person on the extreme right, the troops were so posted as to hold this position, and leaving a competent staff officer to direct them in my name, I moved rapidly to the extreme right. Here I found a strong force, consisting of three parts, without a common head – Brigadier General Breckinridge, with his reserve division, pressing the enemy; Brigadier General Withers, with his splendid division, greatly exhausted and taking a temporary rest, and Major General Cheatham, with his division, of Major General Polk's corps, to their left and rear. These troops were soon put in motion, responding with great alacrity to the command of "Forward! Let every order be forward."

It was now probably past 4:00 p.m., the descending sun warning us to press our advantage and finish the work before night should compel us to desist. Fairly in motion, these commands again, with a common head and a common purpose, swept all before them. Neither battery nor battalion could withstand their onslaught. Passing through camp after camp, rich in military spoils of every kind, the enemy was driven headlong from every position and thrown in confused masses upon the river bank, behind his heavy artillery and under cover of his gunboats at the Landing. He had left nearly the whole of his light artillery in our hands and some 3,000 or more prisoners, who were cut off from their retreat by the closing in of our troops on the left under Major General Polk, with a portion of his reserve corps, and Brigadier General Ruggles, with Anderson's and Pond's brigades of his division

The prisoners were dispatched to the rear under a proper guard, all else being left upon the field that we might press our advantage. The enemy had fallen back in much confusion and was crowded in unorganized masses on the river bank, vainly striving to cross. They were covered by a battery of heavy guns, well served, and their two gunboats, which now poured a heavy fire upon our supposed positions, for we were entirely hid by the forest. Their fire, though terrific in sound and producing some consternation at first, did us no damage, as the shells all passed over and exploded far beyond our positions.

As soon as our troops could be again formed and put in motion the order was given to move forward at all points and sweep the enemy from the field. The sun was about disappearing, so that little time was left us to finish the glorious work of the day, a day unsurpassed in the history of warfare for its daring deeds, brilliant achievements, and heavy sacrifices.

Our troops, greatly exhausted by twelve hours' incessant fighting, without food, mostly responded to the order with alacrity, and the movement commenced with every prospect of success, though a heavy battery in our front and the gunboats on our right seemed determined to dispute every inch of ground.

Just at this time an order was received from the commanding general to withdraw the forces beyond the enemy's fire. As this was communicated, in many instances, direct to brigade commanders, the troops were soon in motion, and the action ceased. The different commands, mixed and scattered, bivouacked at points most convenient to their positions and beyond the range of the enemy's guns. All firing, except a half hour shot from the gunboats, ceased, and the whole night was passed by our exhausted men in quiet. Such as had not sought shelter in the camps of the enemy were again drenched before morning by one of those heavy rainstorms which seemed to be our portion for this expedition.

Such was the nature of the ground over which we had fought, and the heavy resistance we had met, that the commands of the whole army were very much shattered. In a dark and stormy night commanders found it impossible to find or assemble their troops, each body or regiment bivouacking where night overtook them.

In this condition morning found us, confronting a large and fresh army, which had arrived during the night, and for the first time the enemy advanced to meet us. He was received by our whole line with a firm and bold front, and the battle again raged.

> *As soon as our troops could be again formed and put in motion, the order was given to move forward at all points and sweep the enemy from the field.*

From this hour until 2:00 p.m. the action continued with great obstinacy and varying success. Our troops, exhausted by days of incessant fatigue, hunger, and want of rest, and ranks thinned by killed, wounded, and stragglers, mounting in the whole to nearly half our force, fought bravely, but with the want of that animation and spirit which characterized them the preceding day. Many instances of daring and desperate valor, deserving of better success, failed for want of numbers.

My personal services were confined during this day to the extreme left of our line, where my whole time was incessantly occupied. The troops in my front consisted of Ruggles' division, Colonel Trabue's brigade, of Breckinridge's reserve, and other detachments of different corps, all operating to the left of Shiloh Church.

This force advanced in the early morning and pressed the enemy back for nearly a mile, securing for our left flank an eminence in an open field near Owl Creek, which we held until near the close of the conflict against every effort the enemy could make. For this gallant and obstinate defense of our left flank, which the enemy constantly endeavored to force, we were indebted to Colonel Trabue's small brigade, in support of Captain Byrne's battery.

Against overwhelming numbers this gallant command maintained its position from the commencement of the action until about 12:00, when, our forces on the right were left, entirely without support, far in front of our whole army. Safety required it to retire.

During this time the right and center were actively engaged. Withers' division, in conjunction with portions of Hardee's and Breckinridge's commands, obstinately disputed every effort of the enemy. But his overwhelming numbers, a very large portion being perfectly fresh troops, the prostration of our men, and the exhaustion of our ammunition, not a battalion being supplied, rendered our position most perilous, and the commanding general ordered a retrograde movement, to commence on the right. This was gradually extended to the left, now held by Ketchum's battery. The troops fell back generally in perfect order and formed line of battle on a ridge about half a mile in the rear, Ketchum retiring slowly as the rear guard of the whole army. The enemy evinced no disposition to pursue.

After some half hour our troops were again put in motion and moved about a mile farther, where line was formed and final arrangements made for the march to our camp at Corinth, the enemy not making the slightest demonstration upon us. This orderly movement, under the circumstances, was as creditable to the troops as any part of the brilliant advance they had made.

Of the missing, a few were ascertained to have fallen into the hands of the enemy, mostly wounded. The others were no doubt left dead on the field. The heavy loss sustained by the command will best indicate the obstinacy of the resistance met and the determination with which it was overcome.

For the part performed by the different portions of the corps reference is made to the reports of subordinate commanders.

The division of Brigadier General J.M. Withers was gallantly led by that officer from the first gun to the close of the action, and performed service rarely surpassed by any troops on any field.

Brigadier General A.H. Gladden, First Brigade of this division, fell early in the action, mortally wounded, while gallantly leading his command in a successful charge. No better soldier lived. No truer man or nobler patriot ever shed his blood in a just cause.

Later in the day Colonel D.W. Adams, Louisiana infantry, who had succeeded to this splendid brigade, was desperately wounded while gallantly leading it, and later still Colonel Z.C. Deas, 22nd Alabama Volunteers, fell pierced by several balls.

Brigadier General James R. Chalmers, at the head of his gallant Mississippians filled – he could not have exceeded – the measure of my expectations. Never were troops and commander more worthy of each other and of their State.

Brigadier General J.K. Jackson did good service with his Alabama Brigade on the first day, but, becoming much broken; it was not unitedly in action thereafter. The excellent regiment of Colonel Joseph Wheeler, however, joined and did noble service with Gladden's brigade.

Brigadier General D. Ruggles, commanding Second Division, was conspicuous throughout both days for the gallantry with which he led his troops. Brigadier General Patton Anderson, commanding a brigade of this division, was also among the foremost where, the fighting was hardest, and never failed to overcome whatever resistance was opposed to him. With a brigade composed almost entirely of raw troops his personal gallantry and soldierly bearing supplied the place of instruction and discipline.

It would be a pleasing duty to record the deeds of many other noble soldiers of inferior grade, but as subordinate commanders have done so it, their reports a repetition is unnecessary. I shall be pardoned for making an exception in case of Captain R.W. Smith, commanding a company of Alabama cavalry, which served as my personal escort during the action. For personal gallantry and intelligent execution of orders, frequently under the heaviest fire, his example has rarely been equaled. To him, his officers, and his men I feel a deep personal as well as official obligation.

By the officers of my staff, I was most faithfully, laboriously, and gallantly served throughout both days, as well as on the marches before and after the action. A record of their names is an acknowledgment but justly due:

Major George G. Garner, assistant adjutant-general (horse wounded on Sunday); Captain H.W. Walter, assistant adjutant-general; Captain G.B. Cooke, assistant adjutant-general; First Lieutenant Towson Ellis, regular aide: First Lieutenant F.S. Parker, regular aide: Lieutenant Colonel F. Gardner; Lieutenant Colonel W.K. Beard, Florida Volunteers, acting inspector-general (wounded on Monday); Major J.H. Hallonquist, Provisional Army, chief of artillery; Captain W.O. Williams, Provisional Army, assistant to chief of artillery; Captain S.H. Lockerr, Engineers; Captain H. Oladowski, chief of ordnance; Major J.J. Walker, Provisional Army, chief of subsistence; Major L.F. Johnston, Provisional Army, chief quartermaster; Major O.P. Chaffee, Provisional Army, assistant quartermaster; Surgeon A.J. Foard, medical director; Surgeon J.C. Nott, Provisional Army, medical inspector; Dr. Robert O. Butler, of Louisiana, volunteer for the occasion, rendered excellent service in our field hospitals. Lieutenant Colonel David Urquhart, aide to the Governor of Louisiana, served me with great intelligence and efficiency as volunteer aide.

Several other officers during the engagement, temporarily separated from their own commands, did me the favor to act on my staff and served me efficiently.

Privates H. Montague and M. Shehan, Louisiana infantry, and Private John Williams, Tenth Regiment Mississippi Volunteers, orderlies in attendance on myself and staff, though humble in position, rendered services so useful and gallant, that their names are fully entitled to a mention in this report. They encountered the same dangers, and when necessary performed nearly the same duties, as officers of my staff, without the same incentives. In rallying troops, bringing up stragglers, and enforcing orders against refugees they were especially active, energetic, and efficient.

It may not be amiss to refer briefly to the causes it is believed operated to prevent the complete overthrow of the enemy, which we were so near accomplishing, and which would have changed the entire complexion of the war.

The want of proper organization and discipline, and the inferiority in many cases of our officers to the men they were expected to command, left us often without system or order; and the large proportion of stragglers resulting weakened our forces and kept the superior and staff officers constantly engaged in the duties of file closers. Especially was this the case after the occupancy of each of the enemy's camps, the spoils of which served to delay and greatly to demoralize our men. But no one cause probably contributed so largely to our loss of time – which was the loss of success – as the fall of the commanding general. At the moment of this irreparable disaster the plan of battle was being rapidly and successfully executed under his immediate eye and lead on the right.

For want of a common superior to the different commands on that part of the field great delay occurred after this misfortune, and that delay prevented the consummation of the work so gallantly and successfully begun and carried on until the approach of night induced our new commander to recall the exhausted troops for rest and recuperation before a crowning effort on the next morning.

The arrival during the night of a large and fresh army to reinforce the enemy, equal in numbers at least to our own, frustrated all his well grounded expectations, and, after a long and bloody contest with superior forces, compelled us to retire from the field, leaving our killed, many of our wounded, and nearly all of the trophies of the previous day's victories.

In this result we have a valuable lesson, by which we should profit – never on a battlefield to lose a moment's time, but leaving the killed, wounded, and spoils to those whose special business it is to care for them, to press on with every available man, giving a panic stricken and retreating foe no time to rally, and reaping all the benefits of a success never complete until every enemy is killed, wounded, or captured. No course so certain as this to afford succor to the wounded and security to the trophies.

I am, sir, very respectfully, your obedient servant,
BRAXTON BRAGG,
General, Commanding

THE BATTLE OF SHILOH

By General Ulysses S. Grant. *Personal Memoirs of the U.S. Grant.*

When I reassumed command on the 17th of March, I found the army divided, about half being on the east bank of the Tennessee at Savannah, while one division was at Crump's Landing on the west bank about four miles higher up, and the remainder at Pittsburg Landing, five miles above Crump's. The enemy was in force at Corinth, the junction of the two most important railroads in the Mississippi valley – one connecting Memphis and the Mississippi River with the East, and the other leading south to all the cotton states. Still another railroad connects Corinth with Jackson, in west Tennessee. If we obtained possession of Corinth the enemy would have no railroad for the transportation of armies or supplies until that running east from Vicksburg was reached. It was the great strategic position at the West between the Tennessee and the Mississippi rivers and between Nashville and Vicksburg.

I at once put all the troops at Savannah in motion for Pittsburg landing, knowing that the enemy was fortifying at Corinth and collecting an army there under Johnston. It was my expectation to march against that army as soon as Buell, who had been ordered to reinforce me with the Army of the Ohio, should arrive; and the west bank of the river was the place to start from. Pittsburg is only about twenty miles from Corinth, and Hamburg landing, four miles further up the river, is a mile or two nearer. I had not been in command long before I selected Hamburg as the place to put the Army of the Ohio when it arrived. The roads from Pittsburg and Hamburg to Corinth converge some eight miles out. This disposition of the troops would have given additional roads to march over when the advance commenced, within supporting distance of each other.

> *If we obtained possession of Corinth, the enemy would have no railroad for the transportation of armies or supplies until that running east from Vicksburg was reached.*

Before I arrived at Savannah, Sherman, who had joined the Army of the Tennessee and been placed in command of a division, had made an expedition on steamers convoyed by gunboats to the neighborhood of Eastport, thirty miles south, for the purpose of destroying the railroad east of Corinth. The rains had been so heavy for some time before that the lowlands had become impassable swamps.

Sherman debarked his troops and started out to accomplish the object of the expedition; but the river was rising so rapidly that the back water up the small tributaries threatened to cut off the possibility of getting back to the boats, and the expedition had to return without reaching the railroad. The guns had to be hauled by hand through the water to get back to the boats.

On the 17th of March the army on the Tennessee River consisted of five divisions, commanded respectively by Generals C.F. Smith, McClernand, L. Wallace, Hurlbut and Sherman. General W.H.L. Wallace was temporarily in command of Smith's division, General Smith, as I have said, being confined to his bed. Reinforcements were arriving daily and as they came up they were organized, first into brigades, then into a division, and the command given to General Prentiss, who had been ordered to report to me.

General Buell was on his way from Nashville with 40,000 veterans. On the 19th of March he was at Columbia, Tennessee, eighty five miles from Pittsburg. When all reinforcements should have arrived I expected to take the initiative by marching on Corinth, and had no expectation of needing fortifications, though this subject was taken into consideration. McPherson, my only military engineer, was directed to lay out a line to entrench. He did so, but reported that it would have to be made in rear of the line of encampment as it then ran. The new line, while it would be nearer the river, was yet too far away from the Tennessee, or even from the creeks, to be easily supplied with water, and in case of attack these creeks would be in the hands of the enemy.

The fact is, I regarded the campaign we were engaged in as an offensive one and had no idea that the enemy would leave strong entrenchments to take the initiative when he knew he would be attacked where he was if he remained. This view, however, did not prevent every precaution being taken and every effort made to keep advised of all movements of the enemy.

Johnston's cavalry meanwhile had been well out towards our front, and occasional encounters occurred between it and our outposts. On the 1st of April this cavalry became bold and approached our lines, showing that an advance of some kind was contemplated. On the 2nd Johnston left Corinth in force to attack my army. On the 4th, his cavalry dashed down and captured a small picket guard of six or seven men, stationed some five miles out from Pittsburg on the Corinth road. Colonel Buckland sent relief to the guard at once and soon followed in person with an entire regiment, and General Sherman followed Buckland taking the remainder of a brigade. The pursuit was kept up for some three miles beyond the point where the picket guard had been captured, and after nightfall Sherman returned to camp and reported to me by letter what had occurred.

At this time a large body of the enemy was hovering to the west of us, along the line of the Mobile and Ohio railroad. My apprehension was much greater for the safety of Crump's Landing than it was for Pittsburg. I had no apprehension that the enemy could really capture either place. But I feared it was possible that he might make a rapid dash upon Crump's and destroy our transports and stores, most of which were kept at that point, and then retreat before Wallace could be reinforced. Lew Wallace's position I regarded as so well chosen that he was not removed.

At this time I generally spent the day at Pittsburg and returned to Savannah in the evening. I was intending to remove my headquarters to Pittsburg, but Buell was expected daily and would come in at Savannah. I remained at this point, therefore, a few days longer than I otherwise should have done, in order to meet him on his arrival. The skirmishing in our front, however, had been so continuous from about the 3rd of April that I did not leave Pittsburg each night until an hour when I felt there would be no further danger before the morning.

On Friday the 4th, the day of Buckland's advance, I was very much injured by my horse falling with me, and on me, while I was trying to get to the front where firing had been heard. The night was one of impenetrable darkness, with rain pouring down in torrents; nothing was visible to the eye except as revealed by the frequent flashes of lightning. Under these circumstances I had to trust to the horse, without guidance, to keep the road.

I had not gone far, however, when I met General W.H.L. Wallace and Colonel (afterwards General) McPherson coming from the direction of the front. They said all was quiet so far as the enemy was concerned. On the way back to the boat my horse's feet slipped from under him, and he fell with my leg under his body. The extreme softness of the ground, from the excessive rains of the few preceding days, no doubt saved me from a severe injury and protracted lameness. As it was, my ankle was very much injured, so much so that my boot had to be cut off. For two or three days after I was unable to walk except with crutches.

On the 5th General Nelson, with a division of Buell's army, arrived at Savannah and I ordered him to move up the east bank of the river, to be in a position where he could be ferried over to Crump's Landing or Pittsburg as occasion required. I had learned that General Buell himself would be at Savannah the next day, and desired to meet me on his arrival.

Affairs at Pittsburg Landing had been such for several days that I did not want to be away during the day. I determined, therefore, to take a very early breakfast and ride out to meet Buell, and thus save time. He had arrived on the evening of the 5th, but had not advised me of the fact and I was not aware of it until some time after.

While I was at breakfast, however, heavy firing was heard in the direction of Pittsburg landing, and I hastened there, sending a hurried note to Buell informing him of the reason why I could not meet him at Savannah. On the way up the river I directed the dispatch boat to run in close to Crump's Landing, so that I could communicate with General Lew Wallace. I found him waiting on a boat apparently expecting to see me, and I directed him to get his troops in line ready to execute any orders he might receive. He replied that his troops were already under arms and prepared to move.

Up to that time I had felt by no means certain that Crump's Landing might not be the point of attack. On reaching the front, however, about 8:00 a.m.., I found that the attack on Pittsburg was unmistakable, and that nothing more than a small guard, to protect our transports and stores, was needed at Crump's. Captain Baxter, a quartermaster on my staff, was accordingly directed to go back and order General Wallace to march immediately to Pittsburg by the road nearest the river. Captain Baxter made a memorandum of this order.

About 1:00 p.m., not hearing from Wallace and being much in need of reinforcements, I sent two more of my staff, Colonel McPherson and Captain Rowley, to bring him up with his division. They reported finding him marching towards Purdy, Bethel, or some point west from the river, and farther from Pittsburg by several miles than when he started. The road from his first position to Pittsburg Landing was direct and near the river. Between the two points a bridge had been built across Snake Creek by our troops, at which Wallace's command had assisted, expressly to enable the troops at the two places to support each other in case of need.

Wallace did not arrive in time to take part in the first day's fight. General Wallace has since claimed that the order delivered to him by Captain Baxter was simply to join the right of the army, and that the road over which he marched would have taken him to the road from Pittsburg to Purdy where it crosses Owl Creek on the right of Sherman; but this is not where I had ordered him nor where I wanted him to go.

I never could see and do not now see why any order was necessary further than to direct him to come to Pittsburg Landing, without specifying by what route. His was one of three veteran divisions that had been in battle, and its absence was severely felt. Later in the war General Wallace would not have made the mistake that he committed on the 6th of April, 1862. I presume his idea was that by taking the route he did he would be able to come around on the flank or rear of the enemy, and thus perform an act of heroism that would redound to the credit of his command, as well as to the benefit of his country.

> *When the firing ceased at night, the National line was all of a mile in rear of the position it had occupied in the morning.*

Some two or three miles from Pittsburg Landing was a log meeting house called Shiloh. It stood on the ridge which divides the waters of Snake and Lick creeks, the former emptying into the Tennessee just north of Pittsburg landing, and the latter south. This point was the key to our position and was held by Sherman. His division was at that time wholly raw, no part of it ever having been in an engagement; but I thought this deficiency was more than made up by the superiority of the commander.

McClernand was on Sherman's left, with troops that had been engaged at forts Henry and Donelson and were therefore veterans so far as western troops had become such at that stage of the war. Next to McClernand came Prentiss with a raw division, and on the extreme left, Stuart with one brigade of Sherman's division. Hurlbut was in rear of Prentiss, massed, and in reserve at the time of the onset. The division of General C.F. Smith was on the right, also in reserve. General Smith was still sick in bed at Savannah, but within hearing of our guns. His services would no doubt have been of inestimable value had his health permitted his presence. The command of his division devolved upon

Brigadier General W.H.L. Wallace, a most estimable and able officer; a veteran too, for he had served a year in the War and had been with his command at Henry and Donelson. Wallace was mortally wounded in the first day's engagement, and with the change of commanders thus necessarily effected in the heat of battle the efficiency of his division was much weakened.

The position of our troops made a continuous line from Lick Creek on the left to Owl Creek, a branch of Snake Creek, on the right, facing nearly south and possibly a little west. The water in all these streams was very high at the time and contributed to protect our flanks. The enemy was compelled, therefore, to attack directly in front. This he did with great vigor, inflicting heavy losses on the National side, but suffering much heavier on his own.

The Confederate assaults were made with such a disregard of losses on their own side that our line of tents soon fell into their hands. The ground on which the battle was fought was undulating, heavily timbered with scattered clearings, the woods giving some protection to the troops on both sides. There was also considerable underbrush.

A number of attempts were made by the enemy to turn our right flank, where Sherman was posted, but every effort was repulsed with heavy loss. But the front attack was kept up so vigorously that, to prevent the success of these attempts to get on our flanks, the National troops were compelled, several times, to take positions to the rear nearer Pittsburg Landing. When the firing ceased at night the National line was all of a mile in rear of the position it had occupied in the morning.

In one of the backward moves, on the 6th, the division commanded by General Prentiss did not fall back with the others. This left his flanks exposed and enabled the enemy to capture him with about 2,200 of his officers and men. General Badeau gives 4:00 of the 6th as about the time this capture took place. He may be right as to the time, but my recollection is that the hour was later. General Prentiss himself gave the hour as 5:30.

I was with him, as I was with each of the division commanders that day, several times, and my recollection is that the last time I was with him was about 4:30, when his division was standing up firmly and the general was as cool as if expecting victory. But no matter whether it was four or later, the story that he and his command were surprised and captured in their camps is without any foundation whatever. If it had been true, as currently reported at the time and yet believed by thousands of people, that Prentiss and his division had been captured in their beds, there would not have been an all-day struggle, with the loss of thousands killed and wounded on the Confederate side.

With the single exception of a few minutes after the capture of Prentiss, a continuous and unbroken line was maintained all day from Snake Creek or its tributaries on the right to Lick Creek or the Tennessee on the left above Pittsburg. There was no hour during the day when there was not heavy firing and generally hard fighting at some point on the line, but seldom at all points at the same time.

It was a case of Southern dash against Northern pluck and endurance.

Three of the five divisions engaged on Sunday were entirely raw, and many of the men had only received their arms on the way from their States to the field. Many of them had arrived but a day or two before and were hardly able to load their muskets according to the manual. Their officers were equally ignorant of their duties. Under these circumstances it is not astonishing that many of the regiments broke at the first fire.

In two cases, as I now remember, colonels led their regiments from the field on first hearing the whistle of the enemy's bullets. In these cases the colonels were constitutional cowards, unfit for any military position; but not so the officers and men led out of danger by them. Better troops never went upon a battlefield than many of these, officers and men, afterwards proved themselves to be, who fled panic stricken at the first whistle of bullets and shell at Shiloh.

During the whole of Sunday I was continuously engaged in passing from one part of the field to another, giving directions to division commanders. In thus moving along the line, however, I never deemed it important to stay long with Sherman. Although his troops were then under fire for the first time, their commander, by his constant presence with them, inspired a confidence in officers and men that enabled them to render services on that bloody battlefield worthy of the best of veterans.

McClernand was next to Sherman, and the hardest fighting was in front of these two divisions.

> *It was a case of Southern dash against Northern pluck and endurance.*

McClernand told me on that day, the 6th, that he profited much by having so able a commander supporting him. A casualty to Sherman that would have taken him from the field that day would have been a sad one for the troops engaged at Shiloh. And how near we came to this! On the 6th Sherman was shot twice, once in the hand, once in the shoulder, the ball cutting his coat and making a slight wound, and a third ball passed through his hat. In addition to this he had several horses shot during the day

The nature of this battle was such that cavalry could not be used in front; I therefore formed ours into line in rear, to stop stragglers – of whom there were many. When there would be enough of them to make a show, and after they had recovered from their fright, they would be sent to reinforce some part of the line which needed support, without regard to their companies, regiments or brigades.

On one occasion during the day, I rode back as far as the river and met General Buell, who had just arrived; I do not remember the hour, but at that time there probably were as many as four or five thousand stragglers lying under cover of the river bluff, panic-stricken, most of whom would have been shot where they lay, without resistance, before they would have taken muskets and marched to the front to protect themselves.

This meeting between General Buell and myself was on the dispatch-boat used to run between the landing and Savannah. It was brief, and related specially to his getting his troops over the river. As we left the boat together, Buell's attention was attracted by the men lying under cover of the river bank. I saw him berating them and trying to shame them into joining their regiments. He even threatened them with shells from the gunboats near by. But it was all to no effect. Most of these men afterward proved themselves as gallant as any of those who saved the battle from which they had deserted.

I have no doubt that this sight impressed General Buell with the idea that a line of retreat would be a good thing just then. If he had come in by the front instead of through the stragglers in the rear, he would have thought and felt differently. Could he have come through the Confederate rear, he would have witnessed there a scene similar to that at our own. The distant rear of an army engaged in battle is not the best place from which to judge correctly what is going on in front.

Later in the war, while occupying the country between the Tennessee and the Mississippi, I learned that the panic in the Confederate lines had not differed much from that within our own. Some of the country people estimated the stragglers from Johnston's army as high as 20,000. Of course this was an exaggeration.

The situation at the close of Sunday was as follows: along the top of the bluff just south of the log-house which stood at Pittsburg landing, Colonel J.D. Webster, of my staff, had arranged twenty or more pieces of artillery facing south or up the river. This line of artillery was on the crest of a hill overlooking a deep ravine opening into the Tennessee. Hurlbut with his division intact was on the right of this artillery, extending west and possibly a little north.

McClernand came next in the general line, looking more to the west. His division was complete in its organization and ready for any duty. Sherman came next, his right extending to Snake Creek. His command, like the other two, was complete in its organization and ready, like its chief, for any service it might be called upon to render. All three divisions were, as a matter of course, more or less shattered and depleted in numbers from the terrible battle of the day.

The division of W.H.L. Wallace, as much from the disorder arising from changes of division and brigade commanders, under heavy fire, as from any other cause, had lost its organization and did not occupy a place in the line as a division.

Prentiss' command was gone as a division, many of its members having been killed, wounded or captured, but it had rendered valiant services before its final dispersal, and had contributed a good share to the defense of Shiloh.

The right of my line rested near the bank of Snake Creek, a short distance above the bridge which had been built by the troops for the purpose of connecting Crump's landing and Pittsburg landing. Sherman had posted some troops in a loghouse and out buildings which overlooked both the bridge over which Wallace was expected and the creek above that point. In this last position Sherman was frequently attacked before night, but held the point until he voluntarily abandoned it to advance in order to make room for Lew Wallace, who came up after dark.

There was, as I have said, a deep ravine in front of our left. The Tennessee River was very high and there was water to a considerable depth in the ravine. Here the enemy made a last desperate effort to turn our flank, but was repelled. The gunboats *Tyler* and *Lexington*, Gwin and Shirk commanding, with the artillery under Webster, aided the army and effectually checked their further progress.

Before any of Buell's troops had reached the west bank of the Tennessee, firing had almost entirely ceased; anything like an attempt on the part of the enemy to advance had absolutely ceased. There was some artillery firing from an unseen enemy, some of his shells passing beyond us; but I do not remember that there was the whistle of a single musket-ball heard. As his troops arrived in the dusk General Buell marched several of his regiments part way down the face of the hill where they fired briskly for some minutes, but I do not think a single man engaged in this firing received an injury. The attack had spent its force.

General Lew. Wallace, with 5,000 effective men, arrived after firing had ceased for the day, and was placed on the right. Thus night came, Wallace came, and the advance of Nelson's division came; but none – unless night – in time to be of material service to the gallant men who saved Shiloh on that first day against large odds. Buell's loss on the 6th of April was two men killed and one wounded, all members of the 36th Indiana infantry.

The Army of the Tennessee lost on that day at least 7,000 men. The presence of two or three regiments of Buell's army on the west bank before firing ceased had not the slightest effect in preventing the capture of Pittsburg Landing.

So confident was I before firing had ceased on the 6th that the next day would bring victory to our arms if we could only take the initiative, that I visited each division commander in person before any reinforcements had reached the field. I directed them to throw out heavy lines of skirmishers in the morning as soon as they could see, and push them forward until they found the enemy, following with their entire divisions in supporting distance, and to engage the enemy as soon as found. To Sherman I told the story of the assault at Fort Donelson, and said that the same tactics would win at Shiloh. Victory was assured when Wallace arrived, even if there had been no other support. I was glad, however, to see the reinforcements of Buell and credit them with doing all there was for them to do.

During the night of the 6th the remainder of Nelson's division, Buell's army crossed the river and were ready to advance in the morning, forming the left wing. Two other divisions, Crittenden's and McCook's, came up the river from Savannah in the transports and were on the west bank early on the 7th. Buell commanded them in person. My command was thus nearly doubled in numbers and efficiency.

During the night rain fell in torrents and our troops were exposed to the storm without shelter. I made my headquarters

> *Victory was assured when Wallace arrived, even if there had been no other support.*

under a tree a few hundred yards back from the river bank. My ankle was so much swollen from the fall of my horse the Friday night preceding, and the bruise was so painful, that I could get no rest. The drenching rain would have precluded the possibility of sleep without this additional cause.

Some time after midnight, growing restive under the storm and the continuous pain, I moved back to the log-house under the bank. This had been taken as a hospital, and all night wounded men were being brought in, their wounds dressed, a leg or an arm amputated as the case might require, and everything being done to save life or alleviate suffering. The sight was more unendurable than encountering the enemy's fire, and I returned to my tree in the rain.

The advance on the morning of the 7th developed the enemy in the camps occupied by our troops before the battle began, more than a mile back from the most advanced position of the Confederates on the day before. It is known now that they had not yet learned of the arrival of Buell's command. Possibly they fell back so far to get the shelter of our tents during the rain, and also to get away from the shells that were dropped upon them by the gunboats every fifteen minutes during the night.

The position of the Union troops on the morning of the 7th was as follows: General Lew Wallace on the right; Sherman on his left; then McClernand and then Hurlbut. Nelson, of Buell's army, was on our extreme left, next to the river. Crittenden was next in line after Nelson and on his right, McCook followed and formed the extreme right of Buell's command. My old command thus formed the right wing, while the troops directly under Buell constituted the left wing of the army. These relative positions were retained during the entire day, or until the enemy was driven from the field.

In a very short time the battle became general all along the line. This day everything was favorable to the Union side. We had now become the attacking party. The enemy was driven back all day, as we had been the day before, until finally he beat a precipitate retreat. The last point held by him was near the road leading from the landing to Corinth, on the left of Sherman and right of McClernand.

About 3:00, being near that point and seeing that the enemy was giving way everywhere else, I gathered up a couple of regiments, or parts of regiments, from troops near by, formed them in line of battle and marched them forward, going in front myself to prevent premature or long-range firing.

At this point there was a clearing between us and the enemy favorable for charging, although exposed. I knew the enemy were ready to break and only wanted a little encouragement from us to go quickly and join their friends who had started earlier. After marching to within musket-range I stopped and let the troops pass. The command, Charge, was given, and was executed with loud cheers and with a run; when the last of the enemy broke.

Since writing this chapter I have received from Mrs. W.H.L. Wallace, widow of the gallant general who was killed in the first day's fight on the field of Shiloh, a letter from General Lew Wallace to him dated the morning of the 5th. At the date of this letter it was well known that the Confederates had troops out along the Mobile & Ohio railroad west of Crump's landing and Pittsburg landing, and were also collecting near Shiloh. This letter shows that at that time General Lew Wallace was making preparations for the emergency that might happen for the passing of reinforcements between Shiloh and his position, extending from Crump's Landing westward, and he sends it over the road running from Adamsville to the Pittsburg Landing and Purdy road. These two roads intersect nearly a mile west of the crossing of the latter over Owl Creek, where our right rested.

In this letter General Lew Wallace advises General W.H.L. Wallace that he will send "tomorrow" (and his letter also says "April 5th," which is the same day the letter was dated and which, therefore, must have been written on the 4th) some cavalry to report to him at his headquarters, and suggesting the propriety of General W.H.L. Wallace's sending a company back with them for the purpose of having the cavalry at the two landings familiarize themselves with the road so that they could "act promptly in case of emergency as guides to and from the different camps."

This modifies very materially what I have said, and what has been said by others, of the conduct of General Lew Wallace at the battle of Shiloh. It shows that he naturally, with no more experience than he had at the time in the profession of arms, would take the particular road that he did start upon in the absence of orders to move by a different road.

The mistake he made, and which probably caused his apparent dilatoriness, was that of advancing some distance after he found that the firing, which would be at first directly to his front and then off to the left, had fallen back until it had got very much in rear of the position of his advance. This falling back had taken place before I sent General Wallace orders to move up to Pittsburg Landing and, naturally, my order was to follow the road nearest the river. But my order was verbal, and to a staff officer who was to deliver it to General Wallace, so that I am not competent to say just what order the general actually received.

General Wallace's division was stationed, the First brigade at Crump's Landing, the Second out two miles, and the Third two and a half miles out. Hearing the sounds of battle General Wallace early ordered his First and Third brigades to concentrate on the Second. If the position of our front had not changed, the road which Wallace took would have been somewhat shorter to our right than the River road.

HI-LIGHTS OF A HERO'S LIFE
ALBERT SIDNEY JOHNSTON

★ 1803 – (February 2) Born in Washington, Kentucky.

★ 1826 – Graduated from West Point, 8th in a class of 41. Commissioned as brevet 2nd lieutenant in the 2nd U.S. Artillery.

★ 1829 – Married Henrietta Preston.

★ 1832 – Served in the Black Hawk War.

★ 1834 – Resigned his commission to care for his ailing wife, who died two years later.

★ 1836 – Enlisted in the Texas Army as a private.

★ 1837 – (January 31) Promoted to senior brigadier general.

★ 1838 – (December 22) Appointed Texas Secretary of War.

★ 1840 – Resigned and returned to Kentucky.

★ 1843 – Married Eliza Griffith and returned to Texas.

★ 1846 – Returned to the Texas Army during the Mexican-American War.

★ 1849 – Rejoined the U.S. Army.

★ 1855 – Promoted to colonel and put in command of the 2nd U.S. Calvary.

★ 1857 – Fought in the Utah Campaign.

★ 1858 – Promoted to brevet brigadier general.

★ 1860 – (December) Sailed to California to take command of the Department of the Pacific.

★ 1861 – (May 3) Resigned from the U.S. Army.

★ 1861 – (May 27) Left California.

★ 1861 – (September 1) Arrived in Richmond and received the rank of full general and command of the Western Department.

- 1862 – (April 6) Killed at the Battle of Shiloh.

- 1867 – His body was re-interred in Austin, Texas.

ALBERT SIDNEY JOHNSTON AT SHILOH

By Colonel William Preston Johnston, as published in *Battles and Leaders of the Civil War, Volume 1*, New York: *Century Magazine*, 1887-1888, pp 540-569. (Edited by C.L. Gray)

General Johnston's plan of campaign may be summed up in a phrase. It was to concentrate at Corinth and interpose his whole force in front of the great bend of the Tennessee, the natural base of the Federal army: this effected, to crush Grant in battle before the arrival of Buell. This meant immediate and decisive action. The army he had brought from Nashville was ready for the contest, but Generals Beauregard and Bragg represented to him that the troops collected by them were unable to move without thorough reorganization. Ten days were consumed in this work of reorganization. Moments were precious, but there was the hope of re-enforcement by Van Dorn's army, which might arrive before Buell joined Grant, and which did arrive only a day or two later. But Buell's movements were closely watched, and, hearing of his approach on the 2nd of April, General Johnston resolved to delay no longer, but to strike at once a decisive blow.

General Grant's army had been moved up the Tennessee River by boat, and had taken position on its left bank at Pittsburg Landing. It had been landed by divisions, and Bragg had proposed to Beauregard to attack Grant before he assembled his whole force. Beauregard forbade this, intending to await events, and attack him away from his base if possible, though he now insists that his plan of campaign was offensive. Grant's first object was to destroy the railroads which centered at Corinth, and, indeed, to capture that place if he could. But his advance was only a part of a grand plan for a combined movement of his own and Buell's army. With Pittsburg Landing as a base, this army was to occupy North Mississippi and Alabama, command the entire railroad system of that section, and take Memphis in the rear, while Halleck forced his way down the Mississippi River. General Johnston divined the movement before it was begun, and was there to frustrate it. Indeed, Grant's army was assembled at Pittsburg Landing only one week before Johnston completed the concentration.

> *The essential feature of General Johnston's strategy had been to get at his enemy as quickly as possible and in good order.*

Such was the position on April 2nd, when General Johnston, learning that Buell was rapidly approaching, resolved to advance next day and attack Grant before his arrival. His general plan was very simple in outline. It seems to have been to march out and attack the Federals by columns of corps, to make the battle a decisive test, and to crush Grant utterly or lose all in the attempt; to contend with Buell for the possession of Tennessee, Kentucky, and possibly the Northwest.

General Johnston gave orders about 1:00 on the night of Wednesday, the 2nd of April, for the advance. But much time was spent in their elaboration, and the troops did not receive them from the adjutant-general's office until the next afternoon. Some of the troops did not move until the morning of Saturday, the 5th, owing to a still further delay in the delivery of orders by the adjutant general's office, and all were impeded by the heavy condition of the roads, through a dense forest, and across sloughs and marshes.

The order was to attack at 3:00 on the morning of Saturday, the 5th; but the troops were not in position until late that afternoon. After midnight a violent storm broke upon them as they stood under arms in the pitch darkness, with no shelter but the trees. From detention by the rain, ignorance of the roads, and a confusion produced by the order of march, some divisions failed to get into line, and the day was wasted.

Some skirmishing on Friday between the Confederate cavalry and the Federal outposts, in which a few men were killed, wounded, and captured on both sides, had aroused the vigilance of the Northern commanders to some extent. General Prentiss had thrown forward Colonel Moore, with the 21st Missouri regiment, on the Corinth road. Moore, feeling his way cautiously, encountered Hardee's skirmish-line under Major Hardcastle, and, thinking it an outposts, assailed it vigorously. Thus really the Federals began the fight. The struggle was brief, but spirited. The 8th and 9th Arkansas came up. Moore fell wounded. The Missourians gave way, and Shaver's brigade pursued them. Hindman's whole division moved on, following the ridge and drifting to the right, and drove in the grand guards and outposts until they struck Prentiss's camps. Into these they burst, overthrowing all before them.

The essential feature of General Johnston's strategy had been to get at his enemy as quickly as possible, and in as good order. In this he had succeeded. His plan of battle was as simple as his strategy. It had been made known in his order of battle, and was thoroughly understood by every brigade commander. The orders of the 3d of April were, that "every effort should be made to turn the left flank of the enemy, so as to cut off his line of retreat to the Tennessee River and throw him back on Owl Creek, where he will be obliged to surrender."

The Confederate line pushed rapidly to the front, and as gaps widened in the first lines, they were filled by brigades of the second and third. One of Breckinridge's brigades (Trabue's) was sent to the left to support Cleburne and fought under Polk the rest of the day; and the other two were led to the extreme right, only Chalmers being beyond them. Gladden, who was on Hindman's right, and had a longer distance to traverse to strike some of Prentiss' brigades further to the left, found them better prepared, but, after a sanguinary resistance, drove them from their camps. In this bitter struggle Gladden fell mortally wounded. Chambers' brigade, of Bragg's line, came in on Gladden's right, and his Mississippians drove the enemy with the bayonet half a mile. He was about to charge again, when General Johnston came up, and moved him to the right, and brought John K. Jackson's brigade into the interval. The Federals retreated sullenly, not routed, but badly hammered.

With Hindman as a pivot, the turning movement began from the moment of the overthrow of Prentiss' camps. The Federal reports complain that they were flanked and outnumbered, which is true; for the Confederates were probably stronger at every given point throughout the day except at the center called the Hornets' Nest, where the Federals eventually massed nearly two divisions. The fighting was a grapple and a death-struggle all day long, and, as one brigade after another wilted before the deadly fire of the stubborn Federals, still another was pushed into the combat and kept up the fierce assault. A breathing spell, and the shattered command would gather itself up and resume its work of destruction. These were the general aspects of the battle.

When the battle began Hindman, following the ridge, had easy ground to traverse; but Cleburne's large brigade, on his left, with its supports, moving over a more difficult country, was slower in getting upon Sherman's front. That general and his command were aroused by the long roll, the advancing musketry, and the rush of troops to his left, and he got his division in line of battle and was ready for the assault of Cleburne, which was made about 8:00. General Johnston, who had followed close after Hindman, urging on his attack, saw Cleburne's brigade begin its advance, and then returned to where Hindman was gathering his force for another assault. Hardee said of Cleburne that he moved quickly through the fields, and, though far outflanked by the enemy on our left, rushed forward under a terrific fire from the serried ranks drawn up in front of the camp. A morass covered his front, and, being difficult to pass, caused a break in this brigade. Deadly volleys were poured upon the men from behind bales of hay and other defenses, as they advanced; and after a series of desperate charges they were compelled to fall back.

"They are offering stubborn resistance here. I shall have to put the bayonet to them."

While Sherman was repelling Cleburne's attack, McClernand sent up three Illinois regiments to reinforce his left. But General Polk led forward Bushrod R. Johnson's brigade, and General Charles Clark led Russell's brigade, against Sherman's left, while General Johnston himself put A.P. Stewart's brigade in position on their right. Supported by part of Cleburne's line, they attacked Sherman and McClernand fiercely. Clark and Bushrod R. Johnson fell badly wounded. Hildebrand's Federal brigade was swept from the field.

Wood's brigade, of Hindman's division, joined in this charge on the right. As they hesitated at the crest of a hill, General Johnston came to the front and urged them to the attack. They rushed forward with the inspiring "rebel yell," and, with Stewart's brigade, enveloped the Illinois troops. In ten minutes the latter melted away under the fire, and were forced from the field.

The whole Federal front, which had been broken here and there, gave way under this hammering process on front and flank. Sherman's route of retreat was marked by the thick-strewn corpses of his soldiers. At last, pressed back toward both Owl Creek and the river, Sherman and McClernand found safety by the interposition on their left flank of W.H.L. Wallace's fresh division. Hurlbut and Wallace had advanced about 8:00, so that Prentiss' command found a refuge in the intervals of the new and formidable Federal line, with Stuart on the left and Sherman's shattered division on the right.

General Johnston had pushed Chalmers to the right and front, sweeping down the left bank of Lick Creek, driving in pickets, until he encountered Stuart's Federal brigade on the Pittsburg and Hamburg road. After a hard fight, the Federals were driven back. Chalmers' right rested on the Tennessee River bottom-lands, and he fought down the bank toward Pittsburg Landing. The new line of battle was established before 10:00.

A gigantic contest now began which lasted more than five hours: The Hornet's Nest. Here, behind a dense thicket on the crest of a hill, was posted a strong force of as hardy troops as ever fought, almost perfectly protected by the conformation of the ground, and by logs and other rude and hastily prepared defenses. To assail it an open field had to be passed, enfiladed by the fire of its batteries. No figure of speech would be too strong to express the deadly peril of assault upon this natural fortress. For five hours brigade after brigade was led against it. Then Bragg ordered up Gibson's brigade.

The brigade was four times repulsed, but maintained its ground steadily, until W.H.L. Wallace's position was turned, when, renewing its forward movement in conjunction with Cheatham's command, it helped to drive back its stout opponents. Cheatham, charging with Stephens' brigade on Gibson's right, across an open field, had been caught under a murderous cross-fire, but fell back in good order, and, later in the day, came in on Breckinridge's left in the last assault when Prentiss was captured. This bloody fry a lasted till nearly 4:00, without making any visible impression on the Federal center, but when its flanks were turned, these assaulting columns, crowding in on its front, aided in its capture.

General Johnston was with the right of Statham's brigade, confronting the left of Hurlbut's division, which was behind the crest of a hill, with a depression filled with chaparral in its front. Bowen's brigade was further to the right in line with Statham's, touching it near this point. The Confederates held the parallel ridge in easy musket-range; and "as heavy fire as I ever saw during the war," says Governor Harris, was kept up on both sides for an hour or more. It was necessary to cross the valley raked by this deadly ambuscade and assail the opposite ridge in order to drive the enemy from his stronghold. When General Johnston came up and saw the situation, he said to his staff: "They are offering stubborn resistance here. I shall have to put the bayonet to them." With a mighty shout Bowen's and Statham's brigades moved forward at a charge. A sheet of flame and a mighty roar burst from the Federal stronghold. The Confederate line withered; but there was not an instant's pause. The crest was gained. The enemy was in flight.

General Johnston had passed through the ordeal seemingly unhurt. His horse was shot in four places; his clothes were pierced by missiles; his bootsole was cut and torn by a minie; but if he himself had received any severe wound, he did not know it. By the chance of war, a minie-ball from one of these did its fatal work. As he sat there, after his wound, Captain Wickham says that Colonel O'Hara, of his staff, rode up, and General Johnston said to him, "We must go to the left, where the firing is heaviest," and then gave him an order, which O'Hara rode off to obey. Governor Harris returned, and, finding him very pale, asked him, "General, are you wounded?" He answered, in a very deliberate and emphatic tone: "yes, and, I fear, seriously." These were his last words. Harris and Wickham led his horse back under cover of the hill, and lifted him from it. They searched at random for the wound, which had cut an artery in his leg, the blood flowing into his boot. When his brother-in-law, Preston, lifted his head, and addressed him with passionate grief, he smiled faintly, but uttered no word. His life rapidly ebbed away, and in a few moments he was dead.

After the war, General Gibson lamented: "General Johnston's death was a tremendous catastrophe. There are no words adequate to express my own conception of the immensity of the loss to our country. Sometimes the hopes of millions of people depend upon one head and one arm. The West perished with Albert Sidney Johnston, and the Southern country followed."

To subscribe, send an email to: thestainlessbanner@gmail.com or visit our website at www.thestainlessbanner.com

Subscription is free.

The Stainless Banner

An e-zine dedicated to the armies of the Confederacy

Issue 3, Volume 5
May 2012

THE SAME KIND PROVIDENCE
JACKSON IS VICTORIOUS!

Stonewall Jackson's victories at Front Royal and Winchester in the spring of 1862 not only sent the Federal army under the command of General Nathaniel Banks running to Harper's Ferry for safety but also sent shockwaves throughout the Lincoln Administration. President Lincoln fired off a telegram to General George McClellan's headquarters on the Virginia peninsula urging the young Napoleon to either get on with task of taking Richmond or quit the venture altogether and return the army to Washington to defend the Capital from the unstoppable Jackson. General Irving McDowell, making preparations in Fredericksburg to march his 40,000 man division to the peninsula to aid McClellan, received orders to put 20,000 men on the road to the Valley immediately.

From Winchester, Jackson informed the War Department of his recent victories. In response, Richmond dispatched Jackson's friend, Congressman Alexander Boetler, to Winchester with orders for Jackson to demonstrate against Harper's Ferry and effect an invasion of Maryland. Richmond hoped that the threat of a possible Jackson attack on Washington would help relieve the pressure Joe Johnston was under as he retreated up the Peninsula before McClellan's massive army.

Two days after Jackson left Winchester, General Turner Ashby, Jackson's cavalry leader, sent word to Jackson of two armies invading the Valley. The first, under the command of John Fremont, was threading its way through the mountains toward Strasburg.

What's Inside:	
Jackson's Official Report	7
Ewell's Report on Cross-Keys	15
Taylor's Official Report	18
The Stonewall Brigade	20
Richard Ewell	24
Cross-Keys & Port Republic	26

www.thestainlessbanner.com

The second, under the command of James Shields, was on its way from Fredericksburg. When another courier arrived from Ashby warning Jackson that Fremont and Shields were closing in, Jackson withdrew from the Potomac and marched south, risking an attack from Banks.

The Great Escape

To prevent the uniting of the two Federal armies, Jackson sent Ashby's cavalry to take control of the only road Fremont could use to cross North Mountain and reach Strasburg. The 12th Georgia was ordered to hold the bridges at Front Royal for as long as possible. When it became necessary to retreat, the men were to burn the bridges and secure Jackson's flank and rear.

When Jackson arrived back in Winchester, bad news was waiting for him. The Georgians had panicked at the first sight of Union blue and fled the city, abandoning over $300,000 worth of captured stores and failing to burn the bridges. Shields was now only twelve miles from Strasburg, the Valley Turnpike and a possible juncture with Fremont.

Jackson's only hope of escape was to head south as quickly as possible. He spent the night making arrangements to get the captured wagons from his previous victories to safety. Sometime during the hectic night, Colonel Conner of the 12th Georgia appeared and reported to Jackson. In vivid terms he recounted his harrowing escape from Front Royal. Jackson cut him off. "How many men did you have killed, sir?" Conner had no idea. He had left Front Royal before the fight had begun, so he answered, "none." Jackson stared hard at him. "How many wounded?" None was the reply again. Jackson went ballistic. "Colonel Conner, do you call that much of a fight?" With these angry words, Conner was dismissed.[1] Sandie Pendleton, Jackson's aide-de-camp, found Conner moments later and placed him under arrest for dereliction of duty.

That afternoon, in a torrential downpour, the long wagon trains and prisoners departed Winchester followed by the foot-sore and exhausted Army of Shenandoah. Colonel Crutchfield struggled to push the artillery through roads that had turned into seas of mud. The men were stretched out fifteen miles along the Valley Turnpike.

God's favor shined down on Jackson. Fremont halted his advance four miles from the Turnpike and the vulnerable Confederates. At Front Royal, Shields decided to wait for his supporting division to come up before proceeding. This allowed Jackson to ride into Strasburg unmolested.

Before leaving Winchester, Jackson had sent his cartographer, Jedidiah Hotchkiss, to Charles Town to retrieve the Stonewall Brigade. When the brigade did not rejoin the army before it reached Strasburg, Jackson made the decision to wait for the Brigade to arrive before

> *"No, Colonel. Shoot them all. I don't want them to be brave."*

---

[1] James I. Robertson, Jr. *Stonewall Jackson: The Man, the Soldier, the Legend.* (New York: Macmillan Publishing 1997) 418.

journeying on. The trains continued south to New Market, and Richard Ewell headed west to aid Ashby just in case Fremont became trouble.

June 1st dawned, and the Stonewall Brigade still had not arrived. The boom of Ewell's guns could be heard as they engaged in a long dual range with Fremont's artillery. A probe by Fremont at Cedar Creek was sharply contested and won. Finally, in the afternoon and to Jackson's great relief, the Stonewall Brigade filed into Strasburg. General Charles Sidney Winder had performed a masterful feat of marching. Leaving Charles Town the day before with four regiments, he had marched thirty-five miles through a heavy downpour while the 2nd Virginia, entrenched on Loudoun Heights, had marched forty-two.

The Brigade was given no time to rest. At sunset, Jackson evacuated Strasburg. Before departing, the cavalry was ordered to burn the bridges spanning the South Fork of the Shenandoah River at White House, Columbia and Conrad's Store. This would prevent Shields from crossing the flooded river and attacking Jackson's men as they toiled through the driving rain and thick mud.

Marching all night and the next day, the army arrived safely at Hawkinsville. At the rear of the column, the 21st Virginia skirmished with Federal cavalry. Colonel John Patton arrived later that night to make his report and remorsefully recounted the heroics of three Union horsemen who had charged right into the regiment and had been shot for their bravery. "Colonel, why wouldn't you have shot them?" Jackson queried. "I should have spared them," Patton said regretfully, "because they were men who had gotten into a desperate situation." Jackson disagreed. "No, Colonel. Shoot them all. I don't want them to be brave."[2]

At Meem's Bottom, Jackson's men crossed the North Fork of the Shenandoah River, burning the bridge behind them. Before Fremont could build an improvised bridge and follow, the river rose ten feet in four hours and washed away Fremont's hope of running the Confederates to ground. Meanwhile, Shields was bogged down in the mud and strung out for twenty miles in the Luray Valley.

Fifteen miles south of Conrad's Store lay the small hamlet of Port Republic nestled in the angle formed by the North River and the South River as they flowed together to become the South Fork of the Shenandoah River. A bridge spanned the North River at the northern end of the village. On the opposite end of the village were two fords of the South River. On the same side of the village were two roads – one leading to Brown's Gap and the other to Swift Run Gap. From Brown's Gap, Jackson could slip from the Valley, board his troops on the Virginia Central Railroad and head south to Richmond.

Behind Port Republic lay a high ridge and the road to Staunton. On the south side of the North River a ridge line commanded Port Republic. From the ridge, the ground flattened out to meadows and fields. Artillery placed on the ridge would command the whole area.

Port Republic was a strong defensive position, but it was also dangerous one. Concentrating an army to attack Fremont might allow Shields to seize the ridge

---

[2] Ibid., 424

along the North River. But if Jackson chose to attack Shields, then Fremont could seize the high ground. Adding to Jackson's dilemma was if he met Shields first, he would have to burn the bridge over the North River as a precaution. He would then be unable to recross the river and confront Fremont.

The Death of Turner Ashby

As Jackson marched toward Port Republic and Ewell's small division toward Cross Keys, Ashby's cavalry was watching Fremont's progress on the east side of Harrisonburg. A sharp skirmish turned back a small detachment of Federal cavalry, but reinforcements had hurried up. Outnumbered, Ashby retreated to Chestnut Hill and called for infantry support.

Ashby was soon engulfed in a desperate battle. After his horse was shot out from underneath him, he jumped to his feet and ordered his men to charge. A bullet pierced his heart, killing him instantly. His men took up the standard from their fallen leader and won the day.

Jackson was overwhelmed when he heard the news of Ashby's death. The cavalry leader's body was brought to Port Republic and after a brief funeral, Jackson returned to the room where Ashby lay and "remained for time in silent communion with the dead, then left him, with a solemn and elevated countenance."[3]

---
[3] Ibid., 429.

Cross Keys – June 8

Two miles southeast of the village of Cross Keys, midway between Harrisonburg and Port Republic, Ewell and 5,000 men waited for Fremont. The Stonewall Brigade and regiments from Allegheny Johnson's brigade were sent to reinforce Ewell, while artillery was placed on the dominating ridges.

Colonel Samuel Carroll, 150 troopers of the 1st Virginia (U.S.) and four guns of Battery L of the 1st Ohio Artillery and infantry had been organized by Shields as a strike force to secure the bridge at Conrad's Store. When Carroll failed in that mission, he learned of the bridge at Port Republic and headed south. As the sun rose on the 8th, Carroll's cavalry regiment and horse artillery galloped down the banks of the South River, scattered the pickets guarding the fords, rode unopposed into Port Republic and almost captured Jackson.

Jackson fled to the other side of the river. He quickly found William Poague of the Rockbridge Artillery and ordered him to open fire on the Union troopers. In Port Republic, Captain James Carrington, Charlottesville Artillery, hurried a cannon into the street and fired on the Union troopers as they attempted to secure the bridge. The 37th Virginia charged down the street. Carroll had had enough. He gathered up his men and made good his escape. Jackson ordered the Stonewall Brigade to Bogota to prevent any further surprises.

At Cross-Keys, Ewell deployed his division behind Mill Creek. He placed the

> *Trimble waited until the Federals were sixty paces away before ordering his men to fire.*

15th Alabama a mile in advance at Union Church. Isaac Trimble's brigade was deployed across the Port Republic Road and Arnold Elzey's men held the center along the high bluffs. Ewell strengthened his center with a concentration of artillery.

Believing he faced Jackson's entire army, Fremont's advance along the Keezeltown Road was slow and cautious. From his position on the Port Republic Road, Trimble watched with anticipation as General Julius Stahel's brigade advanced toward him. Anxious to fight, Trimble hurried his troops to Victory Hill and placed them behind the crest. In parade ground fashion, Stahel's men marched up the hill. Trimble waited until the Federals were sixty paces away before ordering his men to fire. The Federals recoiled in confusion and regrouped on the heights opposite Victory Hill. Stahel turned the battle over to his artillery.

Trimble sent the 15th Alabama up a ravine to flank and silence the guns. Ewell hurried the 13th and 25th Virginia to reinforce Trimble's right. Stahel's batteries were in command of the battle until the 15th Alabama emerged from the ravine and scampered up the hill. Trimble ordered the rest of his men forward. The Federals quickly limbered up their cannon and withdrew.

Still not content with his gains, Trimble launched a counter-offensive, chasing the Federals for a full mile. Believing he could rout the enemy, he halted his men and went to ask Ewell for reinforcements. Ewell refused, so Trimble hot-footed it to Jackson. Even though Jackson appreciated Trimble's fighting spirit, he sent Trimble back to Ewell for direction.

General Robert Milroy advancing on Stahel's right approached Ewell's center behind Mill Creek and opened fire. Ewell's line easily held. Milroy did not stay long. Stahel's retreat had left Milroy's left flank vulnerable and Fremont sent orders to withdraw immediately.

Fremont now seemed to be at a loss as to what to do. General Robert Schenck's attack on Ewell's left near Union Church was easily rebuked. The battle was over. The Federals retreated back up Keezeltown Road.

> *Impatient to get about beating Shields so he could recross the South River and deal with Fremont, Jackson ordered the Stonewall Brigade to attack.*

Battle of Port Republic

With the bridge at Conrad's Store destroyed, Shields' options were limited. If he wanted to unite with Fremont, he would have to return to Luray and rebuild the bridge over the South Fork or try for the bridge at Port Republic. Since retreating to Luray was not an option, Shields decided to attack Jackson at Port Republic.

Shields' decision meant another sleepless night for Jackson. As soon as Fremont withdrew, Jackson went to work. He stationed Taliaferro's brigade inside Port Republic to protect the city from ambush and recalled the Stonewall Brigade from Bogota. The pioneers worked through the night building a bridge of wagons across the South River.

Since Trimble was in such a fighting mood, Jackson left his brigade and a portion of Patton's 21st Virginia at Cross Keys to oppose Fremont should the Pathfinder decide to come to Shields' aid. The rest of Ewell's division marched to Port Republic.

There was trouble from the start. As the Stonewall Brigade crossed the wagon bridge at dawn, the swift moving South River began to pull it apart. The men tiptoed tentatively over the bridge in single-file, slowing the movement of the army from Port Republic to the other side of the river where Shields was waiting to receive battle. Instead of facing the Federals in strength, the slowly disintegrating bridge forced Jackson to feed his army into the battle piecemeal and almost cost him a victory.

As Winder's skirmishers fanned out, they faced a battlefield that favored the defense. On the left was the South Fork of the Shenandoah. The center consisted of rolling wheatfields. To the right was a spur jutting out from the Blue Ridge Mountains. On the crest of the spur was The Coaling – a flat area void of trees that had been cut down long ago to make charcoal. Shields artillery was there and it threatened to destroy Jackson's assault before it even began.

Impatient to get about beating Shields so he could recross the South River and deal with Fremont, Jackson ordered the Stonewall Brigade to attack. Winder's first objective was to ascend The Coaling and neutralize the artillery. He ordered his men forward but it was no contest. As predicted, Union cannon wrecked havoc on the Brigade. Winder called for the 33rd Virginia, but the 33rd Virginia was still on the other side of the river. The wagon bridge had collapsed and the men were forced to wade across the river.

Jackson ordered Taliaferro's brigade into action, while he sent his chief of artillery, Colonel Crutchfield, racing to the rear for Taylor's Louisianans. Trimble was ordered from Cross Keys and told to burn the bridge over the North River once he was across. With that order, Jackson gave up his plans of defeating Fremont.

Jedidiah Hotchkiss met Taylor's brigade as they come on the field and guided them to a place from which the Louisianans could flank The Coaling and silence the guns.

On the battlefield things were rapidly deteriorating. A courier sped to bring Ewell up. Suddenly, providentially Jackson believed, the Union cannon on The Coaling fell silent. Taylor's men had won the day after a terrible hand-to-hand battle. The sudden loss of The Coaling sapped the courage from Shields. He withdrew his order to attack Jackson's center.

Winder reassembled the Stonewall Brigade, gathered up Taliaferro's brigade, and the first of Ewell's regiments arriving on the field into a long line. The charge was sounded and the line swept forward.

The Federals wanted no part of the surging Confederates. They withdrew from the field. Jackson was not about to let them get away so easily. He ordered both infantry and artillery to give chase. He kept up the pursuit for four miles until the forest gobbled up the road and made pursuit impossible. The battle was over.

Jackson was the sole architect of a strategy that swept the Federal armies from most of the Valley, froze McDowell at Fredericksburg, made the ever cautious McClellan even more cautious and immortalized forever the name of Stonewall Jackson.

JACKSON'S REPORT ON THE BATTLE OF CROSS KEYS AND PORT REPUBLIC

Headquarters Second Corps, Army of Northern Virginia
April 14, 1863

General: I have the honor herewith to submit to you a report of the battle of Port Republic, fought on June 8 and 9, 1862:

Having through the blessing of an ever kind Providence passed Strasburg before the Federal armies under Generals Shields and Fremont effected the contemplated junction in my rear, as referred to in the report of the battle of Winchester, I continued to move up the Valley turnpike, leaving Strasburg on the evening of June 1. The cavalry under Brigadier General George H. Steuart brought up the rear.

Fremont's advance, which had been near us during the day, soon ascertained that our retreat had been resumed, and, pursuing after dark, succeeded in approaching so near our rear guard as to attack it. The 6th Virginia Cavalry, being nearest the enemy, was thrown into confusion and suffered some loss. Disorder was also to some extent communicated to the Second Virginia Cavalry, but its commander, Colonel Munford, soon reformed it and gallantly drove back the Federals and captured some of their number. From information received respecting Shields' movements, and from the fact that he had been in possession of Front Royal for over forty-eight hours and had not succeeded in effecting a junction with Fremont, as originally designed, I became apprehensive that he was moving via Luray for the purpose of reaching New Market, on my line of retreat, before my command should arrive there. To avoid such a result, I caused White House Bridge, which was upon his assumed line of march, over the South Fork of the Shenandoah River, to New Market, to be burned, and also Columbia Bridge, which was a few miles farther up the river.

On June 2, the enemy's advance came within artillery range of and commenced shelling our rear guard, which caused most of the cavalry and that part of its artillery nearest the enemy to retreat in disorder. This led General Ashby to one of those acts of personal heroism and prompt resource which strikingly marked his character. Dismounting from his horse, he collected from the road a small body of infantry from those who from

> *In this affair, General Turner Ashby was killed.*

fatigue were straggling behind their commands, and posting them in a piece of wood near the turnpike he awaited the advance of the Federal cavalry, now pushing forward to reap the fruits of the panic produced by the shells. As they approached within easy range, he poured such an effective fire into their ranks as to empty a number of saddles and check their farther pursuit for that day. Having transferred the 2nd and 6th Virginia Cavalry to Ashby, he was placed in command of the rear guard.

On the 3rd, after my command had crossed the bridge over the Shenandoah near Mount Jackson, General Ashby was ordered to destroy it, which he barely succeeded in accomplishing before the Federal forces reached the opposite bank of the river. Here his horse was killed by the enemy, and he made a very narrow escape with his life.

We reached Harrisonburg at an early hour on the morning of the 5th, and passing beyond that town turned toward the east in the direction of Port Republic.

On the 6th, General Ashby took position on the road between Harrisonburg and Port Republic, and received a spirited charge from a portion of the enemy's cavalry, which resulted in the repulse of the enemy and the capture of Colonel Wyndham and sixty-three others. Apprehending that the Federals would make a more serious attack, Ashby called for an infantry support. The brigade of Brigadier General George H. Steuart was accordingly ordered forward.

In a short time the 58th Virginia Regiment became engaged with a Pennsylvania regiment called the Bucktails, when Colonel Johnson, of the 1st Maryland Regiment, coming up in the hottest period of the fire, charged gallantly into its flank and drove the enemy with heavy loss from the field, capturing Lieutenant Colonel Kane, commanding.

In this skirmish our infantry loss was seventeen killed, fifty wounded, and three missing. In this affair General Turner Ashby was killed.

An official report is not an appropriate place for more than a passing notice of the distinguished dead, but the close relation which General Ashby bore to my command for most of the previous twelve month, will justify me in saying that as a partisan officer I never knew his superior; his daring was proverbial; his powers of endurance almost incredible; his tone of character heroic, and his sagacity almost intuitive in divining the purposes and movements of the enemy.

The main body of my command had now reached the vicinity of Port Republic. This village is situated in the angle formed by the junction of the North and South Rivers, tributaries of the South Fork of the Shenandoah. Over the larger and deeper of those two streams, the North River, there was a wooden bridge connecting the town with the road leading to Harrisonburg. Over the South River there was a passable ford. The troops were immediately under my own eye and were encamped on the high ground north of the village, about a mile from the river.

General Ewell was some four miles distant, near the road leading from Harrisonburg to Port Republic. General Fremont had arrived with his forces in the vicinity of Harrisonburg, and General Shields was moving up the east side of the South Fork of the Shenandoah, and was then at Conrad's Store, some fifteen miles below Port Republic, my position being about equal distance from both hostile armies. To prevent a junction of the two Federal armies, I had caused the bridge over the South Fork of the Shenandoah at Conrad's Store to be destroyed.

Intelligence having been received that General Shields was advancing farther up the river, Captain Sipe with a small cavalry force was sent down during the night of the 7th to verify the report and gain such other information respecting the enemy as he could. Captain G. W. Myers, of the cavalry, was subsequently directed to move with his company in the same direction, for the purpose of supporting Captain Sipe, if necessary.

The next morning Captain Myers' company came rushing back in disgraceful disorder, announcing that the Federal forces were in close pursuit. Captain Chipley and his company of cavalry, which was in town, also shamefully fled. The brigades of Generals Taliaferro and Winder were soon under arms and ordered to occupy positions immediately north of the bridge. By this time the Federal cavalry, accompanied by artillery, were in sight, and after directing a few shots toward the bridge, they crossed South River and, dashing into the village, they planted one of their pieces at the southern entrance of the bridge.

In the meantime the batteries of Weeding, Poague, and Carpenter were being placed in position, and General Taliaferro's brigade, having reached the vicinity of the bridge, was ordered to charge across, capture the piece and occupy the town. While one of Poague's pieces was returning the fire of that of the enemy at the far end of the bridge, the 37th Virginia Regiment, Colonel Fulkerson commanding, after delivering its fire, gallantly charged over the bridge, captured the gun, and, followed by the other regiments of the brigade, entered the town and dispersed and drove back the Federal cavalry. Another piece of artillery with which the Federal cavalry had advanced was abandoned and subsequently fell into our hands.

About this time, a considerable body of infantry was seen advancing up the same road. Our batteries opened with marked effect upon the retreating cavalry and advancing infantry. In a short time the infantry followed the cavalry, falling back to Lewis', three miles down the river, pursued for a mile by our batteries on the opposite bank, when the enemy disappeared in the wood around a bend in the road.

This attack of General Shields had hardly been repulsed before Ewell was seriously engaged with Fremont, moving on the opposite side of the river. The enemy pushed forward, driving in the 15th Alabama, Colonel Cantey commanding, from their post on picket. This regiment made a gallant resistance, which so far checked the Federal advance as to afford to General Ewell time for the choice of his position at leisure.

His ground was well selected, on a commanding ridge, a rivulet and large field of open ground in front, woods on both flanks, and his line intersected near its center by the road leading to Port Republic. General Trimble's brigade was posted on the right, somewhat in advance of his center. The batteries of Courtney, Lusk, Brockenbrough, and Raine in the center; General Steuart's brigade on the left, and General Elzey's brigade in rear of the center, and in position to strengthen either wing. Both wings were in the wood.

About 10:00 the enemy threw out his skirmishers and shortly after posted his artillery opposite to our batteries. The artillery fire was kept up with great animation and spirit on both sides for several hours. In the meantime a brigade of Federal forces advanced, under cover, upon the right, occupied by General Trimble, who reserved his fire until they reached the crest of the hill, in easy range of his musketry, when he poured a deadly fire from his whole front, under which they fell back.

Observing a battery about being posted on the enemy's left, half a mile in front, General Trimble, now supported by the 13th and 25th Virginia Regiments, of Elzey's brigade, pushed forward for the purpose of taking it, but found it withdrawn before he reached the spot, having in the meantime some spirited skirmishing with its infantry supports. General Trimble had now advanced more than a mile from his original position, while the Federal advance had fallen back to the ground occupied by them in the morning.

General Taylor, of the 8th Brigade of Louisiana troops, having arrived from the vicinity of the bridge at Port Republic, toward which he had moved in the morning, reported to General Ewell about 2:00 p.m. and was placed in rear. Colonel Patton, with the 42nd and 48th Virginia Regiments and 1st Battalion of Virginia Regulars, also joined, and with the remainder of General Elzey's brigade was added to the center and left, then supposed to be threatened. General Ewell – having been informed by Lieutenant Hinrichs of the Engineer Corps, who had been sent out to reconnoiter, that the enemy was moving a large column on his left – did not advance at once, but subsequently ascertaining that no attack was designed by the force referred to, he advanced, drove in the enemy's skirmishers, and when night closed was in position on ground previously held by the enemy. During this fighting Brigadier-Generals Elzey and Steuart were wounded and disabled from command.

> *Taylor emerged with his command from the woods just as the loud cheers of the enemy had proclaimed their success in front.*

This engagement with Fremont has generally been known as the battle of Cross Keys, in which our troops were commanded by General Ewell. I had remained at Port Republic during the principal part of the 8th, expecting a renewal of the attack. As no movement was made by General Shields to renew the action that day, I determined to take the initiative and attack him the following morning. Accordingly, General Ewell was directed to move from his position at an early hour on the morning of the 9th toward Port Republic, leaving General Trimble's brigade, supported by Colonel Patton and the 42nd Virginia Infantry and the 1st Battalion of Virginia Regulars, to hold Fremont in check, with instructions, if hard pressed, to retire across the North River and burn the bridge in their rear.

Soon after 10:00 p.m. General Trimble, with the last of our forces, had crossed the North River and the bridge was destroyed.

In the meantime, before 5:00 in the morning, General Winder's brigade was in Port Republic and having crossed the South Fork by a temporary wagon bridge placed there for the purpose, was moving down the River Road to attack the forces of General Shields. Advancing one mile, he encountered the Federal pickets and drove them in. The enemy had judiciously selected his position for defense. Upon a rising ground, near the Lewis house, he had planted six guns, which commanded the road from Port Republic and swept the plateau for a considerable distance in front.

As General Winder moved forward his brigade, a rapid and severe fire of shell was opened upon it. Captain Poague, with two Parrott guns, was promptly placed in position on the left of the road to engage and if possible dislodge the Federal battery. Captain Carpenter was sent to the right to select a position for his battery, but finding it impracticable to drag it through the dense undergrowth, it was brought back and part of it placed near Poague. The artillery fire was well sustained by our batteries but found unequal to that of the enemy.

In the meantime, Winder being now reinforced by the 7th Louisiana Regiment, Colonel Hays command, seeing no mode of silencing the Federal battery or escaping its destructive missiles but by a rapid charge and the capture of it, advanced with great boldness for some distance, but encountered such a heavy fire of artillery and small arms as greatly to disorganize his command, which fell back in disorder. The enemy advanced across the field, and by a heavy musketry fire forced back our infantry supports, in consequence of which our guns had to retire.

The enemy's advance was checked by a spirited attack upon their flank by the 58th and 54th Virginia Regiments, directed by General Ewell and led by Colonel Scott, although his command was afterward driven back to the woods with severe loss. The batteries were all safely withdrawn, except one of Captain Poague's six pound guns, which was carried off by the enemy.

While Winder's command was in this critical condition, the gallant and successful attack of General Taylor on the Federal left and rear diverted attention from the front and led to a concentration of their force upon him. Moving to the right along the mountain acclivity through a rough and tangled forest, and much disordered by the rapidity and obstructions of the march, Taylor emerged with his command from the wood just as the loud cheers of the enemy had proclaimed their success in front, and, although assailed by a superior force in front and flank, with their guns in position, within point-blank range, the charge was gallantly made, and the battery, consisting of six guns, fell into our hands. Three times was this battery lost and won in the desperate and determined efforts to capture and recover it.

After holding the battery for a short time a fresh brigade of the enemy, advancing upon his flank, made a vigorous and well-conducted attack upon him, accompanied by a galling fire of canister from a piece suddenly brought into position at a distance of about 350 yards. Under this combined attack Taylor fell back to the skirt of the wood near which the captured battery was stationed, and from that point continued his fire upon the advancing enemy, who succeeded in recapturing one of the guns, which he carried off, leaving both caisson and limber.

The enemy, now occupied with Taylor, halted his advance to the front. Winder made a renewed effort to rally his command, and, succeeding, with the 7th Louisiana, under Major Penn (the colonel and lieutenant-colonel having been carried from the field wounded), and the 5th Virginia Regiment, Colonel Funk, he placed part of Poague's battery in the position previously occupied by it, and again opened upon the enemy, who were moving against Taylor's left flank, apparently to surround him in the woods.

Chew's battery now reported and was placed in position and did good service. Soon after guns from the batteries of Brockenbrough, Courtney, and Rains were brought forward and placed in position. While these movements were in progress, on the left and front Colonel Scott, having rallied his command, led them under the orders of General Ewell to the support of General Taylor, who, pushing forward with the reinforcements just received, and assisted by the well-directed fire of our artillery, forced the enemy to fall back, which was soon followed by his precipitate retreat, leaving many killed and wounded upon the field.

General Taliaferro, who the previous day had occupied the town, was directed to continue to do so with part of his troops, and with the remainder to hold the elevated position on the north side of the river, for the purpose of cooperating, if necessary, with General Trimble and preventing his being cut off from the main body of the army by the destruction of the bridge in his rear; but finding the resistance more obstinate than I anticipated, orders were sent to Taliaferro and Trimble to join the main body. Taliaferro came up in time to discharge an effective volley into the ranks of the wavering and retreating enemy. The pursuit was continued some five miles beyond the battlefield by Generals Taliaferro and Winder with their brigades and portions of the batteries of Wooding and Caskie. Colonel Munford, with cavalry and some artillery, advanced about three miles beyond the other troops.

Our forces captured in the pursuit about 450 prisoners, some wagons, one piece of abandoned artillery, and about 800 muskets. Some 275 wounded were paroled in the hospitals near Port Republic.

While the forces of Shields were in full retreat and our troops in pursuit, Fremont appeared on the opposite bank of the South Fork of the Shenandoah with his army, and opened his artillery upon our ambulances and parties engaged in the humane labors of attending to our dead and wounded and the dead and wounded of the enemy. The next day withdrawing his forces, he retreated down the valley.

On the morning of the 12th, Munford entered Harrisonburg, where, in addition to wagons, medical stores, and camp equipage, he captured some 200 small arms. At that point there also fell into our hands about 200 of Fremont's men, many of them severely wounded on the 8th, and most of the others had been left behind as sick. The Federal surgeons attending them were released and those under their care paroled. The official reports of the casualties of the battle show a loss of sixteen officers killed, sixty-seven wounded, and two missing; 117 non-commissioned officers and privates killed, 862 wounded, and thirty-two missing, making a total loss of 1,096, including skirmishes on the 6th. Since evacuation of Winchester, 1,167; also one piece of artillery.

If we add to the prisoners captured on the 6th and 9th those who were paroled at Harrisonburg and in the hospitals in the vicinity of Port Republic, it will make the number of the enemy who fell into our possession about 975, exclusive of his killed and such of his wounded as he removed. The small arms taken on the 9th and at Harrisonburg numbered about 1,000. We captured seven pieces of artillery, with their caissons, and all of their limbers except one.

The conduct of officers and men during the action merits high praise. During the battle I received valuable assistance in the transmission of orders from the following members of my staff: Colonel Abner Smead, assistant inspector-general; Major R.L. Dabney, assistant adjutant-general; First Lieutenant A.S. Pendleton, aide-de-camp; First Lieutenant H.K. Douglas, assistant inspector-general; First Lieutenant J.K. Boswell, chief engineer, and Colonel William L. Jackson, volunteer aide-de-camp. The medical director of the army, Dr. Hunter McGuire, gave special attention to the comfort and treatment of the wounded. Major W.J. Hawks, chief commissary, and Major J.A. Harman, chief quartermaster, had their departments in good condition.

For further information respecting the conduct of officers and men who distinguished themselves, as well as for a more detailed account of the movements of the troops, I would respectfully refer you to the accompanying official reports of other officers.

I forward herewith two maps by Mr. J. Hotchkiss, one giving the route of the enemy during the retreat from Strasburg to Port Republic and the other of the battlefield.

On the 12th the troops recrossed South River and encamped near Weyer's Cave.

For the purpose of rendering thanks to God for having crowned our arms with success and to implore His continued favor, divine service was held in the army on the 14th.

The army remained near Weyer's Cave until the 17th, when, in obedience to instructions from the commanding general of the department, it moved toward Richmond.

I am, general, very respectfully, your obedient servant,
T. J. JACKSON,
Lieutenant General

Do You Have a Book on the War You Are Longing to Publish?

THE STAINLESS BANNER PUBLISHING COMPANY
A Full-Service Small Press Dedicated to the
Preservation of Southern Heritage and History

History ✭ Memoir ✭ Biography ✭ Novel ✭ Alternate History

What does The Stainless Banner Publishing Company offer you?

- ✭ Professional editing
- ✭ Dynamic covers
- ✭ Hard cover or paperback
- ✭ Competitively priced books
- ✭ High royalties paid monthly
- ✭ Your book on Amazon, Barnes & Noble and many other outlets
- ✭ Free advertising

What is my investment?

Nothing! **The Stainless Banner Publishing Company** is a full service small press. Once your book is published, your only investment will be for any inventory you wish to keep on hand.

Where do I send my query letter?

No need for an agent or a query letter. Just send an email telling us about your book and include the first three chapters in the body of the email. You will receive a prompt reply, usually within 48 hours. Send your email to books@thestainlessbanner.com

The Stainless Banner Publishing Company
www.thestainlessbanner.com

Ewell's Report of Cross Keys and Port Republic

Headquarters, Third Division, Valley District
June 16, 1862.
Major R. L. Dabney, AAG, Valley District.

Major: I have the honor to submit the following report of the action of the 8th instant at Cross Keys between the division commanded by me and the forces under Major General Fremont:

I was ordered on the 7th by the general commanding to occupy the advance, and my division encamped for that night near Union Church. The enemy made a reconnaissance in the afternoon, and going forward, I found General Elzey drawing up his own and General Taylor's brigades in position. I at once determined to meet the enemy on the ground selected by General Elzey.

On the morning of the 8th the enemy advanced, driving in the 15th Alabama, Colonel Cantey, from their post on picket. The regiment made a gallant resistance, enabling me to take position at leisure. The campfires left by the regiment – no tents or anything else – were the camps from which the enemy reported to have driven us.

At this time I had present Elzey's, Trimble's, and Steuart's brigades, short of 5,000 men, Taylor's having been ordered to Port Republic.

The general features of the ground were a valley and rivulet in my front, woods on both flanks, and a field of some hundreds of acres where the road crossed the center of my line, my side of the valley being more defined and commanding the other. General Trimble's brigade was posted a little in advance of my center on the right, General Elzey in rear of the center, and General Steuart on the left; the artillery was in the center. Both wings were in woods. The center was weak, having open ground in front, where the enemy was not expected. General Elzey was in position to strengthen either wing.

About 10:00, the enemy felt along my front with skirmishers, and shortly after posted his artillery, chiefly opposite mine. He advanced under cover on General Trimble with a force, according to his own statement, of two brigades, which were repulsed with such signal loss that they did not make another determined effort. General Trimble had been reinforced by the 13th and 25th Virginia Regiments, Colonel Walker and Lieutenant Colonel Duffy, of General Elzey's brigade. These regiments assisted in the repulse of the enemy. General Trimble in turn advanced and drove the enemy more than a mile, and remained on his flank ready to make the final attack.

General Taylor, with the 8th Brigade, composed of Louisiana troops, reported about 2:00 p.m., and was placed in rear. Colonel Patton, with the 42nd and 48th Regiments and Irish Battalion, Virginia Volunteers, also joined, and with the remainder of General Elzey's brigade was added to the center and left, then threatened. I did not push my successes at once, because I had no cavalry, and it was reported, and reaffirmed by Lieutenant Hinrichs, topographical engineer, sent to reconnoiter, that the enemy was moving a large column two miles to my left. As soon as I could determine this was not to be an attack, I advanced both my wings, drove in the enemy's skirmishers, and when night closed was in position on the ground previously held by the enemy, ready to attack him at dawn.

My troops were recalled to join in the attack at Port Republic. The enemy's attack was decided by 4:00 p.m., and it being principally directed against General Trimble, and, though from their own statement, they outnumbered us on that flank two to one, it had signally failed. General Trimble's command, including the two regiments on his right, under Colonel Walker, is entitled to the highest praise for the gallant manner in which it repulsed the enemy's main attack. His brigade captured one of their colors.

As before mentioned, the credit of selecting the position is due to General Elzey. I availed myself frequently during the action of that officer's counsel, profiting largely by his known military skill and judgment. He was much exposed. His horse was wounded early in the action, and at a later period of the day was killed by a rifle-ball, which, at the same time, inflicted upon the rider a wound that forced him to retire from the field. He was more particularly employed in the center, directing the artillery. General George H. Steuart was severely wounded, after rendering valuable aid in command of the left.

I had Courtney's, Brockenbrough's, Raine's, and Lusk's batteries. The enemy testifies to the efficiency of their fire. Captain Courtney opened the fight, and was for hours exposed to a terrible storm of shot and shell. He and Captain Brockenbrough have been under my observation since the campaign opened, and I can testify to their efficiency on this as on former occasions. The loss in all the batteries shows the warmth of the fire. I was well satisfied with them all.

The history of the Maryland regiment, gallantly commanded by Colonel Bradley T. Johnson, during the campaign of the valley, would be the history of every action from Front Royal to Cross Keys.

On the 6th instant, near Harrisonburg, the 58th Virginia Regiment was engaged with the Pennsylvania Bucktails, the fighting being close and bloody. Colonel Johnson came up with his regiment in the hottest period of the affair, and by a dashing charge in flank drove the enemy off with heavy loss, capturing the lieutenant-colonel (Kane) commanding. In commemoration of their gallant conduct, I ordered one of the captured bucktails to be appended as a trophy to their flag.

The gallantry of the regiment on this occasion is worthy of acknowledgment from a higher source, more particularly as they avenged the death of the gallant General Ashby, who fell at the same time.

Two color-bearers were shot down in succession, but each time the colors were caught before reaching the ground, and were finally borne by Corporal Shanks to the close of the action.

On the 8th instant, at Cross Keys, they were opposed to three of the enemy's regiments in succession.

My staff at Cross Keys consisted of Lieutenant Colonel J.M. Jones and Major James Barbour, Adjutant-General's Department; Lieutenants. G. Campbell Brown and T.T. Turner, aides, and Captain Hugh M. Nelson, volunteer aide. These officers were much exposed during the day, and were worked hard over an extensive field. Their services were valuable and were rendered with zeal and ability.

I append a list of casualties, showing forty-two killed, and 287 killed, wounded, and missing. I buried my dead and brought off all the wounded except a few, whose mortal agonies would have been uselessly increased by any change of position.

Some of the enemy's wounded were brought off and arrangements made for moving them all, when I was ordered to another field. There are good reasons for estimating their loss at not less than 2,000 in killed, wounded, and prisoners. On one part of the field they buried 101 at one spot, fifteen at another, and a house containing some of their dead was said to have been burned by them, and this only a part of what they lost. They were chiefly of Blenker's division, notorious for months on account of their thefts and dastardly insults to women and children in that part of the State under Federal domination.

The order of march of General Fremont was found on a staff officer left in our hands. It shows seven brigades of infantry, besides numerous cavalry. I had three small brigades during the greater part of the action, and no cavalry at any time. They made no bayonet charge, nor did they commit any particular ravages with grape or canister, although they state otherwise. Colonel Mercer and the 21st Georgia tried to close with them three times, partly succeeding in overtaking them once. That officer is represented to have handled his regiment with great skill, and, with the Sixteenth Mississippi, Colonel Posey, was the closest engaged.

Brigadier General Trimble, 7th Brigade, had the brunt of the action, and is entitled to most thanks. Colonel Bradley T. Johnson (1st Maryland), Colonel Carnot Posey (16th Mississippi), Colonel J.T. Mercer (21st Georgia), Captain Courtney (of the Courtney Battery) are officers who were enabled to render highly valuable service.

I regret that I cannot go more into details of those lower in rank, whose gallant services are recompensed by the esteem of their comrades and their own self-approval; after all, the highest and most enduring record.

I enclose a copy of General Fremont's order of march on the day of battle, and detailed reports of the killed and wounded, names and regiments of the officers killed and wounded, and tabular statements of the same according to regiments; also the official report of Col. J.A. Walker, commanding the Fourth Brigade.
Respectfully,
R. S. EWELL, Major General

TAYLOR'S REPORT ON THE BATTLE OF CROSS KEYS AND PORT REPUBLIC

Headquarters 8th Brigade
June 11, 1862

Major Barbour
Assistant Adjutant-General, Third Division

Major: I have the honor to submit the following report of the 8th Brigade as connected with the actions of the 8th and 9th instant:

On the morning of the 8th, I received orders to march the brigade to Port Republic to assist in repelling the attack commenced on the bridge at that point by Shields' forces. When within one miles of the bridge the column was halted, by order of Major General Jackson, to await further orders. These were shortly received – in effect to return to the front and act as a reserve to the troops there engaged against Fremont. Here the brigade became separated, two regiments, the 7th and 8th Louisiana, being ordered to Major-General Ewell to the support of a battery in the center or on the left, of our line, while I marched the remaining two regiments and Wheat's battalion to the right to support General Trimble's brigade, then much pressed. The display of force caused the enemy to retire still farther from the position to which he had been driven by the vigorous charge of Trimble's command.

The brigade, though not actually in action on this day, was much exposed to the enemy's shell, and suffered a loss of one private killed, one officer (Captain Green, 7th Louisiana) and seven privates and non-commissioned officers wounded.

On the 9th I marched from camp near Dunkard's Church, according to orders, at daylight, and proceeded across the Port Republic Bridge to the field where General Winder's troops had already engaged the enemy. Here I received orders from the major general commanding to leave one regiment near the position then occupied by himself, and with the main body to make a detour to the right for the purpose of checking a

> *Here I received orders from the major-general commanding to leave one regiment near the position then occupied by himself, and with the main body to make a detour to the right for the purpose of checking a formidable battery planted in that locality.*

formidable battery planted in that locality. The nature of the ground over which we passed necessarily rendered our progress slow. On reaching the position indicated, the charge was made and the battery, consisting of six guns, fell into our hands after an obstinate resistance on the part of its supporters. Our troops were at the same time subjected to a most destructive fire from the enemy's sharpshooters, posted in a wood above the battery.

After holding the battery for a short time, a fresh brigade of the enemy's troops, moving up from their position on my left flank, and where they had been fronting the troops of Winder's brigade, made a determined and well-conducted advance upon us, accompanied by a galling fire of canister from a piece suddenly brought into position at a distance of about 350 yards. Under this combined attack my command fell back to the skirts of the wood near which the captured battery was stationed, and from this point continued their fire upon the advancing enemy, who succeeded in reclaiming only one gun, which he carried off, leaving both caisson and limber. At this moment our batteries in my rear opened fire, and reinforcements coming up, led by Major General Ewell, the battle was decided in our favor, and the enemy precipitately fled.

The 7th Louisiana Regiment, Colonel Hays commanding, being the regiment left in the front by order of General Jackson, was meanwhile engaged in another portion of the field, and suffered heavy loss. The guns captured by the brigade were five in number, and one other – a brass 12-pounder howitzer – was afterward discovered deserted in the woods near the Brown's Gap road by Lieutenant Dushane, quartermaster of Wheat's Battalion, and by him brought off.

The above record is a mere statement of facts, but no language can adequately describe the gallant conduct of the 8th Brigade in the action of the 9th instant. Disordered by the rapidity of their charge through a dense thicket, making the charge itself just as the loud cheers of the enemy proclaimed his success in another part of the field, assailed by a superior force in front and on the flanks with two batteries in position within point-blank range, nobly did the sons of Louisiana sustain the reputation of their State. Three times was the captured battery lost and won, the enemy fighting with great determination.

Colonel Seymour, of the 6th Louisiana, and Major Wheat, of the battalion, on the left; Colonel Stafford, of the 9th, in the center, and Colonel Kelly, of the 8th, on the right, all acted with the most determined gallantry and were as gallantly supported by their officers and men. Members of each of the regiments engaged in the charge were found dead under the guns of the captured battery.

Captain Surget, assistant adjutant general, distinguished himself greatly, and rendered the most important service on the left. Lieutenant Hamilton, aide-de-camp, gave me valuable assistance in rallying and reforming the men when driven back to the edge of the wood, as did Lieutenant Killmartin, of the 7th Louisiana Regiment, temporarily attached to my staff.

Circumstances unfortunately retained the 7th Regiment, under the gallant Colonel Hays, in another part of the field. Its record of 156 killed and wounded – fifty percent of the number carried into action – shows the service it performed.

Respectfully, your obedient servant,
R. TAYLOR,
Brigadier General

WINDER'S REPORT OF THE BATTLE OF CROSS KEYS AND PORT REPUBLIC

Headquarters First Brigade, Valley District
Camp near Weyer's Cave, Virginia
June 15, 1862

Major. R. L. Dabney,
Assistant Adjutant-General,
Headquarters Valley District.

Sir: I have the honor herewith to report the part taken by this brigade in the operations of the 8th and 9th instant near Port Republic, Virginia

While quietly in camp on Sunday morning, the 8th instant, between 8:00 and 9:00, I heard artillery to our right and rear, which I inferred must be that of the enemy. Captain Poague came in at this time and informed me he had ordered his battery to be prepared for action. I approved it, and requested him to transmit to Captain Carpenter, camped just by him, instructions to the same effect. The good judgment of both these officers had anticipated such orders – a most fortunate circumstance indeed, as the enemy were pressing rapidly on our rear.

General Jackson rode to my tent at this time and ordered me to send a regiment to the bridge over the Shenandoah at Port Republic in double-quick time. I at once sent orders to Colonel J.W. Allen, commanding Second Regiment, to conduct his regiment to that point. Mounting my horse, I rode in the direction of the bridge. Passing Poague's battery, I observed a Parrott gun hitched up and ordered it to follow me.

About one-fourth of a mile from camp I discovered the position of a battery of the enemy across the river, it sending shell just across the road, but too high to do any damage. The gun arriving, I turned it to the left, to bear on the aforesaid battery, when General Jackson directed me to send it to him on the right. This I did and awaited the arrival of other guns, which were soon brought up and placed in position on the hill commanding the opposite side of (the) river.

The second shot silenced the enemy's battery, causing it to limber up and move off. Carpenter's battery arriving, I ordered it placed on the left of Poague's, and the eight pieces of the two batteries to be directed on the retreating battery and column of infantry advancing up the road. The guns were rapidly and admirably served, pouring a heavy and destructive fire upon the enemy. His column halted, staggered at so warm a reception, wavered, and then retreated down the road, being signally repulsed by the artillery alone.

I directed pieces to move to the left, keeping up a constant fire upon him so long as he was within range. Two or more guns were moved a mile beyond the original position. Colonel Allen, 2nd Regiment, arriving, I directed him to move to the left (General Taliaferro's brigade having gone to the bridge), throwing out skirmishers, guarding against a flank movement by the enemy. The 4th Regiment, Colonel Ronald commanding, was ordered to support this regiment. The 5th Regiment, Lieutenant-Colonel Funk commanding, supported Poague's battery. The 27th, Colonel Grigsby commanding, supported Carpenter's battery. The 33rd Regiment, Colonel Neff commanding, was advanced on the left and held in position to repel a flank movement, and at night picketed near same point.

Some few unimportant changes occurred during the day, but the enemy did not again advance within range of our guns. So heavy and well directed was our artillery fire he was obliged to abandon a howitzer and two limbers, which were found in the woods on the following day, being a portion of the battery used against us in the morning. I had observed him trying to remove it and succeeded beyond my expectations in forcing him to leave it, though I knew he had not taken it off by the road on which it advanced. The brigade moved to camp at dark just above Port Republic. The total strength of brigade was 1,334 rank and file in action.

On the morning of the 9th instant, at 3:45, I received orders to have my brigade in Port Republic at 4:45. Orders were immediately given, and the head of the brigade reached the point indicated at that hour. I met General Jackson shortly thereafter, who ordered me to move across South River on a temporary foot-bridge being constructed. I sent Lieutenant Garnett to recall Colonel Neff's regiment from picket, and then moved the brigade as indicated. I was ordered to follow the road down the valley. I placed the Second Regiment Colonel Allen, in front, throwing forward two companies as an advance guard. Having proceeded about a mile, the cavalry in front reported the enemy's pickets.

General Jackson being near, I referred the officer to him. I then received orders to drive them in, occupy the woods in front, and attack the enemy. I directed Captain Nadenbousch, commanding advance, to deploy skirmishers on either side of the road and

> *General Jackson rode to my tent at this time and ordered me to send a regiment to the bridge over the Shenandoah at Port Republic in double quick-time.*

move forward; Captain Carpenter to advance two pieces, take post on left of road, and shell the pickets. These orders were rapidly and well executed; the enemy's pickets disappeared and the skirmishers advanced, the line being supported by Colonel Allen.

The enemy here opened a rapid fire of shell with great accuracy on the road and vicinity. I was then ordered to send a regiment through the woods to endeavor to turn their battery, also a battery to get a position above them. I directed Colonel Allen to move with his regiment, he being in advance and near the wood, to accomplish this, and Colonel Ronald, 4th Regiment, to support him; Captain Carpenter to take his battery in same direction to execute the above order. Captain Poague's two Parrott guns I ordered in position on left of road in a wheat field and opened on enemy's battery, the smoke of which only could be seen, the remaining pieces being under cover. Colonel Grigsby commanding the 27th Regiment, I ordered to support this battery. Lieutenant-Colonel Funk's 5th Regiment was placed on left and to rear of 27th Regiment. The 33rd Regiment, Colonel Neff commanding, to take position on right of road, but, being detained in crossing the river, this order never reached him.

The enemy's fire was so well directed I found it necessary to separate Poague's two guns, placing one some distance on left, ordering Funk's regiment to follow the movement. Here the fire was resumed. The enemy soon placed a battery of two pieces in front and in a commanding position. I sent Lieutenant Garnett, and afterward Captain Poague, to look for a position nearer and more desirable, but none could be found unless the enemy was driven off.

I then learned his skirmishers were advancing, and ordered Funk's regiment forward to support extreme left of line, at same time sending to General Jackson for re-enforcements, being greatly outnumbered. Colonel H.T. Hays soon reported to me with the 7th Louisiana Regiment. I directed him to take position on the right of Funk's, and ordered Grigsby's regiment up, placing it on the right of Hays.

> *The pursuit was continued some four miles when I met General Jackson, who was in advance, and by his orders halted all the artillery except two pieces of Chew's battery.*

This line under Hays I ordered to move forward, drive the enemy from his position, and carry his battery at the point of the bayonet. I, at the same time, directed the remainder of Poague's and a section of Carpenter's battery – the latter having reported it impossible to get through the thick woods or find any position – to be advanced. Colonel Hays moved his command forward in gallant style with a cheer.

Seeing his movement I advanced with the artillery, placing the guns in battery just in rear of Hays' line, which I found had been halted behind a fence, the enemy being in such strong force and pouring in such a heavy fire of artillery and rifles. I then sent for reinforcements, but received none. The men stood it boldly for some time and fought gallantly

– many until all their cartridges were gone.

Captain Raine reported with two pieces of artillery, one, however, without any cannoneers; this piece I sent from the field, the other being brought into action. I had directed Captain Poague to move with a Parrott gun to the right, and sent Lieutenant Garnett to Carpenter to endeavor to place his section so as to enfilade the enemy. The 31st Regiment Virginia Volunteers (Colonel Hoffman) arrived about this time to relieve Colonel Hays, who was ordered to join his brigade. This change it was impossible to effect, and I held Colonel Hoffman in rear of the batteries for their security, as the infantry line began to waver under the storm of shot, shell, and balls which was being rained upon them.

The batteries were moved to the rear and I tried to rally the men, placing Hoffman's regiment in line on which to rally; here I partially succeeded, but the enemy so greatly outnumbered us, and, getting within such easy range, thinned our ranks so terribly, that it was impossible to rally them for some time, though I was most ably assisted in my endeavors by my staff, the gallant Hays, Grigsby, Funk, Major Williams (5th Regiment), Captains Nadenbousch (2nd), and Burke (5th Regiment); these came particularly under my observation, though doubtless others did their duty as nobly and bravely. Here one piece of Poague's, I regret to say, fell into the enemy's hands, I having ordered it to halt and fire on his advancing column, where it was disabled, as shown in Poague's report.

I still endeavored to rally the remainder of this force, and succeeded in getting the 7th Louisiana, under Major Penn, the colonel and lieutenant-colonel both being wounded, and 5th Regiment, under Funk. I placed two pieces of Poague's battery in the position previously occupied, and again opened fire on the enemy, he having halted in his advance.

A sharp fire from the wood on (the) right told General Taylor's and Allen's forces were engaged. I directed the Parrott gun on the enemy's battery, which was now turned on those forces. I was gratified to learn from General Taylor this fire was of service to him.

The enemy now moved to his left flank, apparently to surround this command in the woods. Seeing two regiments lying quietly on their arms to the right under the woods, I dispatched Lieutenant Garnett to order them forward rapidly to press the enemy's rear. I then moved forward the artillery with its supports and obtained a far better position. Captain Chew here reported to me and did good execution with his battery, displaying great skill and accuracy in his fire.

I soon met General Jackson and reported my impressions to him, and was told he had ordered up other troops. Lieutenant Colonel Garnett (48th Regiment) came up, reporting for orders. I directed him to follow the road in double-quick, pressing the enemy hotly in rear and driving him from his position. Major Holliday (33rd Regiment) rode up at this time, and through him I sent orders to Colonel Neff to do the same. The batteries arriving, I continued to advance them as rapidly as possible, pouring in a heavy and well-directed fire on the retreating columns of the enemy, who were now driven from the field, routed at every point. A section of Captain Brockenbrough's battery joined me just as the retreat commenced and was ably handled.

The road and woods were shelled and the enemy scattered, in every direction. The pursuit was continued some four miles, when I met General Jackson, who was in advance, and by his orders halted all the artillery except two pieces of Chew's battery. The enemy being again driven from their ambuscade, I followed with my command to a point some eight or nine miles below Port Republic, when I received orders to return and camp with my wagons, which order was executed, my advance reaching camp on the summit of the Blue Ridge at Brown's Gap at midnight and the batteries at daylight.

It again affords me sincere and great gratification to bear testimony to the courage, gallantry, fortitude, and good conduct of the officers and men under my command, and to them I return my heartfelt thanks. They fought gallantly and desperately, as our holy cause urged them to do, and though temporarily repulsed, it was only from overwhelming numbers. Although exposed to such a withering fire, the killed are few in number, a kind Providence having guarded many from the great dangers to which they were exposed. Colonels Allen and Ronald were so far separated from me I must refer to their respective reports for the operations of their regiments. To my staff, Captain O'Brien, Lieutenants Howard and Garnett, I tender my sincere thanks for their assistance in transmitting my orders to different points (though under heavy fire frequently after the fight became general), ever ready and prompt.

The casualties were: Killed – officers, two; privates, eleven. Wounded – officers, six; privates, 148. Missing – privates, thirty-two. Total, 199. The strength of the brigade was 1,313, rank and file.

For detailed accounts of the affair I respectfully refer to the reports of the several commanders herewith transmitted.

I am, sir, very respectfully,
Charles. S. Winder
Brigadier General, Commanding

Hi-Lights of a Hero's Life
Richard S. Ewell

★ 1817 – (February 8) Born in Georgetown in the District of Columbia.

★ 1820 – Moved to Stony Lonesome, an estate near Manassas, Virginia.

★ 1840 – Graduated from the Military Academy, 13th in a class of 42 cadets and commissioned a second lieutenant in the 1st U.S. Dragoons.

- 1845 – Promoted to first lieutenant and served on escort duty along the Santa Fe and Oregon Trails.

- 1848 – Promoted to captain for bravery shown at Contreras and Churubusco during the Mexican-American War.

- 1853 – Explored the newly purchased Gadsden Territory.

- 1859 – Wounded in a skirmish with the Apaches.

- 1860 – Commanded Fort Buchanan but grew ill and returned to West Virginia to recuperate.

- 1861 – (May 7) Resigned his U.S. Army commission.

- 1861 – (May 9) Appointed colonel of cavalry in the Virginia Provisional Army.

- 1861 – (May 31) Wounded in a skirmish at Fairfax Court House.

- 1861 – (June 17) Promoted to brigadier general in the Confederate army.

- 1861 – (July 21) Fought at Manassas.

- 1862 – (January 24) Promoted to major general and assigned to the Shenandoah Valley.

- 1862 – (March 23 to June 9) Fought in the Valley Campaign.

- 1862 – (June 25 to July 1) Participated in the Seven Days Battles, including Gaines Mill and Malvern Hill.

- 1862 – (August 9) Fought at Cedar Mountain.

- 1862 – (August 29) Wounded at Groveton which resulted in his left leg being amputated below the knee.

- 1863 – (May 26) Married Lizinka Campbell Brown.

- 1863 – Promoted to lieutenant general and placed in command of the Second Corps.

- 1863 – Won a major victory at the Second Battle of Winchester.

- 1863 – (July 1-3) Participated in the Gettysburg Campaign.

- 1863 – (November) Wounded at Kelly's Ford.

- 1864 – (May 5-7) Fought in the Battle of Wilderness.

- 1864 – (May 12) Robert E. Lee took command of the Second Corps at the Mule Shoe during the Battle of Spotsylvania Court House.

- 1864 – Relieved of command of the Second Corps. Reassigned to command the garrison of the Department of Richmond.

- 1865 – (April) Abandoned Richmond.

- 1865 – (April 6) Surrendered at Saylor's Creek.

- 1865 – Imprisoned at Fort Warren in Boston Harbor.

- 1865 – Returned to Spring Hill, Tennessee and became a gentleman farmer.

- 1872 – Died of pneumonia.

The Battle Of Cross Keys and Port Republic

By George F.R. Henderson. *Stonewall Jackson and the American Civil War.* (New York: Fawcett Publicatons, Inc. 1962).

By the ignorant and the envious success in war is easily explained away. The dead military lion, and, for that matter, even the living, is a fair mark for the heels of a baser animal. The greatest captains have not escaped the critics. The genius of Napoleon has been belittled on the ground that each one of his opponents, except Wellington, was only second-rate. French historians have attributed Wellington's victories to the mutual jealousy of the French marshals; and it has been asserted that Moltke triumphed only because his adversaries blundered. Judged by this rule few reputations would survive.

In war, however, it is as impossible to avoid error as it is to avoid loss of life; but it is by no means simple either to detect or to take advantage of mistakes. Before both Napoleon and Wellington an unsound maneuver was dangerous in the extreme. None were so quick to see the slip; none more prompt to profit by it. Herein, to a very great extent, lay the secret of their success, and herein lies the true measure of military genius. A general is not necessarily incapable because he makes a false move; both

Napoleon and Wellington, in the long course of their campaigns, gave many openings to a resolute foe and both missed opportunities. Under ordinary circumstances mistakes may easily escape notice altogether or at all events pass unpunished, and the reputation of the leader who commits them will remain untarnished. But if he is pitted against a master of war, a single false step may lead to irretrievable ruin; and he will be classed as beneath contempt for a fault which his successful antagonist may have committed with impunity a hundred times over.

Jackson Escapes from Winchester

So Jackson's escape from Winchester was not due simply to the inefficiency of the Federal generals, or to the ignorance of the Federal President. Lincoln was wrong in dispatching McDowell to Front Royal in order to cut off Jackson. When Shields, in execution of this order, left Fredericksburg, the Confederates were only five miles north of Winchester, and had they at once retreated, McDowell must have missed them by many miles. McDowell, hotly protesting, declared, and rightly, that the movement he had been ordered to execute was strategically false. "It is impossible," he said, "that Jackson can have been largely reinforced. He is merely creating a diversion and the surest way to bring him from the lower Valley is for me to move rapidly on Richmond. In any case, it would be wiser to move on Gordonsville."

His arguments were unavailing. But when Jackson pressed forward to the Potomac, it became possible to intercept him, and the President did all he could to assist his generals. He kept them constantly informed of the movements of the enemy and of each other. He left them a free hand, and with an opponent less able, his instructions would have probably brought about complete success.

Nor were the generals to blame. They failed to accomplish the task that had been set them and they made mistakes. But the task was difficult; and, if at the critical moment the hazard of their situation proved too much for their resolution, it was exactly what might have been expected. The initial error of the Federals was in sending two detached forces, under men of no particular strength of character, from opposite points of the compass, to converge upon an enemy who was believed to be superior to either of them. Jackson at once recognized the blunder and foreseeing the consequences that were certain to ensue resolved to profit by them. His escape, then, was the reward of his own sagacity.

When once the actual position of the Confederates had been determined, and the dread that reinforcements were coming down the Valley had passed away, the vigor of the Federal pursuit left nothing to be desired.

> *By the ignorant and envious, success in war is easily explained away.*

Directly it was found that the Confederates had gone south, on the afternoon of June 1, Shields was directed on Luray, and that night his advanced guard was ten miles beyond Front Royal; on the other side of the Massanuttons, Fremont, with Bayard's cavalry heading his advance, moved rapidly on Woodstock.

The Federal generals, however, had to do with a foe who never relaxed his vigilance. Whilst Ashby and Ewell, on May 31, were engaged with Fremont at Cedar Creek, Jackson had expected that Shields would advance on Strasburg. But not a single infantry soldier was observed on the Front Royal road throughout the day. Such inaction was suspicious and the probability to which it pointed had not escaped the penetration of the Confederate leader. His line of retreat was the familiar route by New Market and Harrisonburg to Port Republic, and thence to the Gaps of the Blue Ridge. There he could secure an unassailable position within reach of the railway and of Richmond.

But, during the movement, danger threatened from the valley of the South Fork. Should Shields adopt that line of advance the White House and Columbia bridges would give him easy access to New Market; and while Fremont was pressing the Confederates in rear, their flank might be assailed by fresh foes from the Luray Gap. And even if the retiring column should pass New Market in safety, Shields, holding the bridges at Conrad's Store and Port Republic, might block the passage to the Blue Ridge. Jackson, looking at the situation from his enemy's point of view, came to the conclusion that a movement up the valley of the South Fork was already in progress and that the aim of the Federal commander would be to secure the bridges. His conjectures hit the mark.

Before leaving Front Royal, Shields ordered his cavalry to march rapidly up the valley of the South Fork and seize the bridge at Conrad's Store; the White House and Columbia bridges he intended to secure himself. But Jackson was not to be so easily overreached.

On the night of June 2, the Federal cavalry reached Luray to find that they had come too late. The White House and Columbia bridges had both been burned by a detachment of Confederate cavalry, and Shields was thus cut off from New Market. At dawn on the 4th, after a forced night march, his advanced guard reached Conrad's Store to find that bridge also gone and he was once more foiled.

On his arrival at Luray, the sound of cannon on the other side of the Massanuttons was plainly heard. It seemed probable that Jackson and Fremont were already in collision; but Shields, who had written a few hours before to Mr. Stanton that with supplies and forage he could "stampede the enemy to Richmond," was unable to stir a foot to assist his colleague.

Jackson's Use of the Valley's Topography

Once again Jackson had turned to account the strategic possibilities of the Massanuttons and the Shenandoah; and, to increase General Shields' embarrassment, the weather had broken. Heavy and incessant rainstorms submerged the Virginia roads. He was ahead of his supplies; much hampered by the mud; and the South Fork of the Shenandoah, cutting him off from Fremont, rolled a volume of rushing water which it was impossible to bridge without long delay.

    Meanwhile, west of the great mountain, the tide of war, which had swept with such violence to the Potomac, came surging back. Fremont, by the rapidity of his pursuit, made full amends for his lack of vigor at Cedar Creek. A cloud of horsemen filled the space between the hostile columns. Day after day the quiet farms and sleepy villages on the Valley turnpike heard the thunder of Ashby's guns. Every stream that crossed the road was the scene of a fierce skirmish and the ripening corn was trampled under the hoofs of the charging squadrons.

    On June 2, the first day of the pursuit, between Strasburg and Woodstock, the Federals, boldly led by Bayard, gained a distinct advantage. A dashing attack drove in the Confederate rearguard swept away the horse artillery and sent Ashby's and Steuart's regiments, exhausted by hunger and loss of sleep, flying up the Valley. Many prisoners were taken and the pursuit was only checked by a party of infantry stragglers whom Ashby had succeeded in rallying across the road.

    Next day, June 3, the skirmishing was continued; and the Confederates, burning the bridges across the roads, retreated to Mount Jackson.

    On the 4th the bridge over the North Fork was given to the flames, Ashby, whose horse was shot under him, remaining to the last; and the deep and turbulent river placed an impassable obstacle between the armies. Under a deluge of rain, the Federals attempted to launch their pontoons; but the boats were swept away by the rising flood, and it was not till the next morning that the bridge was made.

    The Confederates had thus gained twenty-four hours' respite, and contact was not resumed until the 6th. Jackson, meanwhile, constructing a ferry at Mount Crawford, had sent his sick and wounded to Staunton, thus saving them the long detour by Port Republic; and dispatching his stores and prisoners by the more circuitous route, had passed through Harrisonburg to Cross Keys, a clump of buildings on Mill Creek, where, on the night of the 5th, his infantry and artillery, with the exception of a brigade supporting the cavalry, went into bivouac.

The Death of Ashby

    On the afternoon of the 6th the Federal cavalry followed Ashby. Some three miles from Harrisonburg is a tract of forest, crowning a long ridge; and within the timber the Confederate squadrons occupied a strong position. The enemy, 800 strong, pursued without precaution, charged up a gentle hill, and was repulsed by a heavy fire. Then Ashby let loose his mounted men on the broken ranks, and the Federals were driven back to within half a mile of Harrisonburg, losing four officers and thirty men.

    Smarting under this defeat, Fremont threw forward a still stronger force of cavalry, strengthened by two battalions of infantry. Ashby had already called up a portion of the brigade which supported him and met the attack in a clearing of the forest. The fight was fierce. The

Confederates were roughly handled by the Northern riflemen and the ranks began to waver. Riding to the front where the opposing lines were already at close range Ashby called upon his infantry to charge.

As he gave the order his horse fell heavily to the ground. Leaping to his feet in an instant, again he shouted, "Charge, men! For God's sake, charge!" The regiments rallied, and inspired by his example swept forward from the wood. But hardly had they left the covert when their leader fell, shot through the heart. He was speedily avenged. The men who followed him, despite the heavy fire, dashed at the enemy in front and flank, and drove them from their ground. The cavalry, meanwhile, had worked round in rear; the horse artillery found an opportunity for action; and under cover of the night the Federals fell back on Harrisonburg.

The losses of the Union troops were heavy; but the Confederate victory was dearly purchased. The death of Ashby was a terrible blow to the Army of the Valley. From the outbreak of the war, he had been employed on the Shenandoah and from Staunton to the Potomac, his was the most familiar figure in the Confederate ranks. His daring rides on his famous white charger were already the theme of song and story; and if the tale of his exploits, as told in camp and farm, sometimes bordered on the marvelous, the bare truth, stripped of all exaggeration, was sufficient in itself to make a hero. His reckless courage, his fine horsemanship, his skill in handling his command, and his power of stimulating devotion, were not the only attributes which incited admiration.

"With such qualities," it is said, "were united the utmost generosity and unselfishness, and a delicacy of feeling equal to a woman's."

His loss came home with especial force to Jackson. After the unfortunate episode in the pursuit from Middletown, he had rated his cavalry leader in no measured terms for the indiscipline of his command; and for some days their intercourse, usually most cordial, had been simply official. Sensitive in the extreme to any reflection upon himself or his troops, Ashby held aloof; and Jackson, always stern when a breach of duty was concerned, made no overtures for a renewal of friendly intercourse.

Fortunately, before the fatal fight near Harrisonburg, they had been fully reconciled; and with no shadow of remorse Jackson was able to offer his tribute to the dead. Entering the room in Port Republic, whither the body had been brought, he remained for a time alone with his old comrade; and in sending an order to his cavalry, added, "Poor Ashby is dead. He fell gloriously – one of the noblest men and soldiers in the Confederate army." A more public testimony was to come. In his official report he wrote: "The close relation General Ashby bore to my command for most of the previous twelve months will justify me in saying that as a partisan officer I never knew his superior. His daring was proverbial, his powers of endurance almost incredible, his character heroic, and his sagacity almost intuitive in divining the purposes and movements of the enemy."

On the 6th and 7th the Confederate infantry rested on the banks of Mill Creek, near Cross Keys. The cavalry, on either flank of the Massanuttons, watched both Fremont's camps at Harrisonburg and the slow advance of Shields; and on the southern peak of the mountains a party of signalers, under a staff officer, looked down upon the roads which converged on the Confederate position.

June 7 was passed in unwonted quiet. For the first time for fifteen days, since the storming of Front Royal, the boom of the guns was silent. The glory of the summer brooded undisturbed on hill and forest; and as the escort which followed Ashby to his grave passed down the quiet country roads, the Valley lay still and peaceful in the sunshine. Not a single Federal scout observed the melancholy cortege.

Fremont's pursuit had been roughly checked. He was uncertain in which direction the main body of the Confederates had retreated; and it was not till evening that a strong force of infantry, reconnoitering through the woods, struck Jackson's outposts near the hamlet of Cross Keys. Only a few shots were exchanged.

Shields in Pursuit

Shields, meanwhile, had concentrated his troops at Columbia Bridge on the 6th, and presuming that Jackson was standing fast on the strong position at Rude's Hill, was preparing to cross the river. Later in the day a patrol, which had managed to communicate with Fremont, informed him that Jackson was retreating, and the instructions he thereupon dispatched to the officer commanding his advanced guard are worthy of record:

"The enemy passed New Market on the 5th; Blenker's division on the 6th in pursuit. The enemy has flung away everything, and their stragglers fill the mountain. They need only a movement on the flank to panic-strike them, and break them into fragments. No man has had such a chance since the war commenced. You are within thirty miles of a broken, retreating enemy, who still hangs together. Ten thousand Germans are on his rear, who hang on like bull-dogs. You have only to throw yourself down on Waynesborough before him, and your cavalry will capture them by the thousands, seize his train and abundant supplies."

In anticipation, therefore, of an easy triumph, and, to use his own words, of "thundering down on Jackson's rear," Shields, throwing precaution to the winds, determined to move as rapidly as possible on Port Republic. He had written to Fremont urging a combined attack on "the demoralized rebels," and he thought that together they "would finish Jackson." His only anxiety was that the enemy might escape, and in his haste he neglected the warning of his Corps commander. McDowell, on dispatching him in pursuit, had directed his attention to the importance of keeping his division well closed up.

Jackson's predilection for dealing with exposed detachments had evidently been noted. Shields' force, however, owing to the difficulties of the road, the mud, the quick-sands, and the swollen streams, was already divided into several distinct fractions. His advanced brigade was south of Conrad's Store; a second was some miles in rear, and two were at Luray, retained at that point in consequence of a report that 8,000 Confederates were crossing the Blue Ridge by Thornton's Gap. To correct this faulty formation before advancing he thought was not worth while. On the night of June 7 he was sure of his prey.

The situation at this juncture was as follows: Shields was stretched out over five-and-twenty miles of road in the valley of the South Fork; Fremont was at Harrisonburg; Ewell's division was near Cross Keys, and the main body of the Valley Army near Port Republic.

During his retreat Jackson had kept his attention fixed on Shields. That ardent Irishman pictured his old enemy flying in confusion, intent only on escape. He would have been much astonished had he learned the truth. From the moment Jackson left Strasburg, during the whole time he was retreating, with the "bulldogs" at his heels, he was meditating a counter stroke and his victim had already been selected. When Shields rushed boldly up the valley of the South Fork it seemed that an opportunity of avenging Kernstown was about to offer.

On June 4, the day that the enemy reached Luray, Ewell was ordered to provide his men with two days' cooked rations and to complete their ammunition "for active service." The next day, however, it was found that Shields had halted. Ewell was ordered to stand fast, and Jackson wrote despondently to Lee: "At present I do not see that I can do much more than rest my command and devote its time to drilling."

On the 6th, however, he learned that Shields' advanced guard had resumed its march; and, like a tiger crouching in the jungle, he prepared to spring upon his prey. But Fremont was close at hand, and Shields and Fremont between them mustered nearly 25,000 men. They were certainly divided by the Shenandoah; but they were fast converging on Port Republic; and in a couple of marches, if not actually within sight of each other's camps, they would come within hearing of each other's guns. Yet, notwithstanding their numbers, Jackson had determined to deal with them in detail.

Jackson Chooses to Fight at Port Republic

A few miles from the camp at Port Republic was a hill honeycombed with caverns, known as the Grottoes of the Shenandoah. In the heart of the limestone Nature has built herself a palace of many chambers, vast, silent, and magnificent. But far beyond the beauty of her mysterious halls was the glorious prospect which lay before the eyes of the Confederate sentries. Glimmering aisles and dark recesses, where no sunbeam lurks nor summer wind whispers, compared but ill with those fruitful valleys, watered by clear brown rivers, and steeped in the glow of a Virginian June. To the north stood the Massanuttons, with their forests sleeping

> *No battlefield boasts a fairer setting than Port Republic; but, lover of Nature as he was, the region was attractive to Jackson for reasons of a sterner sort.*

in the noonday; and to the right of the Massanuttons, displaying, in that transparent atmosphere, every shade of that royal color from which it takes its name, the Blue Ridge loomed large against the eastern sky. Summit after summit, each more delicately penciled than the last, receded to the horizon, and beneath their feet, still, dark, and unbroken as the primeval wilderness, broad leagues of woodland stretched far away over a lonely land.

No battlefield boasts a fairer setting than Port Republic; but, lover of Nature as he was, the region was attractive to Jackson for reasons of a sterner sort. It was eminently adapted for the purpose he had at heart.

1. The South Fork of the Shenandoah is formed by the junction of two streams, the North and South Rivers; the village of Port Republic lying on the peninsula between the two.

2. The bridge crosses the North River just above the junction, carrying the Harrisonburg road into Port Republic; but the South River, which cuts off Port Republic from the Luray Valley, is passable only by two difficult fords.

3. North of the village, on the left bank of the Shenandoah, a line of high bluffs covered with scattered timber completely commands the tract of open country which lies between the river and the Blue Ridge, and across this tract ran the road by which Shields was marching.

4. Four miles northwest of Port Republic, near the village of Cross Keys, the road to Harrisonburg crosses Mill Creek, a strong position for defense.

By transferring his army across the Shenandoah, and burning the bridge at Port Republic, Jackson could easily have escaped Fremont, and have met Shields in the Luray Valley with superior force. But the plain where the battle must be fought was commanded by the bluffs on the left bank of the Shenandoah; and should Fremont advance while an engagement was in progress, even though he could not cross the stream, he might assail the Confederates in flank with his numerous batteries.

In order, then, to gain time in which to deal with Shields, it was essential that Fremont should be held back, and this could only be done on the left bank. Further, if Fremont could be held back until Shields' force was annihilated, the former would be isolated. If Jackson could hold the bridge at Port Republic, and also prevent Fremont reaching the bluffs, he could recross when he had done with Shields, and fight Fremont without fear of interruption.

To reverse the order and to annihilate Fremont before falling upon Shields was out of the question. Whether he advanced against Fremont or whether he stood still to receive his attack, Jackson's rear and communications, threatened by Shields, must be protected by a strong detachment. It would be thus impossible to meet Fremont with superior or even equal numbers, and an army weaker on the battlefield could not make certain of decisive victory.

Jackson had determined to check Fremont at Mill Creek. But the situation was still uncertain. Fremont had halted at Harrisonburg, and it was possible that he might advance no further. So the Confederates were divided, ready to meet either adversary; Ewell remaining at Cross Keys, and the Stonewall division encamping near Port Republic.

Jackson Almost Captured

On the morning of June 8, however, it was found that Fremont was moving. Ewell's division was already under arms. At 8.30 a.m. his pickets, about two miles to the front, became engaged, and the Confederate regiments moved leisurely into position.

The line ran along the crest of a narrow ridge, commanding an open valley, through which Mill Creek, an insignificant brook, ran parallel to the front. The further slopes open and unobstructed except for scattered trees and a few fences rose gently to a lower ridge, about a mile distant. The ground held by the Confederates was only partially cleared, and from the Port Republic road in the center at a distance of six hundred yards on either flank were woods of heavy timber enclosing the valley and jutting out towards the enemy. The ridge beyond the valley was also thickly wooded; but here, too, there were open spaces on which batteries might be deployed; and the forest in rear, where Ashby had been killed, standing on higher ground, completely concealed the Federal approach.

The pickets, however, had given ample warning of the coming attack; and when, at 10:00 a.m., the hostile artillery appeared on the opposite height, it was received with a heavy fire. "Eight and a half batteries," says Fremont, "were brought into action within thirty minutes." Against this long array of guns the Confederates massed only five batteries; but these commanded the open ground, and were all in action from the first.

Ewell had with him no more than three brigades. The Louisiana regiments had bivouacked near Port Republic and were not yet up. The whole strength of the troops which held the ridge was no more than 6,000 infantry, and perhaps 500 cavalry. Fremont had at least 10,000 infantry, twelve batteries, and 2,000 cavalry.

> *It was then against overwhelming numbers that Ewell was asked to hold his ground, and the remainder of the army was four miles in the rear.*

It was then against overwhelming numbers that Ewell was asked to hold his ground, and the remainder of the army was four miles in the rear. Jackson himself was still absent from the field. The arrangements for carrying out his ambitious plans had met with an unexpected hitch. In the Luray Valley, from Conrad's Store northwards, the space between the Blue Ridge and the Shenandoah was covered for the most part with dense forest, and through this forest ran the road. Moving beneath the spreading foliage of oak and hickory, Shields' advanced brigade was concealed from the observation of the Confederate cavalry; and the signalers on the mountain, endangered by Fremont's movement, had been withdrawn. North of Port Republic, between the foot-hills of the Blue Ridge and the Shenandoah, lies a level tract of arable and meadow, nearly a mile wide, and extending for nearly three miles in a northerly direction.

On the plain were the Confederate pickets, furnished by three companies of

Ashby's regiment, with their patrols on the roads towards Conrad's Store; and there seemed little chance that Shields would be able to reach the fords over the South River, much less the Port Republic bridge, without long notice being given of his approach. The cavalry, however, as had been already proved, were not entirely to be depended on. Jackson, whose headquarters were within the village, had already mounted his horse to ride forward to Cross Keys, when there was a distant fire, a sudden commotion in the streets, and a breathless messenger from the outposts reported that not only had the squadrons on picket been surprised and scattered, but that the enemy was already fording the South River.

Between the two rivers, southwest of Port Republic, were the Confederate trains, parked in the open fields. Here was Carrington's battery, with a small escort; and now the cavalry had fled there were no other troops, save a single company of the 2nd Virginia, on this side the Shenandoah. The squadron which headed the Federal advanced guard was accompanied by two guns. One piece was sent towards the bridge; the other, unlimbering on the further bank, opened fire on the church, and the horsemen trotted cautiously forward into the village street.

Jackson, warned of his danger, had already made for the bridge, and crossing at a gallop escaped capture by the barest margin of time. His chief of artillery, Colonel Crutchfield, was made prisoner, with Dr. McGuire and Captain Willis and his whole staff was dispersed save Captain Pendleton a sterling soldier, though hardly more than a boy in years.

And the danger was not over. With the trains was the whole of the reserve ammunition, and it seemed that a crushing disaster was near at hand. The sudden appearance of the enemy caused the greatest consternation amongst the teamsters; several of the wagons went off by the Staunton road; and, had the Federal cavalry come on, the whole would have been stampeded. But Carrington's battery was called to the front by Captain Moore, commanding the company of infantry in the village. The picket, promptly put into position, opened with a well-aimed volley, and a few rounds checked the enemy's advance; the guns came rapidly and effectively into action, and at this critical moment Jackson intervened with his usual vigor.

From the left bank of the North River he saw a gun bearing on the bridge, the village swarming with blue uniforms, and more artillery unlimbering across the river. He had already sent orders for his infantry to fall in, and a six-pounders was hurrying to the front. "I was surprised," said the officer to whose battery this piece belonged, "to see a gun posted on the opposite bank. Although I had met a cavalry man who told me that the enemy was advancing up the river, still I did not think it possible they could have brought any guns into the place in so short a time. It thereupon occurred to me that the piece at the bridge might be one of Carrington's, whose men had new uniforms something like those we saw at the bridge.

Upon suggesting this to the general, he reflected a moment, and then riding a few paces to the left and front, he called out, in a tone loud enough to be heard by the enemy, 'Bring that gun up here!' but getting no reply, he raised himself in his stirrups, and in a most authoritative and seemingly angry tone he shouted, 'Bring that gun up here, I say!' At this they began to move the trail of the gun so as to bring it to bear on us, which, when the general perceived, he turned quickly to the officer in charge of my gun, and said in his sharp, quick way, 'Let 'em have it!' The words had scarcely left his lips when Lieutenant Brown, who had his piece charged and aimed, sent a shot right among them, so disconcerting them that theirs in reply went far above us."

The Confederate battalions, some of which had been formed up for inspection or for the Sunday service, when the alarm was given had now come up, and the 87th Virginia was ordered to capture the gun, and to clear the village. Without a moment's hesitation the regiment charged with a yell across the bridge, and so sudden was the rush that the Federal artillerymen were surprised. The gun was double-shotted with canister, and the head of the column should have been swept away. But the aim was high and the Confederates escaped.

Then, as the limber came forward, the horses, terrified by the heavy fire and the yells of the charging infantry, became unmanageable; and the gunners, abandoning the field-piece, fled through the streets of Port Republic. The 87th rushed forward with a yell. The hostile cavalry, following the gunners, sought safety by the fords; and as the routed army dashed through the shallow water, the Confederate batteries, coming into action on the high bluffs west of the Shenandoah, swept the plain below with shot and shell.

The hostile artillery beyond the stream was quickly overpowered; horses were shot down wholesale; a second gun was abandoned on the road; a third, which had only two horses and a driver left, was thrown into a swamp; and a fourth was found on the field without either team or men.

The Federal infantry was not more fortunate. Carroll's brigade of four regiments was close in rear of the artillery when the Confederate batteries opened fire. Catching the contagion from the flying cavalry, it retreated northward in confusion. A second brigade (Tyler's) came up in support; but the bluffs beyond the river were now occupied by Jackson's infantry; a stream of fire swept the plain; and as Shields' advanced guard, followed by the Confederate cavalry, fell back to the woods whence it had emerged, five miles away on the other flank was heard the roar of the cannonade which opened the battle of Cross Keys.

> "Fortune," it has been said, "loves a daring suito,r and he who throws down the gauntlet may always count on his adversary to help him."

Battle of Cross Keys

From the hurried flight of the Federals it was evident that Shields' main body was not yet up; so, placing two brigades in position to guard the bridge, Jackson sent the remainder to Ewell, and then rode to the scene of action.

Fremont, under cover of his guns, had made his preparations for attack; but the timidity which he had already displayed when face to face with Jackson had once more taken possession of his faculties. Vigorous in pursuit of a flying enemy, when that enemy turned at bay his courage vanished. The Confederate position was undoubtedly strong, but it was not impregnable. The woods on either flank gave access under cover to the central ridge. The superior weight of his artillery was sufficient to cover an advance across the open; and although he was without maps or guide, the country was not so intersected as to render maneuvering impracticable.

In his official report Fremont lays great stress on the difficulties of the ground; but reading between the lines it is easy to see that it was the military situation which overburdened him. The vicious strategy of converging columns, where intercommunication is tedious and uncertain, once more exerted its paralyzing influence. It was some days since he had heard anything of Shields. That general's dispatch, urging a combined attack, had not yet reached him: whether he had passed Luray or whether he had been already beaten, Fremont was altogether ignorant; and, in his opinion, it was quite possible that the whole of the Confederate army was before him.

A more resolute commander would probably have decided that the shortest way out of the dilemma was a vigorous attack. If Shields was within hearing of the guns – and it was by no means improbable that he was – such a course was the surest means of securing his co-operation; and even if no help came, and the Confederates maintained their position, they might be so crippled as to be unable to pursue. Defeat would not have been an irreparable misfortune. Washington was secure. Banks, Saxton, and McDowell held the approaches; and if Fremont himself were beaten back, the strategic situation could be in no way affected. In fact a defeat, if it had followed an attack so hotly pressed as to paralyze Jackson for the time being, would have been hardly less valuable than a victory.

"Fortune," it has been well said, "loves a daring suitor, and he who throws down the gauntlet may always count upon his adversary to help him." Fremont, however, was more afraid of losing the battle than anxious to win it. "Taking counsel of his fears," he would run no risks. But neither could he abstain from action altogether. An enemy was in front of him who for seven days had fled before him, and his own army anticipated an easy triumph.

So, like many another general who has shrunk from the nettle danger, he sought refuge in half-measures, the most damning course of all. Of twenty-four regiments present on the field of battle, five only, of Blenker's Germans, were sent forward to the attack. Their onslaught was directed against the Confederate right; and here, within the woods, Trimble had posted his brigade in a most advantageous position. A flat-topped ridge, covered with great oaks, looked down upon a wide meadow, crossed by a stout fence; and beyond the hollow lay the woods through which the Federals, already in contact with the Confederate outposts, were rapidly advancing. The pickets soon gave way, and crossing the meadow found cover within the thickets, where Trimble's three regiments lay concealed. In hot pursuit came the

Federal skirmishers, with the solid lines of their brigade in close support. Steadily moving forward, they climbed the fence and breasted the gentle slope beyond.

A few scattered shots, fired by the retreating pickets, were the only indications of the enemy's presence; the groves beyond were dark and silent. The skirmishers had reached the crest of the declivity, and the long wave of bayonets, following close upon their tracks, was within sixty paces of the covert, when the thickets stirred suddenly with sound and movement. The Southern riflemen rose swiftly to their feet. A sheet of fire ran along their line, followed by a crash that resounded through the woods; and the German regiments, after a vigorous effort to hold their ground, fell back in disorder across the clearing. Here, on the further edge, they rallied on their reserves, and the Confederates, who had followed up no further than was sufficient to give impetus to the retreat, were once more withdrawn.

A quarter of an hour passed, and as the enemy showed no inclination to attempt a second advance across the meadow, where the dead and wounded were lying thick, Trimble, sending word to Ewell of his intention, determined to complete his victory. More skilful than his enemies, he sent a regiment against their left, to which a convenient ravine gave easy access, while the troops among the oaks were held back till the flank attack was fully developed. The unexpected movement completely surprised the Federal brigadier. Again his troops were driven in, and the Confederates, now reinforced by six regiments which Ewell had sent up, forced them with heavy losses through the woods, compelled two batteries, after a fierce fight, to limber up, routed a brigade which had been sent by Fremont to support the attack, and pressing slowly but continuously forward, threw the whole of the enemy's left wing, consisting of Blenker's eleven regiments, back to the shelter of his line of guns. Trimble had drawn the "bulldog's" teeth.

> *Ewell's division bivouacked within sight of the enemy's watch fires and within hearing of his outposts...*

The Confederates had reached the outskirts of the wood. They were a mile in advance of the batteries in the center; and the Federal position commanding a tract of open ground was strong in itself and strongly held. A general counterstroke was outside the scope of Jackson's designs. He had still Shields to deal with. The Federal left wing had been heavily repulsed, but only a portion of Fremont's force had been engaged; to press the attack further would undoubtedly have cost many lives, and even a partial reverse would have interfered with his comprehensive plan.

In other quarters of the battlefield, the fighting had been unimportant. The Confederate guns, although heavily outnumbered, held their ground gallantly for more than five hours; and when they eventually retired it was from want of ammunition rather than from loss of moral. The wagons which carried their reserve had taken a wrong road, and at the critical moment there were no means of replenishing the supply. But so timid were Fremont's tactics that the blunder passed unpunished.

While the battle on the left was raging fiercely he had contented himself elsewhere with tapping feebly at the enemy's lines. In the center of the field his skirmishers moved against Ewell's batteries but were routed by a bayonet charge; on the right, Milroy and Schenck, the two generals who had withstood Jackson so stubbornly at McDowell, advanced on their own initiative through the woods. They had driven in the Confederate skirmishers, and had induced Ewell to strengthen this portion of his line from his reserve when they were recalled by Fremont, alarmed by Trimble's vigorous attack, to defend the main position.

The Southerners followed slowly. The day was late and Ewell, although his troops were eager to crown their victory, was too cool a soldier to yield to their impatience; and, as at Cedar Creek, where also he had driven back the "Dutch" division, so at Cross Keys he rendered the most loyal support to his commander. Yet he was a dashing fighter, chafing under the restraint of command, and preferring the excitement of the foremost line. "On two occasions in the Valley," says General Taylor, "during the temporary absence of Jackson, he summoned me to his side, and immediately rushed forward amongst the skirmishers, where sharp work was going on. Having refreshed himself, he returned with the hope that "Old Jack would not catch him at it."

Ewell

How thoroughly Jackson trusted his subordinate may be inferred from the fact that, although present on the field, he left Ewell to fight his own battle. The only instructions he gave showed that he had fathomed the temper of Fremont's troops. "Let the Federals," he said, "get very close before your infantry fire; they won't stand long." It was to Ewell's dispositions, his wise use of his reserves, and to Trimble's ready initiative, that Fremont's defeat was due. Beyond sending up a couple of brigades from Port Republic, Jackson gave no orders. His ambition was of too lofty a kind to appropriate the honors which another might fairly claim; and, when once battle had been joined, interference with the plan on which it was being fought did not commend itself to him as sound generalship.

He was not one of those suspicious commanders who believe that no subordinate can act intelligently. If he demanded the strictest compliance with his instructions, he was always content to leave their execution to the judgment of his generals; and with supreme confidence in his own capacity, he was still sensible that his juniors in rank might be just as able. His supervision was constant, but his interference rare; and it was not till some palpable mistake had been committed that he assumed direct control of his divisions or brigades. Nor was any peculiar skill needed to beat back the attack of Fremont.

Nothing proves the Federal leader's want of confidence more clearly than the tale of losses. The Confederate casualties amounted to 288, of which nearly half occurred in Trimble's counterstroke. The Federal reports show 684 killed, wounded, and missing, and of these Trimble's riflemen accounted for nearly

500, one regiment, the 8th New York, being almost annihilated; but such losses, although at one point severe, were altogether insignificant when compared with the total strength; and it was not the troops who were defeated but the general.

Ewell's division bivouacked within sight of the enemy's watch fires, and within hearing of his outposts; and throughout the night the work of removing the wounded, friend and foe alike, went on in the somber woods.

There was work, too, at Port Republic. Jackson, while his men slept, was all activity. His plans were succeeding admirably. From Fremont, cowering on the defensive before inferior numbers, there was little to be feared. It was unlikely that after his repulse he would be found more enterprising on the morrow; a small force would be sufficient to arrest his march until Shields had been crushed; and then, swinging back across the Shenandoah, the soldiers of the Valley would find ample compensation, in the rout of their most powerful foe, for the enforced rapidity of their retreat from Winchester. But to fight two battles in one day, to disappear completely from Fremont's ken, and to recross the rivers before he had time to seize the bridge, were maneuvers of the utmost delicacy, and needed most careful preparation.

It was Jackson's custom, whenever a subordinate was to be entrusted with an independent mission, to explain the part that he was to play in a personal interview. By such means he made certain, first that his instructions were thoroughly understood; and, second, that there was no chance of their purport coming to the knowledge of the enemy. Ewell was first summoned to headquarters, and then Patton, whose brigade, together with that of Trimble, was to have the task of checking Fremont the next day. "I found him at 2 a.m.," says Patton, "actively engaged in making his dispositions for battle.

He immediately proceeded to give me particular instructions as to the management of the men in covering the rear, saying: 'I wish you to throw out all your men, if necessary, as skirmishers, and to make a great show, so as to cause the enemy to think the whole army are behind you. Hold your position as well as you can, then fall back when obliged; take a new position, hold it in the same way, and I will be back to join you in the morning.' "

Colonel Patton reminded him that his brigade was a small one and that the country between Cross Keys and the Shenandoah offered few advantages for protracting such maneuvers. He desired, therefore, to know for how long he would be expected to hold the enemy in check. Jackson replied, "By the blessing of Providence, I hope to be back by 10:00."

These interviews were not the only business which occupied the commanding general. He arranged for the feeding of his troops before their march next day, for the dispositions of his trains and ammunition wagons; and at the rising of the moon, which occurred about midnight, he was seen on the banks of the South River, superintending the construction of a bridge to carry his infantry dryshod across the stream.

June 9

An hour before daybreak he was roused from his short slumbers. Major Imboden, who was in charge of a mule battery, looking for one of the staff, entered by mistake the general's room.

"I opened the door softly, and discovered Jackson lying on his face across the bed, fully dressed, with sword, sash, and boots all on. The low-burnt tallow candle on the table shed a dim light, yet enough by which to recognize him. I endeavored to withdraw without waking him. He turned over, sat upon the bed, and called out, 'Who is that?'

"He checked my apology with, 'That is all right. It's time to be up. I am glad to see you. Were the men all up as you came through camp?'

"'Yes, General, and cooking.'

"'That's right; we move at daybreak. Sit down. I want to talk to you.'

"I had learned never to ask him questions about his plans, for he would never answer such to anyone. I therefore waited for him to speak first. He referred very feelingly to Ashby's death, and spoke of it as an irreparable loss. When he paused I said, 'General, you made a glorious winding-up of your four weeks with yesterday.' He replied, 'Yes, God blessed our army again yesterday, and I hope with His protection and blessing we shall do still better today.'" Then followed instructions as to the use of the mule battery in the forests through which lay Shields' line of advance.

Before 5:00 a.m. the next morning the Stonewall Brigade had assembled in Port Republic and was immediately ordered to advance. On the plain beyond, still dark in the shadow of the mountains, where the cavalry formed the outposts, the fire of the pickets, which had been incessant throughout the night, was increasing in intensity. The Federals were making ready for battle.

Winder had with him four regiments, about 1,200 strong, and two batteries. In rear came Taylor with his Louisianians; and Jackson, leaving Major Dabney to superintend the passage of the river, rode with the leading brigade. The enemy's pickets were encountered about a mile and a half down the river, beyond a strip of woods, on either side of the Luray road. They were quickly driven in, and the Federal position became revealed. From the foot-hills of the Blue Ridge, clothed to their crests with undergrowth and timber, the plain, over a mile in breadth, extended to the Shenandoah. The ground was terraced; the upper level, immediately beneath the mountain, was densely wooded, and fifty or sixty feet above the open fields round the Lewis House.

Shields Comes Up

Here was the hostile front. The Federal force was composed of two brigades of infantry and sixteen guns, not more than 4,000 all told, for Shields, with the remainder of the division, was still far in rear. The right rested on the river; the left on a ravine of the upper level, through which a shallow stream flowed down from the heights above. On the northern shoulder of this ravine was established a battery of seven guns, sweeping every yard of the ground beneath, and a country road, which led directly to the Shenandoah, running between stiff banks and strongly fenced, was lined with riflemen. Part of the artillery was on the plain, near the Lewis House, with a section near the river; on the hillside, beyond the seven guns, two regiments were concealed within the forest, and in rear of the battery was a third.

The position was strong, and the men who held it were of different caliber from Blenker's Germans, and the leaders of stauncher stuff than Fremont. Six of the seven battalions had fought at Kernstown. Tyler, who on that day had seen the Confederates retreat before him, was in command; and neither general nor soldiers had reason to dread the name of Stonewall Jackson. In the sturdy battalions of Ohio and West Virginia the Stonewall Brigade were face-to-face with foemen worthy of their steel; and when Jackson, anxious to get back to Fremont, ordered Winder to attack, he set him a formidable task.

It was first necessary to dislodge the hostile guns. Winder's two batteries were insufficient for the work, and two of his four regiments were ordered into the woods on the terrace, in order to outflank the battery beyond the stream. This detachment, moving with difficulty through the thickets, found a stronger force of infantry within the forest; the guns opened with grape at a range of one hundred yards, and the Confederates, threatened on either flank, fell back in some confusion.

The remainder of Winder's line had meanwhile met with a decided check. The enemy along the hollow road was strongly posted. Both guns and skirmishers were hidden by the embankment; and as the mists of the morning cleared away, and the sun, rising in splendor above the mountains, flooded the valley with light, a long line of hostile infantry, with colors flying and gleaming arms, was seen advancing steadily into battle.

The Federal Commander, observing his opportunity, had, with rare good judgment, determined on a counterstroke. The Louisiana brigade was moving up in support of Winder, but it was still distant. The two regiments which supported the Confederate batteries were suffering from the heavy artillery fire, and the skirmishers were already falling back.

"Below," says General Taylor, "Ewell was hurrying his men over the bridge; but it looked as if we should be doubled up on him ere he could cross and develop much strength. Jackson was on the road, a little in advance of his line, where the fire was hottest, with the reins on his horse's neck. Summoning a young officer from his staff, be pointed up the mountain. The head of my approaching column was turned short up the slope, and within the forest came speedily to a path which came upon the gorge opposite the battery.

But, as Taylor's regiments disappeared within the forest, Winder's brigade was left for the moment isolated, bearing up with difficulty against overwhelming numbers. Ewell's division had found great difficulty in crossing the South River. The bridge, a construction of planks laid on the running gear of wagons, had proved unserviceable. At the deepest part there was a step of two feet between two axletrees of different height; and the boards of the higher stage, except one, had broken from their fastenings. As the men passed over, several were thrown from their treacherous platform into the rushing stream, until at length they refused to trust themselves except to the center plank. The column of fours was thus reduced to single file; men, guns, and wagons were huddled in confusion on the river banks; and the officers present neglected to secure the footway, and refused, despite the order of Major Dabney, to force their men through the breast-high ford.

So, while his subordinates were trifling with the time, which, if Fremont was to be defeated as well as Shields, was of such extreme importance, Jackson saw his old brigade assailed by superior numbers in front and flank. The Federals, matching the rifles of the Confederate marksmen with weapons no less deadly, crossed over the road and bore down upon the guns. The 7th Louisiana, the rear regiment of Taylor's column, was hastily called up, and dashed forward in a vain attempt to stem the tide.

The Battle

A most determined and stubborn conflict now took place, and, as at Kernstown, at the closest range. The Ohio troops repelled every effort to drive them back.

Winder's line was thin. Every man was engaged in the firing line. The flanks were scourged by bursting shells. The deadly fire from the road held back the front. Men and officers were falling fast. The stream of wounded was creeping to the rear; and after thirty minutes of fierce fighting, the wavering line of the Confederates, breaking in disorder, fell back upon the guns. The artillery, firing a final salvo at a range of two hundred yards, was ordered to limber up. One gun alone, standing solitary between the opposing lines, essayed to cover the retreat; but the enemy was within a hundred yards, men and horses were shot down; despite a shower of grape, which rent great gaps in the crowded ranks, the long blue wave swept on, and leaving the captured piece in rear, advanced in triumph across the fields.

In vain two of Ewell's battalions, hurrying forward to the sound of battle, were thrown against the flank of the attack. For an instant the Federal left recoiled, and then, springing forward with still fiercer energy, dashed back their new antagonists as they had done the rest. In vain Jackson, galloping to the front, spurred his horse into the tumult, and called upon his men to rally. Winder's line, for the time being at least, had lost all strength and order; and although another regiment had now come up, the enemy's fire was still so heavy that it was impossible to reform the defeated troops, and two fresh Federal regiments were now advancing to strengthen the attack. Tyler had ordered his left wing to reinforce the center and it seemed that the Confederates would be defeated piecemeal. But at this moment the lines of the assailant came to a sudden halt; and along the slopes of the Blue Ridge a heavy crash of musketry, the rapid discharges of the guns, and the charging yell of the Southern infantry, told of a renewed attack upon the battery on the mountain side.

The Louisianians had come up in the very nick of time. Pursuing his march by the forest path, Taylor had heard the sounds of battle pass beyond his flank, and the cheers of the Federals proved that Winder was hard pressed. Rapidly deploying on his advanced guard, which, led by Colonel Kelley, of the 8th Louisiana, was already in line, he led his companies across the ravine. Down the broken slopes, covered with great boulders and scattered trees, the men slipped and stumbled, and then, splashing through the stream, swarmed up the face of the bank on which the Federal artillery was in action. Breaking through the undergrowth they threw themselves on the guns.

> **By Jackson's brilliant maneuvers, McDowell had been lured westward at the very moment he was about to join McClellan.**

The attention of the enemy had been fixed upon the fight that raged over the plain below, and the thick timber and heavy smoke concealed the approach of Taylor's regiments. The surprise, however, was a failure. The trails were swung round in the new direction, the canister crashed through the laurels, the supporting infantry rushed forward, and the Southerners were driven back. Again, as reinforcements crowded over the ravine, they returned to the charge, and with bayonet and rammer the fight surged to and fro within the battery.

For the second time the Federals cleared their front; but some of the Louisiana companies, clambering up the mountain to the right, appeared upon their flank, and once more the stormers, rallying in the hollow, rushed forward with the bayonet. The battery was carried, one gun alone escaping, and the Federal commander saw the key of his position abandoned to the enemy.

Not a moment was to be lost. The bank was nearly a mile in rear of his right and center, and commanded his line of retreat at effective range. Sending his reserves to retake the battery, he directed his attacking line, already pressing heavily on Winder, to fall back at once. But it was even then too late. The rest of Ewell's division had reached the field. One of his brigades had been ordered to sustain the Lousianians; and across the plain a long column of infantry and artillery was hurrying northwards from Port Republic.

The Stonewall Brigade, relieved of the pressure in front, had already rallied; and when Tyler's reserves, with their backs to the river, advanced to retake the battery, Jackson's artillery was once more moving forward. The guns captured by Taylor were turned against the Federals— Ewell, it is said, indulging to the full his passion for hot work, serving as a gunner—and within a short space of time Tyler was in full retreat, and the Confederate cavalry were thundering on his traces.

It was 10:30. For nearly five hours the Federals had held their ground, and two of Jackson's best brigades had been severely handled. Even if Trimble and Patton had been successful in holding Fremont back, the Valley soldiers were in no condition for a rapid march and a vigorous attack, and their commander had long since recognized that he must rest content with a single victory.

Before 9:00, about the time of Winder's repulse, finding the resistance of the enemy more formidable than be had anticipated, he had recalled his brigades from the opposite bank of the Shenandoah, and had ordered them to burn the bridge. Trimble and Patton abandoned the battlefield of the previous day, and fell back to Port Republic. Hardly a shot was fired during their retreat, and when they took up their march only a single Federal battery had been seen. Fremont's advance was cautious in the extreme. He was actually aware that Shields had two brigades beyond the river, for a scout had reached him, and from the ground about Mill Creek the sound of Tyler's battle could be plainly heard. But he could get no direct information of what was passing.

The crest of the Massanuttens, although the sun shone bright on the cliffs below, was shrouded in haze, completely forbidding all observation; and it was not till near noon, after a march of seven miles, which began at dawn and was practically unopposed, that Fremont reached the Shenandoah. There, in the charred and smoking timbers of the bridge, the groups of Federal prisoners on the plain, the Confederates gathering the wounded, and the faint rattle of musketry far down the Luray Valley, he saw the result of his timidity.

Massing his batteries on the western bluffs, and turning his guns in impotent wrath upon the plain, he drove the ambulances and their escort from the field. But the Confederate dead and wounded had already been removed, and the only effect of his spiteful salvoes was that his suffering comrades lay under a drenching rain until he retired to Harrisonburg. By that time many, whom their enemies would have rescued, had perished miserably, and "not a few of the dead, with some perchance of the mangled living, were partially devoured by swine before their burial."

The pursuit of Tyler was pressed for nine miles down the river. The Ohio regiments, dispersed at first by the Confederate artillery, gathered gradually together, and held the cavalry in check. Near Conrad's Store, where Shields, marching in desperate haste to the sound of the cannonade, had put his two remaining brigades in position across the road, the chase was stayed. The Federal commander admits that he was only just in time. Jackson's horsemen, he says, were enveloping the column; a crowd of fugitives was rushing to the rear, and his own cavalry had dispersed.

The Confederate army, of which some of the brigades and nearly the whole artillery had been halted far in rear, was now withdrawn; but, compelled to move by circuitous paths in order to avoid the fire of Fremont's batteries, it was after midnight before the whole had assembled in Brown's Gap. More than one of the regiments had marched over twenty miles and had been heavily engaged.

Port Republic was the battle most costly to the Army of the Valley during the whole campaign. Out of 5,900 Confederates engaged 804 were disabled. The Federal losses were heavier. The killed, wounded, and missing (including 450 captured) amounted to 1,001, or one-fourth of Tyler's strength.

The success which the Confederates had achieved was undoubtedly important. The Valley army, posted in Brown's Gap, was now in direct communication with Richmond. Not only had its pursuers been roughly checked, but the sudden and unexpected counterstroke, delivered by an enemy whom they believed to be in full flight had surprised Lincoln and Stanton as effectively as Shields and Fremont.

On June 6, the day Jackson halted near Port Republic, McCall's division of McDowell's Army Corps, which had been left at Fredericksburg, had been sent to the Peninsula by water; and two days later McDowell himself, with the remainder of his force, was directed to join McClellan as speedily as possible overland. Fremont, on the same date, was instructed to halt at Harrisonburg, and Shields to march to Fredericksburg. But before Stanton's dispatches reached their destination both Fremont and Shields had been defeated, and the plans of the Northern Cabinet were once more upset.

Instead of moving at once on Fredericksburg, and in spite of McDowell's remonstrances, Shields was detained at Luray, and Ricketts, who had succeeded Ord, at Front Royal; while Fremont, deeming himself too much exposed at Harrisonburg, fell back to Mount Jackson. It was not till June 20 that Ricketts and Shields were permitted to leave the Valley, ten days after the order had been issued for McDowell to move on Richmond. For that space of time, then, his departure was delayed; and there was worse to come. The great strategist at Richmond had not yet done with Lincoln. There was still more profit to be derived from the situation; and from the subsidiary operations in the Valley we may now turn to the main armies.

By Jackson's brilliant maneuvers McDowell had been lured westward at the very moment he was about to join McClellan. The gap between the two Federal armies had been widened from five to fifteen marches, while Jackson at Brown's Gap was no more than nine marches distant from Richmond. McClellan, moreover, had been paralyzed by the vigor of Jackson's blows.

On May 16, as already related, he had reached White House on the Pamunkey, twenty miles from the Confederate capital. Ten miles south, and directly across his path, flowed the Chickahominy, a formidable obstacle to the march of a large army.

To subscribe, send an email to:
thestainlessbanner@gmail.com
or visit our website at www.thestainlessbanner.com

Subscription is free.

The Stainless Banner

An e-zine dedicated to the armies of the Confederacy

Volume 3, Issue 6
June/July 2012

A READY DEFENSE
STUART AT GETTYSBURG

After Robert E. Lee's death in 1870, the leadership of the Army of Northern Virginia, including Lee's aides, published countless articles in newspapers and other periodicals in an effort to deflect responsibility for the defeat at Gettysburg from Lee unto others. One of their favorite scapegoats was Jeb Stuart. Lee's aides were predominantly responsible for advancing the theory that Stuart's absence from the main body during the critical days before the battle blinded Lee to the movements of the Army of the Potomac and forced him into battle before he was ready. This theory, now widely promulgated throughout the Gettysburg's historiography, is simply not true. A careful examination of Stuart's orders reveals a very different story.

On June 22nd, Stuart met with both Lee and General James Longstreet and advanced his plan to pass by the rear of the enemy in order to destroy a large portion of the Army of the Potomac's communications and transportation. Lee told Stuart that he would think about it.

Once the meeting broke up, Lee sent a dispatch to General Richard Ewell waiting at Hagerstown, Maryland, and ordered the Second Corps to proceed into Pennsylvania and, if possible, capture Harrisburg. Whether or not the rest of the army would follow depended "upon the quantity of supplies obtained in that country… There may be enough for your command, but not for the others. Every

What's Inside:

Stuart's Orders	17
Official Reports:	
Lee's Outline	20
Lee's Official	27
Jeb Stuart	42
Charles Marshall's Accuses	44
Mosby For the Defense:	
Against Marshall	63
Against Longstreet	67

exertion should, therefore, be made to locate and secure them. Beef we can drive with us, but bread, we cannot carry, and must secure it in the country."[1]

Sometime during the afternoon, Lee made up his mind about Stuart's plan because he sent another dispatch to Ewell notifying him that Stuart would be "marching with three brigades across the Potomac, and (would) place himself on your right and be in communication with you, keep you advised of the movements of the enemy, and assist in collecting supplies for the army."[2] Since there would be no enemy in front of Ewell, once Stuart reached the Second Corps, the cavalry would have nothing to do but forage for supplies.

That same night, Lee sent orders giving Stuart permission to ride around the Union army. "If you find that he (Hooker) is moving northward and that two brigades can guard the Blue Ridge and take care of your rear, you can move with the other three into Maryland, and take position on General Ewell's right, place yourself in communication with him, guard his flanks, keep him informed of the enemy's movement, and collect all the supplies you can for the use of the army."[3]

Lee's orders were conditional. The first condition depended on whether or not Stuart believed two brigades could provide a sufficient force to guard the Blue Ridge and continue the screen that had so frustrated the Federals. If that force was sufficient, Stuart could then take three brigades of cavalry, move into Maryland, and join Ewell's march. But the movement to Ewell was conditioned on whether or not Stuart found the enemy moving northward. If the Federals were not moving toward the Potomac, then Stuart was not permitted to pass by the rear of the enemy.

Nowhere in the orders was Stuart told to communicate with Lee. He was directed to communicate with Ewell and to keep Ewell informed of the enemy's movement.

Lee sent the orders first to Longstreet since Longstreet needed to determine if he could spare Stuart from his front. Longstreet forwarded the orders to Stuart along with a letter urging Stuart to leave via the Hopewell Gap and pass by the rear of the enemy. Longstreet was concerned that if Stuart crossed the Potomac by the army's rear, he would disclose Confederate plans to the Federals. "You had better not leave us,

> *"You had better not leave us, therefore, unless you take the proposed route in the rear of the enemy."*

---

[1] Robert E. Lee, General. "Dispatch to General Richard S. Ewell, June 22, 1863." *The War of The Rebellion: a Compilation of the Official Records of the Union and Confederate Armies.* Volume 27, Part III, 914.

[2] Robert E. Lee, General. "Dispatch to General Richard S. Ewell, June 22, 1863." *The War of The Rebellion: a Compilation of the Official Records of the Union and Confederate Armies.* Volume 27, Part III, 915.

[3] Robert E. Lee, General. "Dispatch to General J.E.B. Stuart, June 22, 1863." *The War of The Rebellion: a Compilation of the Official Records of the Union and Confederate Armies.* Volume 27, Part III, 913.

therefore, unless you can take the proposed route in the rear of the enemy."[4]

Longstreet gave Stuart specific directions. The terminology to pass by the rear of the enemy only meant one thing. Stuart was to cross the Bull Run Mountains, go around the Union army, and place his command with Ewell. Longstreet mentioned the Hopewell Gap, which is a gap in that mountain chain. Since the Hopewell Gap is not mentioned in Lee's orders, it must have been spoken about in the morning meeting.

What is not stated is this: was the Hopewell Gap the only route Stuart could take? Or could it be, since Lee had not mentioned a specific gap or route, he left the decision on where to pass through the Union columns to Stuart's discretion.

In the letter's postscript, Longstreet warned Stuart that if he should come into the Valley and cross the Potomac west of the Blue Ridge, it would disclose the army's plan to move into Pennsylvania. Longstreet admonished, "You have better not leave us, therefore, unless you take the proposed route in the rear of the enemy."[5]

Longstreet notified Lee that he had "forwarded (Lee's) letter to General Stuart with the suggestion that he pass by the enemy's rear if he thinks that he may get through."[6]

The next day, Stuart received another set of orders because Lee directed his aide, Colonel Charles Marshall, to "repeat it" (the June 22nd order).[7] The second set of orders bear no resemblance to the first set. Whereas in the first set Stuart was ordered to pass by the rear of the enemy only if the Union army was marching north, in the second set that condition changed. Now Stuart could leave two brigades to watch the Federals and withdraw with the three others if Hooker's army remained inactive.

Withdraw where? The orders do not say.

Another condition was introduced. If Stuart found the enemy not moving northward, he was ordered to withdraw behind the Blue Ridge on the evening of June 24th, cross the Potomac at Shepherdstown on June 25th, and then move over to Fredericktown.

Is not moving northward just another way to say inactive? No, not within the context of the orders. The orders read: "If General Hooker's army remains inactive, you can leave two brigades to watch him and withdraw with the three others, but should he appear not to be moving northward..."[8] These were two very different conditions which had two distinct courses of action.

This phrase "not moving northward" could have meant that Richmond's fears

---

[4] James Longstreet, Lieutenant General. "Dispatch to General J.E.B. Stuart, June 22, 1863. *The War of The Rebellion: a Compilation of the Official Records of the Union and Confederate Armies.* Volume 27, Part III, 915.

[5] Ibid.

[6] James Longstreet, Lieutenant General. "Dispatch to General Lee, June 22, 1863." *The War of The Rebellion: a Compilation of the Official Records of the Union and Confederate Armies.* Volume 27, Part III, 915.

[7] I (Marshall) remember saying to the General that it could hardly be necessary to repeat the order, as General Stuart had had the matter fully explained to himself verbally and my letter had been very full and explicit. I had retained a copy of my letter in General Lee's confidential letter book. General Lee said that he felt anxious about the matter and desired to guard against the possibility of error, and desired me to repeat it which I did, and dispatched the second letter. Charles Marshall, Colonel. *An Aide-De-Camp of Lee.* (Boston: Little, Brown, and Company, 1927), 207.

[8] Robert E. Lee, General. "Dispatch to General J.E.B. Stuart, June 23, 1863." *The War of The Rebellion: a Compilation of the Official Records of the Union and Confederate Armies.* Volume 27, Part III, 915.

had been realized, and the Federals had taken advantage of Lee's absence to attack the Confederate capital. This threat was very real. On June 10th, Hooker had written Lincoln "…will it not promote the true interest of the cause for me to march to Richmond at once?"[9]

Richmond politicians may have feared that Hooker would sweep down on them, but it does not appear Lee considered this a possibility. For if Hooker was truly moving on Richmond, it seems improbable that Lee would have ordered Stuart to withdraw to the western side of the Blue Ridge, cross the Potomac, and ride toward Fredericktown. Therefore, the phrase "not moving northward" must have another meaning. Yet, it is not clear to what that meaning is.

The problem with the second set of orders is that they are unclear, illogical, and lack any specificity as to what Stuart should do. If, as Marshall claimed, he was to repeat the first set of orders, he did a very poor job.

So, how does one reconcile the two orders? If you combine the first set of orders, Longstreet's endorsement, and what Lee, Longstreet and Stuart had already discussed, it appeared Lee was giving Stuart another condition under which he could circle around the enemy. Before, he could only pass by the rear of the enemy if the Federals were moving north. Now, he could do so if the Army of the Potomac was inactive (or remained in their camps). The only time Stuart was to come back to the army in the Valley was if the Federals were not marching northward.

The differences between northward, inactive, and not northward were very important because Stuart's actions depended upon them.

> *In both orders Stuart was ordered to communicate with Ewell and not Lee.*

Lee also informed Stuart that it would be best if he did not cross the Potomac until June 25th and then he should do so as quickly as he could. If Stuart was to cross the river with the main body, would Lee have needed to give him a specific date in which to do so?

Stuart was also given discretion to decide whether or not he could pass around the enemy without hindrance. This was the only discretion he was given by Lee. If he thought he could pass by, then he was to do so and damage the enemy all he could.

In both orders Stuart was ordered to communicate with Ewell and not Lee.

One more set of orders would reach Stuart. They arrived late night on the 23rd. These orders actually clarified what Stuart was supposed to do. He was ordered to place himself with Jubal Early, who would be at York, which was also given as the possible concentration point of the army. For the first time Stuart was given a time frame. He must move on "as speedily as possible."[10] In direct contradiction to the orders received earlier in the day, Stuart was warned that the roads from Shepherdstown and

---

[9] Joseph Hooker, Major General. "Dispatch to Abraham Lincoln, June 10, 1863." *The War of The Rebellion: a Compilation of the Official Records of the Union and Confederate Armies.* Volume 27, Part I, 34-35.

[10] Henry B. McClellan, Major. *I Rode With Jeb Stuart.* (New York: Da Capo Press. 1994), 317.

Williamsport were packed with men, artillery, and trains and passing around the enemy would be the quicker route.

On June 24th, Stuart sent orders to Beverly Robertson to keep watch on the Federals and move on the rear and the right of A.P. Hill and Longstreet as they marched into Pennsylvania.

Early on June 25th, Stuart gathered his brigades at Salem Depot. He was aware that there were two divisions of Winfield Scott Hancock's 2nd Corps at Thoroughfare Gap, so he marched to Glasscock Gap near Haymarket. Upon reaching the gap, he found Hancock's other division moving northward. His orders now dictated that he was to pass by the rear of the enemy. Stuart sent Fitz Lee to Gainesville (2-1/2 miles away) to find out if the way was clear. Fitz reported that it was. Stuart then exercised the discretion given to him and passed by the rear of the enemy.

Even though Stuart was not ordered to communicate with Lee, he sent at least two messages, once on the 25th and again on the 27th[11] alerting Lee that the Federals were on the march north.

A Tale of Two Reports

On July 31st, Lee sent his first report of the campaign to General Cooper, Adjutant and Inspector General of the Confederate Army. It was not an official report, but an "outline"[12] or a "general description"[13] of the events. The report, written before Lee received campaign reports from Stuart or his three corps commanders, was immediately leaked to the press.

The outline was very critical of Stuart. It stated that Stuart's orders directed him to move into Maryland, cross the Potomac east or west of the Blue Ridge, as, in his judgment should be best, and take position on the right of the column as it advanced."[14] In the context of the report, the right of the column meant Longstreet and Hill as they marched through the Valley and into Maryland. But as we have seen, Stuart's orders sent him to Ewell who was advancing through Pennsylvania.

In describing Stuart's movements, the report declared that Stuart had been left behind in Virginia in order to follow the movements of the Federal army after Hill and Longstreet crossed the Potomac. But in his efforts to impede Hooker's progress, Stuart advanced as far east as Fairfax Courthouse. When he could no longer delay the enemy, he crossed the river at Seneca and marched through Westminster, arriving at Carlisle after Ewell was summoned to Gettysburg.

Unfortunately, Stuart chose a route that allowed the Union army to get between his brigades and the main army, which prevented him from sending word to Lee that the Federals had crossed the Potomac. In the next sentence, the report linked this lack of communication as the reason the march to Gettysburg had been conducted more slowly than it would have been conducted if the whereabouts of the Union army had been known.

---

[11] J.B. Jones. *A Rebel War Clerk's Diary at the Confederate States Capital.* (Philadelphia: J.B. Lippincott & Co. 1866), 366.
[12] Robert E. Lee, General. "Official Report of the Pennsylvania Campaign, July 31, 1863." *The War of The Rebellion: a Compilation of the Official Records of the Union and Confederate Armies.* Volume 27, Part II, 306.

[13] Ibid., 308
[14] Ibid., 307

In none of the three orders Stuart received was he ordered to guard the gaps, impede or delay the enemy, report the enemy's movements to Lee, or to watch Hooker in Virginia while helping Ewell forage in Pennsylvania, yet Lee's outline accused him of these very failures.

The outline's inaccuracies became all that the soldiers, politicians, and citizens knew about Stuart's actions until Lee sent his official report to the War Department on January 20, 1864, which corrected most of the misstatements made in the first report about Stuart's actions.

In this official report, Stuart was to pass by the rear of the enemy "as soon as he should perceive the enemy (was) moving northward."[15] Though the report still claimed Stuart was ordered to delay the enemy[16] in crossing the river, the report was very clear that Stuart had the discretion to enter Maryland either east or west of the Blue Ridge and "place himself on the right of General Ewell,"[17] as he moved on Harrisburg. More importantly, the statement that censured Stuart for placing himself east of the Federal army was removed.

The reason for the slow pace of the army's march was also revised. No longer was it due to the lack of intelligence of the movement of the enemy. Now, the army marched at a leisurely pace due to the "inclement weather"[18] and "with a view to the comfort of the troops."[19] However, the absence of the cavalry remained the reason Lee had to concentrate his army east of the mountains.[20]

The most quoted portion of report read that the "movements of the army preceding the battle of Gettysburg had been much embarrassed by the absence of cavalry."[21] But the next few lines addressed the fact that Lee had to send for Robertson and the two brigades left behind by Stuart to "rejoin the army without delay."[22]

In the same paragraph that censured Robertson for not marching with the army, Stuart's absence was thoroughly explained. "In the exercise of the discretion given when Longstreet and Hill marched into Maryland, General Stuart determined to pass around the rear of the Federal Army with three brigades and cross the Potomac, between it (the Federal Army) and Washington, believing that he would be able, by that route, to place himself on our right flank in time to keep us properly advised of the enemy's movements."[23] Again, in the context of the report, the right flank was Longstreet and Hill.

The report expressed the expectation that Stuart would send Lee word when

> *In none of the three orders Stuart received was he ordered to guard the gaps, impede or delay the enemy, report the enemy's movements to Lee, or to watch Hooker in Virginia while helping Ewell forage in Pennsylvania, yet Lee's outline accused him of these very failures.*

---

[15] Lee. "Official Campaign Report, January 20, 1864," 316.
[16] Ibid.
[17] Ibid.
[18] Ibid., 317
[19] Ibid.
[20] Ibid., 316.
[21] Ibid., 321.
[22] Ibid.
[23] Ibid.

the Federal army crossed the Potomac. When no message was received, "it was inferred that the enemy had not yet left Virginia."[24] This inference seems almost impossible to believe. According to a June 19th dispatch Lee sent Ewell, he hoped Ewell's presence north of the Potomac would provide Hooker with enough incentive to cross the river. In the June 22nd orders Lee sent to Stuart, he worried that Hooker would steal a march on him and get over the Potomac before Longstreet and Hill could march to the support of Ewell. On June 23rd, Lee sent a dispatch to Jefferson Davis alerting the president to the fact that the Federals were preparing to cross the Potomac and had laid down a pontoon bridge at Edward's Ferry. Furthermore, if Stuart was ordered to pass by the rear of the enemy if he found the enemy moving northward, then the fact that Lee had received no messages from Stuart should have reinforced the fact that Hooker was following the Army of Northern Virginia across the Potomac. The report does not make mention of the two messages Stuart sent.

By leaving in the report the belief that Stuart was to inform Lee when the Federal army had crossed the Potomac, the report severely damaged Stuart's reputation. For the report never explained how Stuart was to accomplish this almost impossible feat. If Stuart had not been delayed by Hancock's march, he would have crossed the Potomac by the evening of the 25th. This would have placed him in front of the Union army with his command moving further away from the Potomac as he rode north to Ewell. There was no way Stuart would have been in a position to know when the Army of the Potomac crossed the river. He left that task for Robertson to do.

When the controversy erupted, the generals and aides who wrote articles explaining the loss used both the outline and the official report to bolster their arguments on what Stuart did or did not do after he left the main body. They filled in the reports' vagueness and blaring contradictions with their own interpretations and insights. Stuart's orders no longer mattered. The reports became the seminal source when discussing Stuart's actions in the Pennsylvania campaign.

> *Not once did Longstreet take responsibility for the orders he forwarded to Stuart on June 22nd or the letter of endorsement he sent along with those orders. Nor did he bring up the fact that Stuart was ordered to join Ewell on the Susquehanna and not join the main body as it marched through Maryland.*

Longstreet Goes Public with the Causes of Defeat

In 1866, William Swinton published *Campaigns of the Army of the Potomac*. Swinton wrote Lee and asked if Lee would contribute to the chapter on Gettysburg. Lee directed Swinton to his January 1864 report. If Swinton had any questions about the battle, he would find the answers there.

---

[24] Ibid., 316

Longstreet was a different story however. Since the army's return from Pennsylvania, he had chafed under the spoken and unspoken accusations that he somehow bore the lion's share of responsibility for the loss. He believed the accusations were not only unfair but were created to shield Lee from blame. When Swinton asked him to provide insight into the Army of Northern Virginia's strategy and movements during the campaign, Longstreet eagerly agreed. It was his first opportunity to publicly defend himself, and he used Swinton's book to deflect blame away from himself and onto the two people he believed most culpable for the loss: Lee and Stuart.

When Swinton asked Longstreet why Stuart was not with the army as it marched into Pennsylvania, Longstreet's explanation sounded very much like Lee's July 31st report. Longstreet actually lifted entire phrases and sentences from the report and used them to indict Stuart for not communicating with Lee about the movements of the Federal army.

Longstreet explained to Swinton that after Stuart passed into Maryland, he "was to take position on the right of the advancing column."[25] In the context of the book, the advancing column was Longstreet and Hill. If Stuart had done this, he would have been in the "proper place to watch the Union cavalry thrown out on the left of the Army of the Potomac."[26] Stuart, however, had advanced too far east in his effort to impede the enemy, and, when he crossed the Potomac, he found the Federal army between his brigades and Lee. This forced him to make a wide sweep through Pennsylvania in order to join the army at Chambersburg.[27] Stuart's absence caused the army's march to be "conducted much more slowly than was usual."[28] Longstreet also attributed Lee's ignorance of the movements of the Federal army to Stuart's absence.[29]

Not once did Longstreet take responsibility for the orders he forwarded to Stuart on June 22nd or the letter of endorsement he sent along with those orders. Nor did he bring up the fact that Stuart was ordered to join Ewell on the Susquehanna and not join the main body as it marched through Maryland. This would not be the last time Longstreet would try to wash his hands of any responsibility for approving Stuart's movement around the rear of the enemy.

Longstreet's cooperation with Swinton became the opening salvo of the Gettysburg controversy. All that was needed for the controversy to explode into a raging battle was for Lee's restraining hand to be removed.

History is Revised

In 1877, the Count of Paris sent a letter to the Southern Historical Society Papers (SHSP) seeking an explanation on why the army had lost at Gettysburg. Dr. J. William Jones, the Society's secretary, mailed copies of the letter to the generals of the Army of Northern Virginia and to Lee's aides with a personal plea to respond. Each reply was then published

---

[25] William Swinton. *Campaigns of the Army of the Potomac.* (New York: Charles B. Richardson 1866), 337-338.
[26] Ibid. (footnote 338)
[27] Ibid.
[28] Ibid., 337-338.
[29] Ibid.

in the Society's papers. Due to the space restrictions of this article, we won't be able to study all the replies received, but we will examine the ones that contributed to the belief that Stuart disobeyed his orders.

General Henry Heth was one of the first to respond. He sent Jones an article he had written for the *Philadelphia Times*. Heth declared that the "failure to crush the Federal Army in Pennsylvania in 1863…can be expressed in five words – *the absence of our cavalry.*"[30]

The basis of Heth's dramatic conclusion was a letter Lee supposedly sent to Stuart before Ewell crossed the Potomac. The letter instructed Stuart to seize the gaps in the Blue Ridge once the First Corps had vacated them and then protect the corps while it crossed the Potomac. Once the corps was across the river, Stuart was to follow on the right flank, keep watch on the enemy, provide Lee with intelligence, and collect supplies.[31]

By stating that the letter was written before Ewell crossed the river, Heth was asserting that on June 17th, Lee had already determined to invade Pennsylvania. Yet, on June 22nd, a dispatch sent to Ewell revealed that Lee still had not made up his mind on whether or not Longstreet and Hill would follow the Second Corps into Pennsylvania. That decision would be based on whether Ewell thought the countryside could sustain the army. Another dispatch sent to Ewell later that day informed Ewell that Lee was sending Stuart to help with the collection of supplies.

Heth was also the author of the narrative of Lee as a giant, stumbling blind through enemy country because the eyes of his army, the cavalry, were joyriding around the Union army. Heth wrote that "every officer who conversed with Lee…well remembers having heard such expressions as these: 'Can you tell me where General Stuart is?' 'Where is my cavalry?' 'Have you any news of the enemy's movements?' 'What is the enemy going to do?' 'If the enemy does not find us, we must try and find him, in the absence of our cavalry, as best we can.'"[32]

Is it feasible that Lee, who originated the entire campaign in order to transfer hostilities north of the Potomac, watched as Hooker followed him from the Rappahannock to the Potomac, knew Hooker had put down pontoons at Edwards Ferry, received couriers from Stuart alerting him that the Union army was moving north, now lurched from general to general begging them for news about the enemy and his cavalry? Is that really the portrait Heth and the rest of the generals wanted to paint of Lee? It does not matter, for it is the portrait that remains.

At the end of the article, Heth declared that his opinion regarding Stuart's failure was shared by "all the officers of the Army of Northern Virginia."[33]

Lee's adjutant, Walter Taylor, wrote that whereas discretion had been given to Stuart on where he would cross the river, once across, "he was to connect at once with General Lee, keep on his flank, and

---

[30] Henry Heth, Major General. "Letter from Major General Henry Heth of A.P. Hill's Corps, A.N.V.," *Southern Historical Society Papers, Volume 4*, (1877), 151-160.
[31] Ibid.
[32] Ibid.
[33] Ibid.

advise him of the enemy's movements."[34] Unfortunately for Lee, Stuart got distracted and wasted time pursuing and capturing a train of wagons.[35] Taylor then consulted Lee's 1863 report and declared that by time Stuart turned west to join the army, he found General Hooker interposed between him and General Lee, and so was compelled to make the circuit of the Federal Army.[36]

In an article written for the *Philadelphia Times*, Taylor declared that Stuart was "especially directed to keep the Commanding General informed of the movements of the Federal army."[37] Through the years, Taylor never backed off this claim. When Stuart's defenders questioned Taylor on why Stuart's orders told him to communicate with Ewell and not Lee, or how Stuart could communicate with Lee with the Union army and two mountain ranges between them, Taylor refused to answer. Instead, he just kept insisting that Stuart was directed to keep Lee informed of the movements of the enemy.

In his response, Lee's military secretary, General A.L. Long, stated that even before the campaign began, Lee had "carefully considered every contingency that could mar success, except the possibility of tactical blunders of those who had always maintained his confidence by a prompt and intelligent execution of instructions."[38] Stuart's blunder was the fatal one which lost the battle. Long also upheld Heth's claim that, once across the river, Lee was a blinded giant. Stuart's separation from the army at such a critical time, forced Lee to grope his way through Pennsylvania in the dark. Without proper warning to the whereabouts of the enemy, Lee was forced into battle before he was ready.[39]

Long released his biography of Lee in 1886, and in the chapter on Gettysburg turned the narrative on its head. He wrote that while the army was still on the Rappahannock, he was invited into Lee's tent to discuss the movement into Pennsylvania. Lee traced the route the army would take to Chambersburg or Gettysburg. In this meeting, Lee confided to Long that Gettysburg would be the best place for battle. York, on the other hand, was less desirable.[40] What Long did not explain is why Lee wrote Stuart on June 23rd that the army would consolidate at York. Neither did Long explain why Ewell was ordered to capture Harrisburg, and, why, before Longstreet's spy came into headquarters, Lee had sent orders to Longstreet and Hill to join Ewell at Harrisburg. One last glaring inconsistency that Long never addressed was this: If, along the Rappahannock, Lee desired to do battle at Gettysburg, why did he not order Early to hold the town when Early's men passed through it days before the battle began.

---

[34] Walter H. Taylor, Colonel. "Memorandum by Colonel Walter H. Taylor, of General Lee's Staff." *Southern Historical Society Papers*, Volume 4 (1877), 83.
[35] Ibid.
[36] Ibid.
[37] Walter H. Taylor, Colonel. "Second Paper by Colonel Walter H. Taylor of General Lee's Staff." *Southern Historical Society Papers*, Volume 4 (1877), 125-126.
[38] A.L. Long, Brigadier General. "Letters from General A.L. Long, Military Secretary to General R.E. Lee, April 1877," *Southern Historical Society Papers*, Volume 4 (1877), 122.
[39] Ibid.
[40] A.L. Long, Brigadier General. *Memoirs of Robert E. Lee* (London: Sampson Low, Marston, Searle, and Rivington, 1886), 268.

If one reads all the replies to the Count of Paris' questions, it does not take long before the reader discovers a familiar narrative. Southern independence was lost because Stuart, angry over critical press received after the Battle of Brandy Station, disobeyed his order, crossed the Potomac east of the Blue Ridge, and allowed the Union army to get between his command and Lee. With no intelligence coming from his cavalry, Lee lurched blindly through the Pennsylvania countryside. When Longstreet's spy informed Lee that the Union army had crossed the river, Lee hurriedly consolidated his army at Gettysburg, and was forced to wage the battle at a severe disadvantage he could not overcome.

Stuart's Harshest Critic

In 1896, Colonel Charles Marshall, one of Lee's aide-de-camps, gave a speech before the Confederate Veterans Association in Washington D.C in celebration of Lee's birthday. Marshall used the opportunity to "correct the impression that had prevailed to some extent that the movement of the cavalry (at Gettysburg) was made by General Lee's orders..."[41] It was not. That error belonged to Stuart alone.

In the speech, Marshall pursued two contradictory points. The first: Stuart had discretion to cross the river east or west of the Blue Ridge, but if he found the enemy moving northward, he was to place his command on the right of Hill and Longstreet.[42] The second: Stuart was to place his brigades on the right of Ewell in Maryland. Which point Marshall expounded on depended on which set of orders he was interpreting.

Marshall read the entire June 22nd order to the audience. He asserts that while Stuart was instructed to move with three brigades and join Ewell in Maryland, the specific movement to pass by the rear of the enemy originated with Longstreet and not with Lee. This is pure spin on Marshall's part. Longstreet's endorsement is quite clear that the idea for the movement came from Lee. Longstreet wrote Stuart that Lee "speaks of you leaving via Hopewell Gap and passing by the rear of the enemy."[43]

Furthermore, Marshall ignored the fact that Lee took responsibility for the orders that sent Stuart around the rear of the enemy in his official report on the campaign.

In his analysis of the order, Marshall kept insisting that Ewell was in Maryland, west of the Blue Ridge. Except,

> *In his analysis of the order, Marshall kept insisting that Ewell was in Maryland, west of the Blue Ridge. Except, according to the very dispatches Marshall just read, by the time Stuart left Salem Depot, Ewell's divisions were near Chambersburg.*

---

[41] Charles Marshall, Colonel. "Address of Colonel Charles Marshall, January 1896." *Richmond Dispatch*, January 26 and February 2, 1896.

[42] Ibid.
[43] James Longstreet, Lieutenant General. "Dispatch to General J.E.B. Stuart, June 22, 1863. *The War of The Rebellion: a Compilation of the Official Records of the Union and Confederate Armies.* Volume 27, Part III, 915.

according to the very dispatches Marshall just read, by the time Stuart left Salem Depot, Ewell's divisions were near Chambersburg.

Marshall concluded his analysis of the June 22nd orders by telling his audience that even though the orders sent Stuart to Ewell, it was Lee's expectations that Stuart would cross the Potomac west of the Blue Ridge (or in the Valley) and join the army (Longstreet and Hill) in its march to Pennsylvania.[44]

Marshall moved on to the June 23rd orders. His explanation was as convoluted as the order. He executes an 180° turnaround. No longer was it Longstreet who ordered Stuart to pass by the rear of the enemy, but Lee. He justifies Lee's order by saying that even though the order gave Stuart permission to move around the Federal army, Lee still expected Stuart to cross at Shepherdstown.

This contradiction is so egregious that it is beyond belief. Why would Lee order Stuart to pass by the rear of the enemy and still expect him to cross with the army at Shepherdstown? No explanation of how this impossibility was to occur was ever forthcoming from Marshall, Taylor, Long, or any of Stuart's critics.

According to Marshall, the end result of Stuart's movement around the rear of the enemy was that General Lee was left without any information as to the movement of the Union army. Stuart's silence caused Lee to move his army to Gettysburg not to engage the enemy but in order to save his communications with Virginia.

Again, this is not true. On June 25th, the same day Stuart left Salem Depot, Lee wrote Jefferson Davis that he was forced to abandon his communications since he did not have sufficient force to sustain them.[45] Marshall concluded his analysis of the June 23rd orders by declaring that Stuart's absence took away Lee's options to fight the enemy on the ground of his choosing.

> *If, along the Rappahannock, Lee desired to do battle at Gettysburg, why did he not order Early to hold the town when Early's men passed through it days before the battle began.*

One June 24th, Hill's corps crossed the Potomac at Charles Town in "plain view of the Federal signal station on Maryland Heights."[46] That signal station telegraphed to Hooker that the Confederates were on the move. Lee said he was forced to put Hill on the road because Ewell was moving swiftly through Pennsylvania, and Lee did not want him to get beyond support of the rest of the army. Hooker put his army in motion to make sure that it stayed between the Confederates and Washington. This was the reason Stuart found Hancock's corps on the roads at Haymarket. If Lee had just sat still for one more day, Mosby said, Stuart could have been over the Potomac by sunset on June

---

[44] Marshall, "Address of Colonel Charles Marshall, January 1896."

[45] Robert E. Lee, General. "Dispatch to Jefferson Davis, June 25, 1863." *The War of The Rebellion: a Compilation of the Official Records of the Union and Confederate Armies.* Volume 27, Part III, pages 931-932.

[46] John Mosby, Colonel. *Mosby's Memoirs.* (Nashville: J.S. Sanders & Company, 1995), 220.

25th. In his many articles, Mosby asked the most important question of the controversy: How could Stuart be on the Potomac watching Lee's flank and on the Susquehanna watching Ewell's flank at the same time? It was a question that Stuart's critics refused to answer. Instead, they just kept insisting that Stuart's orders did not matter. Stuart should have known that Lee would need him at Gettysburg and, therefore, should not have deserted him.

Marshall closed his speech by saying that "the result of General Stuart's actions was that two armies invaded Pennsylvania in 1863, instead of one. One of those armies had no cavalry; the other had nothing but cavalry. One was commanded by General Lee, the other by General Stuart."[47]

Mosby Goes on the Offense

After the war, John Mosby served as Consul to Hong Kong. When he returned to Washington in the winter of 1886-1887 to settle his accounts, he became aware of an article Longstreet authored charging Stuart with insubordination. Incensed to find Stuart under attack, Mosby went to the National Archives and read all the correspondence, orders, and reports of the Pennsylvania campaign.

He joined the debate, writing articles that were published throughout the county. He did so with one purpose: to defend Stuart against the charges that he had disobeyed orders and left Lee blind. Mosby took on all comers. His defense, based on his research, was spirited, passionate, and, at times, acrimonious.

Mosby thought the narrative that Lee was caught by surprise when he heard that the Federal army had crossed the Potomac was particularly insulting to Lee's military acumen. Lee initiated the campaign to move the hostilities north of the Potomac? Hooker had faithfully mirrored Lee's march from the Rappahannock to the counties just south of the Potomac. Lee knew that Hooker could not allow the Army of Northern Virginia to cross the Potomac and invade the North without following. According to the dispatch he sent Ewell on June 19th, Lee hoped Ewell's presence north of the Potomac would provide Hooker with enough incentive to cross the river, and in the order sent to Stuart on June 22nd, he worried Hooker would steal a march on him and get over the Potomac before he could send Longstreet and Hill to support Ewell.

Mosby believed the major source of criticism against Stuart stemmed from the "idea that Gettysburg was General Lee's objective point."[48] Since Stuart was absent from the first day battle, "he must have, therefore, been in default."[49] Well, if Stuart was in default then so was Lee for Lee was not at Gettysburg on the first day of battle either.

> *What Taylor, Marshall and Talcott refused to admit is no matter where Stuart crossed the Potomac – east or west of the Blue Ridge – "he would not have been with General Lee or anywhere near Gettysburg, but away off on the Susquehanna."*

---

[47]Marshall, "Address of Colonel Charles Marshall, January 1896."

[48]Ibid.
[49]Ibid.

When Lee sent orders to recall his scattered forces, he ordered them to Cashtown, a small town about eight miles from Gettysburg. He also ordered his corps commanders not to bring on a general engagement until the army was together. July 1st found Lee at Greenwood, ten miles west of Gettysburg. But with the army consolidating at Cashtown, Lee was soon on the move. He sent General Imboden a message stating that "my headquarters for the present will be at Cashtown, east of the mountain."[50] As he traveled toward Cashtown, he heard the sound of battle. Not understanding its source, he rode at full speed toward the fighting.

Ewell, on his return from Carlisle, had camped with Rodes' and Early's divisions a few miles north of Gettysburg. On the night before the battle, he received a note from A.P. Hill warning him that Hill's corps would be moving on Gettysburg in the morning. It is a good thing Hill sent that note. Without Ewell's aid, Hill might have found himself pushed back to Cashtown.

Until Heth brought on the battle, Lee had not intended to fight at Gettysburg. Colonel Arthur Fremantle, an observer from Her Majesty's Cold Stream Guards and Longstreet's guest, confirmed this. His diary entry for July 1st states, "I have the best reason for supposing that the fight came off prematurely, and that neither Lee nor Longstreet intended that it should have begun that day. I also think that their plans were deranged by the events of the first"[51]

As evening fell, Lee was compelled to order up the remainder of the army and deliver battle on ground he had not chosen, or fall back to Cashtown, leaving his dead and wounded on the field and giving the enemy the prestige of victory. It was clear that the want of cavalry had nothing to do either with precipitating the battle or losing it. "Lee, Longstreet, and Stuart were all absent on the first day for the same reason because the army had not been ordered to Gettysburg and it was not their duty to be there. They were in their proper places – Hill and Heth were not."[52]

Mosby's assertions did not go unchallenged. Colonel T.M.R. Talcott, an aide-de-camp of Lee, read Mosby's book, *Stuart's Cavalry in the Gettysburg Campaign* and wrote Walter Taylor to find out Taylor's thoughts on Mosby's arguments. Taylor told Talcott that he did not "attach much importance to his (Mosby) statements."[53] Taylor restated that Stuart "was admonished all the while to keep in touch with our main army and to keep General Lee informed as to the movements of the enemy. […] It is not a good defense of General Stuart to say that it was impossible for him to communicate with General Lee when he had himself put himself in a position where it was impossible, although admonished all the while not to do this."[54]

---

[50] Ibid., 237, 238.
[51] Arthur Fremantle, Colonel. *Three Months in the Southern States*. (New York: Little Brown & Company, 1954), 256.
[52] John Mosby, Colonel. *Stuart's Cavalry in the Gettysburg Campaign*. (New York: Moffat, Yard & Company, 1908), 138, 139.
[53] T.M.R. Talcott, Colonel. "Stuart's Calvary in the Gettysburg Campaign." *Southern Historical Society Papers*, Volume 37), 21-37.
[54] Ibid.

Armed with Taylor's letter, Talcott published an article in the SHSP. He declared that there was "nothing in either order to Stuart, or in General Lee's letter to General Ewell of June 22nd, that justified Colonel Mosby's inferences that Stuart was to move to Pennsylvania and join Ewell on the Susquehanna."[55] Nothing at all except Ewell was in Pennsylvania and had been ordered to Harrisburg, which is on the Susquehanna.

Mosby was not going to let Talcott's critique go unanswered. And in typical Mosby fashion, he bluntly told Colonel Taylor to "point out one word in General Lee's letter to Stuart about keeping 'in touch with the main army,' or keeping General Lee 'informed of the movements of the enemy.'"[56] What Taylor, Marshall, and Talcott refused to admit is no matter where Stuart crossed the Potomac – east or west of the Blue Ridge – "he would not have been with General Lee or anywhere near Gettysburg, but away off on the Susquehanna."[57]

Talcott was not ready to let go of his belief that Stuart was not ordered to move into Pennsylvania. He wrote that after reading Mosby's reply he "scrutinized very carefully General Lee's letters to General Stuart on the 22nd and 23rd of June, with a view to see how they should have been construed by General Stuart."[58] His conclusion was simple. Stuart's crossing into Maryland by the rear of the enemy was "required only in case the enemy was moving northward."[59] Therefore, Stuart was "going contrary to them (his orders) when he left Rectortown on the night of the June 25th, in his attempt to pass through Hooker's army while it was still inactive."[60]

> *If as Marshall claimed, Stuart's orders did not reflect Lee's expectations, then why did Lee not craft orders that did reflect his expectations?*

Talcott's conclusion is mind boggling. Hooker's army was not inactive. Stuart could not get through at Haymarket because Hancock's corps was on the road heading north. And his orders specifically told him to pass by the rear of the army if he found Hooker's army moving northward. Furthermore, even if the Federals had been in the camps when Stuart reached Haymarket, he could still pass by the rear of the enemy because the orders received on June 23rd told him to do so.

Conclusion

After Lee's death, a narrative was carefully created and rehearsed in articles and memoirs that Gettysburg and independence were lost because Stuart disobeyed his orders, tried for glory to erase the criticism he received after Brandy Station, allowed the Union army

---

[55] Ibid.
[56] John Mosby, Colonel. "A Defense of the Cavalry Commander." *Richmond Times-Dispatch*, January 30, 1910.
[57] Ibid.
[58] T.M.R. Talcott, Colonel. "A Reply to the Letter of Colonel John S. Mosby, Published in the Richmond, Virginia Times-Dispatch, January 30, 1910." *Southern Historical Society Papers*, Volume 38, pages 197-210.
[59] Ibid.
[60] Ibid.

to get between himself and Lee, which meant that Lee had no idea when the Federal army crossed the Potomac until Longstreet's spy told him, and, finally, had to ride through Pennsylvania to reunite with the army at Gettysburg where Lee always intended to fight.

In a multitude of articles, speeches, and memoirs, Stuart's critics insisted that it did not matter what Stuart's orders said. They knew Lee's expectations were for Stuart to march on the right of Longstreet and Hill as they marched into Pennsylvania. Stuart should have known Lee's expectations as well and done everything in his power to meet them. But this gives rise to another question. If, as Marshall claimed, Stuart's orders did not reflect Lee's expectations, then why did Lee not craft orders that did reflect his expectations? If after sending Stuart three orders to pass by the rear of the enemy, Lee still believed Stuart would cross with the army at Shepherdstown, why did Lee not order him to do so?

Another misdirection that Stuart's critics employed was to insist that Ewell was in Maryland when Stuart began his ride. In his speech to the Confederate veterans, Marshall even went as far to claim that Ewell was in the Valley. No, Ewell was in Pennsylvania. There is no denying this. Ewell's official report put his troops at Chambersburg when Stuart began his ride according to the timetable Lee had set for him.

Perhaps the final verdict on whether or not Stuart obeyed his orders should be left to Lee. In the years following the surrender, he was loath to talk about the war, but his eyes would always light up whenever Stuart was mentioned. He met with General Wade Hampton, who succeeded Stuart as cavalry commander, and left this lasting tribute to Stuart. "General Stuart was my ideal of a soldier. He was always cheerful under all circumstances, and always ready for any work, and he was always reliable."[61]

Believe you can whip the enemy and you have won half the battle.

Jeb Stuart

---

[61]Mark Nesbitt. *Saber and Scapegoat.* (Pennsylvania, Stackpole Books, 1994), 107.

Stuart's Orders

June 22, 1863

General: I have just received your note of 7:45 this morning to General Longstreet. I judge the efforts of the enemy yesterday were to arrest our progress and ascertain our whereabouts. Perhaps he is satisfied. Do you know where he is and what he is doing? I fear he will steal a march on us, and get across the Potomac before we are aware. **If you find that he (Hooker) is moving northward and that two brigades can guard the Blue Ridge and take care of your rear, you can move with the other three into Maryland, and take position on General Ewell's right, place yourself in communication with him, guard his flanks, keep him informed of the enemy's movement, and collect all the supplies you can for the use for the army.** One column of General Ewell's army will probably move toward the Susquehanna by the Emmitsburg route; another by Chambersburg. Accounts from him last night state that there was no enemy west of Frederick. A cavalry force (about 100) guarded the Monocacy Bridge, which was barricaded. You will, of course, take charge of Jenkins' brigade, and give him necessary instructions. All supplies taken in Maryland must be authorized by staff officers for their respective departments -- by no one else. They will be paid for, or receipts for the same given to owners. I will send you a general order on this subject, which I wish to see it strictly complied with."

Robert E. Lee, General. "Dispatch to General J.E.B. Stuart, June 22, 1863." *The War of the Rebellion: a Compilation of the Official Records of the Union and Confederate Armies.* Volume 27, Part III, 913.

June 22, 1863
7:00 P.M.

General Lee has enclosed this letter for you, to be forwarded to you, provided you can be spared from my front, and provided I think that you can move across the Potomac without disclosing our plans. **He speaks of you leaving via Hopewell Gap and passing by the rear of the enemy. If you can get through by that route, I think you will be less likely to indicate what our plans are than if you should cross to our rear.** I forward the letter of instructions with these suggestions.

Please advise me of the condition of affairs before you leave, and order General Hampton -- whom I suppose you will leave here in command -- to report to me at Millwood, either by letter or in person, as may be most agreeable to him.

N.B. – I think that your passage of the Potomac by our rear at the present moment, will in a measure, disclose our plans. You have better not leave us, therefore, unless you can take the proposed route in rear of the enemy.

James Longstreet, Lieutenant General. "Dispatch to General J.E.B. Stuart, June 22, 1863." *The War of the Rebellion: a Compilation of the Official Records of the Union and Confederate Armies.* Volume 27, Part III, 915.

June 23, 1863
5:00 p.m.

    General: Your notes of 9:00 and 10:30 a.m. today have just been received. As regards to the purchase of tobacco for your men, supposing that Confederate money will not be taken, I am willing for your commissaries or quartermasters to purchase this tobacco and let the men get it from them, but I can have nothing seized by the men.

    If General Hooker's army remains inactive, you can leave two brigades to watch him and withdraw with the three others, but should he appear not to be moving northward, I think you had better withdraw this side of the mountain tomorrow night (6/24), cross at Shepardstown the next day, and move over to Fredericktown.

    You will, however, be able to judge whether you can pass around the army without hindrance, doing all the damage you can and cross the river east of the mountains. In either case, after crossing the river, you must move on and feel the right of Ewell's troops, collecting information, provisions, etc.

    Give instructions to the commander of the brigades left behind, to watch the flank and rear of the army, and (in the event of the enemy leaving their front) retire the mountain west of the Shenandoah, leaving sufficient pickets to guard the passes, and bringing everything clean along the Valley, closing upon the rear of the army... I think the sooner you cross into Maryland, after tomorrow, the better.

Robert E. Lee, General. "Dispatch to General J.E.B. Stuart, June 23, 1863." *The War of the Rebellion: a Compilation of the Official Records of the Union and Confederate Armies*. Volume 27, Part III, 915.

Late Night June 23/June 24

    The letter discussed at considerable length the plan of passing around the enemy's rear. It informed General Stuart that General Early (commanding a division in the Second Corps) would move upon York, Pa., and that he was desired to place his cavalry as speedily as possible with that, the advance of Lee's right wing.

    The letter suggested that, as the roads leading northward from Shepherdstown and Williamsport were already encumbered by the infantry, the artillery, and the transportation of the army, the delay which would necessarily occur in passing by these would, perhaps be greater than would ensue if General Stuart passed around the enemy's rear. **The letter further informed him that, if he chose the latter route, General Early would receive instructions to look out for him and endeavor to communicate with him; and York, Pa., was designated as the point in the vicinity of which he was expected to hear from Early, and as the possible (if not probable) point of concentration of the army.** The whole tenor of the letter gave evidence that the commanding general approved of the proposed movement, and thought it might be productive of the best results, while the responsibility of the decision was placed upon General Stuart himself.

Henry B. McClellan, Major. *I Rode With Jeb Stuart*. (New York: Da Capo Press. 1994), 317.

June 24, 1863

General, Your own and General Jones' brigades will cover the front of the Ashby's and Snicker's Gaps, yourself, as senior officer, being in command.

Your object will be to watch the enemy; deceive him as to our designs; and harass his rear if you find he is retiring. Be always on the alert; let nothing escape your observation and miss no opportunity which offers to damage the enemy.

After the enemy has moved beyond your reach, leave sufficient pickets in the mountains, withdraw to the west side of the Shenandoah, place a strong and reliable picket to watch the enemy at Harper's Ferry, cross the Potomac and follow the army, keeping on its right and rear.

As long as the enemy remains in your front in force, unless ordered by General R.E. Lee, Lieutenant General Longstreet, or myself, hold the Gaps with a line of pickets reaching across the Shenandoah by Charlestown to the Potomac.

If, in the contingency mentioned, you withdraw, sweep the Valley clear of what pertains to the army, and cross the Potomac at the different points crossed by it.

You will instruct General Jones from time to time as the movements progress, or events may require, and report anything of importance to Lieutenant General Longstreet, with whose position you will communicate by relays through Charlestown.

I will send instructions for General Jones, which please read. Avail yourself of every means in your power to increase the efficiency of your command, and keep it up to the highest number possible. Particular attention should be paid to shoeing horses and to marching off of the turnpike.

In case of advance of the enemy, you will offer such resistance as will be justifiable to check him and discover his intentions and, if possible, you will prevent him from gaining possessions of the Gaps.

In case of a move by the enemy upon Warrenton, you will counteract it as much as you can, compatible with previous instructions.

You will have with you two batteries of horse artillery.

Very respectfully,
Your obedient servant,
J.E.B. Stuart
Major General, Commanding

P.S. Do not change your present line of pickets until daylight tomorrow morning, unless compelled to do so.

J.E.B. Stuart, Major General. "Dispatch to General Beverly Robertson, June 24, 1863". *The War of the Rebellion: a Compilation of the Official Records of the Union and Confederate Armies*. Volume 27, Part III, 927.

ROBERT E. LEE'S REPORT ON THE PENNSYLVANIA CAMPAIGN, JULY 31, 1863

Headquarters, Army of Northern Virginia
July 31, 1863.

General S. COOPER,
Adjutant and Inspector General, Richmond, Va.

GENERAL: I have the honor to submit the following outline of the recent operations of this army, for the information of the Department:

The position occupied by the enemy opposite Fredericksburg – being one in which he could not be attacked to advantage, it was determined to draw him from it. The execution of this purpose embraced the relief of the Shenandoah Valley from the troops that had occupied the lower part of it during the winter and spring, and, if practicable, the transfer of the scene of hostilities north of the Potomac. It was thought that the corresponding movements on the part of the enemy to which those contemplated by us would probably give rise, might offer a fair opportunity to strike a blow at the army then commanded by General Hooker, and that in any event that army would be compelled to leave Virginia, and, possibly, to draw to its support troops designed to operate against other parts of the country. In this way it was supposed that the enemy's plan of campaign for the summer would be broken up, and part of the season of active operations be consumed in the formation of new combinations, and the preparations that they would require. In addition to these advantages, it was hoped that other valuable results might be attained by military success. Actuated by these and other important considerations that may hereafter be presented, the movement began on June 3. McLaws' division, of Longstreet's corps, left Fredericksburg for Culpeper Courthouse, and Hood's division, which was encamped on the Rapidan, marched to the same place. They were followed on the 4th and 5th by Ewell's corps, leaving that of A.P. Hill to occupy our lines at Fredericksburg.

The march of these troops having been discovered by the enemy on the afternoon of the 5th, on the following day, he crossed a force amounting to about one army corps to the south side of the Rappahannock, on a pontoon bridge laid down near the mouth of Deep Run. General Hill disposed his command to resist their advance, but as they seemed intended for the purpose of observation rather than attack, the movements in progress were not arrested.

The forces of Longstreet and Ewell reached Culpeper Courthouse by the 8th, at which point the cavalry, under General Stuart, was also concentrated.

On the 9th, a large force of Federal cavalry, strongly supported by infantry, crossed the Rappahannock at Beverly and Kelly's Fords, and attacked General Stuart. A severe engagement ensued, continuing from early in the morning until late in the afternoon, when the enemy was forced to recross the river with heavy loss, leaving 400 prisoners, three pieces of artillery, and several colors in our hands.

General Jenkins, with his cavalry brigade, had been ordered to advance to Winchester, to cooperate with the infantry in the proposed expedition into the lower Valley, and at the same time General Imboden was directed with his command to make a demonstration in the direction of Romney, in order to cover the movement against Winchester, and prevent the enemy at that place from being reinforced by the troops on the line of the Baltimore and Ohio Railroad. Both of these officers were in position when General Ewell left Culpeper Courthouse on the 10th. Crossing the Shenandoah near Front Royal, he detached Rodes' division to Berryville, with instructions, after dislodging the force stationed there, to cut off the communication between Winchester and the Potomac. With the divisions of Early and [Edward] Johnson, General Ewell advanced directly upon Winchester, driving the enemy into his works around the town on the 13th. On the same day the troops at Berryville fell back before General Rodes, retreating to Winchester.

On the 14th, General Early stormed the works at the latter place, and the whole army of General Milroy was captured or dispersed. Most of those who attempted to escape were intercepted and made prisoners by General Johnson. Their leader fled to Harpers Ferry with a small party of fugitives.

General Rodes marched from Berryville to Martinsburg, entering the latter place on the 14th, where he took 700 prisoners, five pieces of artillery, and a considerable quantity of stores. These operations cleared the Valley of the enemy, those at Harpers Ferry withdrawing to Maryland Heights. More than 4,000 prisoners, twenty-nine pieces of artillery, 270 wagons and ambulances, with 400 horses, were captured, besides a large amount of military stores. Our loss was small.

On the night that Ewell appeared at Winchester, the Federal troops in front of A.P. Hill at Fredericksburg recrossed the Rappahannock, and the next day disappeared behind the hills of Stafford. The whole army of General Hooker withdrew from the line of the Rappahannock, pursuing the roads near the Potomac, and no favorable opportunity was offered for attack. It seemed to be the purpose of General Hooker to take a position which would enable him to cover the approaches to Washington City. With a view to draw him farther from his base, and at the same time to cover the march of A.P. Hill, who, in accordance with instructions, left Fredericksburg for the Valley as soon as the enemy withdrew from his front, Longstreet moved from Culpeper Court-

House on the 15th, and, advancing along the east side of the Blue Ridge, occupied Ashby's and Snicker's Gaps. His force had been augmented, while at Culpeper, by General Pickett, with three brigades of his division. The cavalry under General Stuart was thrown out in front of Longstreet, to watch the enemy, now reported to be moving into Loudoun.

On the 17th, his cavalry encountered two brigades of ours under General Stuart, near Aldie, and was driven back with loss. The next day the engagement was renewed, the Federal cavalry being strongly supported by infantry, and General Stuart was, in turn, compelled to retire. The enemy advanced as far as Upperville, and then fell back. In these engagements, General Stuart took about 400 prisoners and a considerable number of horses and arms.

In the meantime, a part of General Ewell's corps had entered Maryland, and the rest was about to follow. General Jenkins, with his cavalry, who accompanied General Ewell, penetrated Pennsylvania as far as Chambersburg. As these demonstrations did not have the effect of causing the Federal Army to leave Virginia, and as it did not seem disposed to advance upon the position held by Longstreet, the latter was withdrawn to the west side of the Shenandoah, General Hill having already reached the Valley. General Stuart was left to guard the passes of the mountains and observe the movements of the enemy, whom he was instructed to harass and impede as much as possible, should he attempt to cross the Potomac. In that event, General Stuart was directed to move into Maryland, crossing the Potomac east or west of the Blue Ridge, as, in his judgment, should be best, and take position on the right of our column as it advanced.

By the 24th, the progress of Ewell rendered it necessary that the rest of the army should be within supporting distance, and Longstreet and Hill marched to the Potomac. The former crossed at Williamsport and the latter at Shepherdstown. The columns reunited at Hagerstown, and advanced thence into Pennsylvania, camping near Chambersburg on the 27th.

> *Early's division was detached for this purpose and proceeded as far east as York, while the remainder of the corps proceeded to Carlisle.*

No report had been received that the Federal Army had crossed the Potomac, and the absence of the cavalry rendered it impossible to obtain accurate information. In order, however, to retain it on the east side of the mountains, after it should enter Maryland, and thus leave open our communication with the Potomac through Hagerstown and Williamsport, General Ewell had been instructed to send a division eastward from Chambersburg to cross the South Mountain. Early's division was detached for this purpose and proceeded as far east as York, while the remainder of the corps proceeded to Carlisle.

General Imboden, in pursuance of the instructions previously referred to, had been actively engaged on the left of General Ewell during the progress of the latter into Maryland. He had driven off the forces guarding the Baltimore and

Ohio Railroad, destroying all the important bridges on that route from Cumberland to Martinsburg, and seriously damaged the Chesapeake and Ohio Canal. He subsequently took position at Hancock, and, after the arrival of Longstreet and Hill at Chambersburg, was directed to march by way of McConnellsburg to that place.

Preparations were now made to advance upon Harrisburg; but, on the night of the 28th, information was received from a scout that the Federal Army, having crossed the Potomac, was advancing northward, and that the head of the column had reached the South Mountain. As our communications with the Potomac were thus menaced, it was resolved to prevent his farther progress in that direction by concentrating our army on the east side of the mountains. Accordingly, Longstreet and Hill were directed to proceed from Chambersburg to Gettysburg, to which point General Ewell was also instructed to march from Carlisle.

General Stuart continued to follow the movements of the Federal Army south of the Potomac, after our own had entered Maryland, and, in his efforts to impede its progress, advanced as far eastward as Fairfax Courthouse. Finding himself unable to delay the enemy materially, he crossed the river at Seneca, and marched through Westminster to Carlisle, where he arrived after General Ewell had left for Gettysburg. By the route he pursued, the Federal Army was interposed between his command and our main body, preventing any communication with him until his arrival at Carlisle. The march toward Gettysburg was conducted more slowly than it would have been had the movements of the Federal Army been known.

The leading division of Hill met the enemy in advance of Gettysburg on the morning of July 1. Driving back these troops to within a short distance of the town, he there encountered a larger force, with which two of his divisions became engaged. Ewell, coming up with two of his divisions by the Heidlersburg road, joined in the engagement. The enemy was driven through Gettysburg with heavy loss, including about 5,000 prisoners and several pieces of artillery. He retired to a high range of hills south and east of the town. The attack was not pressed that afternoon, the enemy's force being unknown, and it being considered advisable to await the arrival of the rest of our troops. Orders were sent back to hasten their march, and, in the meantime, every effort was made to ascertain the numbers and position of the enemy, and find the most favorable point of attack. It had not been intended to fight a general battle at such a distance from our base, unless attacked by the enemy, but, finding ourselves unexpectedly confronted by the Federal Army, it became a matter of difficulty to withdraw through the mountains with our large trains. At the same time, the country was unfavorable for collecting supplies while in the presence of the enemy's main body, as he was enabled to restrain our foraging parties by occupying the passes of the mountains with regular and local troops. A battle thus became, in a measure, unavoidable.

Encouraged by the successful issue of the engagement of the first day, and in view of the valuable results that would ensue from the defeat of the army of

General Meade, it was thought advisable to renew the attack. The remainder of Ewell's and Hill's corps having arrived, and two divisions of Longstreet's, our preparations were made accordingly. During the afternoon, intelligence was received of the arrival of General Stuart at Carlisle, and he was ordered to march to Gettysburg and take position on our left. A full account of these engagements cannot be given until the reports of the several commanding officers shall have been received, and I shall only offer a general description.

The preparations for attack were not completed until the afternoon of the 2nd. The enemy held a high and commanding ridge, along which he had massed a large amount of artillery. General Ewell occupied the left of our line, General Hill the center, and General Longstreet the right. In front of General Longstreet the enemy held a position from which, if he could be driven, it was thought our artillery could be used to advantage in assailing the more elevated ground beyond, and thus enable us to reach the crest of the ridge. That officer was directed to endeavor to carry this position, while General Ewell attacked directly the high ground on the enemy's right, which had already been partially fortified. General Hill was instructed to threaten the center of the Federal line, in order to prevent reinforcements being sent to either wing, and to avail himself of any opportunity that might present itself to attack. After a severe struggle, Longstreet succeeded in getting possession of and holding the desired ground. Ewell also carried some of the strong positions which he assailed, and the result was such as to lead to the belief that he would ultimately be able to dislodge the enemy. The battle ceased at dark.

These partial successes determined me to continue the assault next day. Pickett, with three of his brigades, joined Longstreet the following morning, and our batteries were moved forward to the positions gained by him the day before. The general plan of attack was unchanged, excepting that one division and two brigades of Hill's corps were ordered to support Longstreet.

The enemy, in the meantime, had strengthened his lines with earthworks. The morning was occupied in necessary preparations, and the battle recommenced in the afternoon of the 3rd, and raged with great violence until sunset. Our troops succeeded in entering the advanced works of the enemy, and getting possession of some of his batteries, but our artillery having nearly expended its ammunition, the attacking columns became exposed to the heavy fire of the numerous batteries near the summit of the ridge, and, after a most determined and gallant struggle, were compelled to relinquish their advantage, and fall back to their original positions with severe loss.

The conduct of the troops was all that I could desire or expect, and they deserve success so far as it can be deserved by heroic valor and fortitude. More may have been required of them than they were able to perform, but my admiration of their noble qualities and confidence in their ability to cope successfully with the enemy has suffered no abatement from the issue of this protracted and sanguinary conflict.

Owing to the strength of the enemy's

position, and the reduction of our ammunition, a renewal of the engagement could not be hazarded, and the difficulty of procuring supplies rendered it impossible to continue longer where we were. Such of the wounded as were in condition to be removed, and part of the arms collected on the field, were ordered to Williamsport.

The army remained at Gettysburg during the 4th, and at night began to retire by the road to Fairfield, carrying with it about 4,000 prisoners. Nearly 2,000 had previously been paroled, but the enemy's numerous wounded that had fallen into our hands after the first and second days' engagements were left behind.

Little progress was made that night, owing to a severe storm, which greatly embarrassed our movements. The rear of the column did not leave its position near Gettysburg until after daylight on the 5th. The march was continued during that day without interruption from the enemy, excepting an unimportant demonstration upon our rear in the afternoon when near Fairfield, which was easily checked. Part of our train moved by the road through Fairfield and the rest by way of Cashtown, guarded by General Imboden.

In passing through the mountains in advance of the column, the great length of the trains exposed them to attack by the enemy's cavalry, which captured a number of wagons and ambulances, but they succeeded in reaching Williamsport without serious loss.

They were attacked at that place on the 6th by the enemy's cavalry, which was gallantly repulsed by General Imboden. The attacking force was subsequently encountered and driven off by General Stuart, and pursued for several miles in the direction of Boonsborough. The army, after an arduous march, rendered more difficult by the rains, reached Hagerstown on the afternoon of July 6 and morning of the 7th.

The Potomac was found to be so much swollen by the rains that had fallen almost incessantly since our entrance into Maryland as to be unfordable. Our communications with the south side were thus interrupted, and it was difficult to procure either ammunition or subsistence, the latter difficulty being enhanced by the high waters impeding the working of the neighboring mills. The trains with the wounded and prisoners were compelled to await at Williamsport the subsiding of the river and the construction of boats, as the pontoon bridge left at Falling Waters had been partially destroyed. The enemy had not yet made his appearance, but as he was in condition to obtain large reinforcements, and our situation, for the reasons above mentioned, was becoming daily more embarrassing, it was deemed advisable to recross the river. Part of the pontoon bridge was recovered and new boats built, so that by the 13th a good bridge was thrown over the river at Falling Waters.

The enemy in force reached our front on the 12th. A position had been previously selected to cover the Potomac from Williamsport to Falling Waters, and an attack was awaited during that and the succeeding day. This did not take place, though the two armies were in close proximity, the enemy being occupied in fortifying his own lines. Our preparations being completed, and the river, though still deep, being pronounced fordable, the army commenced to withdraw to the south side on the night of the 13th. Ewell's corps forded the river at Williamsport. Those of Longstreet and Hill crossed upon the bridge. Owing to the condition of the roads, the troops did not reach the bridge until after daylight on the 14th, and the crossing was not completed until 1:00 p.m., when the bridge was removed. The enemy offered no serious interruption, and the movement was attended with no loss of materiel excepting a few disabled wagons and two pieces of artillery, which the horses were unable to move through the deep mud. Before fresh horses could be sent back for them, the rear of the column had passed.

During the slow and tedious march to the bridge, in the midst of a violent storm of rain, some of the men lay down by the way to rest. Officers sent back for them failed to find many in the obscurity of the night, and these, with some stragglers, fell into the hands of the enemy.

Brigadier General Pettigrew was mortally wounded in an attack made by a small body of cavalry, which was unfortunately mistaken for our own, and permitted to enter our lines. He was brought to Bunker Hill, where he expired a few days afterward. He was a brave and accomplished officer and gentleman, and his loss will be deeply felt by the country and the army.

The following day the army marched to Bunker Hill, in the vicinity of which it encamped for several days. The day after its arrival, a large force of the enemy's cavalry, which had crossed the Potomac at Harpers Ferry, advanced toward Martinsburg. It was attacked by General Fitz Lee, near Kearneysville, and defeated with heavy loss, leaving its dead and many of its wounded on the field.

Owing to the swollen condition of the Shenandoah, the plan of operations which had been contemplated when we recrossed the Potomac could not be put into execution, and before the waters had subsided, the movements of the enemy induced me to cross the Blue Ridge and take position south of the Rappahannock, which was accordingly done.

> *The movements of the enemy induced me to cross the Blue Ridge and take position south of the Rappahannock.*

As soon as the reports of the commanding officers shall be received, a more detailed account of these operations will be given, and occasion will then be taken to speak more particularly of the conspicuous gallantry and good conduct of both officers and men.

It is not yet in my power to give a correct statement of our casualties, which were severe, including many brave men, and an unusual proportion of distinguished and valuable officers. Among them I regret to mention the following general officers: Major-Generals Hood, Pender, and Trimble severely, and Major-General Heth slightly wounded.

General Pender has since died. This lamented officer had borne a distinguished part in every engagement of this army, and was wounded on several occasions while leading his command with conspicuous gallantry and ability. The confidence and admiration inspired by his courage and capacity as an officer were only equaled by the esteem and respect entertained by all with whom he was associated for the noble qualities of his modest and unassuming character.

Brigadier Generals Barksdale and [R. B.] Garnett were killed, and Brigadier General Semmes mortally wounded, while leading their troops with the courage that always distinguished them. These brave officers and patriotic gentlemen fell in the faithful discharge of duty, leaving the army to mourn their loss and emulate their noble examples.

Brigadier Generals Kemper, Armistead, Scales, G. T. Anderson, Hampton, J. M. Jones, and Jenkins were also wounded. Brigadier General Archer was taken prisoner.

General Pettigrew, though wounded at Gettysburg, continued in command until he was mortally wounded, near Falling Waters.

The loss of the enemy is unknown, but from observation on the field, and his subsequent movements, it is supposed that he suffered severely.

Respectfully submitted,
R. E. LEE,
General

ROBERT E. LEE'S REPORT ON THE PENNSYLVANIA CAMPAIGN JANUARY 20, 1864

Headquarters, Army of Northern Virginia
January 20, 1864

GENERAL: I have the honor to submit a detailed report of the operations of this army from the time it left the vicinity of Fredericksburg, early in June, to its occupation of the line of the Rapidan, in August.

Upon the retreat of the Federal Army, commanded by Major General Hooker, from Chancellorsville, it reoccupied the ground north of the Rappahannock, opposite Fredericksburg, where it could not be attacked excepting at a disadvantage. It was determined to draw it from this position, and, if practicable, to transfer the scene of hostilities beyond the Potomac. The execution of this purpose also embraced the expulsion of the force under General Milroy, which had infested the lower Shenandoah Valley during the preceding winter and spring. If unable to attain the valuable results which might be expected to follow a decided advantage gained over the enemy in Maryland or Pennsylvania, it was hoped that we should at least so far disturb his plan for the summer campaign as to prevent its execution during the season of active operations.

The commands of Longstreet and Ewell were put in motion, and encamped around Culpeper Courthouse on June 7. As soon as their march was discovered by the enemy, he threw a force across the

Rappahannock, about two miles below Fredericksburg, apparently for the purpose of observation. Hill's corps was left to watch these troops, with instructions to follow the movements of the army as soon as they should retire.

The cavalry, under General Stuart, which had been concentrated near Culpeper Courthouse, was attacked on June 9 by a large force of Federal cavalry, supported by infantry, which crossed the Rappahannock at Beverly and Kelly's Fords. After a severe engagement, which continued from early in the morning until late in the afternoon, the enemy was compelled to recross the river with heavy loss, leaving about 500 prisoners, three pieces of artillery, and several colors in our hands.

General Imboden and General Jenkins had been ordered to cooperate in the projected expedition into the Valley, General Imboden by moving toward Romney with his command, to prevent the troops guarding the Baltimore and Ohio Railroad from reinforcing those at Winchester, while General Jenkins advanced directly toward the latter place with his cavalry brigade, supported by a battalion of infantry and a battery of the Maryland Line.

General Ewell left Culpeper Courthouse on June 10. He crossed the branches of the Shenandoah near Front Royal, and reached Cedarville on the 12th, where he was joined by General Jenkins. Detaching General Rodes with his division and the greater part of Jenkins' brigade, to dislodge a force of the enemy stationed at Berryville, General Ewell, with the rest of his command, moved upon Winchester. Johnson's division advancing by the Front Royal road, Early's by the Valley turnpike, which it entered at Newtown, where it was joined by the Maryland troops.

> *It was determined to draw the Federal army from this position, and, if practicable, to transfer the scene of hostilities beyond the Potomac.*

Battle of Winchester

The enemy was driven in on both roads, and our troops halted in line of battle near the town on the evening of the 13th. The same day the force which had occupied Berryville retreated to Winchester on the approach of General Rodes. The following morning, General Ewell ordered General Early to carry an entrenched position northwest of Winchester, near the Pughtown road, which the latter officer, upon examining the ground, discovered would command the principal fortifications.

To cover the movement of General Early, General Johnson took position between the road to Millwood and that to Berryville, and advanced his skirmishers toward the town. General Early, leaving a portion of his command to engage the enemy's attention, with the remainder gained a favorable position without being perceived, and, about 5:00 p.m., twenty pieces of artillery, under Lieutenant Colonel H.P. Jones, opened suddenly upon the entrenchments. The enemy's guns were soon silenced. Hays' brigade then advanced to the assault, and carried the works by storm, capturing six rifled pieces, two of which were turned upon and dispersed a column which was forming to retake the position. The enemy immediately abandoned the works on the left of those taken by Hays, and retired into his main fortifications, which General Early prepared to assail in the morning. The loss of the advanced works, however, rendered the others untenable, and the enemy retreated in the night, abandoning his sick and wounded, together with his artillery, wagons, and stores.

Anticipating such a movement, as soon as he heard of Early's success, General Ewell directed General Johnson to occupy, with part of his command, a point on the Martinsburg road, about two and a half miles from Winchester, where he could either intercept the enemy's retreat, or aid in an attack should further resistance be offered in the morning. General Johnson marched with Nicholls' and part of Steuart's brigades, accompanied by Lieutenant-Colonel R.S. Andrews with a detachment of his artillery, the Stonewall Brigade being ordered to follow. Finding the road to the place indicated by General Ewell difficult of passage in the darkness, General Johnson pursued that leading by Jordan Springs to Stephenson's Depot, where he took a favorable position on the Martinsburg road, about five miles from Winchester. Just as his line was formed, the retreating column, consisting of the main body of General Milroy's army, arrived, and immediately attacked him. The enemy, though in superior force, consisting of both infantry and cavalry, was gallantly repulsed, and, finding all efforts to cut his way unavailing, he sent strong flanking parties simultaneously to the right and left, still keeping up a heavy fire in front. The party on the right was driven back and pursued by the Stonewall Brigade, which opportunely arrived. That on the left was broken and dispersed by the Second and Tenth Louisiana Regiments, aided by the artillery, and in a short time nearly the whole infantry force, amounting to more than 2,300 men, with eleven stand of colors, surrendered, the cavalry alone escaping. General Milroy, with a small party of fugitives, fled to Harpers Ferry. The number of prisoners taken in this action exceeded the force engaged under General Johnson, who speaks in terms of well deserved praise of the conduct of the officers and men of his command.

In the meantime, General Rodes marched from Berryville to Martinsburg, reaching the latter place in the afternoon of the 14th. The enemy made a show of resistance, but soon gave way, the cavalry and artillery retreating toward Williamsport, the infantry toward Shepherdstown, under cover of night. The route taken by the latter was not known until it was too late to follow; but the former were pursued so rapidly, Jenkins' troops leading, that they were forced to abandon five of their six pieces of artillery. About 200 prisoners were taken, but the enemy destroyed most of his stores.

These operations resulted in the expulsion of the enemy from the Valley; the capture of 4,000 prisoners, with a corresponding number of small-arms; twenty-eight pieces of superior artillery, including those taken by Generals Rodes and Hays; about 300 wagons and as many horses, together with a considerable quantity of ordnance, commissary, and quartermaster's stores.

Our entire loss was forty-seven killed, 219 wounded, and three missing.

March into Pennsylvania

On the night of Ewell's appearance at Winchester, the enemy in front of A.P. Hill, at Fredericksburg, recrossed the Rappahannock, and the whole army of General Hooker withdrew from the north side of the river. In order to mislead him as to our intentions, and at the same time protect Hill's corps in its march up the Rappahannock, Longstreet left Culpeper Courthouse on the 15th, and, advancing along the eastern side of the Blue Ridge, occupied Ashby's and Snicker's Gaps. He had been joined, while at Culpeper, by General Pickett, with three brigades of his division. General Stuart, with three brigades of cavalry, moved on Longstreet's right, and took position in front of the Gaps. Hampton's and W.E. Jones' brigades remained along the Rappahannock and Hazel Rivers, in front of Culpeper Court-House, with instructions to follow the main body as soon as Hill's corps had passed that point.

On the 17th, Fitz Lee's brigade, under Colonel Munford, which was on the road to Snicker's Gap, was attacked near Aldie by the Federal cavalry. The attack was repulsed with loss, and the brigade held its ground until ordered to fall back, its right being threatened by another body, coming from Hopewell toward Middleburg. The latter force was driven from Middleburg, and pursued toward Hopewell by Robertson's brigade, which arrived about dark. Its retreat was intercepted by W.H.F. Lee's brigade, under Colonel Chambliss, Jr., and the greater part of a regiment captured.

During the three succeeding days there was much skirmishing, General Stuart taking a position west of Middleburg, where he awaited the rest of his command.

General Jones arrived on the 19th, and General Hampton in the afternoon of the following day, having repulsed, on his march, a cavalry force sent to reconnoiter in the direction of Warrenton.

On the 21st, the enemy attacked with infantry and cavalry, and obliged General Stuart, after a brave resistance, to fall back to the gaps of the mountains. The enemy retired the next day, having advanced only a short distance beyond Upperville.

In these engagements, the cavalry sustained a loss of 510 killed, wounded, and missing. Among them were several valuable officers, whose names are mentioned in General Stuart's report. One piece of artillery was disabled and left on the field. The enemy's loss was heavy. About 400 prisoners were taken and several stand of colors.

The Federal Army was apparently guarding the approaches to Washington, and manifested no disposition to assume the offensive.

> *In the meantime, the progress of Ewell, who was already in Maryland, with Jenkins' cavalry, advanced into Pennsylvania as far as Chambersburg, rendered it necessary that the rest of the army should be within supporting distance...*

In the meantime, the progress of Ewell, who was already in Maryland, with Jenkins' cavalry advanced into Pennsylvania as far as Chambersburg, rendered it necessary that the rest of the army should be within supporting distance, and Hill, having reached the Valley, Longstreet was withdrawn to the west side of the Shenandoah, and the two corps encamped near Berryville.

General Stuart was directed to hold the mountain passes with part of his command as long as the enemy remained south of the Potomac, and, with the remainder, to cross into Maryland, and place himself on the right of General Ewell. Upon the suggestion of the former officer that he could damage the enemy and delay his passage of the river by getting in his rear, he was authorized to do so, and it was left to his discretion whether to enter Maryland east or west of the Blue Ridge; but he was instructed to lose no time in placing his command on the right of our column as soon as he should perceive the enemy moving northward.

On the 22nd, General Ewell marched into Pennsylvania with Rodes' and Johnson's divisions, preceded by Jenkins' cavalry, taking the road from Hagerstown, through Chambersburg, to Carlisle, where he arrived on the 27th. Early's division, which had occupied Boonsborough, moved by a parallel road to Greenwood, and, in pursuance of instructions previously given to General Ewell, marched toward York.

On the 24th, Longstreet and Hill were put in motion to follow Ewell, and, on the 27th, encamped near Chambersburg.

General Imboden, under the orders before referred to, had been operating on Ewell's left while the latter was advancing into Maryland. He drove off the troops guarding the Baltimore and Ohio Railroad, and destroyed all the important bridges on that route from Martinsburg to Cumberland, besides inflicting serious damage upon the Chesapeake and Ohio Canal. He was at Hancock when Longstreet and Hill reached Chambersburg, and was directed to proceed to the latter place by way of McConnellsburg, collecting supplies for the army on his route.

The cavalry force at this time with the army, consisting of Jenkins' brigade and E.V. White's battalion, was not greater than was required to accompany the advance of General Ewell and General Early, with whom it performed valuable service, as appears from their reports. It was expected that as soon as the Federal Army should cross the Potomac, General Stuart would give notice of its movements, and nothing having been heard from him since our entrance into Maryland, it was inferred that the enemy had not yet left Virginia. Orders were, therefore, issued to move upon Harrisburg. The expedition of General Early to York was designed in part to prepare for this undertaking by breaking the railroad between Baltimore and Harrisburg, and seizing the bridge over the Susquehanna at Wrightsville. General Early succeeded in the first object, destroying a number of bridges above and below York, but on the approach of the troops sent by him to Wrightsville, a body of militia stationed at that place fled across the river and burned the bridge in their retreat. General Early then marched to rejoin his corps.

The advance against Harrisburg was arrested by intelligence received from a scout on the night of the 28th, to the effect that the army of General Hooker had crossed the Potomac, and was approaching the South Mountain. In the absence of the cavalry, it was impossible to ascertain his intentions; but to deter him from advancing farther west, and intercepting our communication with Virginia, it was determined to concentrate the army east of the mountains.

Battle of Gettysburg

Hill's corps was accordingly ordered to move toward Cashtown on the 29th, and Longstreet to follow the next day, leaving Pickett's division at Chambersburg to guard the rear until relieved by Imboden. General Ewell was recalled from Carlisle, and directed to join the army at Cashtown or Gettysburg, as circumstances might require. The advance of the enemy to the latter place was unknown, and the weather being inclement, the march was conducted with a view to the comfort of the troops. Heth's division reached Cashtown on the 29th, and the following morning Pettigrew's brigade, sent by General Heth to procure supplies at Gettysburg, found it occupied by the enemy. Being ignorant of the extent of his force, General Pettigrew was unwilling to hazard an attack with his single brigade, and returned to Cashtown.

General Hill arrived with Pender's division in the evening, and the following morning (July 1) advanced with these two divisions, accompanied by Pegram's and McIntosh's battalions of artillery, to ascertain the strength of the enemy, whose force was supposed to consist chiefly of cavalry. The leading division, under General Heth, found the enemy's vedettes about three miles west of Gettysburg, and continued to advance until within a mile of the town, when two brigades were sent forward to reconnoiter. They drove in the advance of the enemy very gallantly, but subsequently encountered largely superior numbers, and were compelled to retire with loss, Brigadier General Archer, commanding one of the brigades, being taken prisoner. General Heth then prepared for action, and as soon as Pender arrived to support him, was ordered by General Hill to advance. The artillery was placed in position, and the engagement opened with vigor. General Heth pressed the enemy steadily back, breaking his first and second lines, and attacking his third with great resolution. About 2.30 p.m. the advance of Ewell's corps, consisting of Rodes' division, with Carter's battalion of artillery, arrived by the Middletown road, and, forming on Heth's left, nearly at right angles with his line, became warmly engaged with fresh numbers of the enemy. Heth's troops, having suffered heavily in their protracted contest with a superior force, were relieved by Pender's, and Early,

> *Early's division was detached for this purpose and proceeded as far east as York, while the remainder of the corps proceeded to Carlisle.*

coming up by the Heidlersburg road soon afterward, took position on the left of Rodes, when a general advance was made.

The enemy gave way on all sides, and was driven through Gettysburg with great loss. Major General Reynolds, who was in command, was killed. More than 5,000 prisoners, exclusive of a large number of wounded, three pieces of artillery, and several colors were captured. Among the prisoners were two brigadier-generals, one of whom was badly wounded. Our own loss was heavy, including a number of officers, among whom were Major General Heth, slightly, and Brigadier General Scales, of Pender's division, severely, wounded. The enemy retired to a range of hills south of Gettysburg, where he displayed a strong force of infantry and artillery.

It was ascertained from the prisoners that we had been engaged with two corps of the army formerly commanded by General Hooker, and that the remainder of that army, under General Meade, was approaching Gettysburg. Without information as to its proximity, the strong position which the enemy had assumed could not be attacked without danger of exposing the four divisions present, already weakened and exhausted by a long and bloody struggle, to overwhelming numbers of fresh troops. General Ewell was, therefore, instructed to carry the hill occupied by the enemy, if he found it practicable, but to avoid a general engagement until the arrival of the other divisions of the army, which were ordered to hasten forward. He decided to await Johnson's division, which had marched from Carlisle by the road west of the mountains to guard the trains of his corps, and consequently did not reach Gettysburg until a late hour.

In the meantime the enemy occupied the point which General Ewell designed to seize, but in what force could not be ascertained, owing to the darkness. An intercepted dispatch showed that another corps had halted that afternoon four miles from Gettysburg. Under these circumstances, it was decided not to attack until the arrival of Longstreet, two of whose divisions (those of Hood and McLaws) encamped about four miles in the rear during the night. Anderson's division of Hill's corps came up after the engagement.

It had not been intended to deliver a general battle so far from our base unless attacked, but coming unexpectedly upon the whole Federal Army, to withdraw through the mountains with our extensive trains would have been difficult and dangerous. At the same time we were unable to await an attack, as the country was unfavorable for collecting supplies in the presence of the enemy, who could restrain our foraging parties by holding the mountain passes with local and other troops. A battle had, therefore, become in a measure unavoidable, and the success already gained gave hope of a favorable issue.

The enemy occupied a strong position, with his right upon two commanding elevations adjacent to each other, one southeast and the other, known as Cemetery Hill, immediately south of the town, which lay at its base. His line extended thence upon the high ground along the Emmitsburg road, with a steep ridge in rear, which was also occupied. This ridge was difficult of ascent, particularly the two hills above mentioned as forming its northern extremity, and a third at the other end, on which the enemy's left rested. Numerous stone and rail fences along the slope served to afford protection to his troops and impede our advance. In his front, the ground was undulating and generally open for about three-quarters of a mile.

General Ewell's corps constituted our left, Johnson's division being opposite the height adjoining Cemetery Hill, Early's in the center, in front of the north face of the latter, and Rodes upon his right. Hill's corps faced the west side of Cemetery Hill, and extended nearly parallel to the Emmitsburg road, making an angle with Ewell's, Pender's division formed his left, Anderson's his right, Heth's, under Brigadier General Pettigrew, being in reserve. His artillery, under Colonel R.L. Walker, was posted in eligible positions along his line.

It was determined to make the principal attack upon the enemy's left, and endeavor to gain a position from which it was thought that our artillery could be brought to bear with effect. Longstreet was directed to place the divisions of McLaws and Hood on the right of Hill, partially enveloping the enemy's left, which he was to drive in.

General Hill was ordered to threaten the enemy's center, to prevent reinforcements being drawn to either wing, and cooperate with his right division in Longstreet's attack.

General Ewell was instructed to make a simultaneous demonstration upon the enemy's right, to be converted into a real attack should opportunity offer.

About 4:00 p.m. Longstreet's batteries opened, and soon afterward Hood's division, on the extreme right, moved to the attack. McLaws followed somewhat later, four of Anderson's brigades, those of Wilcox, Perry, [A.R.] Wright, and Posey supporting him on the left, in the order named. The enemy was soon driven from his position on the Emmitsburg road to the cover of a ravine and a line of stone fences at the foot of the ridge in his rear. He was dislodged from these after a severe struggle, and retired up the ridge, leaving a number of his batteries in our possession. Wilcox's and Wright's brigades advanced with great gallantry, breaking successive lines of the enemy's infantry, and compelling him to abandon much of his artillery. Wilcox reached the foot and Wright gained the crest of the ridge itself, driving the enemy down the opposite side; but having become separated from McLaws and gone beyond the other two brigades of the division, they were attacked in front and on both flanks, and compelled to retire, being unable to bring off any of the captured artillery. McLaws' left also fell back, and, it being now nearly dark, General Longstreet determined to await the arrival of General Pickett. He disposed his command to hold the ground gained on the right, withdrawing his left to the first position from which the enemy had been driven.

Four pieces of artillery, several hundred prisoners, and two regimental flags were taken. As soon as the engagement began on our right, General Johnson opened with his artillery, and about two hours later advanced up the hill next to Cemetery Hill with three brigades, the fourth being detained by a demonstration on his left. Soon afterward, General Early attacked

Cemetery Hill with two brigades, supported by a third, the fourth having been previously detached. The enemy had greatly increased by earthworks the strength of the positions assailed by Johnson and Early.

The troops of the former moved steadily up the steep and rugged ascent, under a heavy fire, driving the enemy into his entrenchments, part of which was carried by Steuart's brigade, and a number of prisoners taken. The contest was continued to a late hour, but without further advantage. On Cemetery Hill, the attack by Early's leading brigades--those of Hays and Hoke, under Colonel I.E. Avery – was made with vigor. Two lines of the enemy's infantry were dislodged from the cover of some stone and board fences on the side of the ascent, and driven back into the works on the crest, into which our troops forced their way, and seized several pieces of artillery.

A heavy force advanced against their right, which was without support, and they were compelled to retire, bringing with them about 100 prisoners and four stand of colors. General Ewell had directed General Rodes to attack in concert with Early, covering his right, and had requested Brigadier General Lane, then commanding Pender's division, to cooperate on the right of Rodes. When the time to attack arrived, General Rodes, not having his troops in position, was unprepared to cooperate with General Early, and before he could get in readiness, the latter had been obliged to retire for want of the expected support on his right. General Lane was prepared to give the assistance required of him, and so informed General Rodes, but the latter deemed it useless to advance after the failure of Early's attack.

In this engagement our loss in men and officers was large. Major Generals Hood and Pender, Brigadier Generals J.M. Jones, Semmes, G.T. Anderson, and Barksdale, and Colonel Avery, commanding Hoke's brigade, were wounded, the last two mortally. Generals Pender and Semmes died after their removal to Virginia.

> *About 1:00 p.m., at a given signal, a heavy cannonade was opened and continued for about two hours with marked effect upon the enemy.*

The result of this day's operations induced the belief that, with proper concert of action, and with the increased support that the positions gained on the right would enable the artillery to render the assaulting columns, we should ultimately succeed, and it was accordingly determined to continue the attack. The general plan was unchanged. Longstreet, reinforced by Pickett's three brigades, which arrived near the battlefield during the afternoon of the 2nd, was ordered to attack the next morning, and General Ewell was directed to assail the enemy's right at the same time. The latter, during the night, reinforced General Johnson with two brigades from Rodes' and one from Early's division.

General Longstreet's dispositions were not completed as early as was expected, but before notice could be sent to General Ewell, General Johnson had already become engaged, and it was too late to recall him. The enemy attempted to recover the works taken the preceding evening, but was repulsed, and General Johnson attacked in turn.

After a gallant and prolonged struggle, in which the enemy was forced to abandon part of his entrenchments, General Johnson found himself unable to carry the strongly fortified crest of the hill. The projected attack on the enemy's left not having been made, he was enabled to hold his right with a force largely superior to that of General Johnson, and finally to threaten his flank and rear, rendering it necessary for him to retire to his original position about 1:00 p.m.

General Longstreet was delayed by a force occupying the high, rocky hills on the enemy's extreme left, from which his troops could be attacked in reverse as they advanced. His operations had been embarrassed the day previous by the same cause, and he now deemed it necessary to defend his flank and rear with the divisions of Hood and McLaws. He was, therefore, reinforced by Heth's division and two brigades of Pender's, to the command of which Major General Trimble was assigned. General Hill was directed to hold his line with the rest of his command, afford General Longstreet further assistance, if required, and avail himself of any success that might be gained.

A careful examination was made of the ground secured by Longstreet, and his batteries placed in positions, which, it was believed, would enable them to silence those of the enemy. Hill s artillery and part of Ewell's was ordered to open simultaneously, and the assaulting column to advance under cover of the combined fire of the three. The batteries were directed to be pushed forward as the infantry progressed, protect their flanks, and support their attacks closely.

About 1:00 p.m., at a given signal, a heavy cannonade was opened, and continued for about two hours with marked effect upon the enemy. His batteries replied vigorously at first, but toward the close their fire slackened perceptibly, and General Longstreet ordered forward the column of attack, consisting of Pickett's and Heth's divisions, in two lines, Pickett on the right. Wilcox's brigade marched in rear of Pickett's right, to guard that flank, and Heth's was supported by Lane's and Scales' brigades, under General Trimble.

The troops moved steadily on, under a heavy fire of musketry and artillery, the main attack being directed against the enemy's left center.

His batteries reopened as soon as they appeared. Our own having nearly exhausted their ammunition in the protracted cannonade that preceded the advance of the infantry, were unable to reply, or render the necessary support to the attacking party. Owing to this fact, which was unknown to me when the assault took place, the enemy was enabled to throw a strong force of infantry against our left, already wavering under a concentrated fire of artillery from the ridge in front, and from Cemetery Hill, on the left. It finally gave way, and the right, after penetrating the enemy's lines, entering his advance works, and capturing some of his artillery, was attacked simultaneously in front and on both flanks, and driven back with heavy loss.

The troops were rallied and reformed, but the enemy did not pursue.

A large number of brave officers and men fell or were captured on this occasion. Of Pickett's three brigade commanders, Generals Armistead and [R. B.] Garnett were killed, and General Kemper dangerously wounded.

Major-General Trimble and Brigadier-General Pettigrew were also wounded, the former severely.

The movements of the army preceding the battle of Gettysburg had been much embarrassed by the absence of the cavalry. As soon as it was known that the enemy had crossed into Maryland, orders were sent to the brigades of B.H. Robertson and W.E. Jones, which had been left to guard the passes of the Blue Ridge, to rejoin the army without delay, and it was expected that General Stuart, with the remainder of his command, would soon arrive. In the exercise of the discretion given him when Longstreet and Hill marched into Maryland, General Stuart determined to pass around the rear of the Federal Army with three brigades and cross the Potomac between it and Washington, believing that he would be able, by that route, to place himself on our right flank in time to keep us properly advised of the enemy's movements. He marched from Salem on the night of June 24, intending to pass west of Centreville, but found the enemy's forces so distributed as to render that route impracticable. Adhering to his original plan, he was forced to make a wide *detour* through Buckland and Brentsville, and crossed the Occoquan at Wolf Run Shoals on the morning of the 27th. Continuing his march through Fairfax Courthouse and Dranesville, he arrived at the Potomac, below the mouth of Seneca Creek, in the evening.

He found the river much swollen by the recent rains, but, after great exertion, gained the Maryland shore before midnight with his whole command.

He now ascertained that the Federal Army, which he had discovered to be drawing toward the Potomac, had crossed the day before, and was moving toward Frederick, thus interposing itself between him and our forces.

He accordingly marched northward, through Rockville and Westminster, to Hanover, Pa., where he arrived on the 30th; but the enemy advanced with equal rapidity on his left, and continued to obstruct communication with our main body.

Supposing, from such information as he could obtain, that part of the army was at Carlisle, he left Hanover that night, and proceeded thither by way of Dover.

He reached Carlisle on July 1, where he received orders to proceed to Gettysburg. He arrived in the afternoon of the following day, and took position on General Ewell's left. His leading brigade, under General Hampton, encountered and repulsed a body of the enemy's cavalry at Hunterstown, endeavoring to reach our rear.

General Stuart had several skirmishes during his march, and at Hanover quite a severe engagement took place with a strong force of cavalry, which was finally compelled to withdraw from the town. The prisoners taken by the cavalry and paroled at various places amounted to about 800, and at Rockville a large train of wagons coming from Washington was intercepted and captured. Many of them were destroyed, but 125, with all the animals of the train, were secured.

The ranks of the cavalry were much reduced by its long and arduous march, repeated conflicts, and insufficient supplies of food and forage, but the day after its arrival at Gettysburg it engaged the enemy's cavalry with unabated spirit, and effectually protected our left.

In this action, Brigadier General Hampton was seriously wounded, while acting with his accustomed gallantry.

Robertson's and Jones' brigades arrived on July 3, and were stationed upon our right flank. The severe loss sustained by the army and the reduction of its ammunition, rendered another attempt to dislodge the enemy inadvisable, and it was, therefore, determined to withdraw.

The trains, with such of the wounded as could bear removal, were ordered to Williamsport on July 4, part moving through Cashtown and Greencastle, escorted by General Imboden, and the remainder by the Fairfield road.

The army retained its position until dark, when it was put in motion for the Potomac by the last-named route.

A heavy rain continued throughout the night, and so much impeded its progress that Ewell's corps, which brought up the rear, did not leave Gettysburg until late in the forenoon of the following day. The enemy offered no serious interruption, and, after an arduous march, we arrived at Hagerstown in the afternoon of the 6th and morning of July 7.

The great length of our trains made it difficult to guard them effectually in passing through the mountains, and a number of wagons and ambulances were captured. They succeeded in reaching Williamsport on the 6th, but were unable to cross the Potomac on account of the high stage of water. Here they were attacked by a strong force of cavalry and artillery, which was gallantly repulsed by General Imboden, whose command had been strengthened by several batteries and by two regiments of infantry, which had been detached at Winchester to guard prisoners, and were returning to the army.

> *In the exercise of the discretion given him when Longstreet and Hill marched in Maryland, General Stuart determined to pass around the rear of the Federal Army...*

While the enemy was being held in check, General Stuart arrived with the cavalry, which had performed valuable service in guarding the flanks of the army during the retrograde movement, and, after a short engagement, drove him from the field. The rains that had prevailed almost without intermission since our entrance into Maryland, and greatly interfered with our movements, had made the Potomac unfordable, and the pontoon bridge left at Falling Waters had been partially destroyed by the enemy. The wounded and prisoners were sent over the river as rapidly as possible in a few ferry-boats, while the trains awaited the subsiding of the waters and the construction of a new pontoon bridge.

On July 8, the enemy's cavalry advanced toward Hagerstown, but was repulsed by General Stuart, and pursued as far as Boonsborough.

With this exception, nothing but occasional skirmishing occurred until the 12th, when the main body of the enemy arrived. The army then took a position previously selected, covering the Potomac from Williamsport to Falling Waters, where it remained for two days, with the enemy immediately in front, manifesting no disposition to attack, but throwing up entrenchments along his whole line.

By the 13th, the river at Williamsport, though still deep, was fordable, and a good bridge was completed at Falling Waters, new boats having being constructed and some of the old recovered. As further delay would enable the enemy to obtain reinforcements, and as it was found difficult to procure a sufficient supply of flour for the troops, the working of the mills being interrupted by high water, it was determined to await an attack no longer.

Orders were accordingly given to cross the Potomac that night, Ewell's corps by the ford at Williamsport, and those of Longstreet and Hill on the bridge.

The cavalry was directed to relieve the infantry skirmishers, and bring up the rear.

The movement was much retarded by a severe rain storm and the darkness of the night. Ewell's corps, having the advantage of a turnpike road, marched with less difficulty, and crossed the river by 8:00 the following morning. The condition of the road to the bridge and the time consumed in the passage of the artillery, ammunition wagons, and ambulances, which could not ford the river, so much delayed the progress of Longstreet and Hill, that it was daylight before their troops began to cross. Heth's division was halted about a mile and a half from the bridge, to protect the passage of the column. No interruption was offered by the enemy until about 11:00 a.m., when his cavalry, supported by artillery, appeared in front of General Heth.

A small number in advance of the main body was mistaken for our own cavalry retiring, no notice having been given of the withdrawal of the latter, and was suffered to approach our lines. They were immediately destroyed or captured, with the exception of two or three, but Brigadier General Pettigrew, an officer of great merit and promise, was mortally wounded in the encounter. He survived his removal to Virginia only a few days.

The bridge being clear, General Heth began to withdraw. The enemy advanced, but his efforts to break our lines were repulsed, and the passage of the river was completed by 1:00 p.m. owing to the extent of General Heth's line, some of his men most remote from the bridge were cut off before they could reach it, but the greater part of those taken by the enemy during the movement (supposed to amount in all to about 500) consisted of men from various commands who lingered behind, overcome by previous labors and hardships, and the fatigue of a most trying night march. There was no loss of *materiel* excepting a few broken wagons and two pieces of artillery, which the horses were unable to draw through the deep mud. Other horses were sent back for them, but the rear of the column had passed before their arrival.

The army proceeded to the vicinity of Bunker Hill and Darkesville, where it halted to afford the troops repose. The enemy made no effort to follow excepting with his cavalry, which crossed the Potomac at Harper's Ferry, and advanced toward Martinsburg on July 16.

They were attacked by General Fitz Lee, with his own and Chambliss' brigades, and driven back with loss. When the army returned to Virginia, it was intended to move into Loudoun, but the Shenandoah was found to be impassable. While waiting for it to subside, the enemy crossed the Potomac east of the Blue Ridge and seized the passes we designed to use. As he continued to advance along the eastern slope, apparently with the purpose of cutting us off from the railroad to Richmond, General Longstreet was ordered, on July 19, to proceed to Culpeper Courthouse, by way of Front Royal. He succeeded in passing part of his command over the Shenandoah in time to prevent the occupation of Manassas and Chester Gaps by the enemy, whose cavalry had already made its appearance.

As soon as a pontoon bridge could be laid down, the rest of his corps crossed the river, and marched through Chester Gap to Culpeper Courthouse, where it arrived on the 24th. He was followed without serious opposition by General A.P. Hill.

General Ewell having been detained in the Valley by an effort to capture a force of the enemy guarding the Baltimore and Ohio Railroad west of Martinsburg, Wright's brigade was left to hold Manassas Gap until his arrival. He reached Front Royal on the 23rd, with Johnson's and Rodes' divisions, Early's being near Winchester, and found General Wright skirmishing with the enemy's infantry, which had already appeared in Manassas Gap. General Ewell supported Wright with Rodes' division and some artillery, and the enemy was held in check.

Finding that the Federal force greatly exceeded his own, General Ewell marched through Thornton's Gap, and ordered Early to move up the Valley by Strasburg and New Market. He encamped near Madison Courthouse on July 29.

The enemy massed his army in the vicinity of Warrenton, and, on the night of July 31, his cavalry, with a large supporting force of infantry, crossed the Rappahannock at Rappahannock Station and Kelly's Ford.

The next day they advanced toward Brandy Station, their progress being gallantly resisted by General Stuart with Hampton's brigade, commanded by Colonel L.S. Baker, who fell back gradually to our lines, about two miles south of Brandy. Our infantry skirmishers advanced, and drove the enemy beyond Brandy Station. It was now determined to place the army in a position to enable it more readily to oppose the enemy should he attempt to move southward, that near Culpeper Courthouse being one that he could easily avoid. Longstreet and Hill

> **Longstreet succeeded in passing part of his command over the Shenandoah in time to prevent the occupation of Manassas and Chester Gaps by the enemy, whose cavalry had already made its appearance.**

were put in motion August 3, leaving the cavalry at Culpeper.

Ewell had been previously ordered from Madison, and, by the 4th, the army occupied the line of the Rapidan.

The highest praise is due to both officers and men for their conduct during the campaign. The privations and hardships of the march and camp were cheerfully encountered, and borne with a fortitude unsurpassed by our ancestors in their struggle for independence, while their courage in battle entitles them to rank with the soldiers of any army and of any time. Their forbearance and discipline, under strong provocation to retaliate for the cruelty of the enemy to our own citizens, is not their least claim to the respect and admiration of their countrymen and of the world.

I forward returns of our loss in killed, wounded, and missing. Many of the latter were killed or wounded in the several assaults at Gettysburg, and necessarily left in the hands of the enemy. I cannot speak of these brave men as their merits and exploits deserve. Some of them are appropriately mentioned in the accompanying reports, and the memory of all will be gratefully and affectionately cherished by the people in whose defense they fell.

The loss of Major General Pender is severely felt by the army and the country. He served with this army from the beginning of the war, and took a distinguished part in all its engagements. Wounded on several occasions, he never left his command in action until he received the injury that resulted in his death. His promise and usefulness as an officer were only equaled by the purity and excellence of his private life.

Brigadier Generals Armistead, Barksdale, Garnett, and Semmes died as they had lived, discharging the highest duty of patriots with devotion that never faltered and courage that shrank from no danger.

I earnestly commend to the attention of the Government those gallant officers and men whose conduct merited the special commendation of their superiors, but whose names I am unable to mention in this report.

The officers of the general staff of the army were unremittingly engaged in the duties of their respective departments. Much depended on their management and exertion. The labors of the quartermaster's, commissary, and medical departments were more than usually severe. The inspectors-general were also laboriously occupied in their attention to the troops, both on the march and in camp, and the officers of engineers showed skill and judgment in expediting the passage of rivers and streams, the swollen condition of which, by almost continuous rains, called for extraordinary exertion.

The chief of ordnance and his assistants are entitled to praise for the care and watchfulness given to the ordnance trains and ammunition of the army, which, in a long march and in many conflicts, were always at hand and accessible to the troops.

My thanks are due to my personal staff for their constant aid afforded me at all times, on the march and in the field, and their willing discharge of every duty.

There were captured at Gettysburg nearly 7,000 prisoners, of whom about 1,500 were paroled, and the remainder brought to Virginia. Seven pieces of artillery were also secured.

I forward herewith the reports of the corps, division, and other commanders mentioned in the accompanying schedule, together with maps of the scene of operations, and one showing the routes pursued by the army.

Respectfully submitted.
R. E. LEE,
General.

General S. COOPER,
Adjutant and Inspector General.

HI-LIGHTS OF A HERO'S LIFE
JAMES EWELL BROWN STUART

★ 1833 – (February 6) Born at Laurel Hill in Patrick County, Virginia.

★ 1848 – Entered Emory and Henry College.

★ 1848 – Tried to enlist in the army to fight in the Mexican War but was rejected due to his age.

★ 1850 – Graduated from Emory and Henry College.

★ 1850 – Appointed to West Point.

★ 1854 – Achieved the cadet rank of second captain of the corps and designated as honorary "cavalry officer" for his skills in horsemanship.

★ 1854 – Graduated 13th in a class of 46, commissioned a brevet second lieutenant and assigned to the U.S. Mounted Rifles in Texas.

★ 1855 – Transferred to the newly formed 1st Cavalry Regiment at Fort Leavenworth, Kansas.

★ 1855 – Promoted to first lieutenant.

★ 1855 – (November 14) Married Flora Cooke.

★ 1857 – (July 29) Wounded in a skirmish with the Cheyenne along Solomon River in Kansas.

★ 1859 – Received patent for his saber attachment and licensed it to the U.S. Army.

★ 1859 – Accompanied Colonel Robert E. Lee to Harpers Ferry as aide-de-camp during John Brown's raid.

- 1861 – (April 22) Promoted to captain.

- 1861 – (May 2) Resigned from the Army.

- 1861 – (May 10) Commissioned as lieutenant colonel in the Virginia infantry.

- 1861 – (July 4) Reassigned to command all cavalry companies in the Army of the Shenandoah.

- 1861 – (July 16) Promoted to colonel.

- 1861 – (July 21) Participated in the First Battle of Manassas.

- 1861 – (September 24) Promoted to brigadier general.

- 1862 – (June 12 - 15) Rode around the Army of the Potomac.

- 1862 – (August 28 - 30) Fought at the Second Battle of Manassas.

- 1862 – (September) Participated in the Maryland Campaign and the Battle of Sharpsburg.

- 1862 – (October) Rode around the Army of the Potomac again.

- 1862 – (December 11-15) Protected Jackson's flank during the Battle of Fredericksburg.

- 1862 (December) Conducted a raid north of the Rappahannock River.

- 1863 – (April 30 - May 6) Assumed command of the Second Corps at Chancellorsville.

- 1863 – (June 9) Victorious at the Battle of Brandy Station.

- 1863 – (June - July) Participated in the Pennsylvania Campaign.

- 1863 – (October - November) Participated in the Bristoe Campaign.

- 1864 – (May 5-7) Fought in the Battle of the Wilderness.

- 1864 – (May 8) Fought at Spotsylvania Courthouse.

- 1864 – (May 11) Wounded at the Battle of Yellow Tavern.

- 1862 – (May 12) Died of his wounds and was buried in Hollywood Cemetery in Richmond.

Charles Marshall's Address on the Battle of Gettysburg

By Colonel Charles Marshall. From his speech before the Confederate Veteran Association of Washington D.C., January 1896.

In casting about for a subject on which to address you on this occasion, it seemed to me that I could select nothing more interesting than an account of the movements of General Lee's army which resulted in the battle of Gettysburg. I shall not attempt to describe the battle itself, but I think the movements and events which I shall narrate will be found to have had a controlling influence not only in bringing on the engagement, but in determining the result, so far as that result was affected by the circumstances under which the battle was fought. Although it is true that "the battle is not always to the strong," it is equally true that no force, however strong, can dispense with the precautions that will enable it to put forth its entire strength, and to avail itself of all the aid it can get from advantages of position and of the mode of attack or defense.

I propose to consider the subject in the light of the knowledge possessed by the actors in the events I shall describe, and not in the light of our present knowledge, and shall endeavor to confine myself to the contemporaneous reports and correspondence of those who took leading parts, in the latter of which especially can be found an authentic and trustworthy record of the reasons and motives that controlled their conduct, and of the knowledge of facts upon which their judgments were formed. In other words, I desire to present to you the facts, not as they actually were, but as they appeared at the time to those who were called upon to direct the affairs of which I shall speak.

All who have read what has been written by some of those who took a prominent part in the events of that time will not fail to observe how much the writers are influenced in their judgment of the conduct of others, not to say in their accounts of what they themselves did or advised, by after-acquired information of the facts. Indeed, some of these writers, especially when they are

> *Let us then consider the history of the movements that culminated in the battle of Gettysburg, in the light of the facts as they were known and appeared to General Lee at the time, in order that we may form a judgment of his conduct which will be more just to him than if that conduct be judged as if he knew what we now know.*

writing their memoirs, have developed a degree of military capacity, judgment, and skill, when writing in the light of their present knowledge of facts, which has astounded those who knew them when they were obliged to act upon information derived from the picket-line, from reconnaissances, from scouts, from citizens, from deserters, and other sources of knowledge upon which those in charge of military movements are often obliged to depend.

Those who enjoy the great advantage of a full knowledge of facts in writing of what they advised or did, it will be seen, are usually very positive, and are always right; but so far as what is called the truth of history is concerned, their narratives of what they advised or planned or of what they did, it must be confessed, sometimes do violence to the actual facts.

These writers remind me of something that General Lee once said to me.

While the Confederate army lay on the Rapidan, in the winter of 1863, a report reached General Lee that a change had been made in the disposition of his troops by the enemy on the other side of the river, opposite the extreme right of our line, which, if true, required a corresponding change on our part. He sent me to General Ewell, who commanded on our right, to inform him of the report, and instruct him to make a change in the disposition of the troops to meet that reported on the part of the enemy.

It was a long ride, as General Ewell had heard the same report and had gone to our extreme right, several miles below his headquarters. But when I found him he told me that he had already heard the report, but had discovered that it was incorrect, and that the enemy had made no change. Of course, I did not give him General Lee's order as to changing the location of his troops.

A Lesson in Obedience

I reached our camp about dark and reported what General Ewell had told me and said that I had withheld General Lee's order about changing the position of the troops. General Lee expressed his satisfaction, and told me to get ready for dinner as there were one or two foreign officers to dine with us. I sat at the lower end of a long table in the mess tent, and after dinner conversation became general, and the subject of the report I have mentioned and of my expedition to General Ewell was referred to.

General Lee, with an amused expression, suddenly called to me from his end of the table:

"Colonel Marshall, did you know General Twiggs?"

I replied that I had never met General Twiggs, but that I knew something of him from the history of the Mexican war.

General Lee then said: "General Twiggs had a way of instilling instruction that was very effective, and no one ever forgot a lesson taught by him. When he went to Mexico he had a number of young officers connected with his staff who were without experience but very zealous and desirous to do their duty thoroughly. Sometimes they undertook to change General Twiggs' orders, and would fail to do what he told them to do, or would do it not as the general had ordered it to be done. If General Twiggs remarked upon such liberties being taken with his orders, these gentlemen were always ready to show that they were right and that General Twiggs' order was wrong.

"The General bore with this without complaint or rebuke for some time, but one day a young officer came to report his execution of an order General Twiggs had given him, and reported that when he reached the place where the thing ordered by General Twiggs was to be done, he had found that circumstances were so entirely different from what General Twiggs had supposed that he thought that the General would not have given the order had he known the facts, and was proceeding to satisfy General Twiggs that what the young officer had done was the best under the circumstances. But General Twiggs interrupted him by saying: 'Captain, I know you can prove that you are right, and that my order was wrong, in fact you gentlemen are always right, but for God's sake do wrong sometimes.'"

Although General Lee was satisfied with what I had done on this occasion, he wished to impress the lesson of a literal obedience to orders on my mind, and you may be sure that I never forgot it, when it was possible to refer any doubtful matter back to him for further instructions.

So I think if some of the writers of whom I am speaking would put themselves in the position in which they were when the things of which they write occurred, they would not be perhaps as infallible and as far-seeing as they now make themselves appear, but the truth of history would suffer less if they would "do wrong sometimes."

Let us then consider the history of the movements that culminated in the battle of Gettysburg, in the light of the facts as they were known and appeared to General Lee at the time, in order that we may form a judgment of his conduct which will be more just to him than if that conduct be judged as if he knew what we now know.

Of course, this involves the inquiry as to the accuracy of his knowledge, as to the means he took to inform himself, and as to the discernment he showed in arriving at the truth from a consideration of such facts as were brought to his attention. I think one of the most striking traits of General Lee's mind was his ability to form a correct judgment from all the facts and circumstances that came to his knowledge. This was strikingly illustrated in several important movements. For example, he decided the critical question as to the withdrawal of the Confederate army from Richmond after the battles around that city, in 1862, leaving the large army of General McClellan almost within cannon-shot of the city, trusting to the correctness of his interpretation of a single circumstance and of his estimate of the enterprise of his opponent.

When General McClellan was forced to abandon his fortified position on the Chickahominy and retire to Harrison's landing, on the James, his army was too strong to be left within thirteen miles (as the crow flies) from Richmond, while the army that defended the city moved northward, if there was any reason to apprehend that the Federal commander intended to renew the attempt to capture the place. Immediately after the withdrawal of General McClellan from the Chickahominy to the James, General Lee had dispatched General Jackson, with his own command and that of General Ewell, followed by that of General A.P. Hill, northward to meet the army of General Pope, then advancing along the line of the Orange and Alexandria railroad. Jackson was instructed to cross the Rapidan and attack Pope's advance.

Among other consequences of the defeat of General McClellan before Richmond, Federal troops had been drawn to his support from various other parts of the country, and among them was a large part of the force under General Burnside, on the North Carolina coast. These troops arrived in Hampton Roads and lay there in transports. Upon them the attention of General Lee was immediately concentrated. Their movements would decide his. If they sailed up the James to reinforce McClellan, the latter, being reinforced, intended to renew the attack on Richmond, and General Lee must remain there. If, on the other hand, Burnside sailed up the Chesapeake, McClellan, not being reinforced, did not intend to renew his attempt, but the real attack on Richmond must be looked for from the army of General Pope.

Lee's Accurate Interpretation

Our scouts reported at last that the transports of Burnside had sailed up the Chesapeake, and that night the troops of Longstreet left Richmond and moved northward to the Rapidan, leaving General McClellan at Harrison's landing, with the confident expectation on the part of General Lee that the northward movement of his army would lead to the withdrawal of the Federal army from the James. How accurate General Lee's interpretation of Burnside's movement was we now know, and from that time until some time after the Second Battle of Manassas he practically directed the movements of the Federal army by his own. Another instance of his wonderful capacity in penetrating the intentions of the enemy occurred at Fredericksburg before the Battle of Chancellorsville in 1863.

> *I fear he will steal a march on us and get across the Potomac before we are aware."*

The enemy displayed a large force in our front on the Stafford side of the river, and at the same time another force with infantry and artillery was reported to be on the Rappahannock above Fredericksburg, in our rear. For several days it was doubtful from which quarter the attack would come, but on the afternoon of April 30th, General Lee, after a long examination of the large force displayed on the opposite hills of Stafford, suddenly closed his field glass and remarked, "The main attack will come from above." Within a few hours Jackson's corps was marching towards the illustrious field of Chancellorsville, and its great leader to his last and crowning victory.

I will now proceed to give an account of the movements which began on the 3rd of June, 1863.

The Federal army was opposite Fredericksburg, where it could not be attacked, except at a disadvantage, and we are told by General Lee that the object of his movement was to draw that army from its position, and, if practicable, to transfer the scene of hostilities beyond the Potomac. He also says that "the execution of this purpose embraced the expulsion of the force under General Milroy, which had infested the lower Shenandoah Valley during the preceding winter and spring. If unable to obtain the valuable results which might be expected to follow a decided advantage gained over the enemy in Maryland or Pennsylvania, it was hoped that we should at least so far disturb his plan for the summer campaign as to prevent its execution during the season of active operations."

The commands of Longstreet and Ewell were put in motion on the 3rd of June in the direction of Culpeper Courthouse. On the 5th of June, as soon as their march was discovered by the enemy, he threw a small force across the Rappahannock about two miles below Fredericksburg, and it was thought prudent to halt the command of General Ewell until the object of that movement could be ascertained, but the movement itself, as General Lee says in a letter dated June 7, 1863 "was so devoid of concealment" that he supposed that its object was to ascertain what troops remained near Fredericksburg, and after watching the enemy during the next day, and finding that no advance was made, and that the force displayed on the Stafford side of the river was not larger than could be dealt with in case it should cross by the corps of A.P. Hill, General Ewell was directed to resume his march, and he and Longstreet on the 7th encamped around Culpeper Courthouse.

Orders to Ewell

Knowing by past experience the sensitiveness of the government of the United States to any demonstration in the direction of Washington by way of the Valley, he then ordered General Ewell to move from Culpeper Courthouse to Winchester, to attack the enemy in the Valley, and drive him across the Potomac. The appearance of Ewell in the Valley and his attack on the enemy at Winchester and Berryville resulted, as General Lee had expected, in the disappearance of the Federal army from the Rappahannock, opposite Fredericksburg, and A.P. Hill, in accordance with his instructions, immediately took up his march to join General Lee.

In order to cover Hill's movement, Longstreet, with his corps, was directed to advance along the east side of the Blue Ridge, threatening Washington, with a view to induce the enemy to place his army in a position to cover that city, and to divert him from A.P. Hill.

Longstreet left Culpeper Courthouse on the 15th of June, and occupied Ashby's and Snicker's Gaps, in the Blue Ridge. General Stuart, with three brigades of cavalry, moved on Longstreet's right, and took possession in front of the two gaps. The cavalry brigades of Hampton and W.E. Jones remained along the Rappahannock and Hazel rivers, in front of Culpeper Courthouse, with instructions to follow the main body of the army as soon as Hill's Corps had passed that point.

There was much skirmishing between the cavalry of the two armies during the next three days, General Stuart taking a position west of Middleburg, where he awaited the rest of his command. General Jones arrived on the 19th, and General Hampton on the afternoon of the following day.

On the 21st, Stuart was attacked by infantry and cavalry and forced to fall back to the gaps of the mountains. The enemy retired the next day, having advanced only a short distance beyond Upperville. The Federal army was apparently guarding the approaches to Washington and manifested no disposition to assume the offensive. In the meantime the progress of Ewell, who was already in Maryland with Jenkins's Cavalry Brigade, advanced into Pennsylvania as far as Chambersburg, rendered it necessary that the rest of the army should be within supporting distance, and Hill, having arrived in the Valley, Longstreet was withdrawn to the west side of the Shenandoah, and the two corps encamped near Berryville. General Stuart was directed to hold the mountain passes with part of his command as long as the enemy remained south of the Potomac, and with the remainder to cross into Maryland and place himself on the right of General Ewell, as he moved northward.

General Stuart, having suggested that he could delay the enemy in crossing the Potomac by going in his rear, he was authorized to do so, and it was left to his discretion whether to enter Maryland east or west of the Blue Ridge, but he was instructed to lose no time in placing his command on the right of our column as soon as he perceived that the enemy was moving northward.

As the movement of the cavalry at this time has been much discussed, and perhaps had more to do with the events that immediately followed than any other circumstance, I shall confine myself in stating those movements to the contemporaneous orders and correspondence.

> *I think your best course will be towards the Susquehanna, taking the routes by Emmettsburg, Chambersburg, and McConnellsburg.*

A Great Error

That a great error was committed in the movements of General Stuart cannot be questioned. The object of the movement proposed by him in the rear of the enemy was to strike the line of the latter, who was then marching towards the Potomac from opposite Fredericksburg, his line of march being east of the Bull Run Mountains, and it will be observed that while General Stuart had the discretion to cross the Potomac river, either east or west of the Blue Ridge, his instructions to lose no time in placing his command on the right of our column as soon as he should perceive the enemy moving northward were imperative.

The Federal army was assembling in Loudoun, and for the purpose of ascertaining our movements, strong reconnaissances were made by his cavalry, sometimes supported by infantry.

After the affair at Upperville, on the 21st of June, Stuart remained on the east of the Blue Ridge, in front of Longstreet, one division of whose corps had been recalled from the west of the Shenandoah

River, to aid the cavalry at the time of the attack at Middleburg.

General Longstreet remained on the east of the Blue Ridge, while the headquarters of the army were moved to the west of the Shenandoah, near Berryville. The following letter from General Lee to General Stuart, written on the 22nd of June, will explain the condition of affairs at that time:

Headquarters, June 22nd, 1863.
Major-General J.E.B. Stuart

General, I have just received your note of 7:45 this morning to General Longstreet. I judge the efforts of the enemy yesterday were to arrest our progress, and ascertain our whereabouts. Perhaps he is satisfied. Do you know where he is, and what he is doing? I fear he will steal a march on us and get across the Potomac before we are aware. If you find that he is moving northward, and that two brigades can guard the Blue Ridge and take care of your rear, you can move with the other three into Maryland and take position on General Ewell's right. Place yourself in communication with him, guard his flank, keep him informed of the enemy's movements, and collect all the supplies you can for the use of the army. One column of General Ewell's army will probably move toward the Susquehanna by the Emmettsburg route, another by Chambersburg. Accounts from him last night stated that there was no enemy west of Frederick. A cavalry force (about 100) guarded the Monocacy bridge, which was barricaded. You will, of course, take charge of A.G. Jenkins's Brigade, and give him necessary instructions. All supplies taken in Maryland must be by authorized staff officers for their respective departments – by no one else. They will be paid for, or receipts for same given to the owners. I will send you a general order on this subject, which I wish you to see is strictly complied with.

I am, very respectfully,
R.E. Lee, General

Letters to Ewell

On the same day General Lee wrote the following letter to General Ewell, who had crossed the Potomac at Shepardstown:

June 22, 1863
Lieutenant General R. S. Ewell:

General, Your letter of 6:00 p.m. yesterday has been received. If you are ready to move you can do so. I think your best course will be towards the Susquehanna, taking the routes by Emmettsburg, Chambersburg, and McConnellsburg. Your trains had better be, as far as possible, kept on the center route. You must get command of your cavalry, and use it in gathering supplies, obtaining information, and protecting your flanks. If necessary, send a staff officer to remain with General Jenkins. It will depend upon the quantity of supplies obtained in that country whether the rest of the army can follow. There may be enough for your command, but none for the others. Every exertion should, therefore, be made to locate and secure them. Beef we can drive with us, but bread we cannot carry, and must secure it in the country. I send you copies of a general order on this subject, which, I think, is based on rectitude and sound policy, and the spirit of which I wish you to see enforced in your command. I am much gratified at the success that has attended your movements, and feel assured that if they are conducted with the same energy and circumspection it will continue. Your progress and direction will, of course, depend upon the development of circumstances. If Harrisburg comes within your means capture it. General A.P. Hill arrived yesterday in the vicinity of Berryville. I shall move him on today, if possible. Saturday, Longstreet withdrew from the Blue Ridge. Yesterday the enemy pressed our cavalry with infantry and cavalry on the Upperville road, so that McLaws had to be sent back to hold Ashby's Gap. I have not yet heard from there this morning. General Stuart could not ascertain whether it was intended as a real advance towards the Valley or to ascertain our position.

I am, with great respect,
R. E. Lee, General

Later on the same day General Lee wrote the following letter to General Ewell:

Headquarters, June 22, 1863 3:30 p.m.
General, I have just received your letter of this morning from opposite Shepherdstown. Mine of today authorizing you to move towards the Susquehanna, has reached you ere this. After dispatching my letter, learning that the enemy had not renewed his attempt of yesterday to break through the Blue Ridge, I directed General R.H. Anderson's Division to commence its march towards Shepherdstown. It will reach there tomorrow. I also directed General Stuart, should the enemy so far have retired from his front as to permit of the departure of a portion of the cavalry, to march with three brigades across the Potomac, and place himself on your right and in communication with you, keep you advised of the movements of the enemy, and assist in collecting supplies for the army. I have not heard from him since. I also directed Imboden, if opportunity occurred, to cross the Potomac and perform the same offices on your left.

I am, most respectfully, yours,
R. E. Lee, General

The letter of General Lee to General Stuart of the 22d of June, 1863, giving him specific directions as to his movements, which directions are communicated to General Ewell in General Lee's second letter to that officer of the same date, which I have quoted, was sent by, General Lee through General Longstreet, who was on the east side of the Blue Ridge, and under whose immediate command General Stuart was.

I have not a copy of the letter from General Lee to General Longstreet enclosing General Lee's letter to General Stuart, but I have a copy of the letter from General Longstreet to General Lee acknowledging the receipt of the letter of the latter to General Stuart, containing the order which I have mentioned. It is as follows:

Headquarters, June 22, 1863--7:30 p.m.
General R. E. Lee, Commanding
General, Yours of 4 o'clock this afternoon was received. I have forwarded your letter to General Stuart, with the suggestion that he pass to the enemy's

rear, if he thinks he may get through. We have nothing of the enemy today.

    Most respectfully,
    James Longstreet,
    Lieutenant General Commanding

You will observe that the letter of General Lee to General Stuart, which I have quoted, and which General Stuart received through General Longstreet, contained an order to the former, in case he found that the enemy was moving northward, and that he could protect his rear with two brigades of his force, to move the other three into Maryland and take position on General Ewell's right, place himself in communication with him, guard his flank, and keep him informed of the enemy's movements. This order was sent through General Longstreet, under whose immediate command General Stuart then was, leaving General Longstreet to decide whether the cavalry could be spared to execute the order, and also to direct how it should best move to carry it out in view of the state of things existing when the order was delivered to General Stuart.

What Lee Expected

The letter of General Lee to General Stuart, however, shows that when it was written General Lee expected that General Stuart would pass with all his cavalry, except two brigades, to the west of the Blue Ridge, and cross the Potomac on that side of the mountains, leaving two brigades in the gaps to guard his rear as long as the enemy threatened to attempt to penetrate through the gaps into the Valley.

The letter of General Lee to General Ewell informing that officer of the order General Lee had given to General Stuart, if General Longstreet decided that Stuart could be spared, shows very clearly that the movement that General Lee assumed would be made by General Stuart was to cross into Maryland, and put himself on the right of General Ewell.

The letter of General Longstreet to General Lee, which I have quoted, acknowledging the receipt of General Lee's letter to General Stuart, states that General Longstreet had forwarded that letter with the suggestion that the latter should pass to the enemy's rear, "if he thinks he can get through."

What General Longstreet calls a "suggestion" was, in effect, an order, as will be seen. It was as follows:

Millwood, June 22, 1863 – 7 p.m.
    Major General J.E.B. Stuart, Commanding Cavalry:
    General, General Lee has enclosed to me this letter for you, to be forwarded to you, provided you can be spared from my front, and provided I think you can move across the Potomac without disclosing our plans. He speaks of your leaving via Hopewell Gap, and passing by the rear of the enemy. If you can get through by that route, I think you will be less likely to indicate what our plans are than if you should cross by passing to our rear. I forward the letter of instructions with these suggestions. Please advise me of the condition of affairs before you leave, and order General Hampton, whom, I suppose, you will leave here in command, to report to me at Millwood, either by letter or in person, as may be most agreeable to him.

Most respectfully,
James Longstreet,
Lieutenant General

N.B. – I think your passage of the Potomac by our rear at the present moment will, in a measure, disclose our plans. You had better not leave us, therefore, unless you can take the proposed route in the rear of the enemy.

In effect, General Longstreet tells General Stuart that he had better not leave the army unless he could take the proposed route in the "rear of the enemy," and his "suggestion" substantially amounted to an order to Stuart not to leave the army for the purpose of crossing into Maryland, as directed by General Lee's letter, unless he could do so by that route.

It will be seen that the order of General Longstreet to General Stuart, contained in the letter of the former, which I have just read, appears to be controlled entirely by the idea that General Stuart was to cross the Potomac in such a way as would best conceal the movements of the Confederate army, but it does not notice the positive instruction contained in General Lee's letter to General Stuart, should the latter cross the Potomac, to place himself as speedily as possible, after the enemy begun to move northward, upon General Ewell's right.

You will remember that the order of General Longstreet to General Stuart at the time he sent him General Lee's letter was that he should proceed by way of the enemy's rear to reach the Potomac and cross into Maryland. Now, it must be borne in mind that this suggestion contemplated the possibility of the entire detachment of the cavalry from the rest of the army. To obey the order Stuart had to pass through the Bull Run Mountains across the enemy's line of march from opposite Fredericksburg to the Potomac river, if the way was open. That line of march was east of the Bull Run mountains. The cavalry under Stuart was on the east side of the Blue Ridge, and the enemy was already known to be assembling on the Potomac, in Loudoun, so that General Stuart's march ordered by General Longstreet would take the cavalry east of the Bull Run mountains and bring it to the Potomac river, below where the enemy's army was being concentrated. Of course this might readily prove to be inconsistent with the chief aim of the movement ordered by General Lee, which was that General Stuart should place himself on the right of General Ewell after crossing the river, and there was evident danger that if General Stuart acted under the order of General Longstreet, and the enemy should cross the Potomac before General Stuart, the latter would be separated from General Ewell, who was moving west of the Blue Ridge.

> **To obey the order Stuart had to pass through the Bull Run Mountains across the enemy's line of march from opposite Fredericksburg to the Potomac River, if the way was open.**

Lee to Stuart Again

But there is another letter from General Lee to General Stuart, dated on the 23d of June, at 5:00 p.m., which is as follows:

Headquarters Army Northern Virginia,
June 23, 1863 – 5:00 p.m.
Major General J.E.B. Stuart, Commanding Cavalry:

General, Your notes of 9 and 10:30 a.m. today have just been received. As regards the purchase of tobacco for your men, supposing that Confederate money will not be taken, I am willing for your commissaries or quartermasters to purchase this tobacco and let the men get it from them, but I can have nothing seized by the men. If General Hooker's army remains inactive, you can leave two brigades to watch him, and withdraw the three others; but should he not appear to be moving northward I think you had better withdraw this side of the mountains tomorrow night, cross at Shepherdstown next day, and move over to Fredericktown. You will, however, be able to judge whether you can pass around their army without hindrance, doing them all the damage you can and cross the river east of the mountains. In either case, after crossing the river, you must move on and feel the right of Ewell's troops, collecting information, provisions, etc. Give instructions to the commanders of the brigades left behind to watch the flank and rear of the army, and in the event of the enemy leaving their front, retire from the mountains west of the Shenandoah, leaving sufficient pickets to guard the passes and bring in everything clean along the valley, closing up on the rear of the army. As regards the movements of the two brigades of the enemy moving towards Warrenton, the commander of the brigades to be left in the mountains must do what he can to counteract them, but I think the sooner you cross into Maryland after tomorrow the better. The movements of Ewell's Corps are as stated in my former letter. Hill's First Division will reach the Potomac today, and Longstreet will follow tomorrow. Be watchful and circumspect in your movements.

I am, very respectfully and truly yours,
R. E. Lee, General

This letter was written and received after General Longstreet's letter to General Stuart of the 22nd of June, enclosing that of General Lee, with the suggestion or order of General Longstreet as to the movement of General Stuart, of which I have spoken, and is General Lee's last direction to General Stuart before the army left Virginia. It was written and received before General Stuart started on his march "around the rear of the enemy."

It covers the case of the Federal commander remaining inactive, and also of his not moving northward. In the former event Stuart was to leave two brigades to watch him, and with the other three to withdraw, and in the latter event Stuart's whole command was to be withdrawn tomorrow night (the 24th), "this side of the mountain," cross the Potomac at Shepherdstown, and move towards Fredericktown the next day.

It also leaves Stuart to decide whether he can move around the Federal army (in either of the events mentioned) without hindrance, doing him all the damage he can, and cross east of the mountains. In either case, after crossing the river, Stuart is directed to move on and feel the right of Ewell's Corps, collecting information, etc.

You will see that whether Stuart should cross the Potomac at Shepherdstown, as General Lee directed, or in the exercise of the discretion given him to pass around the rear of the enemy and cross the Potomac east of the mountains, he was ordered, unconditionally, "after crossing the river," to move on and "feel the right of Ewell's troops, collecting information," etc.

This explicit order precluded any movement by Stuart that would prevent him from "feeling the right of Ewell's troops," after crossing the Potomac, and it was the last order General Stuart received before leaving Virginia.

It will also be observed that General Stuart was not permitted to make this movement around the enemy's rear *unless he could pass around the Federal army without hindrance*, and there was the same conditions annexed to the order of General Stuart, as I have shown. In any case, General Stuart, after crossing the Potomac, was to put himself on the right flank of General Ewell, and that any movement on the part of the former which tended to prevent this was entirely inconsistent with General Lee's reiterated instructions.

So, that, under this instruction, General Stuart was practically instructed not to cross the Potomac east of the Federal army, and thus interpose that army between himself and the right of General Ewell.

> *He lost much valuable time in pursuing and capturing trains coming from that city to General Hooker's army, but as he moved northward the Federal army was also moving northward on his left and separating him from the right of the Confederate army, where it was all important that the cavalry should be.*

There were places where the Potomac could be crossed between the enemy's army, at or near Edward's Ferry, and the Blue Ridge, east of the latter, and General Stuart had discretion to use the fords east of the Blue Ridge, but he had no discretion to use any ford that would place the enemy's army between him and the troops of General Ewell.

A Misconstruction

The report of General Stuart of his operations in this campaign states that he had submitted to General Lee a plan of leaving a brigade or two, to use his own language, "in my present front, and passing through Hopewell, or some other gap in Bull Run Mountains, attain the enemy's rear, pass between his main body and Washington, and cross into Maryland, joining our army north of the Potomac.

"The commanding general wrote me, authorizing this move, if I deemed it practicable, and also what instructions should be given the officer in command of the two brigades left in front of the army. He also notified me that one column would move via Gettysburg, the other by Carlisle, towards the Susquehanna, and directed me, after crossing, to proceed with all dispatch to join the right (Early) in Pennsylvania."

There is no such letter as is mentioned by General Stuart contained in the book, in which are found copies of all the other letters of General Lee to him, which I have cited, and it is inconsistent with the other letters I have quoted on the same subject, written by General Lee to him about the same time. But the report of General Stuart evidently refers to the letter of General Lee of June 23rd, which I have read. That letter contains the instructions to be given "to the officer in command of the two brigades to be left in front of the enemy," mentioned in General Stuart's report as being contained in General Lee's letter to him, which he refers to in his report. It also contains the information as to Ewell's movement referred to in the report, and there can be no doubt that General Lee's letter of June 23d, which I have read, is the letter to which General Stuart refers in his report, and that he construed that letter to mean what he there states. If General Lee wrote another letter, in which he gives the same directions as to the instructions to be given the officer in command of the two brigades left in front of the enemy, and in which he informs General Stuart of the movements of Ewell, and which was also inconsistent with his other letters to Stuart, written about the same time, it would be very strange, and the inference is irresistible that General Lee's letter of June 23d is the one to which General Stuart refers in his report, and that he construed that letter to mean what he there states.

That construction, however, is not justified by the, letter itself.

General Stuart's report then proceeds as follows: "Accordingly, three days' rations were prepared, and on the night of the 24th the following brigades – Hampton's, Fitz Lee's, and W.H.F. Lee's, rendezvoused secretly near Salem depot. We had no wagons or vehicles, except six pieces of artillery, caissons, and ambulances. Robertson's and Jones' Brigades, under command of the former, were left in observation of the enemy on the usual front, with full instructions as to following up the enemy in case of withdrawal, and rejoining our main army. Brigadier-General Fitz Lee's Brigade had to march from north of Snicker's Gap to the place of rendezvous.

At 1:00 at night the brigades, with noiseless march, moved out. This precaution was necessary on account of the enemy's having possession of the Bull Run mountains, which in the daytime commanded a view of every movement in consequence of that location. Hancock's Corps occupied Thoroughfare Gap. Moving to the right, we passed through Glasscock's Gap without serious difficulty and marched for Haymarket. I had previously sent Major Mosby, with some picked men, through to gain the vicinity of Dranesville, and bring intelligence to me, near Gum Spring, today." (You will bear in mind that Haymarket is in Prince William county, east of the Bull Run Mountains, and that was the first point to which General Stuart directed his march, using Glasscock's Gap in the mountains,

Glasscock's Gap being further to the south than Hopewell.)

"As we neared Haymarket we found that Hancock's Corps was en route through Haymarket for Gum Spring, his infantry well distributed through his trains. As Hancock had the right of way on my road, I sent Fitz Lee's Brigade to Gainesville to *reconnoitre*, and devoted the remainder of the day to grazing our horses, the only forage procurable in the country. The best of our information represented the enemy still at Centreville, Union Mills, and Wolf Run Shoals. I sent a dispatch to General Lee concerning Hancock's movement and moved back to Buckland to deceive the enemy. It rained heavily that night. To carry out my original design of passing west of Centreville would have involved so much detention on account of the presence of the enemy that I determined to cross Bull Run further down and pass through Fairfax for the Potomac the next day.

The sequel shows this to have been the only practical course. We marched through Brentsville to the vicinity of Wolf Run Shoals, and had to halt again to graze our horses, which hard-marching, without grain, was fast breaking down. We met no enemy today (the 26th). On the following morning (27th), having ascertained that on the night previous the enemy had disappeared entirely from Wolf Run Shoals, a strongly-fortified position on the Occoquan, I marched to that point, and thence directly to Fairfax station, sending General Fitz Lee to the right to cross by Burke station and effect a junction at Fairfax Courthouse, or further on, according to circumstances.

Reaching Fairfax Courthouse, a communication was received from Brigadier-General Fitz Lee from Avondale. At these two points there were evidences of very recent occupation, but the evidence was conclusive that the enemy had left this point entirely, the mobilized army having the day previous moved over towards Leesburg, while the locals had retired to the fortifications near Washington. I had not heard yet from Major Mosby, but the indications favored my successful passage in the rear of the enemy's army. After a halt of a few hours to rest and refresh the command, which regaled itself on stores left by the enemy in the place, the march was resumed at Dranesville late in the afternoon. The campfires at Sedgwick's (Sixth) Corps, just west of the town, were still burning, it having left that morning.

General Hampton's Brigade was still in the advance, and was ordered to move directly for Rowser's Ford on the Potomac, Chambliss' Brigade being held at Dranesville until Brigadier-General Fitz Lee could close up. As General Hampton approached the river, he fortunately met a citizen who had just forded the river, who informed us that there were no pickets on the other side, and that the river, though fordable, was two feet higher than usual. Hampton's Brigade crossed early in the night, but reported to me that it would be utterly impossible to cross the artillery at that ford. In this the residents were also very positive that vehicles could not cross. A ford lower down was examined, and found quite as impracticable, from quicksand, rocks, and rugged banks.

I determined, however, not to give it up without trial, and before 12:00 that night, in spite of the difficulties, to all appearances insuperable, indomitable energy and resolute determination triumphed. Every piece was brought safely over, and the entire command bivouacked on Maryland soil."

Difficult to Occupy

I shall not quote further from the report of General Stuart what I have read already, showing that he crossed the Potomac east of the army of General Hooker, so as to render it extremely difficult, if not impossible, for him to comply with the repeated injunctions he had received from General Lee to place himself on Ewell's right as soon as he entered Maryland. The report states that General Stuart, on reaching the Maryland side, ascertained that General Hooker had already crossed the Potomac, and that on the day before (June 27th) his army was at Poolesville, moving towards Fredericktown.

General Stuart appears to have thought that his movement was intended to threaten Washington. He lost much valuable time in pursuing and capturing trains coming from that city to General Hooker's army, but as he moved northward the Federal army was also moving northward on his left, and separating him from the right of the Confederate army, where it was all important that the cavalry should be.

The report says, speaking of the capture of a large train coming from Washington: "The capture and securing of this train had for the time scattered the leading brigade. I calculated that before the next brigade could march this distance and reach the defenses of Washington it would be after dark. The troops there would have had time to march to positions to meet attack on this road. To attack at night with cavalry, particularly unless certain of surprise, would have been extremely hazardous. To wait until morning would have lost much time from my march to join General Lee, without the probability of compensating results. I therefore determined, after getting the wagons under way, to proceed directly north so as to cut the Baltimore and Ohio railroad (now becoming the enemy's main war artery) that night. I found myself now encumbered by about 400 prisoners, many of whom were officers."

He then proceeds to state how he marched northward, cutting the Baltimore and Ohio railroad at several points, and remained in possession of that road nearly all that day, the 28th. Finding that the enemy was moving north through Frederick City, and it being important for him to reach General Lee's army with as little delay as possible "to acquaint the commanding general with the nature of the enemy's movement, as well as to place with his column my cavalry force," he proceeded, following a ridge road to Westminster, which he reached at 5 p.m.

At this place he had a skirmish with a detachment of Federal cavalry, which he pursued a long distance on the Baltimore

> *The movement of General Stuart, as will be perceived, left the army which had passed into Maryland with no cavalry, except the brigade of Jenkins' and White's battalion, which accompanied General Ewell.*

road, a pursuit that took him further away from the army of General Lee.

The line of march taken by General Stuart on the right of the enemy brought on several skirmishes, which consumed much more time, the consequences of the loss of which will be presently described.

Considerable delay was also caused in an effort to save the captured wagon train. Not being able to learn exactly where the Confederate army was, General Stuart proceeded as far north as Carlisle. It was not until the night of the 1st of July that he was informed that General Lee's army was at Gettysburg, and had been engaged that day with the enemy's advance. He reached Gettysburg on the 2nd of July.

The movement of General Stuart, as will be perceived, left the army which had passed into Maryland with no cavalry, except the brigade of Jenkins's and White's battalion, which accompanied General Ewell. It could not look for supplies in a hostile country, except by the use of artillery and wagon-horses, of which, of course, but a small number could be spared for that purpose, and it was, as we shall see, entirely without knowledge of the enemy's movements.

Let us now return to the movements of the main body of the army.

On the 22nd of June General Ewell marched into Pennsylvania with Rodes' and Johnson's Divisions, preceded by Jenkins's Cavalry, taking the road from Hagerstown through Chambersburg to Carlisle, where he arrived on the 27th. Early's Division moved by a parallel road to Greenwood, and, in pursuance of instructions previously given to General Ewell, marched towards York. On the 24th Longstreet and Hill were put in motion to follow Ewell, and on the 27th encamped near Chambersburg. General Imboden's command, which had been directed to cross the Potomac and take position on General Ewell's left, as he moved northward, reached Hancock, while Longstreet and Hill were at Chambersburg, and was directed to proceed to the latter place.

Implicit Confidence in Stuart

General Lee had the most implicit confidence in the vigilance and enterprise of General Stuart. He had not heard from him since the army left Virginia, and was confident from that fact, in view of the positive orders that Stuart had received, that General Hooker's army had not yet crossed the Potomac. He remained at Chambersburg from the 27th to the 29th, and repeatedly observed while there that the enemy's army must still be in Virginia, as he had heard nothing from Stuart.

Assuming that such was the fact, and that the movements of the Confederate army into Pennsylvania had failed to withdraw that of General Hooker from Virginia, contrary to his confident expectation, General Lee began to become uneasy as to the purpose of the Federal commander, and to fear that he contemplated a strong movement against Richmond.

He remarked that such a proceeding on the part of the enemy would compel the immediate return of his own army to Virginia, if it could, indeed, reach Richmond in time to defend the city. The possession of Richmond was absolutely necessary at that time to preserve communication with the South, and its loss would have led to the evacuation of the whole of Eastern Virginia, at least as far South as the Roanoke.

I heard General Lee express this apprehension more than once while we lay at Chambersburg, and the apprehension was due entirely to his hearing nothing from General Stuart. Under these circumstances he determined to take such action as would compel the enemy to leave Virginia, and deter him from any attempt upon Richmond. General Longstreet's Corps was at Chambersburg with the commanding general. General A.P. Hill's Corps was about four miles east of Chambersburg on the road to Gettysburg. General Ewell was then at Carlisle. On the night of the 28th of June I was directed by General Lee to order General Ewell to move directly upon Harrisburg, and to inform him that General Longstreet would move the next morning (the 29th) to his support.

General A.P. Hill was directed to move eastward to the Susquehanna, and, crossing the river below Harrisburg, seize the railroad between Harrisburg and Philadelphia, it being supposed that all reinforcements that might be coming from the North would be diverted to the defense of that city, and that there would be such alarm created by these movements that the Federal Government would be obliged to withdraw its army from Virginia and abandon any plan that it might have for an attack upon Richmond.

Lee's First Information

I sent the orders about 10:00 at night to General Ewell and General Hill, and had just returned to my tent, when I was sent for by the commanding general. I found him sitting in his tent with a man in citizen's dress, whom I did not know to be a soldier, but who, General Lee informed me, was a scout of General Longstreet's, who had just been brought to him.

He told me that this scout had left the neighborhood of Fredericktown that morning, and had brought information that the Federal army had crossed the Potomac, and that its advance had reached Fredericktown, and was moving thence westward towards the mountains.

The scout informed General Lee that General Meade was then in command of the army, and also as to the movements of the enemy, which was the first information that General Lee had received since he left Virginia. He inferred from the fact that the advance of the enemy had turned westward from Frederick that his purpose was to enter the Cumberland Valley south of our army, and obstruct our communication through Hagerstown with Virginia, General Lee said that, while he did not consider that he had complete communication with Virginia, he had all the communication that he needed, as long as the enemy had no considerable force in the Cumberland Valley. His principal need for communicating with Virginia was to procure ammunition, and he thought that he could always do that with an escort, if the valley were free from a Federal force, but should the enemy have a considerable force in the valley this would be impossible.

He considered it of great importance that the enemy's army should be kept east of the mountains, and, consequently, he determined to move his own army to the east side of the Blue Ridge, so as to threaten Washington and Baltimore, and detain the Federal forces on that side of the mountains to protect those cities. He directed me to countermand the orders to General Ewell and General Hill, and to order the latter to move eastward on the road through Cashtown and Gettysburg, and Ewell to march from Carlisle, so as to form a junction with Hill either at Cashtown or Gettysburg, as circumstances might direct. He ordered General Longstreet to prepare to move the next morning, following Hill. The army moved very slowly, and there would have been no difficulty whatever in having the whole of it at Gettysburg by the morning of the 1st of July had we been aware of the movements of the enemy on the other side of the mountains.

You will thus see that the movement to Gettysburg was the result of the want of information, which the cavalry alone could obtain for us, and that General Lee was compelled to march through the mountains from Chambersburg eastward without the slightest knowledge of the enemy's movements, except that brought by the scout. While making this march the only information he possessed led him to believe that the army of the enemy was moving westward from Frederick to throw itself upon his line of communication with Virginia, and the object of the movement, as I have stated, was simply to arrest the execution of this supposed plan of the enemy, and keep his army on the east side of the Blue Ridge.

It would have been entirely within the power of General Lee to have met the army of the enemy while it was moving on the road between Frederick and Gettysburg, or to have remained west of the mountains. It had not been his intention to deliver a battle north of the Potomac, if it could be avoided, except upon his own terms, and yet, by reason of the absence of the cavalry, his own army marching slowly eastward from Chambersburg, and southward from Carlisle, came unexpectedly on the Federal advance on the 1st day of July, a considerable part of the Confederate army having not yet reached the field of battle.

> *The result of General Stuart's action was that two armies invaded Pennsylvania in 1863, instead of one. One of those armies had no cavalry; the other had nothing but cavalry. One was commanded by General Lee, the other by General Stuart.*

How It Was Brought About

I do not propose to enter into the details of the battle of Gettysburg, but only to show you how that battle was brought about, and how it was fought on the first, second, and third days with troops as they arrived, all of whom could readily have been on the ground on the first day.

It has been my object to correct the impression that has prevailed to some extent that the movement of the cavalry was made by General Lee's orders, and that at a critical moment of the campaign he crossed the Potomac River and moved into Pennsylvania, sending the entire cavalry force of his army upon a useless raid. That this is not true I think the evidence I have laid before you abundantly establishes. The suggestion of General Longstreet in communicating the order of General Lee to General Stuart that the latter should pass by the enemy's rear need not have led to the results which I have described.

You will observe that General Longstreet's suggestion to General Stuart was qualified, as was General Lee's letter to Stuart of June 23rd, by saying that the latter should go by the enemy's rear, "if he thinks he may get through." The first movement of General Stuart after leaving Salem Depot early on the morning of the 25th brought him in conflict with General Hancock's Brigade, near Haymarket, and, finding that he could not pass around the rear of the enemy, the discretion so given him by General Longstreet was at an end, and there was yet time for General Stuart to retrace his steps and obey the order that he had received from General Lee in the letter of the 23rd of June, to cross the Potomac west of the Blue Ridge and move on until he felt the right of Ewell's column. But, instead of pursuing this course, General Stuart, as I have already pointed out, moved to Buckland, east of Bull Run Mountain, and proceeded from that place through Brentsville, down to Wolf Run shoals, and thence across the country by way of Fairfax station to the Potomac River. This latter movement was not sanctioned either by the suggestion of General Longstreet or by the positive orders of General Lee, and from the tenor of General Stuart's report it would seem that he entirely mistook the part that he was expected to take in the movement of the army.

He placed himself east of the Federal army, with that army between his command and the Confederate Force. He left General Lee without any information as to the movements of the enemy from the time he crossed the Potomac river until the 2d of July. By his silence, as I have described, he caused General Lee to move his army to Gettysburg, not with the expectation or purpose of meeting the enemy, but simply to prevent a movement which he supposed the enemy was making to obstruct his line of communication with Virginia, and caused him to fight the battle of Gettysburg without having his whole force present except on the third day, when it was equally possible, had General Lee been informed of what the enemy was doing, for him to have fought that battle with his entire force while the enemy's forces were approaching Gettysburg, or to have remained west of the mountains and have met the Federal army on some other field.

The result of General Stuart's action was that two armies invaded Pennsylvania in 1863, instead of one. One of those armies had no cavalry, the other had nothing but cavalry. One was commanded by General Lee, the other by General Stuart.

IN DEFENSE OF STUART

By Colonel John S. Mosby. *San Francisco Dispatch, January 28, 1896.*

I have just read in the *Post* the report of Colonel Charles Marshall's speech at the celebration of the anniversary of General Lee's birthday. It is the argument of an astute advocate and sophist, and utterly destitute of judicial candor. I shall briefly notice and answer the charge he makes that General Stuart, the Chief of Cavalry, violated General Lee's order in the Gettysburg campaign. Fortunately, in this case, the truth does not lie at the bottom of a well:

> How could Stuart be on the Susquehanna and at the same time watch and report Hooker's movements on the Potomac?

1. General Lee expressly says in his report that he gave Stuart authority to cross the Potomac in the rear of the enemy, which is the route he took. Colonel Marshall was a staff-officer of General Lee's, and, of course, knew this fact; yet he did not mention it.

2. He states that Stuart was ordered to place himself on Ewell's right flank, and did not do it. Any one reading the speech would infer that at the date of the instruction Ewell was with General Lee in the Shenandoah Valley, and that Stuart was in default in this respect.

He ignores the important fact that Ewell was then several days' march in advance of General Lee, in Pennsylvania. Of course, Stuart could not be at the same time with General Lee in Virginia and with Ewell in Pennsylvania. He says that Stuart's instructions were to cover the Confederate right as the enemy moved northward. No such instructions were given, but just the reverse. At 5:00 p.m. June 23rd, General Lee wrote to Stuart, who was then east of the Blue Ridge, in Loudoun County:

"If General Hooker's army remains inactive, you can leave two brigades to watch him, and withdraw with the three others; but should he not appear to be moving northward I think you had better withdraw this side of the mountain tomorrow night, cross at Shepherdstown next day (25th), and move over to Fredericktown. You will, however, be able to judge whether you can pass around their army without hindrance, doing them all the damage you can, and cross the river east of the mountains. In either case, after crossing the river you must move on and feel the right of Ewell's troops, collecting information, provisions," etc.

At that time Longstreet's Corps was the rear guard of the army, and Lee's instructions to Stuart were sent through him. On the day before Longstreet had forwarded a similar letter from General Lee, and urged Stuart to go to Ewell by the route around the rear of the enemy. So far from Stuart having been ordered to wait until the enemy moved northward, he was told to go immediately, if they were not moving northward. At that time

Hooker was waiting quietly on General Lee; all of his movements had been subordinate to Lee's. He had moved in a circle *pari passu* with Lee from the Rappahannock to the Potomac so as to cover Washington. When Lee crossed the river, of course Hooker would cross and maintain the same relative position. General Lee knew that it was physically impossible for Stuart to pass the enemy's rear and keep up communication with him; he knew that it would be equally impossible if he crossed the river west of the Blue Ridge at Shepherdstown, and then (in accordance with his orders) moved on over the South Mountain and joined the right of Ewell's column. How could Stuart be on the Susquehanna and at the same time watch and report Hooker's movements on the Potomac?

Marched Day and Night

On June 22nd General Lee had written Stuart, "One column of Ewell's army (under Early) will probably move toward the Susquehanna by the Emmettsburg route – another by Chambersburg." So it was immaterial so far as giving information of Hooker's movements was concerned whether Stuart crossed the Potomac east or west of the Ridge. In either event after crossing he was required to go out of sight of Hooker and to sever communication with General Lee.

Stuart took the most direct route to join the right of Ewell's column, marching continuously day and night to do so. When he reached York he found that Early had been ordered back to Cashtown, the appointed rendezvous of the army. About all this Colonel Marshall says nothing.

3. Colonel Marshall leaves the impression on the reader that Stuart took the whole cavalry corps with him. He knew that Stuart left two brigades of cavalry with Longstreet.

4. Colonel Marshall says that General Lee, at Chambersburg, not having heard from Stuart since he left Virginia, thought that Hooker was still south of the Potomac until on the night of the 28th he learned through a spy that Hooker was moving northward. This is equivalent to saying that General Lee had lost his head, for no rational being could have supposed that Hooker would remain on the south bank of the Potomac while the Confederates were foraging in Pennsylvania. He might as well have disbanded his army.

When General Lee passed Hagerstown on the 26th, he knew that the bulk of Hooker's army was north of the river and holding the South Mountain passes. If Hooker had still been in Virginia there would have been nothing to prevent General Lee from marching direct to Baltimore and Washington. If General Lee had supposed (as Colonel Marshall says he did) that the way was open to capture those cities, he would have marched east, and not north to Chambersburg. General Lee never committed any such military blunder. The spy, therefore, only told General Lee what he knew before.

> *Colonel Marshall ought to study the facts before he makes another speech.*

On the morning of June 28th, at Frederick, Hooker was superseded by Meade. His army remained there that day. Instead of threatening General Lee's communications, as Colonel Marshall says, Meade withdrew the two corps that were holding the mountain passes when General Lee passed through Maryland, and moved his army the next day to the east so as to cover Washington and Baltimore. There was never any interruption of Lee's communications.

5. Colonel Marshall says that General Lee took his army to Gettysburg simply to keep Meade east of the mountain and prevent a threatened movement against his communications. This statement is contradicted by the record. General Lee attached no such importance to his communications – if he had any. The road was open to the Potomac, but it was not a line of supply; his army lived off the country, and took with it all the ammunition it expected to use. On June 25th, after crossing the river, he wrote Mr. Davis: "I have not sufficient troops to maintain my communications, and therefore have to abandon them."

According to Colonel Marshall, he broke up his whole campaign trying to save them. The fact was they were not even threatened, and General Lee knew it. There was continued passing between the army and the river.

6. I deny that General Lee ever ordered his army to Gettysburg, as Colonel Marshall says, or had any intention of going there before the battle began. In an article published in *Belford's Magazine* (October and November, 1891) I demonstrated this fact from the records. Colonel Marshall ought to study them before he makes another speech.

General Heth Quoted

On the morning of June 29th, General Lee ordered a concentration of the army at Cashtown, a village at the eastern base of the mountain, Hill's Corps was in advance; he reached Cashtown June 30th. That night Hill and Heth heard that there was a force of the enemy at Gettysburg; early the next morning Hill, without orders, with Heth's and Pender's Divisions, started down the Gettysburg pike. General Lee was then west of the mountain with Longstreet. Buford's Cavalry was holding Gettysburg as an outpost. Heth was in advance, and soon ran against Buford. There was a pretty stiff fight with the cavalry until Reynolds, who was camped some six miles back, came to his support. Heth says:

"Archer and Davis were now directed to advance, the object being to feel the enemy; to make a forced reconnaissance, and determine in what force the enemy were – whether or not he was moving his forces on Gettysburg. Heavy columns of the enemy were soon encountered." Davis' and Archer's Brigades were soon smashed, and Archer, with a good many of his men, made prisoners. "The enemy," says General Heth, "had now been felt and found to be in heavy force. The division was now formed in line of battle," etc.

The object of a reconnaissance is to get information; after getting the information the attacking force retires. It seems that General Heth ought now to have been satisfied that the enemy was in force, and should have returned to Cashtown – *i.e.*, if he only went to make a reconnaissance. Hill now put in Pender's and Heth's divisions, and says they drove the enemy until they came upon the First and Eleventh Corps that Reynolds had brought up. He says that he went to Gettysburg "to find out what was in my front." He had now found it.

Hill would have been driven back to Cashtown if Ewell had not come to his support. With Rodes' and Early's divisions, he had camped the night before a few miles north of Gettysburg, and had started to Cashtown when he received a note from Hill telling him he was moving to Gettysburg. The battle had then begun. Ewell, not understanding Hill's object in going to Gettysburg, bearing the sound of battle, and no doubt supposing the army was assembling there, turned the head of his column and marched toward Gettysburg. He came up just in time to save Hill.

General Lee was still west of the mountain when he heard the firing. He did not understand it and rode forward at full speed to the battle. He arrived on the field just at the close. The battle had been brought on without his knowledge, and without his orders, and lasted from early in the morning until 4:00 in the evening. It is clear that Hill took the two divisions to Gettysburg just for an adventure.

When General Lee arrived on the field he found about half of his army there. He had been so compromised that he was compelled to accept battle on those conditions and ordered up the rest of his forces. That morning every division of his army was on the march and converging on Cashtown. That night the whole army – infantry, cavalry, and artillery – would have been concentrated at Cashtown, or in supporting distance, if this rash movement on Gettysburg had not precipitated a battle. A British officer – Colonel Freemantle – was present as a spectator, and spent the night of July 1st at General Longstreet's headquarters. In his diary he says:

"I have the best reason for supposing that the fight came off prematurely, and that neither Lee nor Longstreet intended that it should have begun that day. I also think that their plans were deranged by the events of the 1st."

The record shows who is responsible for the loss of the campaign, and that it was not Stuart. There were no orders to make a reconnaissance on July 1st, and no necessity for making one.

The success of the first day, due to the accident of Ewell's arrival on the field when he was not expected was a misfortune to the Southern army. It would have been far better if Ewell had let Hill and Heth be beaten. They had put the Confederates in the condition of a fish that has swallowed a bait with a hook to it.

LONGSTREET AND STUART

By John Mosby. *Richmond Times, February 2, 1896.*

General Longstreet, having acted a great part as a soldier, now appears as the historian of the war. His book will soon be buried in the dust of oblivion, but, fortunately for him, his fame does not rest upon what he has written, but what he has done. No doubt he has had to endure much, as he says, for the sake of his opinions, as every man must who goes in advance of his age, and he has had strong provocation to speak with bitterness of some of his contemporaries, if he spoke of them at all. But his better angel would have told him that much that he has written about his brothers-in-arms would injure his own reputation more than theirs, and that if he had suffered injustice in defending the right, he had the consolation of knowing that

"Only those are crowned and sainted,
Who with grief have been acquainted."

He will not be able to persuade anyone but himself that he was ever the rival of General Lee and Stonewall Jackson, or that Jackson's fame is factitious and due to his being a Virginian. It is not because he was a Virginian that his monument stands on the bank of the "father of waters," and that a great people beyond the sea gave his statue, in bronze, to the State that will cherish his fame as a possession forever.

The Cavalry

I only propose, however, to review that portion of his book that relates to the management of the cavalry in the Gettysburg campaign. He says that on June 19th, "under the impression that the cavalry was to operate with the first corps (Longstreet's) in the general plan, the commander (Stuart) was ordered to follow its withdrawal west of the Blue Ridge and cross the Potomac on its right at Shepherdstown and make his ride towards Baltimore. He claimed that General Lee had given him authority to cross east of the Blue Ridge.

The point at which the cavalry force should cross the river was not determined between the Confederate commander and his chief of cavalry, there being doubt whether the crossing could be made at Point of Rocks between the Union army and Blue Ridge, or between that army and Washington City. That question was left open, and I was ordered to choose between the points named at the moment that my command took up its line of

> *The criticisms of Stuart are all predicated on the idea that Gettysburg was General Lee's objective point; and as Stuart was absent from the first day's battle he must, therefore, have been in default.*

march. So our plans, adopted after deep study, were suddenly given over to gratify the youthful cavalryman's wish for a romantic ride." General Longstreet does not pretend to have any written record or evidence to support his assertion; on the contrary, the record shows that at that time no such plan could have been entertained, or even discussed.

He writes history on the *a priori* principle of the ancient philosophers, who never went outside of their own consciousness to enquire about facts. It is an exercise of imagination, not of memory; if he runs up against a fact then, like a battery or a line of battle that got in his way – so much the worse for the fact. Not that I would insinuate that he has consciously been guilty of invention; but seeing, as he supposes, in the light of events, that certain things ought to have been done, he persuades himself that they were done.

At the above date (June 19th) General Lee had not determined on sending any of his army north of the Potomac, except Ewell's Corps that was in the advance. Only Rodes' and Johnson's Divisions, with Jenkins' Cavalry, had then crossed the river. A.P. Hill's Corps, that had been left at Fredericksburg, had not then reached the Shenandoah Valley. General Lee, with Longstreet's Corps, was about Berryville; Stuart, with the cavalry, was east of the Blue Ridge, guarding the approaches to the gaps; Longstreet on the west, was supporting him. Longstreet was facing east; Hooker in his front, was, of course, facing west.

General Lee's Plans

Now on June 19th, the day that Longstreet says that all their plans of invasion were matured, and Stuart was ordered to follow his corps and cross the Potomac at Shepherdstown, General Lee wrote to Ewell, who, with two of his divisions, was about Hagerstown, Md., Early not having then crossed the river. General Lee says: "I very much regret that you have not the benefit of your whole corps, for with that north of the Potomac you would accomplish as much unmolested as the whole army could perform with General Hooker in its front. … If your advance causes Hooker to cross the Potomac, or separate his army in any way, Longstreet can follow you."

So on June 19th it was uncertain whether Longstreet would cross the river or not. On the 22nd Hill arrived near Charlestown. Ewell was then ordered to enter Pennsylvania with his whole corps; Jenkins' Cavalry was with him. That day (22nd) in a letter to Ewell, General Lee says: "If you are ready to move you can do so. I think your best course will be toward the Susquehanna, taking the routes by Emmettsburg, Chambersburg, and McConnellsburg. It will depend upon the quantity of supplies obtained in that country whether the rest of the army can follow. If Harrisburg comes within your means, capture it." So on the morning of June 22nd it had not been settled that Longstreet and Hill should follow Ewell.

Later in the day (3:30 p.m.) he again writes Ewell: "I also directed General Stuart, should the enemy have so far retired from his front as to permit of the departure of a portion of the cavalry, to march with three brigades across the Potomac, and place himself on your right, and in communication with you, &c. I also directed Imboden, if opportunity offered, to cross the Potomac, and perform the same offices on your left." Ewell marched with two divisions down the Cumberland Valley to Chambersburg: thence to Carlisle, where he halted. Early was detached and sent east through the Cashtown pass in the South Mountain to York.

What the Letters Show

These letters of General Lee's show that Stuart could not have been ordered to march on Longstreet's flank, because (1) Ewell was then in Pennsylvania and Longstreet in Virginia, and (2) Longstreet and Hill had received no orders to march.

The next day General Lee wrote to Mr. Davis: "Reports of movements of the enemy cast of the Blue Ridge cause me to believe that he is preparing to cross the Potomac. A pontoon bridge is said to be laid at Harper's Ferry; his army corps, that he has advanced to Leesburg and the foot of the mountains, appear to be withdrawing. Their attempt to penetrate the mountains has been successfully repelled by General Stuart with the cavalry.

General Ewell's corps is in motion toward the Susquehanna. General A.P. Hill is moving toward the Potomac; his leading division will reach Shepherdstown today. I have withdrawn Longstreet west of the Shenandoah, and if nothing prevents he will follow tomorrow."

General Lee was then satisfied of Hooker's purpose to cross the Potomac. During the time that Stuart was defending the gaps on account of the presence of Longstreet's corps, Stuart was, to some extent, brought under his authority; for convenience, and to preserve concert of action, all of his correspondence with General Lee passed through Longstreet. In this way Lee and Longstreet were both kept informed of the movements of the enemy. On the day that Ewell left Hagerstown (22d), General Lee sent unsealed through Longstreet the following letter of instructions:

Headquarters, June 22, 1863.
Major-General J.E.B. Stuart, Commanding Cavalry

General, I have just received your note of 7:45 this morning to General Longstreet. I judge the efforts of the enemy yesterday were to arrest our progress and ascertain our whereabouts. I fear he will steal a march on us, and get across the Potomac before we are aware. If you find that he is moving northward, and that two brigades can guard the Blue Ridge and take care of your rear, you can move the other three into Maryland, and take position on Ewell's right, place yourself in communication with him, guard his flank, and keep him informed of the enemy's movements, and collect all the supplies you can for the use of the army. One column of General Ewell's army will probably move towards the Susquehanna by the Emmettsburg route, another by Chambersburg.

Stuart is here given discretion as to the route he should go; but the orders to leave Longstreet and go to Ewell are peremptory. Stuart's headquarters were then at Rector's Cross Roads, about twelve miles east of the Ridge. These letters demonstrate how erroneous are the statements of Generals Longstreet and Heth, and of Long, in the romance he published and called the Memoirs of General Lee, that Stuart was ordered to march on the flank of the column with which General Lee was present. He couldn't be on Ewell's flank on the Susquehanna and Longstreet's flank on the Potomac at the same time. Neither would Longstreet have ordered Stuart to remain with him, knowing that General Lee had ordered him to Ewell. All of Stuart's critics have ignored the fact that General Lee ordered Stuart to leave him and go to Ewell. General Longstreet wrote as follows to General Lee:

June 22, 1863 - 7:30 p.m.
General R. E. Lee, Commanding:

General, Yours of 4:00 this afternoon is received. I have forwarded your letter to General Stuart, with the suggestion that he pass by the enemy's rear if he thinks he may get through. We have nothing of the enemy today.
Most respectfully,
James Longstreet,
Lieutenant General, Commanding

Longstreet to Stuart

In the correspondence during this period between Lee, Longstreet, and Stuart this is the first intimation about taking the route in the rear of the enemy, and it seems that General Longstreet suggested it. This is his letter to Stuart:

Millwood, June 22, 1863 7:00 p.m.
Major-General. J.E.B. Stuart,
Commanding Cavalry

General, General Lee has enclosed to me this letter for you, to be forwarded to you, provided you can be spared from my front, and provided I think that you can move across the Potomac without disclosing our plans. He speaks of your leaving via Hopewell Gap, and passing by the rear of the enemy. If you can get through by that route, I think that you will be less likely to indicate what our plans are than if you should cross by passing to our rear. I forward the letter of instructions with these suggestions. Please advise me of the condition of affairs before you leave and order General Hampton, whom I suppose you will leave here in command, to report to me at Millwood, either by letter or in person, as may be the most agreeable to him.
Most respectfully,
James Longstreet,
Lieutenant-General

N.B. I think that your passage of the Potomac by our rear (Shepherdstown), at the present moment, will in a measure disclose our plans. You had better not leave us, therefore, unless you can take the proposed route in rear of the enemy.

In his book General Longstreet says: "The extent of authority with me, therefore, was to decide whether the crossing should be made at the Point of

> *Stuart took the shortest and most direct route to join Early's Division that was then marching east toward York.*

Rocks, or around Hopewell Gap, east of the Union Army." The Point of Rocks is nowhere mentioned in the correspondence, and General Longstreet's own letter is proof that it was not considered as a place for Stuart's crossing. He tells Stuart that it is better to go by the rear of the enemy than by "our rear." Now at that time Longstreet and Hill were in the valley fronting east; the Point of Rocks is twelve miles east of the Blue Ridge; their rear way, then, of course, toward the west.

In crossing at Point of Rocks Stuart would not have been in rear of either army, but in front of both. If, on the contrary, Stuart had come over the Blue Ridge and crossed the Potomac at Shepherdstown, he would have passed in our rear. General Longstreet says: "In the postscript three points are indicated: First, the move along my rear to the crossing at Point of Rocks." As Longstreet was west of the Blue Ridge facing east, and Stuart was east of the Ridge, it is hard to see how he would pass Longstreet's rear in moving to the Point of Rocks. The Point of Rocks is not mentioned in the letter. "Second, my preferred march on my flank to the Shepherdstown crossing."

There is no such preference shown in the letter; just the reverse, as Longstreet urges Stuart not to cross in "our rear," which would have been at Shepherdstown. "Third, the route indicated by General Lee." But in his letter of the 22nd, to Stuart, General Lee indicated no route – he merely ordered Stuart (if General Longstreet could spare him from his front) to join Ewell. Of course he couldn't join Ewell – stay with Longstreet, as they were seventy-five miles apart, and the distance widening.

He further says: "Especially did he (Stuart) know that my orders were that he should ride on the right of my column, as originally designed, to the Shepherdstown crossing." Stuart didn't know anything of the kind – neither did General Longstreet. The record is against him. The very letter that Longstreet forwarded to Stuart from General Lee told him to leave Longstreet and go to Ewell.

Lee's Final Instruction

But General Lee's final instructions to Stuart, dated June 23rd, 5:00 p.m. shows what choice of routes was given to Stuart. General Lee says: "If General Hooker's army remains inactive, you can leave two brigades to watch him and withdraw with the three others; but should he not appear to be moving northward, I think you had better withdraw this side of the mountain to-morrow night, cross at Shepherdstown next day, and move over to Fredericktown. You will, however, be able to judge whether you can pass around their army without hindrance, doing them all the damage you can, and cross the river east of the mountains. In either case, after crossing the river, you must move on, and feel the right of Ewell's troops, collecting information, provisions."

The movements of Ewell's Corps are as stated in my former letter. Hill's first division will reach the Potomac today (23rd), and Longstreet will follow tomorrow." This letter proves that the choice of routes lay between Shepherdstown, and west of the Blue Ridge, or crossing the river in rear of the enemy to the east. It also shows that

Stuart was not to march on the flank of the column with General Lee even if he crossed at Shepherdstown, but to move on through Boonsboro Gap, and put himself on Ewell's right. Stuart took the shortest and most direct route to join Early's Division that was then marching east toward York.

General Longstreet gives himself away when he says: "The first corps was to draw back from the Blue Ridge, and cross the Potomac at Williamsport, to be followed by the cavalry, which was to cross at Shepherdstown, and ride severely towards Baltimore, to force the enemy to eastern concentration." Now Stuart did ride "severely toward Baltimore," and near to the gates of the city. But if he had gone the other way, and crossed at Shepherdstown, and then ridden through Boonsboro Gap to Baltimore, he would have been as far from Longstreet's flank as he was by the route he took in rear of Hooker. He did not, as he says, order Stuart to put Hampton in command of the two brigades that were left behind, for he had no such authority; neither is it true that Robertson was assigned to this command "without orders to report," at his headquarters.

Should Read

Stuart's instructions to Robertson, which, through abundant caution, he repeated to Jones, and all the correspondence to which I have referred, has been published. It may be that he hasn't read it. If he has not, then he ought to stop writing, and go to reading history. The instructions to Robertson says: "you will instruct General Jones, from time to time, as the movements progress or events may require, and report anything of importance to Lieutenant General Longstreet, with whose position you will communicate by relays through Charlestown. I send instructions for General Jones, which please read."

Jones was one of the best outpost officers in the army. Stuart's main reliance was on him. His brigade was at that time much nearer the Potomac than Robertson's. Jones in accordance with Stuart's order places the Twelfth Virginia Cavalry at Charlestown. Longstreet was responsible for the use made of these two brigades, as they were under his orders. It would have been much easier to send a courier back for them from Hagerstown, if the cavalry was needed, than from Chambersburg. He knew that Hooker's army had crossed the river, and was holding the South Mountain passes when he was at Hagerstown. So his spy only told General Lee what he already knew.

It could not have been a surprise to hear at Chambersburg that the Northern army was moving north. There was nothing else for it to do. If when General Lee was at Hagerstown he had supposed that Hooker was still south of the Potomac he would not have moved north, but due east, toward Baltimore and Washington. There is not the slightest evidence to show that in this campaign any injury resulted to the army from want of cavalry. Our communications were never interrupted. General Longstreet speaks of Stuart's movement toward Ewell's right flank as a raid. As I have shown, it was nothing of the kind, but a part of a combined movement of the whole army.

The criticisms of Stuart are all predicated on the idea that Gettysburg was General Lee's objective point; and as Stuart was absent from the first day's battle he must, therefore, have been in default. But General Lee was not present in the battle; he arrived just at the close. On this assumption a plausible theory was invented that the battle was precipitated for want of cavalry. In *Belford's Magazine* (October and November, 1891), in an article on Gettysburg, based on a study of the records, I demonstrated the error; and showed that General Lee never intended to go to Gettysburg, but that Cashtown was his expected point of concentration.

General Heth, General Longstreet, Long and others, had represented Gettysburg to be the strategic point on which General Lee was maneuvering. They forgot that we had held and then abandoned it. Of course, when the base was knocked from under it, the theory fell.

Who was Responsible?

General Longstreet now says that Cashtown was the place where General Lee ordered the concentration. He did not say so in the *Century*. He fails to show the genesis of the battle, and who was responsible for the defeat of General Lee's plans. I will first say that in my opinion General Longstreet was not. Hill, with Heth's and Pender's Divisions, was at Cashtown on the evening of July 30th. General Lee, with Longstreet, was still some distance west of the mountain.

Every division of his army – infantry, cavalry, and artillery – was on the march, and converging on Cashtown on the morning of July 1st. They could all have reached there by night, or in supporting distance. On the evening before (30th), Hill and Heth heard that a body of the enemy had just occupied Gettysburg. Early on the morning of July 1st, Hill, with Heth's and Pender's Divisions, started down without orders to attack them. Before reaching Gettysburg they met Buford's Cavalry on the pike.

Buford held them in check until Reynolds, who had camped some six miles off with two corps, hearing the firing, came to his support. Heth first put two brigades into the fight that were soon knocked to pieces; Archer and most of his brigade were captured. Heth says: "Archer and Davis were now directed to advance, the object being to feel the enemy and to determine in what force the enemy were – whether or not he was massing his forces on Gettysburg. Heavy columns of the enemy were soon encountered. General Davis was unable to hold the position he had gained. The enemy concentrated on his front and flanks in overwhelming force. The 'enemy had now been felt, and found to be in heavy force.'

Hill states substantially the same thing. He put in Heth's other two brigades, and then Pender's Division. He would have been badly beaten, but Ewell, on the march to Cashtown, received a note from Hill, and hearing the firing, came to his rescue. Hill and Heth called the fight, which lasted from about 8:00 a.m. to 4:00 p.m. and in which over 20,000 men were engaged on a side, and five or six thousand killed and wounded on each side, a reconnaissance. If this was a reconnaissance, then what is a battle?

General Lee had not ordered any reconnaissance, and there was no necessity for it was west of the mountain when he heard the firing and did not understand its significance.

It Was a Raid

The object of a *reconnaissance* is to get information, not to fight. Only sufficient force is applied to compel an enemy to develop his strength and display his position. The attacking force then retires. After two of Heth's Brigades had been shattered and heavy columns of the enemy deployed in his front, he knew the enemy was in force and ought to have retired and gone back to Cashtown. The trouble was, Hill had found out too much.

It is plain that this expedition was not a *reconnaissance*, but a raid. A high military authority says: "When once the object of a *reconnaissance* has been gained, a retreat must be sounded even in the middle of a combat." General Lee was in a state of duress when he arrived on the field at the close of the fight. He was compelled to order up the remainder of the army and deliver battle on ground he had not chosen, or fall back to Cashtown, leaving his dead and wounded on the field, and giving the enemy the prestige of victory. It is clear that the want of cavalry had nothing to do either with precipitating the battle or losing it. Stuart was absent on the day it began for the same reason that General Lee was.

This has been written more in sorrow than in anger. It is no pleasure to me to expose the mistakes of others; my motive is to defend the dead, and that arm of the service to which I belonged. It is a sacred duty I owe to the memory of a friend.

To subscribe, send an email to:
thestainlessbanner@gmail.com
or visit our website at www.thestainlessbanner.com

Subscription is free.

The Stainless Banner

An e-zine dedicated to the armies of the Confederacy

Volume 3, Issue 7
August 2012

WHERE THE FIGHTING IS THE HOTTEST
JACKSON AT SECOND MANASSAS

Seven days of fighting around the Confederate capital in late June and early July had thrown the Army of the Potomac back down the Virginia peninsula. Richmond could breathe easier, but General Robert E. Lee, newly appointed Commanding General of the Army of Northern Army, could not. General McClellan had stopped his retreat at Harrison's Landing and once he gathered his courage, he would retrace his steps and threaten Richmond again. More challenges arrived. General Burnside landed 10,000 men at Fort Monroe and moving through northern Virginia were the shattered forces that had faced Stonewall Jackson in the Valley, now united under General John Pope in the newly formed Army of Virginia.

While Jackson's Army of the Valley recuperated from the long months of marching and fighting, word filtered into Richmond that Pope had occupied Culpepper on the Orange and Alexandria Railroad. South of Culpepper was the small town of Gordonsville located on another railroad – the Virginia Central, the only link between Richmond and the fertile Shenandoah Valley. Its protection was paramount to the survival of the army. So, Lee ordered Jackson to occupy Louisa Courthouse, and, once in the Piedmont, to ascertain whether it was safe to occupy Gordonsville.

Jackson transported the Army of the Valley, now the Army of Northern Virginia's Second Corps, to Fredericks Hall via the Virginia Central Railroad. When he disembarked the train, he was surprised to

What's Inside:	
Jackson's Official Report	8
Stuart's Official Report	16
Lee's Campaign Against Pope	23
Isaac Trimble	36
Jackson's Foot Cavalry at Second Manassas	38

www.thestainlessbanner.com

learn that Pope had not yet occupied Gordonsville. The next day, Jackson led a portion of his command into the town. Now, there was nothing to do but wait for Pope to commit his troops.

On August 7th, Pope finally made his move. He sent a portion of his army south. Jackson quickly assembled his troops and marched north searching for battle. He found it and victory at Cedar Mountain.

In the aftermath of Jackson's victory, Lee realized that the danger to his army was now on the Rappahannock and not on the James. Longstreet was ordered to reinforce Jackson. Lee secured the defenses around Richmond then took the train to Gordonsville. He arrived on the evening of the 14th.

The following morning, Lee called a council of war. Jackson informed the commanding general that Pope had made a tactical blunder. In retreating from Cedar Mountain, Pope had encamped his 70,000 men on a peninsula created by the confluence of the Rappahannock and Rapidan Rivers. The bridge at Rappahannock Station was the major escape route off the peninsula. Burn the bridge and trap the Federals.

Speed was of the essence. The Confederates had to move before Pope realized his mistake. Lee ordered Jackson to be prepared to cross the Rapidan the next day, but Longstreet put on the brakes. He was not one who moved quickly, especially if he disagreed with a strategy. And he disagreed now. Lee delayed Jackson's crossing a day, but not because he agreed with Longstreet. His cavalry was scattered throughout the countryside and it would take a day or two to gather them in.

Jackson's divisions left Gordonsville within hours after the meeting ended, marching along the south side of Clark's Mountain to hide his column from Pope's scouts. At nightfall, Jackson arrived at Mt. Pisgah Church on the road leading to Somerville Ford on the Rapidan.

Delays sprang up. First, Stuart was forced to postpone the capture of Rappahannock Station when Fitz Lee sent word he would not arrive at the rendezvous point until the 19th. Second, General R. H. Anderson's division was late arriving from Richmond. The final delay belonged to Longstreet. He was still not ready to march. Lee pushed back Jackson's crossing of the Rapidan until the 19th and then to the 20th.

> *The Confederates had to move before Pope realized his mistake.*

The fortunes of war smiled on Pope. On the evening of the 18th, a group of fugitive slaves warned him that the Confederates were setting a trap. Pope withdrew from the peninsula the next day.

Then the fortunes of war turned against Pope to smile on the Confederates. On the 22nd, Stuart raided Catlett's Station and captured Pope's coat and hat plus valuable information on the disposition of the Union army and the date of McClellan's arrival on the Rappahannock.

That Sunday afternoon, Jackson arrived at Lee's headquarters near Jeffersonton. Lee proposed a daring strategy. Since McClellan planned to disembark his army at Aquia Creek, Pope would be compelled to keep his left flank in contact with his base at Fredericksburg. This opened up the Union right to a flanking movement.

Lee gave Jackson the order to attack the Orange and Alexandria Railroad far in Pope's rear. Longstreet would hold the Army of Virginia in place as long as possible then cross the Rappahannock and join Jackson. This would trap Pope between Lee's two corps.

Jackson had his men on the road early the next morning. Led by his chief engineer, Lieutenant Keith Boswell, who was familiar with the area, the Second Corps marched through Amissville then to Hedgeman's River. Cavalry protected the advance.

For six hours, Jackson marched in secret before Union scouts discovered the long column and hurried to inform Pope, who promptly drew the wrong conclusion. He believed Jackson was headed toward the Valley and away from any confrontation with his army.

Midafternoon, Jackson was at Orleans and as the sun began to set, the head of the column marched into Salem on the Manassas Gap Railroad. The next morning, the column turned east and headed toward the Bull Run Mountains. As Jackson neared Thoroughfare Gap, he sent cavalry up the mountain to make certain that the Gap was not a trap. The cavalrymen reported that the way was clear.

At 3:00 p.m., Stuart, trailed by his cavalry brigades, rode up. "Hello, General!" He shouted at Jackson. "I've got Pope's coat! If you don't believe me, here it is."[1] He waved a blue overcoat at Jackson.

Jackson gave the cavalry leader a salute. "General Stuart, I would much rather you had brought General Pope instead of his coat."[2]

At Gainesville, Jackson turned toward Bristoe Station on the Orange and Alexandria Railroad to cut Pope's communications. As the men flooded into the small station, the sound of an approaching train filled the evening. The men attempted to derail it, but it sped through the station and up the tracks to Manassas where the news of Jackson's march was broadcasted to anyone with ears to hear.

The Confederates were able to stop a second train, but a third train was able to reverse itself in time and get away. Now, news of Jackson's march was broadcasted to all points south of Bristoe Station. Pope's misinterpretation of the Confederate march would soon be corrected.

As Jackson stood among the wreckage of the second train, citizens of the small village told him that the Union army had stockpiled a multitude of stores at Manassas. Jackson knew he had to seize the junction before Pope sent reinforcements to protect his supply depot, but the Second Corps had just marched fifty-six miles in the heat and dust, and the sun had set long ago.

Isaac Trimble volunteered his brigade for the task. Jackson gladly accepted. Trimble along with the 21st North Carolina and 21st Georgia disappeared in the dark. Jackson summoned Stuart and sent him to Manassas to take command of the assault.

When news of the raids at Bristoe Station and Manassas reached the War Department in Washington, the powers that be consulted and decided that the perpetrator was Stuart. To drive Stuart back and secure the bridge over Bull Run Creek until Pope arrived, General Taylor and his New Jersey brigade were dispatched.

Not satisfied with securing the bridge, Taylor threw his men out in a long line of

---

[1] James Robertson, Jr. *Stonewall Jackson: The Man, the Soldier, the Legend.* (New York: MacMillan Publishing, 1997) 551.
[2] Ibid.

battle and advanced toward Manassas to seize the junction. As his men advanced, Taylor was unaware that Jackson was in the area.

When the Federals were 300 yards away from Jackson's position, canister from the Rockbridge Artillery ripped through their ranks. Taylor reorganized his troops and kept coming. It was a slaughter. Finally, Jackson rode out, waving a white handkerchief, shouting for Federals to surrender. The small brigade ran from the field, and the Confederates ran after them, shooting as they chased. With the way now clear, Jackson's men marched into Manassas to feast on the stores Stuart and Trimble had captured.

It had taken some time, but Pope was fully awake to the fact that Jackson had not marched to the Valley but was now astride his communications. He forsook the Rappahannock line, forgot all about Longstreet and headed north to defeat the troublesome Jackson once and for all.

On the afternoon of the 27th, Ewell sent word that the Federals were pressing him hard at Bristoe Station. Jackson evacuated Manassas, burning the stores that had not been transported before he left.

Pope expected Jackson to head west through Thoroughfare Gap and join up with Longstreet, who had left the Rappahannock and was following Jackson's path to Manassas. Instead, Jackson gathered his troops at Groveton and took a position between the turnpike in his front and Bull Run in his rear.

August 28th, noon. A courier found Jackson and alerted him to the fact that a large portion of Pope's army was marching down the turnpike, headed toward Centreville. Jackson wasted no time. He ordered Taliaferro to attack the enemy and ordered Ewell to support the attack. It was a false alarm. The Federals had turned off the turnpike and marched to Manassas.

Two hours later, Rufus King's division, 10,000 strong, marched down the turnpike into Jackson's waiting corps. Taliaferro sent his men forward followed by Trimble. Three batteries rolled into position. At 6:30 p.m., Jackson's men fell on the left flank of the Federals. For the next two and half hours, the battle raged and the casualties mounted. Taliaferro was wounded and Ewell was taken from the field with a severe wound that caused his leg to be amputated. He would be out of action until next summer.

The next morning, Jackson woke to find the Federals in force to the east – between his command and Washington. He drew his lines back into an unfinished rail line along the top of a ridge. Throughout the morning and early afternoon, four substantial Union attacks were launched at Jackson's left at Sudley Springs and the determined men of A.P. Hill's Light Division beat each one back…barely.

As Jackson's men battled, Lee and Longstreet arrived. By early afternoon, the First Corps was deployed. Lee ordered Longstreet to attack immediately and relieve the pressure on Jackson. Longstreet declined. He informed Lee that he needed time to examine the ground and make sure that the Union line did not over lap his to the south.

At Sudley Springs, a desperate Hill told Jackson's aide, Henry Kyd Douglas, that if the Federals came again, he was not sure his men could beat them back.

Two more Federal waves rolled toward the Light Division. Hill's line bent and snapped back only to bend again. Just when it seemed that the line would break into pieces, Jubal Early appeared with six regiments and unleashed a furious assault that tore into the Federals, sapping their strength, courage and will. They fled the field with Early's soldiers at their heels. At the end of the day's battle, Pope made two blunders. He assumed Jackson would use the night to retreat, and he was oblivious that Longstreet was on field.

Jackson did use the night to retreat, but only to the woods for concealment. This sudden disappearance of the Second Corps reinforced Pope's assumption that Jackson had retreated.

In the morning, Federal artillery opened up, this time on Jackson's right, but it was sporadic and half-hearted. As the morning wore on, Jackson began to think the Federals were not serious about attacking.

Pope ordered Fitz John Porter to move around Jackson's right and cut off the Second Corps retreat. Porter's movement was in trouble even before it began. First, Jackson was not retreating and second, Longstreet was in the way.

The sun rose and the day grew hotter. Still, there was no sign of an attack. Finally, at 3:00, a cannon barked from the Federal line signaling the commencement of the long awaited attack. Three lines of Union soldiers, in parade ground formation, marched across the open ground toward the Stonewall Brigade and into the waiting guns of Stephen D. Lee. Blasts of canisters ripped through the lines, but the Federals closed ranks and kept coming.

For the next two hours, the Federals pounded the Confederate defenses.

> **But Lee had emerged and now the Confederates stood not more than thirty miles from Washington City.**

Jackson's men ran out of ammunition and threw rocks to keep the blue lines back. Jackson sent word to Lee for reinforcements. Lee started a division toward Jackson's position and ordered Longstreet to launch his attack. Once again Longstreet said no. This time his reason was that it would take too long to dispatch his men. Instead, he ordered his artillery to fire into the left flank of Pope's assault. Canister ripped into the second and third wave of Pope's men, causing the Federals to flee the field.

Longstreet ordered his men to attack. The First Corps swept down unto the field and into the fray. Jackson was impatient to join the pursuit, but the Second Corps was in a snarl that had to be untangled first. By time the order was given, precious time had been lost. Darkness halted the pursuit before Jackson could seize the familiar ground of Henry House Hill.

In the morning, Lee and Jackson reconnoitered the ground and discovered that the retreating Federals had destroyed the stone bridge over Bull Run. If the Confederates were to pursue, they would have to do so by a large, circuitous route.

The Second Corps plodded down roads that had been turned into mud from last night's rainstorm. As night fell, Jackson called a halt and allowed his exhausted men to get some much needed rest.

The next morning, Stuart rode into Jackson's headquarters and reported a heavy Federal position astride Germantown Road two miles northwest of Fairfax. Jackson put his men on the move. Union cavalry sniped at his flanks warning Jackson that Pope was aware of his advance. As the Second Corps neared Fairfax, Jackson halted to wait for Longstreet to close up. Once Longstreet was within supporting distance, Jackson resumed his march. As the head of the column reached the grounds of the plantation Chantilly and moved on toward Ox Road, Stuart reappeared and informed Jackson that the Federals were just ahead.

Jackson halted again to wait for Longstreet. But the Federals were not about to allow the Army of Northern Virginia to consolidate. They opened the attack just as the skies opened up. The fighting continued even though the powder was wet, and the men could barely see in the driving rain. The Federals withdrew and continued to do so through the rest of the night. The rain made it impossible for the Confederates to continue.

The Battle of Second Manassas was over. At the end of June the Confederates were in dire straits, backed up to the Confederate capital. But Lee had emerged and now the Confederates stood not more than thirty miles from Washington City. Lee would continue his advance into Maryland where the armies would tangle once more on the banks of another creek.

★ ★ ★

I myself see in this war, if the North triumph, a dissolution of the bonds of all society. It is not alone the destruction of our property, but the prelude to anarchy, infidelity, and the ultimate loss of free responsible government on this continent.

Stonewall Jackson

Do You Have a Book on the War You Are Longing to Publish?

THE STAINLESS BANNER PUBLISHING COMPANY
A Full-Service Small Press Dedicated to the
Preservation of Southern Heritage and History

History ✯ Memoir ✯ Biography ✯ Novel ✯ Alternate History

What does The Stainless Banner Publishing Company offer you?

- ✯ Professional editing
- ✯ Dynamic covers
- ✯ Hard cover or paperback
- ✯ Competitively priced books
- ✯ High royalties paid monthly
- ✯ Your book on Amazon, Barnes & Noble and many other outlets
- ✯ Free advertising

What is my investment?

Nothing! **The Stainless Banner Publishing Company** is a full service small press. Once your book is published, your only investment will be for any inventory you wish to keep on hand.

Where do I send my query letter?

No need for an agent or a query letter. Just send an email telling us about your book and include the first three chapters in the body of the email. You will receive a prompt reply, usually within 48 hours. Send your email to books@thestainlessbanner.com

The Stainless Banner Publishing Company
www.thestainlessbanner.com

Jackson's Official Report on the Battle of Second Manassas

Headquarters, Second Corps, Army of Northern Virginia
April 27, 1863

Brigadier General R.H. Chilton,
Assistant Adjutant General,
Headquarters, Department Northern Virginia

General:

I have the honor herewith to submit to you a report of the operations of my command from August 15 to September 5, 1862, embracing the several engagements of Manassas Junction, Bristoe Station, Ox Hill, and so much of the battle of Groveton (on August 28, 29, and 30) as was fought by the troops under my command. On August 15, in obedience to instructions from the commanding General, I left my encampment near Gordonsville and, passing Orange Courthouse, encamped in the evening near Mount Pisgah Church where I remained until the 20th, when, in accordance with my instructions, while General Longstreet was crossing the Rapidan at Raccoon Ford, I crossed the same river at Somerville Ford. The command encamped for the night near Stevensburg.

My command at this time comprised Ewell's, A.P. Hill's, and Jackson's divisions. Ewell's was composed of the brigades of Generals Lawton, Early, Hays (Colonel Forno commanding), and Trimble, with the batteries of William D. Brown, W.F. Dement, J.W. Latimer, W.L. Baithis, and L.E. D'Aquin. A.P. Hill's division was composed of the brigades of Generals Branch, Gregg, Field, Pender, Archer, and Colonel Thomas, with the batteries of C. M. Braxton. H.G. Latham, W.G. Crenshaw, D.G. Mcintosh, Greenlee Davidson, and W.J. Pegram. Jackson's division, commanded by Brig. Gen. William B. Taliaferro, was composed of Winder's brigade, Colonel Baylor commanding; Colonel Campbell's brigade, Major John Seddon commanding; Brigadier General William B. Taliaferro's brigade, Colonel A.G. Taliaferro commanding, and Starke's brigade, with the batteries of Brockenbrough, Wooding, W.T. Poague, Joseph Carpenter, W.H. Caskie and Charles I. Raine.

Major General Stuart, with his cavalry, cooperated during the expedition, and I shall more than once have to acknowledge my obligations for the valuable and efficient aid which he rendered.

Early on the morning of the 21st, the command left its encampment and moved in the direction of Beverly Ford on the Rappahannock, General Taliaferro's command in the lead. On approaching the ford, the enemy was seen on the opposite bank. Batteries of that division, under the direction of Major Shumaker, chief of

> *My command was now in the rear of General Pope's army, separating it from the Federal capital and its base of supply.*

artillery, were placed in position, which, after a short resistance (as reported by General Taliaferro), silenced the enemy's guns and dispersed his infantry. Major General Stuart had crossed with a portion of his cavalry, supported by some pieces of artillery, and after skirmishing with the enemy a few hours, taking some prisoners and arms, returned with the information that the Federal forces were moving in strength upon his position and were close at hand. The enemy soon appeared on the opposite bank and an animated firing was opened and, to a considerable extent, kept up across the river for the rest of the day between the Federal artillery and the batteries of Taliaferro's command.

On the following morning (22nd), the three divisions continued their march up the bank of the Rappahannock, General Ewell in the advance, and crossed Hazel River, one of its tributaries, at Wellford's Mill, near which General Trimble was left with his brigade to protect the flank of our wagon train from the enemy who was moving up the north side of the Rappahannock simultaneously with the advance of our troops on the south side.

About 12:00 p.m., a small party surprised part of the train and captured some ambulances and mules, which were, however, soon recovered and some prisoners taken who gave information that a more considerable Federal force had crossed the river.

About 4:00 p.m., General Trimble, supported by General Hood (who was the advance of Longstreet's command), had a sharp engagement with this force, in which, after gallantly charging and taking a number of prisoners, they drove the residue with severe loss across the river under the protection of the guns of the main body of the Federal Army on the opposite side. In the meantime, the command passed Freeman's Ford, which it found strongly guarded, and moved on to a point opposite the Fauquier White Sulphur Springs where we found the bridge destroyed and other evidence that the enemy was in close proximity.

In the afternoon of the 22nd, the 13th Georgia, Colonel M. Douglass, Brown's and Dement's batteries of four guns each, and Early's brigade crossing over took possession of the Springs and adjacent heights and taking some prisoners and incurring some risk from the rain and sudden rise of the water, which for a few hours cut off communication with the main body. In this critical situation the skill and presence of mind of General Early was favorably displayed. It was deemed advisable not to attempt a passage at that point but to proceed higher up the river. By dawn on the morning of the 24th, General Early, by means of a temporary bridge which had been constructed for his relief, had his troops and artillery safely on the southern side.

On the 24th, there was a fierce cannonade between General Hill's artillery and that of the enemy across the river. In the meantime, General Stuart, who had preceded me, crossed the Rappahannock, striking the enemy in his rear, making his brilliant night attack upon his camp at

> *Learning that the enemy had collected at Manassas Junction, a station about seven miles distant, stores of great value, I deemed it important that no time should be lost in securing them.*

Catlett's Station, capturing many prisoners, personal baggage of General Pope and his dispatch book, containing information of value to us in this expedition. In the evening we moved near Jeffersonton.

Pursuing the instructions of the commanding general, I left Jeffersonton on the morning of the 25th to throw my command between Washington City and the army of General Pope and to break up his railroad communication with the Federal capital. Taking the route by Amissville, crossing Hedgeman River (one of the tributaries of the Rappahannock) at Henson's Mill and moving via Orleans, we reached the vicinity of Salem after a severe day's march and bivouacked there for the night.

On the next day (26th), the march was continued diverging to the right at Salem, crossing the Bull Run Mountain through Thoroughfare Gap, and passing Gainesville reached Bristoe Station on the Orange and Alexandria Railroad after sunset. At Gainesville I was joined by General Stuart, who, after leaving the vicinity of Waterloo Bridge about 2:00 a.m., had by a rapid march come up in time to render all needful assistance. He kept upon my right flank during the residue of the day. My command was now in rear of General Pope's army separating it from the Federal capital and its base of supply.

As we approached Bristoe Station, the sound of cars coming from the direction of Warrenton Junction was heard, and General Ewell divided his force so as to take simultaneous possession of two points of the railroad. Colonel T. T. Munford, with the 2nd Virginia Cavalry, cooperated in this movement. Two trains of cars and some prisoners were captured, the largest portion of the small Federal force at that point making its escape.

Learning that the enemy had collected at Manassas Junction, a station about seven miles distant, stores of great value, I deemed it important that no time should be lost in securing them. Notwithstanding the darkness of the night and the fatiguing march, which would be over thirty miles before reaching the Junction, Brigadier-General Trimble volunteered to proceed there forthwith with the 21st North Carolina, Lieutenant Colonel S. Fulton commanding, and the 21st Georgia, Major T.C. Glover commanding (in all about 500 men) and capture the place. I accepted the gallant offer and gave him orders to move without delay.

In order to increase the prospect of success, Major General Stuart, with a portion of his cavalry, was subsequently directed to move forward, and, as the ranking officer, to take command of the expedition. The duty was cheerfully undertaken by all who were assigned to it and most promptly and successfully executed. Notwithstanding the Federal fire of musketry and artillery, our infantry dispersed the troops placed there for the defense of the place, and captured eight guns, with seventy-two horses, equipments, and ammunition complete, immense supplies of commissary and quartermaster's stores, upward of 200 new tents; and General Trimble also reports the capture of over 300 prisoners and 175 horses, exclusive of those belonging to the artillery, besides recovering over 200 negroes.

The next morning the divisions under command of Generals Hill and Taliaferro moved to Manassas Junction, the division of General Ewell remaining at Bristoe Station. About a mile before reaching the Junction Colonel, W.S.H. Baylor encountered and dispersed a regiment of Federal cavalry. Soon after the advance of the troops from Bristoe Station reached the Junction, they were fired upon by a distant battery of the enemy posted in the direction of the battlefield of Manassas. This artillery was soon driven off and retreated in the direction of Centreville.

Soon after a considerable body of Federal infantry, under Brigadier General Taylor, of New Jersey, came in sight having, it is believed, that morning left Alexandria in the cars and boldly pushed forward to recover the position and stores which had been lost the previous night. The advance was made with great spirit and determination and under a leader worthy of a better cause. Assailed by the batteries of Poague and Carpenter and some of General Hill's division, and apparently seeing that there was danger of its retreat being cut off by our other troops if it continued to move forward, it soon commenced retreating, and being subjected to a heavy fire from our batteries was soon routed, leaving its killed and wounded upon the field. Several brigades of General Hill's division pressed forward in pursuit. In this conflict the Federal commander, General Taylor, was mortally wounded. Our loss was small.

In the afternoon of the same day, heavy columns of the enemy were seen approaching Bristoe Station from the direction of Warrenton Junction and on the right of the railroad. General Ewell promptly made his dispositions to meet them. So soon as the enemy came within range, the batteries of his division opened upon them from their several positions, as did also the 6th and 8th Louisiana and 60th Georgia regiments. By this combined fire, two columns of the enemy, of not less than a brigade each, were driven back; but fresh columns soon supplied their places, and it was obvious that the enemy was advancing in heavy force.

General Ewell's instructions were, if hard pressed, to fall back and join the main command at Manassas Junction, and orders were accordingly given for the withdrawal of his forces north of Broad Run. At the moment of issuing this order, a portion of the troops were actively engaged and the enemy advancing, and yet the withdrawal of the infantry and artillery was conducted with perfect order, General Early closing up the rear. The Federals halted near Bristoe Station, and General Ewell moved without further molestation, Colonel Munford of the 2nd Virginia and Colonel T.L. Rosser of the 5th Virginia Cavalry bringing up his rear to Manassas.

The destruction of the railroad bridge across Broad Run was entrusted to Lieutenant (now Captain) J.K. Boswell of the Engineer Corps under whose superintendence the duty was promptly and efficiently executed.

Orders were given to supply the troops with rations and other articles which they could properly make subservient to their use from the captured property. It was vast in quantity and of great value, comprising 50,000 pounds of bacon, 1,000 barrels of corned beef, 2,000 barrels of salt pork, 2,000 barrels of flour, quartermaster's, ordnance, and sutlers' stores deposited in buildings and filling two trains of cars. Having appropriated all that we could use and unwilling that the residue should again fall into the hands of the enemy who took possession of the place the next day, orders were given to destroy all that remained after supplying the immediate wants of the army. This was done during the night.

General Taliaferro moved his division that night across to the Warrenton and Alexandria turnpike, pursuing the road to Sudley's Mill and crossing the turnpike in the vicinity of Groveton, halted near the battlefield of July 21, 1861. Ewell's and Hill's divisions joined Jackson's on the 28th.

My command had hardly concentrated north of the turnpike before the enemy's advance reached the vicinity of Groveton from the direction of Warrenton. General Stuart kept me advised of the general movements of the enemy, while Colonel Rosser of the cavalry with his command, and Colonel Bradley T. Johnson, commanding Campbell's brigade, remained in front of the Federals and operated against their advance. Dispositions were promptly made to attack the enemy, based upon the idea that he would continue to press forward upon the turnpike toward Alexandria; but as he did not appear to advance in force, and there was reason to believe that his main body was leaving the road and inclining toward Manassas Junction, my command was advanced through the woods, leaving Groveton on the left, until it reached a commanding position near Brawner's house. By this time it was sunset; but as his column appeared to be moving by with its flank exposed, I determined to attack at once, which was vigorously done by the divisions of Taliaferro and Ewell. The batteries of Weeding, Poague, and Carpenter were placed in position in front of Starke's brigade and above the village of Groveton, and, firing over the heads of our skirmishers, poured a heavy fire of shot and shell upon the enemy. This was responded to by a very heavy fire from the enemy, forcing our batteries to select another position.

By this time Taliaferro's command, with Lawton's and Trimble's brigades on his left, was advanced from the woods to the open field and was now moving in gallant style until it reached an orchard on the right of our line and was less than 100 yards from a large force of the enemy. The conflict here was fierce and sanguinary. Although largely reinforced, the Federals did not attempt to advance but maintained their ground with obstinate determination.

Both lines stood exposed to the discharges of musketry and artillery until about 9:00, when the enemy slowly fell back, yielding the field to our troops.

The loss on both sides was heavy and among our wounded were Major General Ewell and Brigadier General Taliaferro. The latter after a few months was able to resume his duties; the former, I regret to say, is still disabled by his wound, and the army thus deprived of his valuable services.

This obstinate resistance of the enemy appears to have been for the purpose of protecting the flank of his column until it should pass the position occupied by our troops. Owing to the difficulty of getting artillery through the woods, I did not have as much of that arm as I desired at the opening of the engagement; but this want was met by Major Pelham with the Stuart Horse Artillery, who dashed forward on my right and opened upon the enemy at a moment when his services were much needed.

Although the enemy moved off under cover of the night and left us in quiet possession of the field, he did not long permit us to remain inactive or in doubt as to his intention to renew the conflict.

The next morning (29th), I found that he had abandoned the ground occupied as the battlefield the evening before and had moved farther to the east and to my left, placing himself between my command and the Federal capital. My troops on this day were distributed along and in the vicinity of the cut of an unfinished railroad (intended as a part of the track to connect the Manassas road directly with Alexandria) stretching from the Warrenton turnpike in the direction of Sudley's Mill. It was mainly along the excavation of this unfinished road that my line of battle was formed on the 29th (Jackson's division, under Brigadier General Starke on the right, Ewell's division, under Brigadier General Lawton in the center, and Hill's division on the left).

In the morning, about 10:00, the Federal artillery opened with spirit and animation upon our right, which was soon replied to by the batteries of Poague, Carpenter, Dement, Brockenbrough, and Latimer under Major Shumaker. This lasted for some time, when the enemy moved around more to our left to another point of attack. His next effort was directed against our left. This was vigorously repulsed by the batteries of Braxton, Crenshaw, and Pegram.

About 2:00 p.m., the Federal infantry in large force advanced to the attack of our left, occupied by the division of General Hill. It pressed forward, in defiance of our fatal and destructive fire, with great determination, a portion of it crossing a deep cut in the railroad track and penetrating in heavy force an interval of nearly 175 yards, which separated the right of Gregg's from the left of Thomas' brigade. For a short time, Gregg's brigade on the extreme left was isolated from the main body of the command; but the 14th South Carolina in reserve, with the 49th Georgia, left of Colonel Thomas, attacked the exultant enemy with vigor and drove them back across the railroad track with great slaughter. General McGowan reports that the opposing forces at one time delivered their volleys into each other at the distance of 10 paces. Assault after assault was made on the left, exhibiting on the part of the enemy great pertinacity and determination, but every advance was most successfully and gallantly driven back.

General Hill reports that six separate and distinct assaults were thus met and repulsed by his division, assisted by Hays' brigade, Colonel Forno commanding.

By this time the brigade of General Gregg, which, from its position on the extreme left was most exposed to the enemy's attack, had nearly expended its ammunition. It had suffered severely in its men, and all its field officers except two were killed or wounded. About 4:00, it had been assisted by Hays' brigade. It was now retired to the rear to take some repose after seven hours of severe service, and General Early's brigade, of Ewell's division, with the 8th Louisiana took its place.

On reaching his position, General Early found that the enemy had obtained possession of the railroad and a piece of wood in front, there being at this point a deep cut, which furnished a strong defense. Moving through a field, he advanced upon the enemy, drove them from the wood and railroad cut with great slaughter, and followed in pursuit some 200 yards; the 13th Georgia at the same time advanced to the railroad and crossed with Early's brigade. As it was not desirable to bring on a general engagement that evening, General Early was recalled to the railroad, where Thomas, Pender, and Archer had firmly maintained their positions during the day. Early kept his position there until the following morning.

Brigadier General Field and Colonel Forno (commanding Hays' brigade) were severely wounded. Brigadier General Trimble was also seriously wounded.

During the day, a force of the enemy penetrated the wood in my rear, endangering the safety of my ambulances and train. Upon being advised of this by General Stuart, I sent a body of infantry to drive them from the wood; but in the meantime, the vigilant Pelham had unlimbered his battery and dispersed that portion of them which had reached the wood. At a later period, Major Patrick of the cavalry, who was by General Stuart entrusted with guarding the train, was attacked, and although it was promptly and effectually repulsed, it was not without the loss of that intrepid officer, who fell in the attack while setting an example of gallantry to his men well worthy of imitation. During the day, the commanding general arrived and also General Longstreet with his command.

On the following day (30th), my command occupied the ground and the divisions the same relative position to each other and to the field which they held the day before forming the left wing of the army, General Longstreet's command forming the right wing. A large quantity of artillery was posted upon a commanding eminence in the center.

After some desultory skirmishing and heavy cannonading during the day, the Federal infantry, about 4:00 in the evening, moved from under cover of the wood and advanced in several lines, first engaging the right, but soon extending its attack to the center and left. In a few moments, our entire line was engaged in a fierce and sanguinary struggle with the enemy. As one line was repulsed, another took its place and pressed forward as if determined by force of numbers and fury of assault to drive us from our positions. So impetuous and well sustained were these onsets as to induce me to send to the commanding general for reinforcements, but the timely and gallant advance of General Longstreet on the right relieved my troops from the pressure of overwhelming numbers and gave to those brave men the chances of a more equal conflict.

> *By the following morning, the Federal Army had entirely disappeared from our view, and it soon appeared, by a report from General Stuart, that it had passed Fairfax Courthouse and had moved in the direction of Washington City.*

As Longstreet pressed upon the right, the Federal advance was checked, and soon a general advance of my whole line was ordered. Eagerly and fiercely did each brigade press forward, exhibiting in parts of the field scenes of close encounter and murderous strife not witnessed often in the turmoil of battle. The Federals gave way before our troops, fell back in disorder, and fled precipitately, leaving their dead and wounded on the field. During their retreat the artillery opened with destructive power upon the fugitive masses. The infantry followed until darkness put an end to the pursuit.

Our loss was heavy; that of the enemy, as shown by the battlefield, of which we were in possession, much heavier. Among the losses was Colonel Baylor, commanding Winder's brigade, who fell in front of his brigade while nobly leading and cheering it on to the charge.

We captured eight pieces of artillery with their caissons, and 6,520 small arms were collected from the battlefield.

It being ascertained next morning that the Federal Army had retreated in the direction of Centreville, I was ordered by the commanding general to turn that position. Crossing Bull Run at Sudley Ford thence pursuing a country road until we reached the Little River turnpike, which we followed in the direction of Fairfax Courthouse until the troops halted for the night.

Early the next morning (September 1), we moved forward, and late in the evening, after reaching Ox Hill, came in contact with the enemy, who were in position on our right and front covering his line of retreat from Centreville to Fairfax Courthouse. Our line of battle was formed, General Hill's division on the right, Ewell's division, General Lawton commanding, in the center, and Jackson's division, General Starke commanding, on the left, all on the right of the turnpike road. Artillery was posted on an eminence to the left of the road. The brigades of Branch and Field, Colonel Brockenbrough commanding the latter, were sent forward to feel and engage the enemy.

A cold and drenching thunder shower swept over the field at this time, striking directly into the faces of our troops. These two brigades gallantly engaged the enemy, but so severe was the fire in front and flank of Branch's brigade as to produce in it some disorder and falling back. The brigades of Gregg, Thomas, and Pender were then thrown into the fight. Soon a portion of Ewell's division became engaged.

The conflict now raged with great fury, the enemy obstinately and desperately contesting the ground until their generals (Kearny and Stevens) fell in front of Thomas' brigade, after which they retired from the field. By the following morning the Federal Army had entirely disappeared from our view, and it soon appeared, by a report from General Stuart, that it had passed Fairfax Courthouse and had moved in the direction of Washington City.

On September 3rd, we left Ox Hill, taking the road by Dranesville and Leesburg, and on the 4th bivouacked near the Big Spring between Leesburg and the Potomac.

The official reports of the casualties of my command in its operations from the Rappahannock to the Potomac will show a loss of seventy-five officers killed and 273 wounded, 730 non-commissioned officers and privates killed, 3,274 wounded and thirty-five missing, making a total loss of 4,387.

Colonel Crutchfield, chief of artillery, discharged his duties well. The conduct of officers and men during the various engagements described was such as to entitle them to great praise. The wounded were skillfully cared for by medical director Dr. Hunter McGuire.

In the transmission of orders I was greatly assisted during the expedition by the following members of my staff: Colonel Smead, assistant inspector-general; Major E.F. Paxton, acting assistant adjutant-general; Captain R.E. Wilbourn, chief signal officer; First Lieutenant H.K. Douglas, assistant inspector-general; First Lieutenant J.G. Morrison, aide-de-camp, and Colonel William L. Jackson, volunteer aide-de-camp. Captain Wilbourn was so severely wounded at the battle of Groveton as to be unable to go farther with the army. The ordnance, quartermaster, and commissary departments were well managed by their respective chiefs, Majors G.H. Bier, J.A. Harman, and W.J. Hawks.

For further information respecting the detailed movements of troops and the conduct of individuals, I would respectfully refer you to the accompanying reports.

For these great and signal victories our sincere and humble thanks are due unto Almighty God. We should in all things acknowledge the hand of Him who reigns in heaven and rules among the armies of men. In view of the arduous labors and great privations the troops were called to endure and the isolated and perilous position which the command occupied while engaged with greatly superior numbers of the enemy we can but express the grateful conviction of our mind that God was with us and gave to us the victory and unto His holy name be the praise.

I am, General, very respectfully, your obedient servant,
T.J. JACKSON,
Lieutenant General

Stuart's Official Report on Second Manassas

Headquarters Stuart's Calvary Division, Army of Northern Virginia
February 28, 1863

Brigadier General R.H. Chilton,
Assistant General, Army of Northern Virginia

General:

I have the honor to furnish the following summary of events in which my command participated immediately preceding and subsequent to the second battle of Manassas, or, as it should be more properly termed, the battle of Groveton Heights, August 30, 1862.

My command had hardly recrossed the Rappahannock, as narrated in my last, when that portion of it left on outpost duty on the river became engaged with the enemy, who had advanced to the opposite bank. It was soon apparent that the enemy meditated the destruction of the Waterloo Bridge, the only bridge over the stream then standing. Appreciating its importance to us, I directed the sharpshooters of the two brigades to be sent to its defense, and the command of this party, numbering about 100 men, devolved by selection upon Colonel T.L. Rosser, 5th Virginia Cavalry, whose judgment in posting his command enabled him to prevent the destruction of the bridge in spite of desperate attempts to reach it, and held possession all day and night against infantry and artillery until the next day, when he turned over his position and the bridge intact to a regiment of infantry sent to relieve him.

During the day I sent Captain J. Hardeman Stuart, my signal officer, to capture the enemy's signal party on View Tree, an eminence overlooking Warrenton and establish his own flag instead; the sequel shows with what success.

Colonel Munford's regiment (2nd Virginia Cavalry) was detached for temporary service with General Jackson.

That night (25th) I repaired to the headquarters of the commanding general and received my final instructions to accompany the movement of Major General Jackson, already began. I was to start at 2:00 a.m. and upon arriving at the brigades that night at 1:00 a.m., I had reveille sounded and preparations made for the march at 2:00. In this way I got no sleep, but continued in the saddle all night. I followed by direction the route of General Jackson through Amissville, across the Rappahannock at Henson's Mill, four miles above Waterloo, proceeded through Orleans, and thence on the road to Salem, until, getting near that place, I found my way blocked by the baggage trains and artillery of General Jackson's command. Directing the artillery and ambulances to follow the road, I left it with the cavalry and proceeded by farm roads and bypaths parallel to General Jackson's route to reach the head of his column, which left Salem and the plains early in the morning for the direction of Gainesville. The country was exceedingly rough, but I succeeded, by the aid of skillful guides, in passing Bull Run Mountain without passing Thoroughfare Gap and without incident worthy of record passed through Hay Market and overtook General Jackson near Gainesville and reported to him. Ewell's division was in advance, and to my command was entrusted guarding the two flanks during the remainder of the pending operations.

On the 26th, as Lee's brigade passed Hay Market, he received information of a train of forage wagons of the enemy and sent out promptly a regiment and captured it.

Having made dispositions above and below Gainesville on the Warrenton road with cavalry and artillery, I kept with the main portion on General Jackson's right crossing Broad Run a few miles above Bristoe and intersecting the railroad to the right (south) of that point. The cavalry now fronted toward the main body of the enemy, still in the direction of the Rappahannock, and covered General Jackson's operations on the railroad bridge, on approaching which Colonel Munford's regiment (2nd Virginia), as advance guard, made a bold dash into the place and secured most of the occupants.

About dusk, and simultaneously with the arrival of the command at the railroad,

> *That night (25th), I repaired to the headquarters of the commanding general and received my final instructions to accompany the movement of Major General Jackson, already began.*

trains of cars came rapidly on from the direction of Warrenton Junction and before obstruction could be made the first passed on though fired into by the infantry. Several subsequent ones followed and were captured by the infantry. Details of these operations will no doubt be given by General Jackson and the division commanders.

As soon as practicable, I reported to General Jackson who desired me to proceed to Manassas and ordered General Trimble to follow with his brigade notifying me to take charge of the whole. The 4th Virginia Cavalry (Colonel Wickham) was sent around to gain the rear of Manassas and with a portion of Robertson's brigade not on outpost duty, I proceeded by the direct road to Manassas. I marched until challenged by the enemy's interior sentinels and received a fire of canister.

As the infantry were near, coming on, I awaited its arrival as it was too dark to venture cavalry over uncertain ground against artillery. I directed General Trimble upon his arrival to rest his center directly on the railroad and advance upon the place with skirmishers well to the front. He soon sent me word it was so dark he preferred waiting until morning, which I accordingly directed he should do. As soon as day broke, the place was taken without much difficulty and with it many prisoners and millions of stores of every kind, which his report will doubtless show. Rosser (5th Virginia Cavalry) was left on outpost duty in front of Ewell at Bristoe, and Brien (1st Virginia Cavalry) above Gainesville. During the 27th detachments of Robertson's and Lee's brigades had great sport chasing fugitive parties of the enemy's cavalry.

General Jackson having arrived early in the day took direction of affairs, and the day was occupied mainly in rationing the command, but several serious demonstrations were made by the enemy during the day from the north side, and in this connection, I will mention the coolness and tact of Mr. Louis F. Terrill, volunteer aide to General Robertson, who extemporized lanyards and with detachments from the infantry as cannoneers turned the captured guns with marked effect upon the enemy. Their general (G.W. Taylor, of New Jersey) was killed during this fire.

> During the 27th, detachments of Robertson's and Lee's brigades had great sport chasing fugitive parties of the enemy's cavalry.

Brigadier General Fitzhugh Lee with the 9th, 4th, and 3rd Virginia Cavalry was detached and sent in rear of Fairfax Courthouse to damage the enemy's communication as much as possible and, if possible, cut off the retreat of this party. Colonels Munford and Rosser brought up the rear of General Ewell and that night, when Manassas was destroyed and evacuated, the cavalry brought up the rear, a portion remaining in the place until daylight.

Captain Pelham, arriving late, was indefatigable in his efforts to get away the captured guns, which duty was entrusted specially to him, a part of the command marching by Centreville and a part directly to the stone bridge over Bull Run. Detachments of cavalry were so arranged as to guard both flanks.

The next morning (28th), the main body of Robertson's brigade rendezvoused near Sudley Church. General Jackson's were massed between the turnpike and Sudley

Ford on Bull Run fronting toward Manassas and Gainesville. Colonel Brien (1st Virginia Cavalry) had to retire being hard pressed by the enemy from the direction of Warrenton and was on the turnpike covering Jackson's front toward Gainesville and Rosser toward Manassas where the enemy had also appeared in force early. The remainder of Lee's brigade was still detached on an expedition towards Alexandria.

Early in the day, a dispatch from the enemy had been intercepted giving the order of march from Warrenton toward Manassas and directing cavalry to report to General Bayard at Hay Market. I proposed to General Jackson to allow me to go up there and do what I could with the two fragments of brigades I still had. I proceeded to that point capturing a detachment of the enemy en route. Approaching the place by a bypath, I saw indications of a large force there prepared for attack. About this time I could see the fight going on at Thoroughfare Gap where Longstreet had his progress disputed by the enemy, and it was to establish communication with him that I was anxious to make this march. I sent a trusty man with the dispatch to the right of Hay Market.

I kept up a brisk skirmish with the enemy without any result until in the afternoon when General Jackson having engaged the enemy, I quietly withdrew and hastened to place my command on his right flank. Not reaching General Jackson's right until dark, the fighting ceased and this command rendezvoused as before, but the cavalry under Colonel Rosser had played an important part in attacking the enemy's baggage train. Captain John Pelham's battery of Horse Artillery acted a conspicuous part on the extreme right of the battlefield dashing forward to his position under heavy fire.

The next morning (29th), in pursuance of General Jackson's wishes, I set out again to endeavor to establish communication with Longstreet from whom he had received a favorable report the night before. Just after leaving the Sudley road, my party was fired on from the woods bordering the road, which was in rear of Jackson's lines and which the enemy had penetrated with small force, it was afterward ascertained, and captured some stragglers.

They were between General Jackson and his baggage at Sudley. I immediately sent to Major Patrick, whose six companies of cavalry were near Sudley, to interpose in defense of the baggage and use all the means at hand for its protection and ordered the baggage at once to start for Aldie. General Jackson, also being notified of this movement in his rear, sent back infantry to clear the woods. Captain Pelham, always at the right place at the right time, unlimbered his battery and soon dispersed that portion in the woods.

Major Patrick was attacked later, but he repulsed the enemy with considerable loss, though not without loss to us, for the gallant Major, himself, setting the example to his men, was mortally wounded. He lived long enough to witness the triumph of our arms, and expired thus in the arms of victory. The sacrifice was noble but the loss to us irreparable.

I met with the head of General Longstreet's column between Hay Market and Gainesville, and there communicated to the commanding general General Jackson's position and the enemy's. I then passed the cavalry through the column so as to place it on Longstreet's right flank and advanced directly toward Manassas, while the column kept directly down the pike to join General Jackson's right. I selected a fine position for a battery on the right, and one having been sent to me, I fired a few shots at the enemy's supposed position, which induced him to shift his position.

General Robertson, who with his command was sent to reconnoiter farther down the road toward Manassas, reported the enemy in his front. Upon repairing to that front, I found that Rosser's regiment was engaged with the enemy to the left of the road and Robertson's vedettes had found the enemy approaching from the direction of Bristoe Station toward Sudley. The prolongation of his line of march would have passed through my position, which was a very fine one for artillery as well as observation, and struck Longstreet in flank.

I waited his approach long enough to ascertain that there was at least an army corps, at the same time keeping detachments of cavalry dragging brush down the road from the direction of Gainesville, so as to deceive the enemy – a ruse which Porter's report shows was successful – and notified the commanding general, then opposite me on the turnpike, that Longstreet's flank and rear were seriously threatened and of the importance to us of the ridge I then held. Immediately upon receipt of that intelligence Jenkins', Kemper's, and D.R. Jones' brigades and several pieces of artillery were ordered to me by General Longstreet and being placed in position fronting Bristoe awaited the enemy's advance. After exchanging a few shots with rifle pieces, this corps withdrew toward Manassas leaving artillery and supports to hold the position until night.

Brigadier General Fitz Lee returned to the vicinity of Sudley after a very successful expedition, of which his official report has not been received and was instructed to co-operate with Jackson's left. Late in the afternoon the artillery on this commanding ridge was to an important degree auxiliary to the attack upon the enemy, and Jenkins' brigade repulsed the enemy in handsome style at one volley as they advanced across a corn field. Thus the day ended, our lines having considerably advanced.

> *The Lord of Hosts was plainly fighting on our side, and the solid walls of Federal infantry melted away before the straggling but nevertheless determined onsets of our infantry columns.*

Captain Pelham's battery was still with the left wing. (See his interesting report of its action on the 28th and 29th, herewith.)

Next morning (30th), it became evident that the enemy had materially retired his left wing. My cavalry reconnoitered to the front gaining at the next house an important point of observation. A large walnut tree being used as an observatory, the enemy was discovered gradually massing his troops in three lines opposite Jackson and his left wing seemed to have entirely shifted. The commanding general was informed of these changes. Captain Throckmorton, 6th Virginia Cavalry, commanding sharpshooters, took position along a stone fence and stoutly defended our observation against the attacks of the enemy's dismounted cavalry.

About 3:00 p.m., the enemy having disclosed his movement on Jackson, our right wing advanced to the attack. I directed

Robertson's brigade and Rosser's regiment to push forward on the extreme right, and, at the same time, all the batteries I could get hold of were advanced at a gallop to take position to enfilade the enemy in front of our lines. This was done with splendid effect. Colonel Rosser, a fine artillerist as well as bold cavalier, having the immediate direction of the batteries. The enemy's lines were distinctly visible and every shot told upon them fearfully. Robertson's brigade was late coming forward, and, consequently, our right flank was at one time somewhat threatened by the enemy's cavalry, but the artillery of Captain Rogers with a few well-directed shots relieved us on that score.

When our cavalry arrived on the field, no time was lost in crowding the enemy, the artillery being kept always far in advance of the infantry lines. The fight was of remarkably short duration. The Lord of Hosts was plainly fighting on our side, and the solid walls of Federal infantry melted away before the straggling but nevertheless determined onsets of our infantry columns.

The head of Robertson's cavalry was now on the ridge overlooking Bull Run and having seen no enemy in that direction, I was returning to the position of the artillery enfilading the Groveton road, when I received intelligence from General Robertson at the point I had just left that the enemy was there in force and asking reinforcements. I ordered the two reserve regiments (7th and 12th) rapidly forward and also a section of artillery, but before the latter could reach the point our cavalry by resolute bravery had put the enemy, under Buford, to ignominious flight across Bull Run and were in full pursuit until our own artillery fire at the fugitives rendered it dangerous to proceed farther.

In this brilliant affair, over 300 of the enemy's cavalry were put hors de combat, they, together with their horses and equipments, falling into our hands. Colonel Brodhead, 1st Michigan, died from his wounds next day. He was cut down by Adjutant Harman, 12th Virginia Cavalry. Major Atwood and a number of captains and lieutenants were among the prisoners.

The further details of this fight will be found in the accompanying reports of Brigadier General Robertson and Col. T.T. Munford. The latter as well as his Lieutenant Colonel Watts, Major Breckinridge and Lieutenants Kelso and Walton were wounded in the action, conspicuously displaying great gallantry and heroism. The 2nd Virginia Cavalry suffered most.

Nothing could have equaled the splendor with which Robertson's regiments swept down upon a force greatly outnumbering them, thus successfully indicating a claim for courage and discipline equal to any cavalry in the world.

Night soon ensued and as the enemy's masses of infantry had not retreated across Bull Run, I was anxious to cut off that retreat. Upon the enemy's position after dark, however, infantry only could move, and I was anxious for Brigadier General Armistead to attack from a position he took after dark directly on the enemy's flank and urged it. He however doubted the policy of night attack with his command, especially as there was danger of collision with our own infantry and I did not feel authorized to order it, particularly as there was time to communicate with the commanding general which was promptly done. The attack was not made.

Before daylight next morning, the cavalry was in the saddle and after the enemy but met with nothing but stragglers until we came within range of the guns at Centreville where his forces appeared to be in position. Twenty or thirty ambulances were captured and sent back with orders to go to work removing our wounded from the battlefield. I have never heard of those ambulances except that they were seized as fresh captures by the Texas Brigade. I think this not improbable as a large number of prisoners I sent to the rear were fired upon by our infantry near the stone bridge. At this time Colonel T.L. Rosser was sent with 100 men and a section of artillery back to recapture Manassas, in which he succeeded. His report of his operations those few days will be found of interest.

At one time on the 30th, I noticed our front lines near Chinn's house giving way and looking back saw the reserve line stationary. I sent word to the general commanding (whose name I did not learn) to move up as he was much needed to support the attack. That order was carried by Captain W.D. Farley, volunteer aide, under circumstances of great personal danger, in which his horse was shot.

Generals Jenkins and Kemper came under my observation as exhibiting good conduct, bravery and coolness.

Brigadier General D.R. Jones was with me part of the time on the extreme right during the battle in which several batteries of his division took part, and I think he left me to bring his infantry into action.

My division surgeon, Talcott Eliason, besides being an adept in his profession, exhibited on this, as on former occasions, the attributes of a cavalry commander.

First Lieutenant R. Channing Price was of invaluable assistance as aide-de-camp.

Major Von Borcke, assistant adjutant-general and Major Hairston and Lieutenant Chiswell Dabney, aide-de-camp, rendered important service throughout the period embraced in this report.

My division quartermaster, Major Samuel Hardin Hairston, in coming on to join me, was put in command of a detachment of cavalry at Salem by the commanding general and sent on an important reconnaissance toward Warrenton, of which his report is appended.

Captain Blackford, Corps of Engineers, was quick and indefatigable in his efforts to detect the designs of the enemy and improve the positions within our reach.

Private Stringfellow displayed great daring and enterprise as a scout.

I append a map of the country embraced in the foregoing operations, drawn by Captain Blackford, Corps of Engineers.

I have to mourn the loss of Capt. J. Hardeman Stuart, signal officer, the particulars of whose death are given below.

I have the honor to be, most respectfully, your obedient servant,

J. E. B. STUART,
Major General, Commanding Cavalry

Lee's Campaign Against Pope in Northern Virginia

Confederate Military History, Volume 2, Chapter 28.

The battle of Cedar Run, as General Lee says in his report, "effectually checked the progress of the enemy for the time," but the pressure from Washington was so great that Pope had to respond with an advance, which he made on August 14th when Reno's arrival increased his force to 50,000. He disposed his army from the crossing of Rappahannock River by the Orange road, to the crossing of the Rapidan at the historic Raccoon ford, across which Wayne led his Pennsylvania brigade to reinforce Lafayette in 1781. Lee, in expectation of this, had, on the 13th of August, ordered Longstreet with his division and two brigades under Hood to move to Gordonsville, and R.H. Anderson to follow him anticipating by a day McClellan's movement from Harrison's landing toward Fort Monroe. At the same time Stuart was ordered to move the main body of his cavalry toward Orange Courthouse covering the right of Longstreet's movement and placing his cavalry upon the right of Lee's army when concentrated in Orange.

Longstreet's troops reached the neighborhood of Gordonsville on the 16th, and the same day Jackson, in advance, moving secretly, put his command behind the outlying Clark's mountain range, east of Orange Courthouse, covering the Raccoon and Somerville fords of the Rapidan.

Lee followed and joined his army in Orange near the middle of August, and, on the 19th, gave orders for an advance having determined to strike Pope and defeat him before the great force under McClellan could join him. Longstreet advised a movement to the left, so that Lee's army, with the Blue Ridge behind it, might fall upon Pope's right, but Lee and Jackson thought it better to turn Pope's left and put the army of Northern Virginia between him and Washington and cutting his line of supplies and retreat. Lee's order of the 19th directed Longstreet to cross the Rapidan at Raccoon ford with the right wing of the army, and move toward Culpeper Courthouse, while Jackson, with the left wing, was to cross at Somerville ford and move in the same direction keeping on Longstreet's left. Anderson's division and S.D. Lee's battalion of artillery were to follow Jackson, while Stuart, crossing at Morton's ford, was to reach the Rappahannock by way of Stevensburg, destroy the railroad bridge, cut Pope's communications, and operate on Longstreet's right.

The men were to carry three days' rations in their haversacks, and the movement was to begin at dawn of the 20th. Jackson desired to attack earlier, but Longstreet was not prepared. The concentrated army was ready to move on the 19th, but Fitz Lee's brigade of Stuart's cavalry, the leading one in the march from

> *Lee had the mortification of seeing from the summit of Clark's mountain Pope's army in full retreat across the plains of Culpeper.*

Richmond, had gone too far to the right in the direction of Fredericksburg and was a day late in joining the army, thus causing another delay.

Pope, on the 19th, ordered a cavalry reconnaissance across the Rapidan, which captured one of Stuart's staff with Lee's order of march on his person. This was quickly furnished to Pope, who hastened to evacuate Culpeper and put the Rappahannock between himself and the now famous Confederate general-in-chief. Lee had the mortification of seeing from the summit of Clark's mountain, the southeastern of "the little mountains of Orange," Pope's army in full retreat across the plains of Culpeper, on the very day that he would have fallen upon it had his strategic orders been promptly and energetically obeyed by his first lieutenant.

Lee's 50,000 men followed his marching orders at dawn of the 20th; but not against Culpeper Courthouse, for Pope had evacuated that the day before. Longstreet, preceded by Fitz Lee's cavalry, marched to Kelly's ford of the Rappahannock, while Jackson marched by way of Stevensburg and Brandy Station toward the Rappahannock bridge, bivouacking for the night near Stevensburg. Stuart, with Robertson's cavalry brigade, had a spirited contest that day with Bayard's cavalry near Brandy station. Forced from that point, Bayard took position between Brandy and the Rappahannock bridge, still guarding the Federal rear from which Stuart again routed him and drove him across the Rappahannock, under cover of Pope's batteries on the high northern bank. The Confederates captured sixty-four prisoners and lost sixteen, killed and wounded.

The morning of the 21st found Lee's 50,000 veterans on the south bank of the Rappahannock with Jackson on the left, extending from the railroad bridge to Beverly ford, across which Robertson's 5th Virginia Cavalry had made a dash, scattering the Federal infantry near by, disabling a battery and spending most of the day on the north side of the river by the aid of Jackson's batteries on the south side. On the approach of a large Federal force, Rosser, by order of Stuart, recrossed. Longstreet extended Lee's line from the Rappahannock bridge to Kelly's ford. Pope's 55,000 men held the commanding ground on the north bank of the Rappahannock and a lively artillery duel was kept up during the day between the confronting armies, but with little or no damage to either.

The undulating Midland plain on which these contending armies had now met was far better fighting ground than was the swampy and densely forested Tidewater country, which was so recently the field of contention. The larger portion of this vicinity of the Rappahannock was cleared and had been under cultivation in large plantations until the opening of the war. At the same time it was a more difficult region for strategic movements to be covered from observation.

> *Lee's military genius and his conferences with Jackson convinced him that the proper movement was one that should turn Pope's right and place the Confederates in his rear, cutting him off from the old time highway that led through the Piedmont country by Warrenton toward Washington.*

It was evident that Pope's concentrated army could not easily be reached by a front attack, while his left was difficult of approach, and receiving the reinforcements steadily coming to him from the direction of Fredericksburg. Lee's military genius and his conferences with Jackson convinced him that the proper movement was one that should turn Pope's right and place the Confederates in his rear, cutting him off from the old time highway that led through the Piedmont country by Warrenton toward Washington. Moreover, "the strength of the hills" lay in that direction; for within sight, looking to the northward and westward, were the outlying ridges of the coast range, the Rappahannock and Bull Run mountains, behind which concealed movements could be made in the desired direction.

The first step in this strategic movement was to get the mobile left wing of his army, under the energetic and always-ready Jackson, behind these covering low mountain ranges, the southwestward extensions of the Bull Run Mountains, without the knowledge of Pope. To accomplish this, Lee adopted a series of novel advances. While Jackson and Stuart were engaging the attention of Pope along the Rappahannock, north of the railroad, he moved Longstreet from his right, by concealed roads, and placed him in Jackson's rear, leaving the latter free to fall back after dark, giving place to Longstreet, and march to a position farther up the river but still holding on to Longstreet's left.

This first exchange of positions was made during the night of the 21st, or rather the early morning of the 22nd, and that day, preceded by cavalry, Jackson reached the neighborhood of Warrenton Springs where the great highway, from Culpeper Courthouse toward Washington crosses the Rappahannock and goes on through Warrenton to Centreville. During that day Longstreet, by a vigorous contention with skirmishers and artillery, engaged Pope's attention in his first position north of the Rappahannock and caused him to add to his force at Beverly ford apprehending that Longstreet was about to force a passage there and attack his center.

Detachments of Federal cavalry and infantry made dashes on Jackson's line of march from a detached column that Pope was moving up the north bank of the river to keep pace with whatever movement Lee might be making to his left. Especially was a bold dash made at Freeman's ford, about noon, as Jackson's rear was passing that point. His rear guard, under Trimble, deployed and awaited the Federal attack. Hood with two of Longstreet's brigades came up about 4:00 in the afternoon, when Trimble, aided by these, vigorously attacked the Federal brigade which had crossed the river, and drove it back with slaughter and in confusion. A third crossing, in pursuit of information, was made at Fant's ford, by cavalry, infantry and artillery, but these soon retired, having learned but little.

When Jackson reached the river opposite the Warrenton Springs and found the ford guarded, he at once began moving his troops to the other side, sending over the 13th Georgia and two batteries, while Early crossed on an old mill dam about a mile further down the river. It began raining while these troops were crossing and an afternoon of showers was followed by a night of heavy downpour and darkness, preventing the crossing of more troops. By morning the river was swollen past fording and Jackson's advance, under Early, was isolated on the further shore.

Pope's main body had continued to hold its position near the railway on the 22nd, as he was unwilling to remove further from his expected reinforcements from Fredericksburg. Apprehensive of an attack from Longstreet, he did not care to move farther to his right to intercept Jackson's movement, concerning which he as yet had no reliable information. Longstreet still held him at bay.

On this same 22nd, Lee initiated one of the boldest of his deceiving strategic movements. During the forenoon he dispatched Stuart with the main body of his cavalry by concealed roads behind his army to Waterloo bridge four miles above Warrenton Springs, held by Jackson, and where the graded highway from Warrenton to Little Washington crosses the Rappahannock. There, Stuart with 1,500 men and two guns crossed the river and began a rapid march for Pope's rear, to break the railway leading to Washington and gather information just as he had recently done in his grand ride around McClellan at Richmond.

With a good road to march on, he reached Warrenton unopposed, in the afternoon. After halting there for a short rest, he continued eastward by Auburn Mills to Catlett's station on the Orange & Alexandria railroad, intending to destroy the bridge over Cedar creek near that place. The downpour that had swelled the Rappahannock caught Stuart on the march, and he reached his objective in the midst of rain and darkness; but an intercepted and captured negro led him to a camp where were the headquarters wagons of General Pope. These Stuart quickly captured with one of the Federal commander's staff and his personal baggage and official papers.

His efforts to destroy the wagon trains and the railroad bridge were but partially successful, in consequence of the rain and the darkness. He began his return march before daylight of the 23rd, bringing off 300 prisoners, and recrossed the Rappahannock in the evening of the same day without molestation after having taught Pope a second lesson on the subject of rear guards and infused an element of fear into the Federal army as to the safety of its lines of retreat, also bringing off the captured correspondence between Pope and Halleck, which informed Lee fully concerning the strength and the plans of his antagonist.

In the afternoon of the 23rd, before Stuart cut the railway and the telegraph at Catlett's station, Pope had telegraphed to Halleck: "Under present circumstances I shall not attempt to prevent his (Lee's) crossing at Sulphur Springs, but will mass my whole force on his flank in the neighborhood of Fayetteville," a crossroads hamlet five miles to the southeast of Sulphur Springs and about the same distance northeast from the right of his position on the Rappahannock. An hour and a half later he telegraphed: "I cannot move against Sulphur Springs just now without exposing my rear to the heavy force in front of me," still looking with alarm across the Rappahannock at Longstreet.

Three hours later, after reporting Jackson's crossing, he again telegraphed: "I must either fall back and meet Heintzelman behind Cedar run or cross the Rappahannock with my whole force and

> "Pope awoke on the morning of the 23rd with no very clear notions of what he intended to do."

assail the enemy's flank and rear. I must do one or the other at daylight; which shall it be?" Halleck approved the suggested bold attack on Lee's rear, and directed the troops approaching from Fredericksburg to march to Stevensburg and Brandy Station on the south side of the river, proposing to unite these with Pope the next day to attack Lee's rear.

General George H. Gordon, who has written so well concerning the Army of Virginia, in which he served, and who fought so bravely at Winchester and Cedar run, says of Pope: "He awoke on the morning of the 23rd with no very clear notions of what he intended to do."

The heavy rain of the night of the 22nd interrupted Jackson's movement and compelled Lee to abandon, for the time being, his intended flank movement. Jackson, by the most persistent efforts, repaired the bridge at the springs in order to extricate Early from the perilous position which he was so boldly holding on the north bank of the Rappahannock, and Pope, knowing that river to be impassable, gave up, no doubt gladly, his scheme of crossing to attack Lee's rear and determined to concentrate against the Confederates on the north side of the river as he had at first proposed.

In the early morning of the 23rd, he turned Sigel toward Sulphur Springs by way of Fayetteville, followed by Banks and Reno. McDowell from his left was ordered to burn the railroad bridge, which up to this time, by the aid of guards and artillery, he had kept intact, and move toward Warrenton. These movements would bring him into line of battle facing any movement of Lee from Sulphur Springs toward Warrenton. Longstreet's batteries gave parting salutes to these backward movements. Reynolds' division of 6,000 men from Aquia Creek reported during the forenoon of the 23rd and followed after McDowell.

Early put on a bold front while awaiting the reconstruction of the bridge in his rear, aided by the swollen condition of the Great Run in his front. He destroyed the bridge over that stream and held the road against Sigel's advance of 25,000 men, which Pope had ordered to make attack and beat the Confederates on the north side of the river. Sigel conceived the idea that Lee's whole army was in front of him and therefore only skirmishing and artillery firing took place during the afternoon and until dark. Sigel, in the meantime, going into camp and advising Pope to withdraw his corps to a better position.

Robertson, with his cavalry and some guns returning from Stuart's expedition in Pope's rear joined Early during the day. As soon as the bridge was made passable, at about nightfall, Lawton's brigade was crossed over to Early's support. Ewell himself went over for a consultation with Early. It was decided in view of the large force before him that it was not expedient to bring on a battle at that place, so orders were given at 3:00 o'clock next morning for Early to withdraw, which he did soon after daylight and removed his men to Jackson's rear, where they broke their fast of two nights and the intervening day.

About 10:00 on the night of the 23rd, Pope, accompanied by the corps of McDowell and the division of Reynolds, reached Warrenton. At that time, more than 50,000 men of the army of Virginia were concentrated along the turnpike road between Jackson at Sulphur Springs and Warrenton. On the morning of the 24th, Pope girded himself to destroy the army of

Lee, which he supposed was still north of the Rappahannock as Sigel had reported. Buford's cavalry was sent to Waterloo whence a good country road led to Warrenton to reconnoiter and to destroy the bridge over the Rappahannock at that point and get in Lee's supposed rear.

Sigel, Banks and Reno were to move toward the same point from opposite Sulphur Springs, while McDowell was placed along the roads leading to Sulphur Springs and to Waterloo to support the movement. As Sigel approached the river, A.P. Hill, who held the Confederate side, opened his batteries and an engagement of artillery was brought on. Sigel continued cautiously, his march up the river annoyed by Hill's batteries, and it was well into the afternoon before Buford learned that there were no Confederates on the north side of the Rappahannock.

It was nearly 4:00 p.m. when Pope telegraphed Halleck that "Sigel is pursuing the enemy in the direction of Waterloo bridge… No force of the enemy has as yet been able to cross except that now enclosed by our forces between Sulphur Springs and Waterloo bridge, which will undoubtedly be captured unless they find some means of escaping."

Sigel occupied most of the 24th in his cautious march of six miles from Sulphur Springs to Waterloo, where he arrived late in the afternoon and found the Confederates on the south side of the river but holding and defending the bridge. The continuing thunder of Lee's guns, from point-to-point of vantage between Sulphur Springs and Waterloo had thoroughly engaged Sigel's attention during the entire day as Lee intended they should, to divert attention from the new flank movement which he had already begun.

Pope was equally ignorant, for, in the afternoon, after learning that there were no Confederates north of the Rappahannock, he dispatched to Halleck that he would "early tomorrow move back a considerable part of my force to the neighborhood of Rappahannock station," evidently disturbed by the long-staying qualities of Longstreet, which he had now been testing for a number of days, while he himself had been zigzagging around in a vain attempt to find the other portions of Lee's army.

Still desiring to strike a telling blow at Pope before McClellan's main body could reach him, Lee ordered from Richmond the divisions of Walker, McLaws and D.H. Hill, which had been held there for prudential reasons and sought a conference with Jackson, to which the latter, a little later, called in his chief engineer, Lieutenant James Keith Boswell for information concerning the roads leading behind the Rappahannock mountains to the line of the Manassas Gap railroad and to Pope's rear, with which he was familiar.

Lee and Jackson having devised a plan of campaign by which Jackson, free from all encumbrances, should move rapidly to Pope's rear, cut his line of communication at Bristoe, destroy his stores back to Manassas Junction, then fall back to the north of the Warrenton and Washington turnpike, and there await the arrival of Lee with Longstreet, who would remain a day longer on the banks of the Rappahannock for the purpose of detaining and perplexing Pope.

During the night of the 24th, Longstreet's batteries took the place of Jackson opposite Warrenton Springs, as did also his troops, leaving Jackson free to begin his movement on the morning of the 25th, which he did at an early hour, leaving his baggage train behind and taking with him only ambulances and ordnance wagons. His troops carried in their haversacks scant rations for three days, Jackson confident of being able to abundantly supply them from the enemy's stores.

Starting from the vicinity of Jeffersonton to which he fell back in giving place to Longstreet, Jackson marched for some distance to the northwestward along the great highway leading to the Valley by way of Chester Gap, and his bronzed veterans were elated with the conviction that they were again bound for the scene of their victories of the preceding spring; but, when a short distance beyond Amissville, their course was turned from the northwest to the northeast, they looked questioningly one to the other as to whither they were going led by Lieutenant Boswell and portions of the noted Black Horse cavalry through their Fauquier homeland. Jackson pressed steadily forward through the long August day without halt, until he had covered twenty-five miles and reached the vicinity of Salem on the Manassas Gap railroad just as the sun sank behind the Blue Ridge to his left.

At dawn of the 26th, Jackson's men were again puzzled on finding themselves marching to the southeast following the line of the Manassas Gap railroad through Thoroughfare Gap to Gainesville, where Stuart joined them with his cavalry and led the way from that hamlet directly to Bristoe Station on the Orange and Alexandria railroad, which they reached about dark after a march of twenty-four miles without having met opposition on the way. Jackson and his 22,000 enthusiastic men and Stuart with wide-awake and jolly cavalry were now in Pope's rear and on his line of communication, which they proceeded to destroy, capturing trains moving toward Washington and breaking up detached Federal encampments along the railway.

> **Apprised of these various movements by his scouts and spies but not comprehending them or their objects or destination, Pope issued orders, which scattered rather than concentrated his large army.**

Not satisfied with this and desiring to not only reap the spoils stored at Manassas but to guard against movements from Washington, Jackson sent Trimble's brigade of infantry and Stuart with a portion of his cavalry, through the darkness, four miles further to Manassas Junction, which they reached and captured after a brief resistance about midnight.

On this same 26th of August, Lee and Longstreet leaving 6,000 men at Waterloo to guard the trains, followed after Jackson and encamped at Orleans.

Apprised of these various movements by his scouts and spies but not comprehending them or their objects or destination, Pope issued orders, which scattered rather than concentrated his large army. He first ordered a concentration on Warrenton. Porter, with 10,000 men, reached Bealeton and Heintzelman, with his 10,000 men reached Warrenton Junction on their way to obey this order. The corps of Sumner, Franklin and Cox, from McClellan's army, were that day marching toward Pope under urgent orders from Alexandria.

Late in the night when the import of Jackson's movement dawned upon him, Pope again changed his orders directing his troops to march on Gainesville to intercept what he supposed would be Jackson's line of retreat. The different portions of his command were headed in that direction but all hindered by a confusion of orders and a resulting mixing of marching columns.

On the 27th, Lee with Longstreet continued his march through Salem and the Plains station on the Manassas Gap railroad, but once interrupted, by the attack of a small body of Federal cavalry, which came near capturing General Lee.

In the early morning of this same day, Jackson marched the divisions of Taliaferro (recently Winder) and of A.P. Hill to Manassas Junction, where during the day they rested and reveled in the vast stores of quartermaster and commissary supplies the Federals had gathered at that important junction. Ewell was left behind, at Bristoe to protect Jackson's rear and oppose any advance from the line of the Rappahannock. There in the afternoon, he had a vigorous combat with Porter, repulsing him then withdrew across Broad Run and late in the day followed on to Manassas Junction.

Longstreet was slow in getting under way on the morning of the 28th and so did not reach Thoroughfare Gap, but seven miles from his camp, until 3:00 in the afternoon, to find that important way, the gate he must pass through to reach Jackson's right at the appointed rendezvous, held by Ricketts and a Federal division. Lee promptly addressed himself to clear the way. Wilcox, with three brigades, was sent three miles to the northward to cross the Bull Run Mountains at Hopewell Gap and flank the right of Ricketts. Law's brigade was ordered to climb the ends of the mountains cut by Broad Run along which the road and the railway followed, while D.R. Jones was to make a direct attack with his brigade through the pass. Law's toughened veterans soon scaled the mountains, fell upon Ricketts' flanks and forced him to retire just as the day closed, when Longstreet led his command through Thoroughfare Gap and encamped east of the Bull Run Mountains and eight miles from the battlefield of Groveton Heights, where Jackson was hotly engaged with King's division of Pope's army, and anxiously awaiting the coming of Lee and Longstreet.

Satisfied by the contention of Hooker with Ewell at Bristoe that Jackson's command was at Manassas Junction, Pope concluded that there was a good opportunity for "bagging the whole crowd," so he issued orders that turning from the ways to Gainesville, his columns should on the morning of the 28th march rapidly on Manassas Junction. Jackson spoiled this third plan of concentration for his capture by not waiting for Pope at Manassas Junction.

On the night of the 27th, he set fire to the stores at Manassas that his men had not appropriated and his wagons could not carry away and hastened to the appointed place for meeting Lee, but by ways that completely baffled his over-confident adversary. Taliaferro's division with the trains was sent northward by the direct road to Sudley church with orders to occupy the forest covered position behind the unfinished Gainesville and Alexandria railroad with which Jackson was thoroughly familiar from having encamped in that region after the First Bull Run battle. A.P. Hill was sent northeastward by the highway across Bull Run to Centreville on the great road leading to Washington, and Ewell was left to follow after him in the same direction.

Porter could not find his way, even with the aid of lighted candles through the darkness of this night, from Warrenton Junction to Manassas, but Jackson's men found the way to their ordered destinations. Hill, on the morning of the 28th, took the big road from Centreville westward, marched across Bull run and took position on Taliaferro's left near Sudley church. Ewell, who had encamped the night before on the south side of Bull Run at Blackburn's ford, crossed over and marching up that stream to the stone bridge, followed after Hill and took position on his right, Taliaferro moving still farther to the right in the direction of Gainesville, so that by the middle of the day, Jackson was concentrated in a strong position, the one the Federals had first occupied at the first battle of Bull Run, looking down upon the stream valley of Young's Branch along which ran the Warrenton and Alexandria Turnpike, his guns in place and his troops ready for action.

That same noonday, Pope having reached Manassas Junction was still seeking for Jackson. The movement of Hill and Ewell toward Centreville, the threatening of Washington by Fitz Lee and his horsemen at Fairfax Courthouse and Burke's station meant Pope knew not what, but he proceeded to issue a third order for concentration. Gainesville and Manassas Junction had failed him, and now, thinking he was after a defeated and retreating foe, he ordered his columns to Centreville.

The leading divisions of McDowell's corps had passed through Gainesville on the way to the junction, early in the day, but King's division did not reach that point until after Pope had ordered a concentration at Centreville, so King on receiving these orders decided to take the direct road from Gainesville to Centreville rather than the circuitous one by Manassas Junction, ignorant of the fact that Jackson lay concealed in the forest flanking the left of this direct road but a short distance from Gainesville. So it came to pass that when, late in the afternoon, he was marching along in front of Jackson's concealed army, the divisions of Taliaferro and Ewell sprang upon him and by a short but fierce and bloody struggle drove him back under cover of the night to Gainesville and to the road to Manassas Junction on which Ricketts' column, retreating from Lee's bold assault at the Thoroughfare gap, overtook him during the night.

On the morning of the 29th, these discomfited divisions of King and Ricketts appeared in the vicinity of the junction and there was now no Federal force to oppose the coming together of the two wings of Lee's army on the famous battlefield of "Groveton Heights," as Jackson named it, that of the first day of the Second Bull Run or Manassas.

Stuart, from Jackson's right, on the 29th, soon opened communication with Lee and Longstreet, who had but eight miles to march to the field of action and extend his lines southward from Jackson's right and cover the roads leading from Centreville and from Manassas Junction. By 10 a.m. of the morning of the 29th, Lee had stationed himself on a commanding knoll near the head of Young's branch on the south side of the turnpike from which he could see his left under Jackson stretching away to the northeast in his strong position on the Sudley ridge for nearly three miles, those of Longstreet reaching to the southward through fields and forests for nearly the same distance, like two gigantic arms outstretched with the fingers of Robertson's cavalry on the right and those of Fitz Lee on the left and ready to close in deadly embrace upon any foe that should venture to come within their far-extending reach.

In the early morning of the 29th, Pope, at Centreville, was issuing orders for a fourth concentration of his troops, which were now scattered anywhere and everywhere within the 12 miles of broken and much afforested country between his headquarters and Bristoe, still believing that he had but Jackson's command before him only seeking an opportunity to escape and ignorant of the position of Longstreet. Pope ordered a vigorous attack on Jackson's left by Sigel's corps supported by Heintzelman, Reno and Reynolds. This attack was bold and vigorous, and from 6:30 to 10:30 there was a fierce contention between A.P. Hill and the Federals, but the latter were repulsed when, just as Lee was leading Longstreet into position, 18,500 men under Heintzelman and Reno were moving in to Sigel's aid.

Pope's men, wearied by the constant marchings and countermarchings of previous days, were slow in moving forward; but at noonday, when Pope himself appeared and took post on Buck Hill, whence his own lines and those of Jackson were visible, he found his 35,000 men in battle order facing Jackson. These he urged to renew the attack from which Sigel had been repulsed. He also ordered McDowell and Porter to advance their 30,000 men from Manassas upon Gainesville, his numerous cavalry hovered about the flanks of the Confederates. Pope did not believe that Lee was yet on the field, so he proposed to hurl his 75,000 against Jackson's 20,000 and win a victory before Longstreet could arrive.

Earnestly watching the battlefield from his well-chosen point of observation, Lee discovered that Longstreet was not far from the left of Pope's line of attack and as that solid mass of Federal veterans marched with quick and resolute step to assault Jackson, Lee urged Longstreet to join in the issue. After overlooking the field, the latter reported the prospect as "not inviting" and greatly disappointed his commander-in-chief by obstinately persisting in his opposition to make an attack.

> *After overlooking the field, Longstreet reported the prospect as "not inviting" and greatly disappointed his commander-in-chief by obstinately persisting in his opposition to make an attack.*

Just then, Stuart, who was on the right and had been reconnoitering toward Manassas Junction, reported the approach of McDowell and Porter, but these soon turned to the northward and marched by the Sudley road to the left of Pope's contention with Jackson. Through all the long day, during ten hours of hotly-contested battle, constantly adding fresh troops and in six vigorous assaults, did Pope force his men against Jackson's position, mainly against A.P. Hill on his left.

The Federal soldiers, well led with the skill of veterans and the courage of brave men, marched to the very front of Jackson's lines, which, by determined efforts, they several times broke and carried, but were every time driven back, once partly with cobblestones picked from the fills of the unfinished railway when the supply of ammunition gave out.

Lee anxiously watched these fierce assaults and desperate repulses and urged his stubborn lieutenant to join in the combat and relieve the pressure upon his other and indomitable lieutenant, who, with another sort of stubbornness held to his lines and drove back the successive waves of Federal assaults.

At 5:00 p.m., when less than two hours of the day remained, Pope massed the divisions of Kearney and Stevens for a last assault upon Jackson's left. Gregg had exhausted his ammunition and sent for more, adding that his Carolinians would hold on with the bayonet, but these were forced backward when the Georgians and the North Carolinians of Branch, dropped in behind them and all, like Indian fighters, took advantage of every rock and tree as the stubborn Federals forced them back. Jackson promptly moved from his center the Virginians of Field and Early, the Georgians of Lawton, and the Louisianians of Hays, and threw these into A.P. Hill's hot contest on his left and routed and dispersed the brave Federal attack, shattering the brigades of Pope's right.

Again Lee, with all the earnestness of his heroic nature, urged Longstreet to participate and help Jackson in meeting this furious attack. But he persisted in his refusal to move, claiming that it was now too late in the day for so doing. But Lee had one force obedient to his commands, or rather his requests, for thus were the orders of that high-toned gentleman expressed. He had massed Hood's batteries on Longstreet's left on commanding ground, and as Pope's left under Reynolds moved forward to attack, a hot fire from these guns drove him back. Just at set of sun when Longstreet yielded for what he called a reconnaissance in force, he turned loose Hood's courageous Texans, who fell upon the Federal center and drove King back with heavy loss, capturing three of his battleflags and one of his guns.

The night closed on this long day of furious and bloody battle, in which the contending armies had each displayed the undaunted courage of their common, fighting, ancestral stock, but the skill of leadership had again asserted itself against the mere power of numbers, and history, in all its annals, nowhere records braver deeds of heroic and daring defense and persistent courage than were exhibited by Jackson's men through all that long day of steady contention against fearful odds. The invincible Stonewall had unflinchingly held the left, confident that the equally invincible Lee was not only watching the contest, but would in the crisis of the day throw his sword into the scale and decide the unequal contest.

The battle over, Jackson's men cared for their wounded, gathered their dead for burial and prepared for another day of conflict, which they well knew was impending, gathered in groups, praying for further aid to the God of Battles, and then, in trusting confidence, slept on their arms awaiting the coming day.

The 30th of August, as the summer neared its end, opened clear and bright with the two armies ready for the renewal of the mighty conflict. The position of Lee's two wings was unchanged except that he had massed thirty-six guns, under Colonel Stephen D. Lee on the commanding watershed swell in the center of his lines, where their lines of fire led down the center of the depression followed by Young's Branch and threaded by the turnpike leading through the midst of the Federal host to the stone bridge over Bull Run. The brigades of Longstreet from the center southward were those of Wilcox, Hood, Kemper and D.R. Jones. R.H. Anderson was in reserve with his 6,000 men on the turnpike to the rear.

Lee then had about 50,000 men at command in his two far-reaching wings, the great jaws of the war monster into which the army of Pope was preparing to move unconscious of the fate that awaited it when these jaws should close and crush it in defeat.

Noticing that the nearby skirmishers of the previous day had disappeared, Pope again rashly concluded that the Confederate army had been defeated by his assaults of the day before and was now in full retreat seeking safety behind the Bull Run Mountains. Therefore he ordered a prompt pursuit along the Warrenton road to Gainesville and then toward the Thoroughfare Gap.

He had brought up Porter's Corps, which had been holding the line of Dawkin's Branch on the road from Manassas Junction to Gainesville and placed it in his center. So it fell to that brave and skillful officer to lead in the supposed pursuit. Recalling Cold Harbor, Porter did not believe as Pope did that Lee and Jackson had given up the contest and were retreating, so he formed his men into a triple line of battle, across the turnpike, and placed King's division to support his right and Reynolds' his left; in his rear followed Sigel's corps and half of Reno's. These dispositions were made in the dense forest along the turnpike and to the east of the Sudley road, and thence Porter was ready to advance on Lee's center.

Pope having had on the previous day, experienced the sharp temper of Jackson's left massed the whole of Heintzelman's and the half of each of the corps of McDowell and Reno, ready to throw them against Jackson with the advance of Porter.

> *This day's advance and retreat cost Pope some 20,000 of his brave men in killed, wounded, and missing.*

In the morning, Heintzelman moved against A.P. Hill with Ricketts' division but soon drew back from the hot reception he met. The skirmishers of Reynolds met the same fate from S.D. Lee's guns when they advanced to feel Lee's center. It was three in the afternoon when Pope was good and ready with his entire army in hand for his grand assault. The signal was given and Porter's men rushed forward, wheeling on their left, and struck the Stonewall brigade now in command of Starke and Lawton's division. The contest was as fierce and earnest as brave men could make it. The lines for some minutes were almost within touch and the dead and dying on both sides strewed the ground.

As Porter closed in across the open field, his left was exposed to S.D. Lee's masked batteries, which now swept through his lines their shot and shell and aided to stagger Porter's attack, while Longstreet opened with three batteries upon his left rear. This unexpectedly received, Porter's men fled in routed masses followed by the men of Jackson's old division from his right, who leaped across their defenses and chased them in hot pursuit. The fierce attacks of Pope on Jackson's left had in the meantime been also repulsed.

Lee now saw that the supreme moment for action had come and he ordered

Longstreet to close in upon the Federal left; but his veteran soldiery now well trained in the art of war had at the same moment reached the same conclusion and without waiting for the word of command, they fairly leaped forward swinging on their left, and with Lee leading in person in the midst of them charged grandly to the front, responding to the movement of all of Jackson's men on the left and hurrying on the rout of the Federal army. The Confederate batteries also joined in the rushing charge and were abreast of their infantry comrades all along the lines where there was opportunity for giving parting shots to the retreating Federals. Stuart on the right on the old Alexandria road heard the well-known shouts of Confederate pursuit and rushed his brigades and batteries far in advance against the Federal left.

Warren's attempt to stem the tide just east of Groveton cost him dearly. Schenck with German tenacity hung on to the Bald Hill on the Federal left but the victory-compelling Confederates swarmed upon his flank and forced him from the summit.

Hood swept the line of the turnpike to the east of the Stone house. Pope's reserves on the Henry Hill the old plateau, which was the center of the fierce fighting of the year, before resisted the tide of victory for a time until Jackson closed down with his left, upon the retreating Federals toward the stone bridge. Darkness put an end to his advance and gave Pope's demoralized brigades an opportunity to follow the crowd of fugitives that crowded over that bridge seeking safety behind Franklin's corps, then advancing from Alexandria and the earthworks at Centreville.

This day's advance and retreat cost Pope some 20,000 of his brave men in killed, wounded and missing. Since Jackson met him at Cedar run, he had lost 30,000 men, 30 pieces of artillery and military stores and small-arms worth millions in value and many thousands in number. This great victory of Groveton Heights cost Lee 8,000 men mostly in Jackson's command, including many of his noblest and bravest officers.

A deluge of rain followed the great battle such as had followed most of those that had preceded it; but through that and the mud that followed it, Stuart rode in the early morning of Sunday, August 31st, across Bull Run to learn what had become of Pope. He found the reinforcements that had the day before come up from Washington holding the formidable entrenchments at Centreville bristling with artillery. Informed of this delay in Pope's retreat, Lee ordered Jackson, who was on his left and nearest Centreville, to cross Bull Run and march to the Little River turnpike, which enters the Alexandria road near Fairfax Courthouse, turn Pope's right and cut off his retreat to Washington.

The rain and mud made the march a difficult one for Jackson's weary and battle worn surviving veterans; but they divined their important mission and eagerly followed their great leader. When Pope learned of Jackson's new flanking movement, although he had in hand 20,000 fresh troops who had not fired a gun, he hastened in retreat to Fairfax Courthouse, after placing Reno's corps across the two converging turnpikes covering the approaches to Fairfax Courthouse from Centreville and Chantilly, with orders to keep back the irrepressible Confederates.

Jackson by continuing his march well into the night took position across the Little River turnpike at Ox Hill in front of

Chantilly. In the midst of a terrific storm of driving rain with almost continuous thunder and lightning, on Monday, September 1st, he met and repulsed a Federal advance under Reno. Jackson ordered the use of bayonets informed that the rain-soaked ammunition could not be used.

Heintzelman supported Reno, but Jackson's well-directed blows forced them both back until darkness ended the contest, when they followed Pope's line of retreat to within the fortifications of the Federal city. Pope's brief career, of less than two months' duration as commander of the Army of Virginia, came to an inglorious end and McClellan again took charge to reorganize the army of the Potomac from the broken Federal forces there gathered.

Longstreet followed Jackson to Chantilly but did not reach there in time to take part in the battle. Lee paused in his onward march, at this noble Chantilly mansion of one of his relatives to give his men much-needed rest and bring forward the supply trains which his rapid marches had left far in the rear.

In four short months the army of Northern Virginia had under his leadership with its 80,000 men, met and driven Banks, Fremont, McDowell, McClellan and Pope with their 200,000 veteran troops from far within the bounds of Virginia in disastrous retreat to beyond its borders with the exception of a small body that still held the line of the Baltimore & Ohio in the lower Valley and the remnant that had found refuge within the fortifications of Washington, on the Virginia side of the Potomac.

HI-LIGHTS OF A HERO'S LIFE
ISAAC TRIMBLE

★ 1802 – (May 15) Born in Culpeper County, Virginia.

★ 1818 – Entered the U.S. Military Academy.

★ 1822 – Graduated 17th in a class of 42 and commissioned as a brevet second lieutenant of artillery.

★ 1831 – Married Maria Catell Presstman and had two sons.

★ 1832 – Resigned from the army and moved to Maryland to become a railroad engineer.

★ 1835 – Maria died. He married Ann Ferguson Presstman.

★ 1849 – Built the President Street Station.

★ 1859 – Superintendent for the Baltimore and Potomac Railroad.

- 1861 – Participated in efforts to restrict movements of Union troops to Washington City by burning bridges north of Baltimore.

- 1861 – (May) When Maryland did not secede, returned to Virginia and joined the Provisional Virginia Army as a colonel in the engineer corps.

- 1861 – (August 9) Promoted to brigadier general and constructed artillery batteries along the Potomac River and the defenses of Norfolk.

- 1862 – Took command of a brigade in the Army of Potomac.

- 1862 – Fought in the Valley Campaign and distinguished himself at the Battle of Cross Keys.

- 1862 – (June 27 & July 1) Fought at Gaines Mills and Malvern Hill.

- 1862 – (August 9) Fought at Cedar Mountain.

- 1862 – (August 28) Joined Stonewall Jackson in the flank march around Pope and helped to seize Manassas Junction.

- 1862 – (August 29) Severely wounded in the leg.

- 1863 – (January 17) Promoted to major general and assigned division command in the Second Corps, but was unable to assume command because of lingering health issues.

- 1863 – (May 28) Assigned as commander of the Valley District.

- 1863 – (June) Visited Robert E. Lee's headquarters in Maryland to await formal assignment. None was forthcoming. He joined Ewell's Second Corps near Harrisburg, Pennsylvania as a supernumerary.

- 1863 – (July 3) Led Pender's Division in what became known as Pickett's Charge and received a wound in the same leg that was injured during Second Manassas. After the leg was amputated, he was left behind when the Army of Northern Virginia retreated.

- 1863 – Imprisoned at Johnson's Island and Fort Warren.

- 1865 – (March) He was moved to City Point, Virginia in order to be exchanged.

- 1865 – (April 16) Paroled in Lynchburg, Virginia.

- 1865 – Returned to Maryland and found employment with the Baltimore and Potomac Railroad.

- 1867 – Resigned as chief engineer of the Baltimore and Potomac Railroad.

- 1888 – (January 2) Died and was buried in Green Mount Cemetery.

Jackson's Foot Cavalry at Second Manassas

By Allen C. Redwood, 55th Virginia Regiment. *Battle and Leaders, Volume 2*, pages 530-538.

In the operations of 1862, in Northern Virginia, the men of Jackson's corps have always claimed a peculiar proprietorship. The reorganization of the disrupted forces of Banks, Frémont, and McDowell under a new head seemed a direct challenge to the soldiers who had made the Valley Campaign and the proclamation of General Pope betokened to the "foot-cavalry" an infringement of their specialty, demanding emphatic rebuke. Some remnant of the old esprit de corps yet survives and prompts this narrative.

After the check to Pope's advance at Cedar Mountain, on the 9th of August, and while we awaited the arrival of Longstreet's troops, A.P. Hill's division rested in camp at Crenshaw's farm. Our brigade (Field's) was rather a new one in organization and experience, most of us having "smelt powder" for the first time in the Seven Days before Richmond. We reached the field at Cedar Mountain too late to be more than slightly engaged, but on the 10th and 11th covered the leisurely retreat to Orange Court House without molestation.

When, about a week later, Pope began to retreat in the direction of the Rappahannock, we did some sharp marching through Stevensburg and Brandy Station, but did not come up with him until he was over the river. While our artillery was dueling with him across the stream, I passed the time with my head in the scant shade of a sassafras bush by the roadside with a chill and fever brought from the Chickahominy low grounds.

For the next few days there was skirmishing at the fords, we moving up the south bank of the river, the enemy confronting us on the opposite side. The weather was very sultry and the troops were much weakened by it, and our rations of unsalted beef eked out with green corn and unripe apples formed a diet unsuited to soldiers on the march and there was much straggling. I fell behind several times but managed to catch up from day to day. Once some cavalry made a dash across the river at our train, I joined a party who, like myself, were separated from their commands, and we fought the enemy until Trimble's brigade, the rearguard, came up.

We were then opposite the Warrenton Springs and were making a great show of crossing, Early's brigade having been thrown over the river where it became smartly engaged. I have since heard that this officer demonstrated more than once at the service required of him, receiving each time in reply a peremptory order from Jackson "to hold his position." He finally retorted: "Oh well, old Jube can die if that's what he wants, but tell General Jackson I'll be d-----d if this position can be held!"

The brigade moved off next morning, leaving me in the grip of the ague, which reported promptly for duty and, thanks to a soaking overnight, got in its work most effectually. The fever did not let go until about sundown, when I made two feeble trips to carry my effects to the porch of a house about one hundred yards distant where I passed the night without a blanket – mine having been stolen between the trips. I found a better one next morning thrown away in a field and soon after came up with the command, in bivouac and breakfasting on some beef which had just been issued. Two ribs on a stump were indicated as my share, and I broiled them on the coals and made the first substantial meal I had eaten for forty-eight hours. This was interrupted by artillery fire from beyond the river, and as I was taking my place in line, my colonel ordered me to the ambulance to recruit. Here I got a dose of Fowler's solution, "in lieu of quinine," and at the wagon camp that day I fared better than for a long time before.

Meanwhile they were having a hot time down at Waterloo Bridge, which the enemy's engineers were trying to burn, while some companies of sharpshooters under Lieutenant Robert Healy of "ours" – whose rank was no measure of his services or merit – were disputing the attempt. A concentrated fire from the Federal batteries failed to dislodge the plucky riflemen, while our guns were now brought up and some hard pounding ensued. But at sunset the bridge still stood, and I "spread down" for the night under the pole of a wagon fully expecting a serious fight on the morrow.

I was roused by a courier's horse stepping on my leg and found this rude waking meant orders to move. With no idea whither, we pulled out at half-past two in the morning and for some time traveled by fields and "new cuts" in the woods, following no road but by the growing dawn evidently keeping up the river. Now Hill's "Light Division" was to earn its name and qualify itself for membership in Jackson's corps.

The hot August sun rose, clouds of choking dust enveloped the hurrying column, but on and on the march was pushed without relenting. Knapsacks had been left behind in the wagons and haversacks were empty by noon, for the unsalted beef spoiled and was thrown away, and the column subsisted itself, without process of commissariat, upon green corn and apples from the fields and orchards along the route, devoured while marching; for there were no stated mealtimes, and no systematic halts for rest.

I recall a sumptuous banquet of "middling" bacon and "collards" which I was fortunate enough to obtain during the delay at Hinson's Mill where we forded the river and the still more dainty fare of tea and biscuits, the bounty of some good maiden ladies at "The Plains," where our ambulance stopped some hours to repair a

> "Oh well, old Jube can die if that's what he wants, but tell General Jackson I'll be d-----d if this position can be held!"

broken axle – the only episodes of the march which now stand out with distinctness.

It was far on in the night when the column stopped, and the weary men dropped beside their stacked muskets and were instantly asleep without so much as unrolling a blanket. A few hours of much needed repose, and they were shaken up again long before "crack of day," and limped on in the darkness only half-awake. There was no mood for speech nor breath to spare if there had been – only the shuffling tramp of the marching feet, the steady rumbling of wheels, the creak and rattle and clank of harness and accouterment. With an occasional order, uttered under the breath and always the same: "Close up! Close up, men!"

All this time we had the vaguest notions as to our objective. At first we had expected to strike the enemy's flank, but as the march prolonged itself, a theory obtained that we were going to the Valley. But we threaded Thoroughfare Gap, heading eastward and in the morning of the third day (August 27th), struck a railroad running north and south – Pope's line of communication and supply. Manassas was ours.

What a prize it was! Here were long warehouses full of stores. Cars loaded with boxes of new clothing en route to General Pope but destined to adorn the backs of his enemies, camps and sutlers' shops. In view of the abundance, it was not an easy matter to determine what we should eat and drink and where withal we should be clothed.

One was limited in his choice to only so much as he could personally transport and the one thing needful in each individual case was not always readily found. However, as the day wore on an equitable distribution of our wealth was effected by barter, upon a crude and irregular tariff in which the rule of supply and demand was somewhat complicated by fluctuating estimates of the imminence of marching orders. A mounted man would offer large odds in shirts or blankets for a pair of spurs or a bridle, and while in anxious guest of a pair of shoes, I fell heir to a case of cavalry half boots, which I would gladly have exchanged for the object of my search. For a change of underclothing, I owe grateful thanks to the major of the 12th Pennsylvania Cavalry. It is with regret that I could not use his library. Whisky was, of course, at a high premium, but a keg of "larger" – a drink less popular then than now – went begging in our company.

But our brief holiday was drawing to a close, for by this time General Pope had some inkling of the disaster which lurked in his rear. When, some time after dark, having set fire to the remnant of the stores, we took the road to Centreville, our mystification as to Jackson's plans was complete. Could we actually be moving on Washington with his shall force, or was he only seeking escape to the mountains?

The glare of our big bonfire lighted up the country for miles and was just dying out when we reached Centreville. The corduroy road had been full of pitfalls and stumbling blocks, to some one of which our cracked

> *What a prize it was! Here were long warehouses full of stores. Cars loaded with boxes of new clothing en route to General Pope but destined to adorn the backs of his enemies, camps, and sutlers' shops.*

axle had succumbed before we crossed Bull Run and being on ahead, I did not know of the casualty until it was too late to save my personal belongings involved in the wreck. Thus suddenly reduced from affluence to poverty just as the gray dawn revealed the features of the forlorn little hamlet, typical of this war-harried region, I had a distinct sense of being a long way from home.

The night's march had seemed to put the climax to the endurance of the jaded troops. Such specters of men they were – gaunt-cheeked and hollow-eyed, hair, beard, clothing, and accouterments covered with dust – only their faces and hands, where mingled soil and sweat streaked and crusted the skin, showed any departure from the whitey-gray uniformity.

The ranks were sadly thinned, too, by the stupendous work of the previous week. Our regiment, which had begun the campaign 1015 strong and had carried into action at Richmond 620, counted off that Thursday morning (August 28th) just 82 muskets! Such were the troops about to deliver battle on the already historic field of Manassas.

We were soon on the road again, heading west; we crossed Stone Bridge, and a short distance beyond, our ambulances halted, the brigade having entered some woods on the right of the road ahead, going into camp, I thought. This pleasing delusion was soon dispelled by artillery firing in front and our train was moved off through the fields to the right, out of range, and was parked near Sudley Church.

Everything pointed to a battle next day. The customary hospital preparations were made, but few if any wounded came in that night and I slept soundly, a thing to be grateful for. My bedfellow and I had decided to report for duty in the morning, knowing that every musket would be needed. I had picked up a good "Enfield" with the proper trappings on the road from Centreville to replace my own left in the abandoned ambulance and, having broken my chills and gained strength from marching unencumbered, was fit for service – as much so as were the rest at least.

Friday morning early, we started in what we supposed to be the right direction, guided by the firing, which more and more betokened that the fight was on. Once we stopped for a few moments at a field-hospital to make inquiries and were informed that our brigade was farther along to the right. General Ewell, who had lost his leg the evening before, was carried by on a stretcher while we were there. Very soon we heard sharp musketry over a low ridge which we had been skirting and almost immediately we became involved with stragglers from that direction – Georgians, I think they were.

It looked as if a whole line was giving way, and we hurried on to gain our own colors before it should grow too hot. The proverbial effect of bad company was soon apparent. We were halted by a Louisiana major, who was trying to rally these fragments upon his own command. My companion took the short cut out of the scrape by showing his "sick-permit" and was allowed to pass. Alas! Mine had been left in my cartridge-box with my other belongings in that unlucky ambulance. The major was courteous but firm. He listened to my story with more attention than I could have expected, but attached my person all the same. "Better stay with us, my boy, and if you do your duty I'll make it right with your company officers when the fight's over. They won't find fault with you when they know you've been in with the

'Pelicans,'" he added, as he assigned me to Company F.

The command was as unlike my own as it was possible to conceive. Such a congress of nations only the cosmopolitan Crescent City could have sent forth, and the tongues of Babel seemed resurrected in its speech: English, German, French, and Spanish, all were represented, to say nothing of Doric brogue and local "gumbo." There was, moreover, a vehemence of utterance and gesture curiously at variance with the reticence of our Virginians.

The battalion did a good deal of counter-marching, and some skirmishing, but most of the time we were acting as support to a section of Cutshaw's battery. The tedium of this last service my companions relieved by games of "seven up," with a greasy, well-thumbed deck, and in smoking cigarettes, rolled with great dexterity, between the deals. Once, when a detail was ordered to go some distance under fire to fill the canteens of the company, a hand was dealt to determine who should go, and the decision was accepted by the loser without demur. Our numerous shifts of position completely confused what vague ideas I had of the situation, but we must have been near our extreme left at Sudley Church, and never very far from my own brigade, which was warmly engaged that day and the day following. Toward evening we were again within sight of Sudley Church.

I could see the light of fires among the trees, as if cooking for the wounded was going on, and the idea occurred to me that there I could easily learn the exact position of my proper command. Once clear of my major and his polyglot "Pelicans," the rest would be plain sailing.

My flank movement was easily effected, and I suddenly found myself the most private soldier on that field. There seemed to be nobody else anywhere near. I passed a farmhouse, which seemed to have been used as a hospital, and where I picked up a Zouave fez. Some cavalrymen were there, one of whom advised "me not to go down there," but as he gave no special reason and did not urge his views, I paid no heed to him, but went on my way down a long barren slope, ending in a small water-course at the bottom, beyond which the ground rose abruptly and was covered by small growth.

The deepening twilight and strange solitude about me, with a remembrance of what had happened a year ago on this same ground, made me feel uncomfortably lonely. By this time I was close to the stream, and while noting the lay of the land on the opposite bank with regard to choice of a crossing-place, I became aware of a man observing me from the end of the cut above. I could not distinguish the color of his uniform, but the crown of his hat tapered suspiciously, I thought, and instinctively I dropped the butt of my rifle to the ground and reached behind me for a cartridge. "Come here!" He called, his accent was worse than his hat.

"Who are you?" I responded as I executed the movement of "tear cartridge." He laughed and then invited me to "come and see." Meanwhile I was trying to draw my rammer, but this operation was arrested by the dry click of several gunlocks, and I found myself covered by half a dozen rifles, and my friend of the steeple-crown, with less urbanity in his intonation, called out to me to "drop that." In our brief intercourse he had acquired a curious influence over me. I did so.

My captors were of Kearny's division, on picket duty. They thought I was deserting until they saw me try to load. I could not account for their being where they were, and when they informed me that they had Jackson surrounded and that he must surrender next day, though I openly discounted the notion, I must own the weight of evidence seemed to be with them.

The discussion of this and kindred topics was continued until a late hour that night with the sergeant of the guard at Kearny's headquarters, where I supped in unwonted luxury on hard-tack and "genuine" coffee, the sergeant explaining that the fare was no better because of our destruction of their supplies at the Junction. Kearny's orderly gave me a blanket, and so I passed the night.

We were astir early in the morning (August 30th), and I saw Kearny as he passed with his staff to the front – a spare, erect, military figure, looking every inch the fighter he was. He fell three days later, killed by some of my own brigade.

Near the Stone Bridge, I found about 500 other prisoners, mostly stragglers picked up along the line of our march. Here my polite provost-sergeant turned me over to other guardians, and after drawing rations, hard-tack, coffee, and sugar, we took the road to Centreville. That thoroughfare was thronged with troops, trains, and batteries, and we had to stand a good deal of chaff on the way at our forlorn appearance. We were a motley crowd enough, certainly, and it did look as if our friends in blue were having their return innings.

> *Jackson had made good the name of "Stonewall" on his baptismal battlefield.*

More than once that day as I thought of the thin line I had left, I wondered how the boys were doing, for disturbing rumors came to us as we lay in a field near Centreville, exchanging rude badinage across the cordon of sentries surrounding us. Other prisoners came in from time to time who brought the same unvarying story, "Jackson hard-pressed – no news of Longstreet yet."

So the day wore on. Toward evening there was a noticeable stir in the camps around us, a continual riding to and fro of couriers and orderlies, and now we thought we could hear more distinctly the deep-toned, jarring growl which had interjected itself at intervals all the afternoon through the trivial buzz about us. Watchful of indications, we noted, too, that the drift of wagons and ambulances was from the battlefield, and soon orders came for us to take the road in the same direction. The cannonading down the pike was sensibly nearer now, and at times we could catch even the roll of musketry, and once we thought we could distinguish, far off and faint, the prolonged, murmurous sound familiar to our ears as the changing shout of the gray people – but this may have been fancy.

All the same, we gave tongue to the cry, and shouts of "Longstreet! Longstreet's at 'em, boys! Hurrah for Longstreet!" went up from our ranks, while the guards trudged beside us in sulky silence.

There is not much more to tell. An all-day march on Sunday through rain and mud brought us to Alexandria, where we were locked in a cotton-factory. Monday, we embarked on a transport steamer and the next evening, were off Fort Monroe, where we got news of Pope's defeat. I was paroled and back in Richmond within ten days of my capture, and then and there learned how completely Jackson had made good the name of "Stonewall" on his baptismal battlefield.

To subscribe, send an email to:
thestainlessbanner@gmail.com
or visit our website at www.thestainlessbanner.com

Subscription is free.

The Stainless Banner

An e-zine dedicated to the armies of the Confederacy

Issue 3, Volume 8
September 2012

CHARGE 'EM BOTH WAYS
FORREST AVOIDS DISASTER AT PARKER'S CROSSROADS

General Ulysses S. Grant, in command of the Army of the Tennessee, was preparing to launch an assault on Vicksburg – the last Southern stronghold on the Mississippi River. Lieutenant General John Pemberton did not have enough men to keep Grant from his goal. An urgent message was dispatched to General Braxton Bragg in Tennessee for reinforcements. Bragg had no troops to spare. What he did have though was a formidable weapon: Nathan Bedford Forrest. Earlier that autumn, Forrest had performed splendidly in a raid behind enemy lines, convincing Union General Don Carlos Buell that the Confederates were planning to attack Nashville even as Bragg hurried his Army of Mississippi through Tennessee and into Kentucky.

Bragg's orders to Forrest presented great difficulties for the cavalry leader. To get at Grant's supply line, Forrest had to cross a rain swollen Tennessee River, where Union gunboats routinely patrolled and protected the river's fords, and travel down roads that winter rain and snow had turned into bottomless pits.

Compounding Forrest's difficulties was the fact that his men lacked ammunition and modern weaponry. Appeals to Richmond for more guns had gone unanswered.

What's Inside:

Forrest's Official Report	
Of the Campaign into Tennessee	4
Of Parker's Crossroads	7
Cummings' Official Report	10
John Hunt Morgan	12
Forrest's Artillery at Parker's Crossroads	14
Battle of Parker's Crossroads	16
Forrest and the Double Envelopment	24

www.thestainlessbanner.com

Forrest crossed the Tennessee on December 17th. News traveled quickly, and from Nashville, General Rosecrans warned Grant that Forrest was now in the army's rear. Grant ordered a concentration of troops at Jackson, Tennessee. General Jeremiah Sullivan was then ordered to hunt down and destroy Forrest. Sullivan gave chase, but he could not stop Forrest, who was having his way in Tennessee.

For two weeks, Forrest destroyed the railroad, bridges and trestles, disrupted Grant's communications, and, more importantly, resupplied his men at the expense of the Union Quartermaster.

As December drew to a close, Forrest decided he had done all the damage he could to Grant's supplies and communications. His men and horses were exhausted. But extracting himself proved hazardous. The Union gunboats were on alert, making crossing the Tennessee even more dangerous the second time. Plus, the forces sent by Grant to capture Forrest were closing in.

Colonel Dunham's brigade, comprised of regiments from the Midwest, was encamped at Clarksburg – four miles from Forrest's own camp at Flake's Store. At Huntingdon, in easy support of Dunham, were Sullivan and General Fuller's Ohio brigade.

> *Study Forrest's life before, during, or after the war, and you quickly come to realize that Forrest never backed down from a fight in his life and, at times, went out of his way to seek them out. But when he fought, he always fought to win. Which he usually did.*

Forrest was only forty miles from the river and could have easily outrun his pursuers, but crossing on the two flat boats that had borne his troops over the swollen river two weeks earlier would allow his foes to catch him with part of his men out of reach on the other side. Forrest decided the best course of action would be to stand his ground and fight each Union force separately. Dunham was the closest, so he would need to be whipped first.

Forrest sent his brother, Captain William Forrest, and his band of independents to locate Dunham. Bill Forrest found the Federals at Parker's Crossroads and sent word back to his older brother, who ordered his men up and in the saddle.

To keep eye on Fuller, Forrest dispatched Captain McLemore and three companies from the 4th Alabama across country toward Huntingdon. McLemore's orders were to skirmish with Fuller in order to slow his approach and notify Forrest when the Ohioans were in striking distance.

As Forrest neared Parker's Crossroads, he found Dunham's men deployed at Hick's Field. To spare his men for the fight with Fuller, Forrest turned to his artillery. Placing his cannon at close range to the Union line of battle, Forrest ordered Lieutenant Morton to "give 'em hell."

Dunham's artillery answered in kind, but the Union commander only had a section of the 7th Wisconsin Light Artillery, and the Confederate cannon swiftly turned the tide. Dunham was driven back toward the Parker house.

Forrest moved his artillery and cavalry forward, flanking Dunham's line to the north. Dunham retreated again, his men finding refuge behind a split-rail fence. Forrest sent Starnes' 4th Tennessee and Russell's 4th Alabama to the left and right of the Union line to get into the Federals' rear. He then ordered a general advance. A half hour later, the Federals broke, crossed the road and entered a cornfield. When gunfire broke out in his rear, Dunham ordered his line to about face and charge, but it was too late. He was surrounded. White flags blossomed along the Union line.

Forrest rode forward to receive Dunham's surrender. While negotiations were going on, gunfire crackled in the Confederate rear. Sullivan and Fuller had arrived from Huntingdon, surprising Forrest's horse holders near the Parker's house. The horses stampeded, leaving 300 Confederates without an escape route. They were captured.

Forrest was none the wiser to Sullivan's arrival. Aide-de-camp Colonel Carroll relayed the news. At first, Forrest refused to believe it. After all, he had sent McLemore to prevent that very thing from happening. He galloped to the rear and beheld a superior force of bluecoats rapidly forming into a line of battle.

Forrest's objective quickly changed from victory to escape.

As he galloped along, a Federal officer called for him to surrender. Forrest replied that he had already done that. Then he wheeled back toward the makeshift hospital at the rear of his line and instructed the wounded to make their escape. Forrest put his remaining forces into motion by the left flank and hastened away from Fuller's forces. Dunham's men, who had just surrendered, picked up their weapons and rejoined the battle.

If Forrest was to escape – bold action was required. Colonel Dibrell and Major Jeffrey Forrest charged Sullivan's batteries, scattering gunners and spooking the horses tethered to the caissons. As those horses came careening into the Confederate line, Forrest seized them and carried them away. Starnes hit Dunham's reorganized line, stopping Dunham in his tracks.

In the chaos, the Confederates captured Dunham's trains and baggage. Forrest secured a small force and charged into the Ohioans, which disrupted the attack and caused Sullivan to pull his forces back into a defensive position. The Union forces froze long enough to allow Forrest to make his escape toward Lexington.

The question remains. What happened to McLemore and why had he allowed Sullivan to get past him unobserved? The answer was a simple one. When McLemore heard the sound of cannon, he simply turned around and headed back to the battle. He got lost along the way and arrived at Parker's Crossroad just as Forrest quit the field. After the war, Forrest blamed poorly written orders which did not fully convey the purpose for McLemore's reconnaissance.

When viewed against the other battles that were fought during the war, this small battle at Parker's Crossroads does not seem very important. Yet, the battle has echoed through history. The reason for this is two-fold.

First, it revealed Nathan Bedford Forrest's inherent ability for tactics and strategy. Forrest was able to take in the entire field and the natural ebb and flow of battle and realize the advent of that golden moment when fortune favors the bold. And Forrest was bold. He was a natural fighter, who fought not just to hold ground, but to defeat the enemy – no matter the size and strength of the foe. Forrest's boldness at the precise moment when boldness was called for probably did more to win victories throughout his career than any other weapon he possessed.

The second reason Parker's Crossroads is celebrated is because of Forrest's instinctive reply when Colonel Carroll questioned what Forrest was going to do now that Sullivan had surprised him. "Charge 'em both ways!" Forrest responded without hesitation. This quote has become legendary not only for its sheer audacity, but because it was a window into Forrest's character.

Study Forrest's life before, during or after the war, and you quickly come to realize that Forrest never backed down from a fight in his life and, at times, he went out of his way to seek them out. But when he fought, he always fought to win. Which he usually did.

Forrest's Official Report – Campaign in Western Tennessee

Brigade Headquarters,
Near Union City, Tennessee, December 24, 1862.

GENERAL: In accordance with your order I moved with my command from Columbia on the 11th instant, reached the river at Clifton on Sunday, the 13th, and after much difficulty, working night and day, finished crossing on the 15th, encamping that night eight miles west of the river.

On the 16th [18th] we met the pickets of the enemy near Lexington and attacked their forces at Lexington, consisting of one section of artillery and 800 cavalry. We routed them completely, capturing the two guns and 148 prisoners including Colonel Ingersoll and Major Kerr of the 11th Illinois Cavalry. We also captured about 70 horses, which were badly needed and immediately put in service in our batteries. The balance of the Federal cavalry fled in the direction of Trenton and Jackson.

We pushed on rapidly to Jackson, and on the evening of the 18th drove in their pickets on all the roads leading out of Jackson. On the same night, I sent Colonel Dibrell on the right of Jackson to tear up the railroad track and destroy the telegraph wires. He captured at Webb's Station 101 Federals, destroying their stockade, and tore up the road, switch, etc., at the turn-out.

At the same time that Dibrell was sent on the right, Colonel Russell, 4th Alabama Cavarly and Major Cox, 2nd Battalion Tennessee Cavalry, with their commands were sent out on the left to destroy bridges and culverts on the railroads from Jackson to Corinth and Bolivar.

The next morning, December 19, I advanced on Jackson with Colonel Woodward's two companies and Colonel Biffle's battalion of about 400 men, with two pieces of artillery from Freeman's battery. About four miles from Jackson skirmishing began with the skirmishers, and the enemy was reported advancing with two regiments of infantry and a battalion of cavalry. We opened on them with the guns, and after a running fight of about an hour drove them into their fortifications.

The enemy had heavily reinforced at Jackson from Corinth, Bolivar, and La Grange, and numbered, from the best information I could obtain, about 9,000 men. I withdrew my forces that evening and moved rapidly on Trenton and Humboldt. Colonel Dibrell's command was sent to destroy the bridge over the Forked Deer River between Humboldt and Jackson. Colonel Starnes was sent to attack Humboldt. Colonel Biffle was sent so as to get in the rear of Trenton, while with Major Cox's command and my bodyguard, commanded by Captain Little, and Freeman's Tennessee battery

I dashed into town and attacked the enemy at Trenton. They were fortified at the depot, but were without artillery. After a short engagement between their sharpshooters and our cavalry, our battery opened on them, and on the third fire from the battery, they surrendered. We lost two men killed and seven wounded; the enemy two killed and over 700 prisoners, with a large quantity of stores, arms, ammunition, and provisions, which for want of transportation, we were compelled to destroy. We captured several hundred horses, but few of them were of any value; those that were of service we took, and the balance I handed over to the citizens, from whom many of them had been pressed or stolen.

> *I dashed into town and attacked the enemy at Trenton.*

Colonel Russell, who was protecting our rear at Spring Creek, found the enemy advancing and following us with 3,000 infantry, two batteries, and several hundred cavalry. He skirmished with them during the evening and the next morning before daylight dismounted half of his command and succeeded in getting within sixty yards of their encampment. They discovered him and formed in line of battle. He delivered a volley as soon as their line was formed and the balance of the regiment charged on horseback.

The enemy became panic-stricken and retreated hastily across Spring Creek, burning the bridge after them. We have heard nothing from them since in that direction. Colonel Starnes took Humboldt, capturing over 100 prisoners. He destroyed the stockade, railroad depot, and burned up a trestle bridge near that point.

Colonel Dibrell's command failed to destroy the bridge over the Forked Deer River, as the enemy were strongly fortified and protected by two creeks on one side of the railroad and a wide, swampy bottom on the other, which rendered the approach of cavalry impossible. He dismounted his men, and while approaching their fort, a train arrived from Jackson with a regiment of infantry. Lieutenant Morton with two guns opened on the train, when it retired, the troops on it gaining the stockade. Owing to the situation of the stockade and the density of the timber and the wet, miry condition of the bottom, the guns could not be brought to bear on it. Night coming on, Colonel Dibrell withdrew and rejoined my command.

We remained in Trenton during the night of the 20th, paroling all the prisoners and selecting from the stores at the depot such as were needed by the command.

On the morning of the 21st, I fired the depot, burning up the remaining supplies, with about 600 bales of cotton, 200 barrels of pork, and a large lot of tobacco in hogsheads, used by the enemy for breastworks. After seeing everything destroyed I moved on in the direction of Union City, capturing at Rutherford Station two companies of Federals and destroying the railroad from Trenton to Kenton Station, at which place we captured Colonel Kinney, of the 122nd Illinois Regiment, and twenty-two men left sick in the hospital.

I took a portion of the command and pushed ahead to Union City, capturing 106 Federals without firing a gun. I destroyed the railroad bridge over the bayou near Moscow and am completing the destruction of the bridges over the North and South Fork of Obion River, with nearly four miles of trestling in the bottom between them. We have made a clean sweep of the Federals and roads north of Jackson, and know of no Federals except at Fort Heiman, Paducah, and Columbus, north of Jackson and west of the Tennessee River.

Reports that are reliable show that the Federals are rapidly sending up troops from Memphis. One hundred and twenty-five transports passed down a few days ago within ten hours, and daily they are passing up loaded with troops. General Grant must either be in a very critical condition or else affairs in Kentucky require the movement.

In closing my report, general, allow me to say that great credit is due to the officers of my command. They have exhibited great zeal, energy, endurance, and gallantry.

Colonel Russell and his command deserve especial notice for their gallantry in the fight at Lexington and Spring Creek. Captain Gurley, 4th Alabama Cavalry, with twelve men charged a gun at Lexington supported by over 100 Federal cavalry. He captured the gun, losing his orderly-sergeant by the fire of the gun when within fifteen feet of its muzzle. My men have all behaved well in action, and as soon as rested a little you will hear from me in another quarter.

We have been so busy and kept so constantly moving that we have not had time to make out a report of our strength, and ask to be excused until the next courier comes over. We send by courier a list of prisoners paroled.

General, I am, very respectfully, your most obedient servant,

N.B. FORREST,
Brigadier-General, Commanding in West Tennessee

If Major Ewing bothers me any further about this matter, I'll come down to his office, tie his long legs into a double bowknot around his neck, and choke him to death with his own shins.

Forrest railing against an order he disagreed with.

Forrest's Official Report – Battle of Parker's Crossroads

BRIGADE HEADQUARTERS,
Clifton, Tennessee, January 3, 1863.

GENERAL: I forwarded you from Middleburg, per Lieutenant Martin, a detailed report of my operations up to the 25th ultimo, which I hope reached you safely.

I left Middleburg on the 25th, proceeding via the Northwestern Railroad to McKenzie's Station, destroying all the bridges and trestles on that road from Union City to McKenzie's Station. From McKenzie's Station, we were compelled to move southward in the direction of Lexington, as the enemy in force occupied Trenton, Humboldt, Huntingdon, and Lexington. After my command left Trenton they commenced reinforcing and moving to the points named with a view of cutting off my command and prevent us from re-crossing the Tennessee.

Understanding a force was moving on me from Trenton in the direction of Dresden, I sent Colonel Biffle, 9th Tennessee Cavalry, in that direction to protect our movements toward Lexington, intending if possible to avoid the enemy and go on and attack the enemy at Bethel Station, on the Mobile and Ohio road, south of Jackson.

We left McKenzie's Station on the morning of December 28, but in crossing the bottom had great difficulty in crossing our artillery and wagons; the bridges proved to be much decayed and gave way, forcing us to drag our artillery and wagons through the bottom and the creeks. It was with great difficulty we got through by working the entire night, and our men and horses were so much fatigued that I was compelled to encamp at Flake's Store, about sixteen miles north of Lexington, when under ordinary circumstances and good roads we ought to have reached Lexington that night, which place had been evacuated by the enemy, believing that I would either cross the Tennessee at Huntingdon or else that I would move northward.

On the morning of the 31st, we moved off in the direction of Lexington, but had not gone more than four miles before we met the skirmishers of the enemy. We engaged and fought six regiments for five hours, driving them back until 3:00 in the evening, when they took shelter in a grove of timber of about sixty acres enclosed by a fence and surrounded by open fields. I had sent four companies to Clarksburg to protect and advise me of any advance from Huntingdon, and finding that we were able to whip the enemy, dismounted a portion of my cavalry to support my artillery and attack in front while I could flank them on each side and get Colonel Russell's

> *Thirty minutes more would have given us the day, when to my surprise and astonishment, a fire was opened on us in our rear, and the enemy in heavy force under General Sullivan advanced on us.*

regiment, 4th Alabama Cavalry, in their rear.

We drove them through the woods with great slaughter and several white flags were raised in various portions of the woods and the killed and wounded were strewn over the ground. Thirty minutes more would have given us the day, when to my surprise and astonishment a fire was opened on us in our rear and the enemy in heavy force under General Sullivan advanced on us.

Knowing that I had four companies at Clarksburg, seven miles from us on the Huntingdon road, I could not believe that they were Federals until I rode up myself into their lines. The heavy fire of their infantry unexpected and unlooked for by all caused a stampede of horses belonging to my dismounted men, who were following up and driving the enemy before them. They also killed and crippled many of the horses attached to our caissons and reserved guns.

I had sent back two miles for more ammunition. My men had been fighting for five hours, and both artillery and small-arm ammunition were well-nigh exhausted. We occupied the battlefield, were in possession of the enemy's dead and wounded and their three pieces of artillery and had demanded a surrender of the brigade, which would doubtless have been forced or accepted in half an hour, the colonel commanding proposing to leave the field entirely and withdraw his force provided we would allow him to bury his dead; but believing I could force, and that in a short time,

demand, the fighting continued, the Federals scattering in every direction.

The stampede of horses and horse holders announced that help was at hand, and finding my command now exposed to fire from both front and rear I was compelled to withdraw, which I did in good order, leaving behind our dead and wounded. We were able to bring off six pieces of artillery and two caissons, the balance, with the three guns we captured, we were compelled to leave, as most of the horses were killed or crippled and the drivers in the same condition, which rendered it impossible to get them out under the heavy fire of the enemy from both front and rear. Our loss in artillery is three guns and eight caissons and one piece which burst during the action.

> *The stampede of horses and horse holders announced that help was at hand, and finding my command now exposed to fire from both front and rear I was compelled to withdraw, which I did in good order, leaving behind our dead and wounded.*

The enemy's loss was very heavy in killed and wounded, and as we had the field and saw them piled up and around the fences had a good opportunity of judging their loss. We gave them grape and canister from our guns at 300 yards, and as they fell back through the timber their loss was terrible. The prisoners say that at least one-third of the command was killed or wounded. From all I could see and learn from my aides and officers they must have lost in killed and wounded from 800 to 1,000 men. The fire of our artillery for accuracy and rapidity was scarcely, if ever, excelled, and their position in the fence corners proved to the enemy, instead of a protection, a source of great loss, as our shot and shell scattered them to the winds, and many were killed by rails that were untouched by balls.

Captain Freeman and Lieutenant Morton of our batteries, with all of their men, deserve special mention, keeping up, as they did, a constant fire from their pieces, notwithstanding the enemy made every effort at silencing their pieces by shooting down the artillerists at the guns. The whole command fought well.

We had about 1,800 men in the engagement and fought six regiments of infantry, with three pieces of artillery, which we charged and took, but were compelled to leave them as the horses were all killed or crippled. We brought off eighty-three prisoners, and they report their respective regiments as badly cut up. They lost three colonels and many company officers. We have on our side to deplore the death of Colonel Alonzo Napier, 10th Tennessee Cavalry, who was killed while leading his men in a charge on foot. He was a gallant officer, and after he fell, his command continued to drive the enemy from their position on the right bank, strewing their path with dead and wounded Federals.

I cannot speak in too high terms of all my commanding officers; and the men, considering they were mostly raw recruits, fought well. I have not been able as yet to ascertain our exact loss, but am of the opinion that sixty killed and wounded and 100 captured or missing will cover it. I saved all my wagons except my ammunition wagons, which, by a mistake of orders, were driven right into the enemy's line. This is seriously to be regretted, as we had captured six wagon loads of it; and when I ordered up one wagon of ammunition and two ambulances, the wagon-master and ordnance officer not knowing exactly what kind was wanted, or misunderstanding the order, brought up all the ammunition, and by the time he reached the point with them where the battle begun that portion of the ground was in possession of the enemy, and the guards, etc., and were forced to abandon them.

> *I had them entirely surrounded and driving them before me and was taking it leisurely and trying as much as possible to save my men.*

We have always been short of shotgun caps, and as we captured nothing but musket-caps, all the men using shot-guns were out, or nearly so, of caps after the action was over. Considering our want of ammunition for small-arms and artillery and the worn down condition of our men and horses I determined at once to recross the Tennessee River and fit up for a return. Had we been entirely successful in the battle of the 31st, I should have attacked Bethel Station on the 2nd instant; had already sent a company to cut wires and bridges and had forage prepared twelve miles south of Lexington for my entire command; but after the fight, and knowing we were followed by Federals in heavy force from Trenton and Huntingdon, and that a force would also move on from Jackson as soon as they learned I had pushed south of Lexington, I deemed it advisable to cross the Tennessee, which I accomplished yesterday and last night in safety.

Colonel Biffle, who I before mentioned as having been sent to Trenton, or in that direction, returned in time to take part in the battle at Parker's Crossroads. He captured and paroled 150 Federals within six miles of Trenton.

The captains of the four companies sent to Clarksburg have not yet reached here with their commands. Had they done their duty by advising me of the approach of the enemy, I could have terminated the fight by making it short and decisive, when without such advice, I was whipping them badly with my artillery, and unless absolutely necessary was not pressing them with my cavalry. I had them entirely surrounded and was driving them before me and was taking it leisurely and trying as much as possible to save my men. The four companies on the approach of the enemy left for Tennessee River and have not yet reported here.

I do not design this, general, as a regular report, but will make one as soon as I can do so. We crossed the river at three points, and the brigade is not yet together, or reports front the different commands have not come in. We have worked, rode, and fought hard, and I hope accomplished to a considerable extent, if not entirely, the object of our campaign, as we drew from Corinth, Grand Junction, and La Grange about 20,000 Federals. Will send you an additional list of paroles, etc., by next courier.

I am, general, very respectfully, your most obedient servant,
N. B. FORREST,
Brigadier General, Commanding Brigade

Lieutenant Colonel GEORGE WILLIAM BRENT, Assistant Adjutant-General

COLONEL CUMMINGS OFFICIAL REPORT ON THE BATTLE OF PARKER'S CROSSROADS

Headquarters, 39th Iowa Infantry
Parker's Crossroads, December 31, 1862.

LIEUTENANT: I have the honor to submit the following report of the part taken by my regiment in the battle of Parker's Crossroads this day:

Upon arriving at the crossroads we were halted, and remained in that position some time, while the 50th Indiana Infantry, deployed as skirmishers and supported by two pieces of artillery, engaged the rebels upon the hill to the right and west of the road. We were then ordered to file to the right, up the lane, to take position in the woods upon the hills, and upon arriving there I was ordered to countermarch and take position about a mile south of the cross-roads, and there formed in front of a few log houses, upon the left of the 122nd Illinois Infantry.

About 11:00 a.m., I changed front forward on first company and moved north about a quarter of a mile, and again formed on the left of the 122nd Illinois behind a fence. Here we were exposed to a murderous fire from two pieces of the enemy's artillery in front and a battery of about six guns upon our right, which enfiladed my entire line; we were also exposed to a heavy musketry fire from the enemy's dismounted cavalry.

My men were in a low skirt of timber, but returned for a long time, with much energy, the fire from their rifles. Notwithstanding the grape, canister, and shell of the enemy were falling thick upon them, wounding many, they behaved admirably and fought with much coolness; and here allow me to remark that they were greatly encouraged by the presence of Colonel Dunham, commanding the brigade, who, amid the thickest of the iron hail, rode in front and rear of them, urging them to do or die for their country.

After fighting for an hour or more in this position some officer came down to my right and gave an order, which several of my officers say to me was "Rally to the rear." Had the officer passed down as far as my colors he would have found me, and I am satisfied I could have had my command heard my voice; have about-faced the regiment, and led them anywhere without confusion; but, being raw troops and imperfectly drilled, they mistook the command for an order to retreat and commenced breaking to the rear from near the right of the regiment, which, despite my efforts, became propagated along the whole line. I hastened toward the right of the retreating men and ordered a halt and the command to form, and had done much toward reforming when we were opened upon by a heavy fire of dismounted men, who had advanced under cover of the thick underbrush to within fifty feet of my men.

They then in more confusion fell back toward the fence, and received standing the fire of the enemy's artillery, and under it and the fire from the rear the confusion became worse. Companies F and D, and several from other companies, formed upon the now right of the 122nd Illinois, which had faced to the rear, and assisted them in driving the rebels back at the point of the bayonet, taking a number of prisoners. Under this fire, so unexpected from both front and rear (and the enemy's cannon seemed to be entirely concentrated upon our left, to save their own force in our rear), about half of my regiment broke to the left of our line as formed behind the fence and crossed the road into the cornfield upon the opposite side.

Assisted by Colonel Dunham, Lieutenant-Colonel Redfield, who was severely wounded; Major Griffiths, who had been struck on the head by a spent grapeshot, and yourself, I attempted to halt and reform the scattered men. The enemy turned their cannon upon us and we were fired upon by their cavalry, and I was unable to form a line until we reached a skirt of timber about a quarter of a mile from where we laid in line.

> *Under this fire, so unexpected from both front and rear (and enemy's cannon seemed to entirely concentrated upon our left, to save their own force in our rear), about half of my regiment broke to the left of our line as formed behind the fence and crossed the road into the cornfield upon the opposite side.*

Here I formed and marched back upon the left again of the 122nd Illinois. Let me say that in this confusion we found a number of the 122nd Illinois and of the 50th Indiana, but they fell in with us and marched back to the battle ground. Shortly afterward, perhaps half an hour, and at about 1:30 p.m., reinforcements arrived and the battle ended.

I have omitted to state that at the crossroads Company A was detached from the regiment and guarded our trains. When we fell back to the ground on which the battle was fought, they, or rather all but fifteen of them, with Company G, of the 122nd Illinois, were stationed at the house in the rear of our line of battle. Here they three times repulsed a regiment of cavalry who attempted to force their way through the lane to reach our main body. The fifteen spoken of were near the trains, and there succeeded in capturing over forty of the rebels.

There were many cases of individual bravery among those under my command; but to particularize would make my report too lengthy. I must, however, say that, from information received through reliable men of my command who were taken prisoners and paroled, I am satisfied that the rebels had men dressed in our uniform so close in our rear that they could see our exact position – knew the numbers of our regiment and strength.

Allow me to add that while I cannot take the room to name the many of my company officers who did their whole duty, I must bear witness to the coolness and bravery of Lieutenant-Colonel Redfield (who ceased his labors only when his wound compelled him), Major Griffiths, Surgeon Woods, and Adjutant Tichenor. They rendered me all the assistance possible.

I am, very respectfully, your obedient servant,
H. J. B. CUMMINGS,
Colonel, Commanding

HI-LIGHTS OF A HERO'S LIFE
JOHN HUNT MORGAN

★ 1825 – (June 1) Born in Huntsville, Alabama.

★ 1831 – Moved to Lexington, Kentucky.

★ 1844 – Suspended from Transylvania College for dueling.

★ 1846 – During the Mexican War, enlisted in the United States Army as a private in the Cavalry.

★ 1846 – Promoted to first lieutenant and saw combat at the Battle of Buena Vista.

- ★ 1848 – Returned to Kentucky and married Rebecca Bruce.

- ★ 1852 – Raised a militia artillery company.

- ★ 1857 – Raised an independent infantry company known as the Lexington Rifles.

- ★ 1861 – (July 21) Rebecca Graves Morgan died.

- ★ 1861 – (September) Traveled to Tennessee and enlisted in the Confederate Army.

- ★ 1862 – (April 4) Promoted to colonel of the 2nd Kentucky Cavalry Regiment.

- ★ 1862 – (April 6-7) Fought at Shiloh.

- ★ 1862 – (July 4) Began a raid deep into Kentucky.

- ★ 1862 – (December 7) Victorious at the Battle of Hartsville.

- ★ 1862 – (December 11) Promoted to brigadier general.

- ★ 1862 – (December 14) Married Martha Ready.

- ★ 1863 – (May 1) Received the gratitude of the Confederate Congress for raids conducted on General Rosecrans' supply line.

- ★ 1863 – (Summer) Disobeyed Braxton Bragg's orders not to cross the Ohio River, conducted a raid across southern Indiana and Ohio.

- ★ 1863 – (July 26) Surrendered at Salineville, Ohio and held prisoner in the Ohio Penitentiary.

- ★ 1863 – (November 27) Escaped from the Ohio Penitentiary.

- ★ 1864 – (August 22) Received command of the Trans-Allegheny Department.

- ★ 1864 – (September 4) Surprised and killed by a Union cavalryman in Greeneville, Tennessee.

- ★ 1864 – Buried in Lexington Cemetery in Kentucky.

Forrest's Artillery at Parker's Crossroads

John Morton. *The Artillery of Nathan Bedford Forrest.*

General Forrest received news that General Dunham was advancing on one side and General Fuller on the other, in an effort to prevent his crossing the river at Clifton; and realizing that with the poor facilities afforded by the two small boats he would be unable to cope with such a force (which would be augmented, in all probability, by the gunboats now patrolling the river), he determined to give battle at once. He therefore dispatched two reconnoitering parties to see where the enemy was. The first of these, commanded by Captain William Forrest, encountered Dunham's Brigade at Parker's Crossroads. The second detachment made a mistake and took the wrong road, leaving the field clear for the approach of Fuller's Brigade, when they came up later and turned the tide of battle for the already defeated Federals.

Moving down the lane, Dunham's troops were found in a strong position, and with his usual vigor General Forrest placed the artillery at close range and ordered Captain Morton to "give 'em hell," a very emphatic order often given to his artillery commander.

Captain Morton has been given the credit, in some quarters, for suggesting to General Forrest that artillery could be fought at as close range as rifles, but this honor does not belong to him. Indeed, he wishes no greater claim to merit than a quick comprehension of, and thorough sympathy with, his commander's ideas. Many of the orders he received were outside of his knowledge of military tactics, but it never occurred to him to doubt their feasibility nor their success.

The author has never forgotten General Forrest's peculiarly penetrating gaze from under the bushy black brows, the dark gray eyes seeming to read the mind before him. The ready conformity he invariably saw in the young artillery commander never drew from him any but the kindest of words. Although witnessing numerous outbreaks of anger and infraction of the Second Commandment, the author was never the recipient of violent language or treatment.

Once, indeed, at the battle of Brice's Crossroads, Captain Morton impulsively advised his chief to get out of danger; and although this advice was followed by a quick apology, he expected a reprimand. To his surprise, however, his commander replied mildly, "Well, John, I will rest a few minutes," and withdrew to shelter under a tree. An incident to be related later will show the only occasion on which Morton met with anything approaching sternness.

The lines were so close that it soon became an artillery duel between the respective sides, the troops of the Confederates being held in reserve to engage Fuller when he should come up. The Union forces made a stubborn fight and

> *What had looked like a victory began to resemble a defeat, as General Sullivan was also seen advancing from the rear at this time.*

defended their position with marked valor, but the Confederate fire was too galling for mortals to stand and they were driven from point to point. By noon they had been repulsed twice, Morton's guns following them with almost equal speed, as the roads had been churned to a deep, muddy paste, making all progress slow. Colonel Starnes was sent to the right and Captain Russell to the left to cut off Dunham's retreat; Biffle's Regiment coming up at the same moment, victory seemed assured.

It was at this point that the brave Colonel Napier fell. In the ardor of the charge he advanced his battalion, without orders and without support, to a position of extreme exposure, and was mortally wounded at once, a number of his intrepid followers falling with him.

Colonel H. J. B. Cummings says the Confederates advanced to within fifty feet of his troops with their artillery, and "about half of my regiment broke and crossed the road into the cornfield."

All of the Confederate guns, with the exception of one which had accidentally exploded, were in fine working order and were kept to the front, together with three guns which had been captured with four horses each. When Colonel Cummings' men fled, General Dunham welcomed a flag of truce which General Forrest sent forward. All firing ceased; and when the leaders met to arrange terms, the soldiers in both armies mingled freely, as was their custom.

Major Strange, confident of surrender, rode alone to the ordnance train of eighteen wagons, and, taking possession began to make an inventory. At this moment, when the surrender was all but consummated, there came the sound of firing from the rear. General Forrest, who was on the lookout for his other detachment, pushed heavily in front, and it was not until the newcomers got at close range that he perceived that Fuller's men had slipped in without warning, and that what had looked like a victory began to resemble a defeat, as General Sullivan was also seen advancing from the rear at this time.

With incomparable address and coolness General Forrest ordered the artillery out between the enfilading lines of fire and rallied his men. Firing as they went and adding to the din and confusion by the "Rebel yell," the bulk of the command escaped, leaving Major Strange, Colonel Cox, and some 300 men, who had dismounted, prisoners, eighteen of these being members of Morton's Battery. A singular fact in this connection lies in the carrying in the Confederate retreat about eighty prisoners – surely it is a doubtful battle where both sides carry away prisoners.

Wyeth says of this movement: "Placing himself at the head of his escort and Dibrell's Regiment, he threw his command as a rear guard between his pet guns and Sullivan's advance. He was not going to give up his artillery without a struggle."

With the exception of the exploded gun, all the Confederate artillery was safely removed by Captain Morton; but the three captured pieces were left behind, as their horses had been killed and there was not time to substitute others. During the retreat Captain Morton rode by the side of General

> *But for a miscarriage of his orders, this most brilliant of tactics would have resulted in the defeating and possible capture of the enemy.*

Forrest at the head of the retiring column. The commanding officer seemed in deep thought, and nothing was said for a time; but as a Minie ball from Sullivan's forces, which had just reached the field in our rear, whizzed by Captain Morton's face, the General dropped his head to his breast. Thinking he was wounded, Captain Morton touched him on the shoulder and inquired: "General, are you hurt much?" General Forrest raised his head, took off his hat, and, noticing that a big hole had been made in the brim, replied: "No, but didn't it come damn close to me?"

While the Confederates were advancing and firing down a narrow lane, before reaching the battlefield, one of Captain Morton's guns was fired with a singular effect. The enemy had piled rails along the road to delay approach, and the Confederates were obliged to turn aside and make a detour through some fields. Captain Morton kept his guns trained on the road, and one of the shells, striking a rail, glanced aside and killed three Federal officers and wounded seven men. This incident was reported to the author by Dr. Henry Long, now a prominent physician of Mt. Pleasant, Tennessee. Dr. Long saw the occurrence and dressed the wounds of the Federal soldiers.

This battle was, perhaps, the supreme test of General Forrest's greatness as a strategist. Knowing that the two divisions of the enemy would meet and, cooperating with the gunboats, prevent his re-crossing the Tennessee, he seized the moment when the two commands were widely separated, and attacked the first with his artillery, holding the infantry for the second attack. But for the miscarriage of his orders, this most brilliant of tactics would have resulted in the defeating and possible capture of the enemy. In the face of disaster he succeeded in withdrawing the greater part of his troops, his artillery, his wagon trains and supplies, as well as some prisoners from the battlefield.

Nevertheless, as everything goes by comparison, the Federals were immensely pleased with the victory and telegrams of a congratulatory nature were showered upon the generals taking part in the achievement. More reinforcements were promised, and on the side of the officers promises of a speedy cleaning out of the Rebels were confidently made. "All the enemies' ferries" were reported destroyed, and the common opinion was that there was nothing left but to "go through the country and pick Forrest and his men up."

In his official report General Forrest commended the action of his artillery. "Captain Freeman and Lieutenant Morton, of our batteries, with all of their men, deserve especial mention, keeping up, as they did, a constant fire from their pieces, notwithstanding the enemy made every effort at silencing them by shooting down the artillerists at their guns."

General Forrest's achievements for the two weeks between and including December 15 and December 31, 1862, may be summed up as follows: The battles of Lexington, Trenton, and Parker's Crossroads, besides daily skirmishes; fifty bridges destroyed on the Mobile and Ohio Railroad; twenty stockades captured and burned; 2,500 of the enemy killed and captured; 10 pieces of artillery captured; fifty wagons and ambulances and their teams, 10,000 stands of excellent small arms, 1,000,000 rounds of ammunition, and 1,800 blankets and knapsacks were captured.

He recrossed the Tennessee River thoroughly armed and equipped, with 500

Enfield rifles to spare and recruits sufficient to cover all losses in men. Many of these recruits were on furlough from other commands and seized the opportunity to join General Forrest's victorious forces.

Battle of Parker's Crossroads

By Thomas Jordan and J.P. Pryor. *The Campaigns of Lieutenant General Nathan Bedford Forrest and of Forrest's Cavalry.*

Encamping for the night some nine miles short of Lexington, the Confederate General detached his brother, Captain Forrest, with his company toward Huntington, to observe the enemy, retard his march as much as possible, and report hostile movements and appearances in that quarter. Captain Forrest encountered a column of the enemy at or near Clarksburg, within six miles of the Confederate camp and moving in that direction. A skirmish ensued, with the loss of several men to the Confederates, but with more casualties to the Federals. This was speedily reported to General Forrest, who, in the jaded condition of his command, regarded it most judicious to remain in position and risk a battle on the following morning with impending odds.

Accordingly, Captain Forrest was instructed through his courier to do all that he could to check the march of the Federal troops, disputing the road as obstinately as practicable, and to make frequent reports of the situation. During this time the men were left undisturbed until 4:00 a.m., when they were quickly roused, ordered to saddle up and prepare for the march.

Speedily in motion toward Parker's Crossroads, (or Red Mound,) about a mile and a half distant, General Forrest, informed of the near approach of his enemy, threw his command in order of battle. About the same time the Federal column made its appearance from the north-eastward and formed promptly to attack.

Dibrell's and Russell's regiments were at once dismounted and thrown forward as skirmishers, the artillery was brought up – six pieces – and placed in a favorable position on a ridge in an open field, within about six hundred yards of the Federal artillery and supports, which were all well posted within a skirt of woods with an open field between them and the Confederates.

The Federal artillery – three or four pieces – meanwhile had been opened upon the Confederate line with spirit. As quickly as possible a vigorous reply was made, and at the first fire a Federal gun was dismounted and several of the gunners and horses killed or disabled; the others were

> **The Confederates were now completely masters of the situation, and a staff officer was sent to where the white flags were shown to receive the capitulation.**

then withdrawn speedily to better cover, their supports falling back also toward Parker's Crossroads. The Confederates followed eagerly, the cavalry – 300 men – disposed equally on either flank of the line of dismounted men.

At the crossroads the Federal force was again formed, apparently in two lines, in an open, undulating field, with their front to the north and at right angles to the highway, a brigade of infantry, a battery of artillery, and a detachment of cavalry, or a total not short of 1800.

Dismounting his men, except about 200, who were distributed equally on his right and left flanks, the Confederate commander disposed them in one line of battle northward of his opponent and nearly parallel to the hostile array, on ground somewhat lower, partly in a peach-orchard and partly in an open field – the two lines about six hundred yards apart – with his artillery occupying three positions: Morton's Battery in the center, and a section of Freeman's on each flank, a few paces in advance of the dismounted men, unlimbered and ready for action. In a few moments the Confederate artillery opened the engagement in earnest, quickly driving that of the enemy under cover of a ridge in their rear.

General Forrest now pushed forward his entire line, as arranged, in battle order, and brought his musketry to bear with a heavy, continuous fire. The enemy fought with stubbornness and spirit, but after an hour were driven back into a skirt of woods eastward of the highway and about half a mile to the southeast of Parker's Crossroads. There they stood their ground stoutly, and, indeed, in turn made a resolute, well led attempt to regain the ridge from which they had been driven, so that, as the Confederates reached the crown of that position, the Federal line was found advanced to within eighty yards of the crest also. But a withering fire was opened with all arms, and the enemy were soon obliged to fall back again under shelter of the woods, leaving two pieces on the ground, their horses killed, as well as a number of officers and men killed and wounded.

General Forrest now concentrated the fire of his artillery upon the Federals, who had retired behind a strong fence in the woods, only about two hundred yards distant, killing and disabling a number of men and horses, and silencing all their other pieces. About 11:00 a.m., they essayed another slight advance, but were repulsed easily. About midday, however, another and very resolute forward movement was made to within sixty paces of our pieces and their supports, though only to be repulsed with slaughter.

It was at this juncture Colonel Napier gallantly, but without the orders or wish of his General, charged with his battalion upon the position of the enemy, up to the very fence behind which they were posted, there to fall mortally wounded with several of his men. Thus an intrepid and promising soldier lost his life through a spirit of martial ardor that unfortunately impelled him forward without due reflection or regard for the instructions and combinations of his commander.

Finding that the enemy were now weakened and doubtless discouraged by their heavy losses, the Confederate commander threw Colonel Russell around by the left to take them in flank and reverse, and meantime had several pieces of artillery so posted as to enfilade both flanks of their line. Apparently observing that the Confederate line had been diminished, the

Federal force made another charge, which was met and foiled by a discharge of grape and canister from the whole Confederate artillery.

At the same time, Colonel Russell dashed forward upon their flank and rear, as General Forrest moved forward in front. Under this stress the enemy's lines gave way, and, breaking, the men ran across the road westward into an open field, leaving many prisoners in our hands. In the meantime, the Confederate lines had fallen into disorder, and the several companies and regiments were intermingled; their General, therefore, found it now necessary to pause and reorganize.

The enemy fleeing, as before said, were cut off and brought to a stand by Colonel Starnes, who at that moment opportunely came up in that quarter of the field, and they resumed some order, but numerous white flags were displayed among them. The Confederates were now completely masters of the situation, and a staff officer was sent to where the white flags were shown to receive the capitulation. Scarcely had this been done, however, when Colonel Charles Carroll of General Forrest's staff dashed up and informed his leader that a fresh and superior force of Federals had reached the field and were forming to attack in his rear.

Ordering the proper disposition of his forces for this unlooked-for and untoward exigency, Forrest galloped to the indicated quarter, and there, indeed, did he find, already in battle order, about to sweep down upon him, two brigades of the enemy. They were, in fact, in possession of the peach orchard and adjacent field, which he had occupied at the outset of the engagement, and he was within eighty yards of their line before he could discover them. Perceived then by a Federal officer, who called, "Halt and surrender!" Forrest promptly replied that he had already done so some time since, but would move up what remained of his command and surrender in form; and with this, wheeling his horse, galloped away in the direction of his troops, notifying, as he passed, the inmates of his hospital of the emergency, that they might make their escape.

Joining his command, he at once put it in motion, by the left flank, at a double-quick. Scarcely had their rapid departure been made, when the enemy, also at a double-quick, came in view of their discomfited comrades, who immediately resumed their arms and renewed their fire.

Major Cox, being at the time with his and Napier's men on the extreme right of the Confederate line, was unable to get off the field, and was captured with about 250 men. The horses of four caissons and of two brass six pounders were disabled as they attempted to withdraw across the open field under the fire of both bodies of the enemy, and had to be abandoned with the loss of a number of drivers and artillerists killed and wounded.

The newcomers upon the field charged onward with spirit, and, from a position on the ridge that Forrest had occupied with such effect that morning, poured into the

> *It will doubtless be asked how it happened that a commander, wary and alert, like Forrest, permitted himself to be taken unaware, as he was by General Sullivan's fresh force...*

Confederate rear a rapid fire with at least two field guns.

Wheeling, with his escort, (75 men,) and a detachment, some fifty strong, of Dibrell's Regiment, under Major Forrest, the Confederate commander now made one of his characteristic dashes at their pieces, dispersed their gunners, and threw their infantry support into such confusion as served materially to aid his command at the moment to regain and mount their horses. The horses of the caissons of three pieces having taken fright and carried them in the direction of the Confederates, the General seized and carried them along with him.

Colonel Starnes, detached during the fight until now, with a mounted force of about 250 men, observing the condition of affairs, happily and boldly fell upon the rear of Dunham's force, and brought the whole Federal army to a halt. This afforded the Confederate troops time to get beyond their immediate reach. Meanwhile, General Forrest had taken post with about two hundred men on an eminence some eight hundred yards eastward of Parker's house, whence he, too, threw himself upon the rear of Dunham's Brigade for a parting blow, capturing his wagon train, with all the baggage of that force, and which were carried safely from the field.

Gathering his whole force now well in hand, the Confederate leader moved off in the direction of Lexington without further molestation, and encamped at that place about 6:00 that evening.

In the several conflicts of the day, between the hours of 6:00 a.m. and 3:00 p.m., the Confederate losses were some twenty-five officers, including Colonel Napier, and men killed, and not to exceed seventy-five wounded, with about 250 captured by the enemy, three pieces and four caissons, five wagons, two ambulances and their teams, with their contents – 7,500 rounds of ammunition

The casualties on the other side were three guns put hors de combat, two caissons, fifteen wagons, two ambulances and their teams, carried off the field, with some 1,800 knapsacks and as many blankets, and about 100 prisoners taken, and subsequently paroled; also, at least, fifty killed, including several prominent officers, and 150 wounded, among whom were one of their Colonels and a Lieutenant Colonel; and fully 100 animals either killed or disabled. All this, be it noted, was achieved on the Confederate side by a force at no moment in the day exceeding 1,200 men, opposed by at least 1,800, whom they vanquished, and finally by two fresh brigades.

So complete was the demoralization wrought on Dunham's Brigade that Major Strange, General Forrest's Adjutant General, unaccompanied by even an orderly, took possession of their ordnance-train and its escort of 22 men, who had surrendered to him just as General Sullivan reached the scene; but soon after which that gallant and able staff officer was himself captured while taking the list and inventory of his own captures.

Colonels Russell, Biffle, and Dibrell, and Major J. E. Forrest bore important parts in the brilliant combats of this field, and gave shining evidences of soldierly capacity. They handled their men with as much resolution as skill.

The lamented Napier displayed an admirable courage, and was able to lead his raw troops into the hottest part of the battle, where he fell a victim to an impetuous, brave soul, eager to do his utmost to win a victory. Captains Freeman and Morton, in command of the artillery, were conspicuous for their coolness, their intelligent, intrepid management of their guns, and their General attributes the larger part of the loss inflicted that day on the enemy to this and the bravery of their companies.

It will doubtless be asked how it happened that a commander, wary and alert, like Forrest, permitted himself to be taken unaware, as he was by General Sullivan's fresh force, and this is a question that must be duly answered. He made the proper provision to guard against such a contingency by ordering the detachment of a battalion of the Fourth Tennessee, under a good officer, to proceed to Clarksburg, on the Huntington road, with the object of holding that approach in close observation, and as a provision against the unannounced advent of any enemy from that direction.

Captain McLemore was accordingly detailed with three companies, about 100 men, for that service; but unfortunately the written instructions given him, at second hand, proved to be vague and inexpressive of the actual purposes of the movement, and failed indeed to indicate any other object than a reconnaissance, a juncture at Clarksburg with Captain Forrest, and their prompt return to the main body.

Moving across the country on byways or through the woods, some seven miles, to Clarksburg, McLemore found that Captain

> *Forrest returned stronger in numbers than when he entered upon the campaign, admirably armed, as before said, with a surplus of 500 Enfield rifles, some 1,800 blankets and knapsacks, and the raw, native courage of his men ripened by battle and the sharp hardships of the expedition into the perfection of cool, confident, soldierly valor, which makes men invincible except in conflict with invincible odds.*

Forrest had been obliged to fall back during the night before a very heavy Federal force, with infantry, cavalry, and artillery that had followed southward before daylight. The roads, too, gave evidence of the recent passage of considerable bodies of troops, and a small detachment of Federal cavalry disappeared in the same direction at a gallop as the Confederates came up.

Meanwhile, hearing the sound of the artillery engaged at Parker's Crossroads, Captain McLemore felt that, having executed his orders, his presence was now needed as soon as possible with his regiment, evidently in conflict at the time with a superior force of the enemy. He, therefore, rapidly retraced his steps, not taking the main road, as he supposed it was occupied by the Federals, but seeking to reach the Confederates by a detour to the right.

In this some time was necessarily lost, from the nature of the country, and he reached the scene only to find that Forrest was quitting the field, and that a large Federal force was interposed between him and his friends. Therefore, looking to the safety of his command, he made another detour northward and eastward and affected a safe crossing of the Tennessee River at Perryville on the following day. But for this misunderstanding of his orders on the part of subordinates, Forrest is confident he must have vanquished, and so disposed of Dunham's Brigade, before the advent of

Sullivan, as to have been in condition to encounter the new column, with strong chances of victory, flushed as the Confederates were with success and discouraged as was their enemy by defeat, while misled in regard to his numbers.

At Lexington – twelve miles from the battlefield – the Confederate force, as before said, was halted, and men and animals were fed, while proper attention was paid to the wounded. This done, that night the column was again in motion toward Clifton. Ten miles in that direction the train and prisoners, sent ahead meanwhile, were overtaken. The prisoners, some 300 in number, were then paroled between that and daybreak and turned adrift – disabled by their parole until exchanged – to find their way back to their comrades.

Early the next morning, the rearguard and scouts having come up, the Confederate commander put his whole force in rapid march for the Tennessee River, a portion of the command having been sent in advance during the night under Major Forrest, a courier from whom met the Commanding General, when ten miles on his way, with the information that a heavy hostile force was confronting him some eight miles from Clifton. About the same time, moreover, a scout brought the intelligence of the approach of about 10,000 infantry and cavalry from the direction of Purdy, and moving on what is known in that region as the Jack's Creek-Clifton road, with the evident purpose of cutting off the Confederates from that crossing of the river.

Giving the order to gallop, and keeping that pace for fifteen miles, Forrest caught up with the main body of his command about eleven miles from the river-crossing. Forming, without loss of a moment, the whole force in order of battle, with a front of one regiment deployed, the others also deployed and following at a distance of two hundred yards – the artillery immediately in the rear and followed by the train, with stringent orders to keep well closed up with the command – General Forrest again advanced about three miles, when the enemy (cavalry) were met drawn up, some 1200 strong, directly across his line of march.

Colonel Dibrell, whose regiment was in advance, was directed to charge. This he did promptly, cutting the Federals in sunder. Next Colonel Starnes was thrown with his regiment upon the leftward fragment, and Colonel Biffle upon the one on the right hand. Dibrell continuing to push that portion retreating before him, the detachments attacked severally by Starnes and Biffle broke and scattered in all directions, with slight show of belligerency.

The road thus cleared, the train and artillery now moved as rapidly as practicable toward the ferry. In this re-encounter the enemy lost some twenty killed and wounded and about fifty prisoners.

Signals were promptly made for skiffs that were on the other bank; they were brought over, and a party was as swiftly sent back to raise the flat-boat which had been sunk to conceal it after the passage on the 15th of the month. All possible haste was used, and the flat was brought to the west bank. Meanwhile, the animals, detached from the vehicles and artillery, were being driven into the river and made to swim across to the eastern bank, as also the horses of the cavalry – a process which was hurried because of the intelligence received that the enemy was moving his whole available forces by forced marches upon the point.

It was a spectacle full of life and movement; quite as many as 1000 animals were at one time in the river, which was about 600 yards broad, with favorable banks. The ferriage of the artillery and wagons was very much slower. Loaded upon the old flat, it was poled up-stream a distance of nearly half a mile, close to the west bank, and, pushed out into the stream, was caught and carried by the current gradually to the other bank some distance below; there discharged, the process was reversed on its return. In the like way did the flat ply to and fro until 8:00 at night, when the work was completed and the men stood cheerfully once more in Middle Tennessee, with five pieces of artillery, six caissons with their horses, sixty wagons, and four ambulances with their teams, which had been successfully ferried in the short space of eight hours.

As will be remembered, it was on the 15th of December that the passage into West-Tennessee had been concluded – that is, a fortnight previously. In the interval, seldom in the annals of war had more hard, swift riding, as many sharp encounters, affluent in results, been crowded in the same short space. That command had averaged over twenty miles a day; it had fought three well-contested engagements, with diurnal skirmishes had destroyed some fifty large and small bridges on the Mobile and Ohio Railroad, and had broken up so much of the trestle work of that road as to make it useless to the enemy for the rest of the war ; had captured and burned eighteen or twenty stockades, and captured or killed 2,500 of the enemy; had taken or disabled ten pieces of field artillery, and carried off fifty wagons and ambulances with their teams; had captured 10,000 stands of excellent small-arms, 1,000,000 rounds of ammunition, and had returned thoroughly armed and equipped, including blankets, after having traversed with artillery and a heavy wagon-train roads which in the country were considered and pronounced impracticable at that season for horsemen, resting undisturbed scarcely one whole night during the fortnight, and all the while subjected, unsheltered, to the most inclement weather of mid-winter.

Crossing the river into West Tennessee with his command wretchedly armed and equipped, and with only ten rounds of percussion caps to his shotguns, Forrest returned stronger in numbers than when he entered upon the campaign, admirably armed, as before said, with a surplus of 500 Enfield rifles, some 1,800 blankets and knapsacks, and the raw, native courage of his men ripened by battle and the sharp hardships of the expedition into the perfection of cool, confident, soldierly valor, which makes men invincible except in conflict with invincible odds.

GENERAL NATHAN BEDFORD FORREST AND THE DOUBLE ENVELOPMENT

By Max Lee Waldrop, Jr., Past Commander General of the Military Order of the Stars and Bars.

Throughout the origins of warfare, the flanking maneuver or flanking attack has been a fundamental basic military tactic. It is defined as an attack made on one or more sides of an enemy force. The double envelopment (pincer movement) is the most effective flanking maneuver. It is a coordinated, simultaneous attack on both sides of the enemy with care taken to avoid friendly fire. The attacking commander uses direct and indirect suppressive fire to "fix and hold" the enemy in place while the maneuvering forces advance to engage the enemy flanks. Once surrounded from several sides and compelled to fight in several directions, the enemy loses the abilities to maneuver and defend and can be possibly destroyed in position. Should both flanking forces link up in the enemy's rear, the enemy becomes encircled leading to surrender or annihilation.

Diagram of the Double Envelopment Maneuver

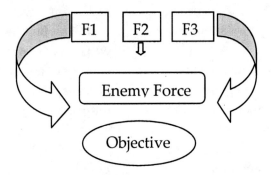

F = Friendly Force

Military commanders such as Hannibal at the Battle of Cannae (216 BC), Julius Caesar, Napoleon, Brigadier General Daniel Morgan at the Battle of Cowpens (1781), General Robert E. Lee and General Thomas Jackson at Chancellorsville, and Nathan Bedford Forrest at Parker's Crossroads have successfully utilized this tactic. In 1991 during the ground campaign of Desert Storm (first Persian Gulf War), the Coalition forces under General Norman Schwarzkopf dramatically used the single envelopment "left hook flanking maneuver" to bypass the Iraqi forces located near the Kuwait-Saudi border. In modern warfare, new variations of envelopment are "vertically" achieved with the use of airborne or air assault troops and "amphibiously" achieved with naval and river forces.

German military strategist, Alfred Graf von Schlieffen – Chief of the German General Staff taught the "Cannae model" of Hannibal's double envelopment of the Roman Army. He is quoted:

> *A battle of annihilation can be carried out today according to the same plan devised by Hannibal in long forgotten times. The enemy front is not the goal of the principal attack. The mass of the troops and the reserves should not be concentrated against the enemy front; the essential is that the flanks be crushed. The wings should not be sought at the advanced points of the front but rather along the entire depth and extension of the enemy formation. The annihilation is completed through an attack against the enemy's rear. To bring about a decisive and annihilating victory requires an attack against the front and against one or both flanks ..."*[1]

Forrest as a Tactician

Nathan Bedford Forrest is described by Shelby Foote in his book *The Civil War* as a natural and brilliant tactician who earned the respect of both sides. He enlisted in 1861 as a private in the Tennessee Mounted Rifles Company, commissioned a lieutenant colonel a month later by Tennessee Governor Isham G. Harris with authority to raise and finance his own force of 600-man cavalry, and achieved the rank of lieutenant general. Foote stated:

> *In his first fight, northeast of Bowling Green, the forty year old Forrest improvised a double envelopment, combined it with a frontal assault – classic maneuvers which he could not identify by name and of which he had most likely never heard...*[2]

Forrest was called "that devil Forrest" by Sherman and was known for his use of mobility, deception, terrain, inventive tactics, and motivating personal courage on the battlefield. The famous Prussian general and military theorist, Carl von Clausewitz defined tactics as the art of using troops in battle, and, in his first battle, Forrest's tactical leadership of obtaining offensive, mass, and surprise through the maneuvering of his cavalry (at times dismounted) and flying artillery (used primarily at Parker's Crossroads) became hallmark characteristics. His often quoted military principle of "Get there first with the most men"[3] succinctly defines this self-made warrior. After this first battle he was described by a subordinate (D.C. Kelley):

> *So fierce did his passion become that he was almost equally dangerous to friend or foe, and, as it seemed to some of us, he was too wildly excitable to be capable of judicious command. Later we became aware that excitement neither paralyzed nor mislead his magnificent military genius.*[4]

Dan Kennerly in his book *Forrest at Parker's Crossroads – The Dawn of Lightening War* wrote:

---

[1] Battle of Cannae (http://.en.wikipedia.org/w/index.php?title=Battle of Cannae)
[2] Shelby Foote, *The Civil War A Narrative – Fort Sumter to Perryville,* 172
[3] Basil W. Duke, *Reminiscences of General Basil W. Duke* (Garden City, New York: Doubleday, 1911), 346
[4] Wyeth, *Life of General Nathan Bedford Forrest,* 35

Unfortunately, Forrest himself did not realize that he had initiated and perfected a new and exciting arm of attack. He had had no formal military training, and therefore could not have comprehended that his intuitive and imaginative use of psychological warfare, artillery maneuver, and small combat commands were the beginnings of armored warfare strategy that would be later utilized in WWII. The Battle of Parker's Crossroads and the campaign preceding it were truly the dawn of the blitzkrieg – the dawn of lightning war.[5]

Forrest's greatest victory was at the Battle of Brice's Crossroads in northern Mississippi on June 10, 1864, in what has often been called the perfect battle. Forrest defeated and routed Major General Samuel D. Sturgis whose 8,000 man army was more than two times the size of the Confederates. Sturgis later described the battle as "What was confusion became chaos…" In this battle, Forrest was able to effectively use terrain against Sturgis as a bridge in the rear of the Union forces served as a roadblock along the escape route.

The Battle of Parker's Crossroads

On November 21, 1862, Forrest was ordered to Columbia, Tennessee by General Braxton Bragg to lead a strike against General Grant's extended rail supply and communication lines. At this time, Grant was heavily involved in his planning and preparation for the attack on Vicksburg, Mississippi. Forrest began his movement north out of Columbia with 1,800 men who were poorly equipped on December 10, 1862.

Beginning December 19th and carrying through December 27th, Forrest's unit engaged Union forces in numerous locations at Lexington, Jackson, Humboldt, Trenton, and Union City, Tennessee.

On December 31, Union forces under Colonel Dunham moved from Clarksburg towards Parker's Crossroads to guard the approaches against the reported proximity of Forrest's cavalry. Forrest placing his horsemen and artillery in motion met the Indiana 50th Infantry Regiment and 18th Illinois Infantry Regiment at Hicks' Field – one mile northwest of the Crossroads. He immediately flanked the Union's right flank while constantly bombarding the Union center with Lieutenant Baxter's grape and canister artillery fire.

Dunham's forces were forced to retreat to the Crossroads where they were kept under intense artillery pressure by Freeman's and Morton's Batteries of seven guns. By early afternoon, Forrest had completely divided and surrounded Dunham by use of the double envelopment as he sent the 4th Alabama and the Kentucky Battalion on a flanking maneuver around the Union's right while simultaneously flanking the Union's left front with the 4th and 19th Tennessee.

Unfortunately, as Forrest was negotiating with Colonel Dunham for the surrender of his brigade, Colonel Fuller's Ohio Brigade approached from the north in the area where Forrest had sent Captain McLemore and four cavalry companies to watch for any Union reinforcements before the battle began.

Captain McLemore, hearing the commencement of the battle, attempted to rejoin his regiment, leaving Forrest without a scouting force and exposed to any additional Union

---

[5] Dan Kennerly, *The Dawn of Lightning War – General Forrest and Parker's Crossroads*, 11

reinforcements.[6] Fuller captured about 300 dismounted Confederate cavalry and completely surprised Forrest, who immediately led a charge on Fuller's left flank with a small reconstituted force. This charge dispersed the enemy's leading edge resulting in an overnight defensive position.

Forrest was then able to make a prudent disengagement and undertaking a dramatic race to and crossing the Tennessee River. Despite the ignominy of the late afternoon surprise reinforcement of Dunham's Brigade by Fuller, Forrest completely dominated the field of battle with his effective use of artillery at short range, dismounted cavalry providing support for the artillery, strategic use of terrain, and the successful encirclement of Dunham's units by maneuvering his forces using the double envelopment.[7]

Forrest's successful raid of Western Tennessee in late 1862 combined with the raid of General Van Dorn on Holly Springs, Mississippi, forced Grant to leave Grenada, Mississippi, and move northward to Memphis, Tennessee, to protect his base of supply. This enabled Vicksburg and Pemberton to delay the ultimate siege for another six months.

To subscribe, send an email to:
thestainlessbanner@gmail.com
or visit our website at www.thestainlessbanner.com

Subscription is free.

---

[6] Official Records, Series I, Part I, 17:580, 595-597
[7] Kennerly, *The Dawn of Lightning War – General Forrest and Parker's Crossroads*, 29-42

The Stainless Banner

An e-zine dedicated to the armies of the Confederacy

Volume 3, Issue 9
October, 2012

MR. LINCOLN GOES TO WAR
THE REASONS BEHIND LINCOLN'S DECISION TO WAGE WAR ON THE CONFEDERACY

In his first Inaugural address, Abraham Lincoln details the specific conditions that would lead to a Northern invasion of the Southern States, but the testimony of a Southern peace representative who spoke with Lincoln on April 4, 1861, in an effort to avert the War Between the States provides keen insight into a side of the issue seldom heard or taught. There surely will be no mention of this significant event on any History Channel specials. It does qualify as an example of Dr. Grady McWhiney's quote (as referenced in *The South Was Right*): "What passes as standard American history is really Yankee history written by New Englanders or their puppets to glorify Yankee heroes and ideals."[1] Many of the modern-day court historians will immediately dismiss the meeting between Abraham Lincoln and Colonel John Baldwin but it did happen, and the account was documented beyond dispute.

Reverend Robert L. Dabney, Chief of Staff to Major-General T.J. "Stonewall" Jackson, met Colonel John B. Baldwin in March of 1865 in Petersburg, Virginia, when the Army of Northern Virginia was under siege. Reverend Dabney gathered valuable information from their conversations.

Before hostilities began, Baldwin was selected to carry out a private mission, at the direction of the Virginia Secession

---

[1] James Ronald Kennedy and Walter Donald Kennedy, *The South Was Right!* Second Edition, Pelican Publishing Company, Gretna, 1994, page 15.

What's Inside:

Colonel Baldwin	7
States Right vs Slavery	18
The Confederate Soldier	19
The Origin of the Late War	22
States Rights	32

convention. His instructions were to meet with Abraham Lincoln in April 1861 and negotiate a peaceful settlement. The South was not anxious to go to war and sought to diffuse the schism between the two regions. The deck was stacked heavily in favor of the North – their population was at least three times that of the South; the North had a decided manufacturing advantage; the North a government, army, navy, etc. Colonel Baldwin found out first hand that there was one issue where Lincoln would not compromise – taxes!

Virginia had already experienced one ill-fated attempt to compromise with Lincoln to try to derail the efforts of certain elements in the North to establish an American Empire. That was the rebuffed attempt by William B. Preston, Alexander H.H. Stuart, and George B. Randolph. Also, the failure of the "Peace Congress," rejection of the Crittenden Amendment, the hint of coercion in Lincoln's inaugural address, and the clandestine arming of the Federal Government for war struck fear and concern in the minds of Southerners.

This concern was echoed by the words of William Preston, who sought compromise and supported the Union as it was originally intended. He said, "If our voices and votes are to be exerted farther to hold Virginia in the Union, *we must know what the nature of the Union is to be. We have valued Union, but we are also Virginians, and we love the Union only as it is based upon the Constitution. If the power of the United States is to be perverted to invade the rights of States and of the people, we would support the Federal government no farther. And now that the attitude of that government was so ominous of usurpation, we must know whither it is going, or we can go with it no farther."[2]

> *Lincoln's response made it clear that slavery was not the issue – he never mentioned slavery, but he was quick to point out that he would lose revenue if the South had low-duty ports.*

Preston voiced serious concerns about the threats of coercion and usurpation, two patently un-American concepts and direct violations of the U.S. Constitution. He also alluded to States' Rights (true federalism) as the central government – created to be the agent of the States – threatened to violate the voluntary relationship of the compact.

Preston's view paralleled that of Robert E. Lee, a proud son of Virginia and a true American hero. Lee also identified the unconstitutional efforts of Lincoln's government, and he correctly refused to commit anything resembling treason (Article 3, Section 3) by commanding the Union Army.

Secretary of State Seward sent a messenger, Allen B. Magruder, to talk with members of the Virginia Convention and request that a Virginia representative come to Washington to talk with Lincoln. This group included Mr. Janney, Convention President, Mr. Stuart, and others. Magruder stated that he was authorized by Seward to say that Fort Sumter would be evacuated on

---

[2]Robert L. Dabney, D.D., *The Origin & Real Cause of the War*, A Memoir of a Narrative Received of Colonel John B. Baldwin, *Reprinted from Discussions, Volume IV*, pages 2 and 3.

the Friday of the ensuing week, and that the *Pawnee* would sail on the following Monday for Charleston to effect the evacuation. Seward said that secrecy was all important, and while it was extremely desirable that one of them should see Lincoln, it was equally important that the public should know nothing of the interview.[3]

Since this was a mutually agreed upon clandestine meeting, Virginia did her part by selecting the relatively unknown Baldwin to send to Washington. Though somewhat reluctant, Baldwin accepted the role as Virginia representative.

Baldwin and Magruder rode in a carriage with raised glasses (for maximum secrecy) to meet Seward. Seward then took Baldwin to the White House, arriving slightly after 9:00 a.m. The porter immediately admitted Baldwin.

Baldwin accordingly followed him (the porter) and Seward into what he presumed was the President's ordinary business room, where he found him in evidently anxious consultation with three or four elderly men, who appeared to wear importance in their aspect.[4] Seward informed Lincoln of his guest's arrival. Lincoln immediately excused himself from the meeting, and then took Baldwin upstairs to a bedroom.

Lincoln greeted him as follows: "Well, I suppose this is Colonel Baldwin of Virginia? I have hearn of you a good deal, and am glad to see you. How d'ye, do sir?"[5]

Baldwin presented his note of credential or introduction, which Lincoln read, sitting upon the edge of the bed, and spitting from time to time on the carpet."[6]

Lincoln admitted that Virginians were good Unionists, but he did not favor their kind of "conditional" Unionism. Baldwin countered with a reaffirmation of Virginia's belief in the Constitution as it was written, and that Virginia would not subscribe to a conflict based on the sectional, free-soil question.

> "You are too late, sir, too late!"

Baldwin then told Lincoln, that as much as Virginia opposed his platform, Virginia would support him as long as he adhered to the Constitution and the laws of the land. Furthermore, if Lincoln would agree on this, Virginia would not only support him, but would also use all possible influence to keep the border States in the Union and convince the already seceded States (seven of them) to come back into the Union.

He made it clear that Virginia would never support the unconstitutional attempt by the Federal Government to coerce any State to remain in the Union against the will of its people.

Lincoln plainly did not like what he heard and tried to paint the South as insincere, as people with hollow words backed by no action. It is difficult to imagine Lincoln being that naïve. He had to have some insight into the Southern psyche since his wife was of Southern ancestry.

Baldwin assured him that coercion of the South would undoubtedly lead to separation and possibly war.

Lincoln awkwardly paced about, and, in obvious dismay, exclaimed: "I ought to have known this sooner! You are too late,

---

[3] Ibid., 3.
[4] Ibid., 2-3.
[5] Ibid., 3.

[6] Ibid.

sir, too late! Why did you not come here four days ago, and tell me all this?"[7]

Baldwin responded: "Why, Mr. President, you did not ask our advice. Besides, as soon as we received permission to tender it, I came by the first train, as fast as steam could bring me."[8]

Lincoln said: "Yes, but you are too late, I tell you, too late!"[9] It was as though the numerous compromise efforts of the South had been blanked out of Lincoln's memory.

Baldwin tried again to convince Lincoln of Virginia's sincerity, but Lincoln's countenance reflected that it was indeed too late. The fate of the Constitutional Union sat squarely on Lincoln's shoulders, for he alone could have saved the Original Republic. Baldwin imparted to him in no uncertain terms that the secession movement could be diffused. All Lincoln had to do was issue "a simple proclamation, firmly pledging the new administration to respect the Constitution and laws, and the rights of the States; to repudiate the power of coercing seceded States by force of arms; to rely upon conciliation and enlightened self interest in the latter to bring them back in the Union, and meantime to leave all questions at issue to be adjudicated by the constitutional tribunals. The obvious ground of this policy was in fact that it was not the question of free-soil which threatened to rend the country in twain, but a well grounded alarm at the attempted overthrow of the Constitution and liberty, by the usurpation of a power to crush States. The question of free-soil had no such importance in the eyes of the people of the border States, nor even of the seceded States, as to become at once a casus belli."[10]

Lincoln claimed secession was unconstitutional, yet saw nothing wrong with coercion, which was and is patently unconstitutional. He felt secession automatically signaled war when it should signify just the opposite since the Constitution was a compact that the States joined voluntarily.

"Only give this assurance to the country, in a proclamation of five lines," said Baldwin, "and we pledge ourselves that Virginia (and with her the border States) will stand by you as though you were our own Washington.

"So sure am I," he added, "of this, and of the inevitable ruin which will be precipitated by the opposite policy, that I would this day freely consent, if you would let me write those decisive lines, you might cut off my head, were my own life my own, the hour after you signed them."[11]

At least on the exterior, Lincoln seemed to be genuinely touched by Baldwin's plea, but he did not like the idea of having the Southern States remain out of the Union until a compromise could be reached. He was concerned about lost revenue.

His reply underscored his deepest concerns: "And open Charleston, etc., as ports of entry, with their ten percent tariff. What, then, would become of my tariff?"[12]

Lincoln dismissed Baldwin and made no promises, but it was crystal clear that the tariff was the crux of the impending dilemma and the decision had already been made to invade the South. The South's ten percent uniform duties would be a severe blow to the Federal economy. This episode

---

[7] Ibid., 6
[8] Ibid.
[9] Ibid.
[10] Ibid.
[11] Ibid., 7-8.
[12] Ibid., 8.

makes one wonder just who the well-dressed men were in Lincoln's earlier meeting. Baldwin had fulfilled his duty and returned to Virginia with the verdict.

Although Lincoln did appear to experience some genuine anguish, he fell back upon the Northern industrialists and railroads that helped get him elected. It was they who had directed him to inflict a punishing tariff on the South so that their own protectionist interests could be preserved and the South could be relegated to permanent subservient status.

Baldwin told Lincoln repeatedly that Virginia (and the South) would not fight over the free-soil (slavery) issue. About 6% of Southerners were slave owners, affecting perhaps twenty-five to thirty percent of Southern families. Clearly, fighting over slavery was not an option. Lincoln's response made it clear that slavery was not the issue – he never mentioned slavery but he was quick to point out that he would lose revenue if the South had low-duty ports.

The account given by Baldwin to Dabney was later confirmed to be accurate by A.H.H. Stuart; it was totally consistent with the account Baldwin had given to the Virginia Commission upon his return.

Dabney sums up the situation as follows: "But what was the decisive weight that turned the scale against peace, and right, and patriotism? It was the interest of a sectional tariff! His single objection, both to the wise advice of Baldwin and Stuart, was: 'Then what would become of my tariffs?' He was shrewd enough to see that the just and liberal free-trade policy proposed by the Montgomery Government would speedily build up, by the help of the magnificent Southern staples, a beneficent foreign commerce through Confederate ports; that the Northern people, whose lawless and mercenary character he understood, could never be restrained from smuggling across the long open frontier of the Confederacy; that thus the whole country would become habituated to the benefits of free-trade, so that when the schism was healed (as he knew it would be healed in a few years by the policy of Virginia), it would be too late to restore the iniquitous system of sectional plunder by tariffs which his section so craved. Hence, when Virginia offered him a safe way to preserve the Union, he preferred to destroy the Union and preserve his (redistributive) tariffs. The war was conceived in duplicity, and brought forth in iniquity.

"The calculated treason of Lincoln's Radical advisers is yet more glaring. When their own chosen leader, Seward, avowed that there was no need for war, they deliberately and malignantly practiced to produce war, for the purpose of overthrowing the Constitution and the Union, to rear their own greedy faction upon the ruins. This war, with all its crimes and miseries, was proximately concocted in Washington City, by Northern men, with malice prepense."[13]

I want to thank John Taylor for contributing the article. John was the previous editor of The *Tallapoosa Rifle*, the newsletter of the Edmund W. Pettus SCV Camp 574. He also edited the *Alabama Confederate*. John's first book: *Consolidated Union at All Cost: Abraham Lincoln's Attack on the South, Liberty and the U.S. Constitution* will be published by The Stainless Banner Publishing Company next year.

---

[13] Ibid., 14.

Do You Have a Book on the War You Are Longing to Publish?

THE STAINLESS BANNER PUBLISHING COMPANY

A Full-Service Small Press Dedicated to the
Preservation of Southern Heritage and History

History ✯ Memoir ✯ Biography ✯ Novel ✯ Alternate History

What does The Stainless Banner Publishing Company offer you?

- ✯ Professional editing
- ✯ Dynamic covers
- ✯ Hard cover or paperback
- ✯ Competitively priced books
- ✯ High royalties paid monthly
- ✯ Your book on Amazon, Barnes & Noble and many other outlets
- ✯ Free advertising

What is my investment?

Nothing! **The Stainless Banner Publishing Company** is a full service small press. Once your book is published, your only investment will be for any inventory you wish to keep on hand.

Where do I send my query letter?

No need for an agent or a query letter. Just send an email telling us about your book and include the first three chapters in the body of the email. You will receive a prompt reply, usually within 48 hours. Send your email to books@thestainlessbanner.com

The Stainless Banner Publishing Company
www.thestainlessbanner.com

Memoir of a Narrative Received from Colonel John B. Baldwin Touching the Origin of the War

By Reverend R.L. Dabney, D.D. *Southern Historical Papers, Volume 1,* pages 443-456.

In March, 1865, being with the army in Petersburg, Virginia, I had the pleasure of meeting Colonel Baldwin at a small entertainment at a friend's house, where he conversed with me some two hours on public affairs. During this time, he detailed to me the history of his private mission, from the Virginia Secession Convention, to Mr. Lincoln in April, 1861.

The facts he gave me have struck me, especially since the conquest of the South, as of great importance in a history of the origin of the war. It was my earnest hope that Colonel Baldwin would reduce them into a narrative for publication, and I, afterwards, took measures to induce him to do so, but I fear without effect. Should it appear that he has left such a narrative, while it will confirm the substantial fidelity of my narrative at second hand, it will also supersede mine, and of this result I should be extremely glad.

Surviving friends and political associates of Colonel Baldwin must have heard him narrate the same interesting facts. I would earnestly invoke their recollection of his statements to them, so as to correct me, if in any point I misconceived the author and to confirm me where I am correct, so that the history may regain, as far as possible, that full certainty of which it is in danger of losing a part by the lamented death of Colonel Baldwin. What I here attempt to do is to give faithfully, in my own language, what I understood Colonel Baldwin to tell me, according to my best comprehension of it. His narration was eminently perspicuous and impressive.

It should also be premised, that the Virginia Convention, as a body, was not in favor of secession. It was prevalently under the influence of statesmen of the school known as the "Clay-Whig." One of the few original secessionists told me that, at first, there were but twenty-five members of that opinion, and that they gained no accessions until they were given them by the usurpations of the Lincoln party. The Convention assembled with a fixed determination to preserve the Union, if forbearance and prudence could do it consistently with the rights of the States. Such, as is well known, were, in the main, Colonel Baldwin's views and purposes.

But Mr. Lincoln's inaugural with its hints of coercion and usurpation, the utter failure of the "Peace-Congress," the rejection of Mr. Crittenden's overtures, the refusal to hear the commissioners from Mr. Davis' Government at Montgomery and the secret arming of the Federal government for attack had now produced feverish apprehensions in and out of the Convention.

> *Lincoln said querulously: "Yes! You Virginia people are good Unionists, but it is always with an if! I don't like that sort of Unionism."*

Colonel Baldwin considered Mr. William Ballard Preston of Montgomery County as deservedly one of the most influential members of that body. This statesman now began to feel those sentiments, which soon after prompted him to move and secure the passage of the resolution to appoint a formal commission of three ambassadors from the Convention to Lincoln's Government, who should communicate the views of Virginia and demand those of Mr. Lincoln. (That commission consisted of William B. Preston, Alexander H. H. Stuart and George W. Randolph. We will refer to its history in the sequel.)

Meantime Mr. Preston, with other original Union men, were feeling thus: "If our voices and votes are to be exerted farther to hold Virginia in the Union, we must know what the nature of that Union is to be. We have valued Union, but we are also Virginians, and we love the Union only as it is based upon the Constitution. If the power of the United States is to be perverted to invade the rights of States and of the people, we would support the Federal government no farther. And now that the attitude of that government was so ominous of usurpation, we must know whither it is going, or we can go with it no farther." Mr. Preston especially declared that if he were to become an agent for holding Virginia in the Union to the destruction of her honor and of the liberty of her people and her sister states, he would rather die than exert that agency.

Meantime Mr. Seward, Lincoln's Secretary of State, sent Allen B. Magruder, Esq., as a confidential messenger to Richmond to hold an interview with Mr. Janney (President of the Convention), Mr. Stuart, and other influential members, and to urge that one of them should come to Washington as promptly as possible, to confer with Mr. Lincoln. Mr. Magruder stated that he was authorized by Mr. Seward to say that Fort Sumter would be evacuated on the Friday of the ensuing week, and that the *Pawnee* would sail on the following Monday for Charleston to effect the evacuation. Mr. Seward said that secrecy was all important, and while it was extremely desirable that one of them should see Mr. Lincoln, it was equally important that the public should know nothing of the interview.

> *We have valued Union, but we are also Virginians, and we love the Union only as it is based upon the Constitution.*

These gentlemen held a conference and determined that as each of them was well known in Washington by person, the required secrecy could not be preserved if either of them went. They therefore asked Colonel Baldwin to go, furnished with the necessary credentials to Mr. Lincoln. He at first demurred, saying that all his public services had been to Virginia, and that he knew nothing of Washington and the Federal politics, but they replied that this was precisely what qualified him because his presence there would not excite remark or suspicion. Colonel Baldwin accordingly agreed to the mission and went with Mr. Magruder the following night, reaching Washington the next morning by the "Acquia Creek route" a little after dawn, and driving direct to the house of Mr. Magruder's brother. (These gentlemen were brothers of General J. B. Magruder of Virginia.)

These prefatory statements prepare the way for Colonel Baldwin's special narrative.

He stated that after breakfasting and attending to his toilet at the house of Captain Magruder, he went with Mr. A.B. Magruder in a carriage with the glasses carefully raised to Seward, who took charge of Mr. Baldwin and went direct with him to the White House, reaching it, he thought, not much after 9:00 a.m. At the door, the man who was acting as usher, or porter, was directed by Colonel Baldwin's companion to inform the President that a gentleman wished to see him on important business. The man replied, as Colonel Baldwin thought with an air of negligence that he would report the application of course, but that it would be useless because the President was already engaged with very important personages. Some card or such missive was given him, and he took it in. He soon returned with a surprised look and said that the gentleman was to be admitted instantly.

Colonel Baldwin accordingly followed him and Mr. Seward into what he presumed was the President's ordinary business room, where he found him in evidently anxious consultation with three or four elderly men who appeared to wear importance in their aspect. Mr. Seward whispered something to the President, who at once arose with eagerness and without making any movement to introduce Colonel Baldwin, said bluntly, in substance: "Gentlemen, excuse me, for I must talk with this man at once. Come this way, sir!" (to Colonel Baldwin).

He then took him up stairs to quite a different part of the house and into what was evidently a private sleeping apartment. There was a handsome bed, with bureau and mirror, washstand, etc. and a chair or two. Lincoln closed the door and locked it. He then said: "Well, I suppose this is Colonel Baldwin of Virginia? I have hearn of you a good deal and am glad to see you. How d'ye, do sir?"

Colonel Baldwin presented his note of credential or introduction, which Lincoln read, sitting upon the edge of the bed and spitting from time to time on the carpet. He then, looking inquiringly at Colonel Baldwin, intimated that he understood he was authorized to state for his friends in the Virginia Convention the real state of opinion and purpose there.

Upon Colonel Baldwin's portraying the sentiments which prevailed among the majority there, Lincoln said querulously: "Yes! Your Virginia people are good Unionists, but it is always with an if! I don't like that sort of Unionism." Colonel Baldwin firmly and respectfully explained, that in one sense no freeman could be more than a conditional Union man, for the value of the Union was in that equitable and beneficent Constitution on which it was founded and if this were lost, Union might become but another name for mischievous oppression.

He also gave Mr. Lincoln assurances that the description which he was making of the state of opinion in Virginia was in perfect candor and fidelity, and that he might rest assured the great body of

> *All Virginians were unanimous in believing that no right existed in the Federal Government to coerce a State by force of arms because it was expressly withheld by the Constitution.*

Virginia, in and out of the Convention, would concur in these views, viz: That although strongly opposed to a presidential election upon a sectional, free-soil platform, which they deplored as most dangerous and unwise, Virginia did not approve of making that, evil as it was, a casus belli, or a ground for disrupting the Union. That much as Virginia disapproved it, if Mr. Lincoln would only adhere faithfully to the Constitution and the laws, she would support him just as faithfully as though he were the man of her choice and would wield her whole moral force to keep the Border States in the Union and to bring back the seven seceded States. But that while much difference of opinion existed on the question whether the right of secession was a constitutional one, all Virginians were unanimous in believing that no right existed in the Federal government to coerce a state by force of arms because it was expressly withheld by the Constitution; that the state of Virginia was unanimously resolved not to acquiesce in the usurpation of that power as had been declared by unanimous joint resolution of her present Legislature, and by the sovereign Convention now sitting, according to the traditionary principles of the State; that if Virginia remained in the Union, the other Border States would follow her example, while, if she were driven out, they would probably go with her, and the whole South would be united in irreconcilable hostility to his. Government; and that the friends of peace desired to have a guarantee that his policy towards the seven seceded states would be pacific and would regard their rights as states; without which guarantee the Convention could not keep the people in the Union even if they would.

Lincoln now showed very plainly that this view was distasteful to him. He intimated that the people of the South were not in earnest in all this. He said that in Washington he was assured that all the resolutions and speeches and declarations of this tenor from the South were but a "game of brag," intended to intimidate the administration party, the ordinary and hollow expedient of politicians; that, in short, when the Government showed its hand, there would "be nothing in it but talk."

Colonel Baldwin assured him solemnly that such advisers fatally misunderstood the South and especially Virginia, and that upon the relinquishment or adoption of the policy of violent coercion, peace or a dreadful war would inevitably turn.

Lincoln's native good sense, with Colonel Baldwin's evident sincerity, seemed now to open his eyes to this truth. He slid off the edge of the bed and began to stalk in his awkward manner across the chamber, in great excitement and perplexity. He clutched his shaggy hair as though he would jerk out handfuls by the roots; he frowned and contorted his features, exclaiming: "I ought to have known this sooner! You are too late, sir, too late! Why did you not come here four days ago and tell me all this?" turning almost fiercely upon Colonel Baldwin.

He replied: "Why, Mr. President, you did not ask our advice. Besides, as soon as we received permission to tender it, I came by the first train, as fast as steam would bring me."

"Yes, but you are too late, I tell you, too late!" Colonel Baldwin understood this as a clear intimation that the policy of coercion was determined on and that within the last four days. He said that he therefore felt impelled by a solemn sense of duty to his country to make a final effort for impressing Lincoln with the truth.

"Never," said he to me, "did I make a speech on behalf of a client, in jeopardy of his life, with such earnest solemnity and endeavor.

"And, there was no simulated emotions; for when he perceived from Lincoln's hints, and from the workings of his crafty and saturnine countenance, the truculence of his purpose, his own soul was filled with such a sense of the coming miseries of the country, and of the irreparable ruin of the Constitution that he felt he would willingly lay down his life to avert them."

He endeavored to make the President feel that Providence had placed the destiny of the country in his hands, so that he might be forever blessed and venerated as the second Washington — the savior of his country — or execrated as its destroyer.

What policy then did the Union men of Virginia advise? We believe, answered Colonel Baldwin, that one single step will be sufficient to paralyze the secession movement and to make the true friends of the Union masters of the situation. This was a simple proclamation, firmly pledging the new administration to respect the Constitution and laws and the rights of the states; to repudiate the power of coercing seceded states by force of arms; to rely upon conciliation and enlightened self-interest in the latter to bring them back into the Union and meantime to leave all questions at issue to be adjudicated by the constitutional tribunals.

The obvious ground of this policy was in the fact that it was not the question of free-soil which threatened to rend the country in twain, but a well grounded alarm at the attempted overthrow of the Constitution and liberty by the usurpation of a power to crush states. The question of free-soil had no such importance in the eyes of the people of the Border States, nor even of the seceded states as to become at once a casus belli. But in the view of all parties in the Border States the claim of coercion had infinite importance.

If, as Mr. Lincoln had argued, secession was unconstitutional, coercion was more clearly so. When attempted, it must necessarily take the form of a war of some states against other states. It was thus the death-knell of constitutional Union and so a thorough revolution of the Federal government. It was the overthrow of the reserved rights of the states, and these were the only bulwark of the liberty of the

> *The obvious ground of this policy was in the fact that it was not the question of free-soil which threatened to rend the country in twain, but a well grounded alarm at the attempted overthrow of the Constitution and liberty by the usurpation of a power to crush states.*

people. This then was the real cause of alarm at the South, and not the claim of free-soil, unjust as was the latter; hence, all that was necessary to reduce the free-soil controversy to harmless and manageable dimensions was to reassure the South against the dreaded usurpation of which free-soil threatened to be made the pretext.

This, Colonel Baldwin showed, could easily be done by a policy of conciliation without giving sanction to what Mr. Lincoln's administration chose to regard as the heresy of secession. The government would still hold the Union and the Constitution as perpetual and the separate attitude of the seceded states as temporary, while it relied upon moderation, justice, self-interest of the Southern people, and the potent mediation of the Border States to terminate it.

"Only give this assurance to the country, in a proclamation of five lines," said Colonel Baldwin, and we pledge ourselves that Virginia (and with her the Border States) will stand by you as though you were our own Washington. So sure am I of this, and of the inevitable ruin which will be precipitated by the opposite policy, that I would this day freely consent, if you would let me write those decisive lines, you might cut off my head were my life my own, the hour after you signed them."

Lincoln seemed impressed by his solemnity and asked a few questions: "But what am I to do meantime with those men at Montgomery? Am I to let them go on?"

"Yes, sir," replied Colonel Baldwin decisively, "until they can be peaceably brought back."

"And open Charleston, etc. as ports of entry with their ten percent tariff. What, then, would become of my tariff?"

This last question he announced with such emphasis, as showed that in his view it decided the whole matter. He then indicated that the interview was at an end and dismissed Colonel Baldwin without promising anything more definite.

> *They were fully aware that neither the Constitution nor laws gave them any right to coerce a state to remain in the Union.*

In order to confirm the accuracy of my own memory, I have submitted the above narrative to the Honorable A. H. H. Stuart, Colonel Baldwin's neighbor and political associate and the only surviving member of the commission soon after sent from the Virginia Convention to Washington. In a letter to me, he says: "When Colonel Baldwin returned to Richmond, he reported to the four gentlemen above named, and to Mr. Samuel Price of Greenbrier, the substance of his interview with Lincoln substantially as he stated it to you."

I asked Colonel Baldwin what was the explanation of this remarkable scene and especially of Lincoln's perplexity. He replied that the explanation had always appeared to him to be this: When the seven Gulf States had actually seceded, the Lincoln faction was greatly surprised and in great uncertainty of what to do; for they had been blind enough to suppose that all Southern opposition to a sectional president had been empty bluster. They were fully aware that neither Constitution nor laws gave them any right to coerce a State to remain in the Union. The whole people, even in the imperious North, knew and recognized this truth. The New York *Tribune*, even, admitted it, violent as it was, and deprecated a Union "pinned together with bayonets." Even General Winfield Scott, the military "Man Friday," of Federal power, advised that the Government should say: "Erring Sisters, go in peace."

So strong was the conviction, even in the Northern mind, that such journals as *Harper's Weekly and Monthly*, shrewdly mercenary in their whole aim, were notoriously courting the secession feeling. New York, the financial capital of America, was well known to be opposed to the faction and to coercion. The previous Congress had expired without daring to pass any coercive measures.

The administration was not at all certain that the public opinion of the American people could be made to tolerate anything so illegal and mischievous as a war of coercion. (Subsequent events and declarations betrayed also how well the Lincoln faction knew at the time that it was utterly unlawful. For instance: when Lincoln launched into that war, he did not dare to say that he was warring against states, and for the purpose of coercing them into a Federal Union of force. In his proclamation calling for the first 75,000 soldiers, he had deceitfully stated that they were to be used to support the laws, to repossess Federal property and places, and to suppress irregular combinations of individuals pretending to or usurping the powers of state governments.

The same was the tone of all the war speakers and war journals at first. They admitted that a state could not be coerced into the Union; but they held that no state really and legitimately desired to go out, or

> *They admitted that a state could not be coerced into the Union; but they held no state really and legitimately desired to go out, or had gone out...*

had gone out – "the great Union-loving majority in the South had been overruled by a factious secession minority, and the Union troops were only to liberate them from that violence, and enable them to declare their unabated love for the Union."

No well informed man was, at first, absurd enough to speak of a state as "committing treason" against the confederation, the creature of the states; the measure was always spoken of as "Secession," the actors were "Secessionists," and even their territory was "Secessia."

It remained for an ecclesiastical body, pretended representative of the Church of the Prince of Peace, in their ignorant and venomous spirit of persecution, to apply the term "treason" first to the movement in favor of liberty.)

The action of the seven states then perplexed the Lincoln faction excessively. On the other hand, the greed and spite of the hungry crew, who were now grasping the power and spoils so long passionately craved, could not endure the thought that the prize should thus collapse in their hands. Hence, when the administration assembled at Washington, it probably had no very definite policy. Seward, who assumed to do the thinking for them, was temporizing. Colonel Baldwin supposed it was the visit, and the terrorizing of the "radical governors," which had just decided Lincoln to adopt the violent policy. They had especially asserted that the secession of the seven states, and the convening and solemn admonitions of State conventions in the others formed but a system of bluster, or, in the vulgar phrase of Lincoln, but a "game of brag;" that the Southern States were neither willing nor able to fight for their own cause, being paralyzed by their fear of servile insurrection.

Thus they had urged upon Lincoln, that the best way to secure his party triumph was to precipitate a collision. Lincoln had probably committed himself to this policy, without Seward's privity within the last four days; and the very men whom Colonel Baldwin found in conclave with him were probably intent upon this conspiracy at the time. But when Colonel Baldwin solemnly assured Lincoln that this violent policy would infallibly precipitate the Border States into an obstinate war, the natural shrewdness of the latter was sufficient to open his eyes, at least partially, and he saw that his factious counselors blinded by hatred and contempt of the South had reasoned falsely; yet, having just committed himself to them, he had not manliness enough to recede. And above all, the policy urged by Colonel Baldwin would have disappointed the hopes of legislative plunder by means of inflated tariffs, which were the real aims for which free-soil was the mask.

Thus far Colonel Baldwin's narrative proceeded. The conversation then turned upon the astonishing supineness (or blindness) of the conservatives, so-called, of the North, to the high-handed usurpations of their own rights, perpetrated by Lincoln and Seward under the pretext of subduing the seceded states, such as the suspension of habeas corpus, the state prisons, the arrests without indictment and the martial law imposed at the beck of the Federal power, in states called by itself "loyal."

> "It I do that, what will become of my revenue? I might as well shut up housekeeping at once!"

I asked: "Can it be possible that the Northern people are so ignorant as to have lost the traditionary rudiments of a free government?" His reply was, that he apprehended the Northern mind really cared nothing for liberty; what they desired was only lucrative arrangements with other states.

The correctness of Colonel Baldwin's surmises concerning the motives of Lincoln's policy receives these two confirmations. After the return of the former to Richmond, the Convention sent the commission, which has been described, composed of Messrs. William B. Preston, A. H.H. Stuart, and George W. Randolph. They were to ascertain definitely what the President's policy was to be.

They endeavored to reach Washington in the early part of the week in which Fort Sumter was bombarded but were delayed by storms and high water, so that they only reached there via Baltimore, Friday, April 12th. They appeared promptly at the White House and were put off until Saturday for their formal interview, although Lincoln saw them for a short time. On Saturday, Lincoln read to them a written answer to the resolutions of Convention laid before him, which was obviously scarcely dry from the pen of a clerk.

"This paper," says Mr. Stuart, "was ambiguous and evasive, but in the main professed peaceful intentions." Mr. Stuart in answer to this paper spoke freely and at large, "urging forbearance and the evacuation of the forts, etc." Lincoln made the objection that all the goods would be imported through the ports of Charleston, etc, and the sources of revenue dried up.

"I remember," says Mr. Stuart, "that he used this homely expression: 'If I do that, what will become of my revenue? I might as well shut up housekeeping at once!' But his declarations were distinctly pacific and he expressly disclaimed all purpose of war."

Mr. Seward and Mr. Bates, Attorney General, also gave Mr. Stuart the same assurances of peace. The next day the commissioners returned to Richmond and the very train on which they traveled carried Lincoln's proclamation calling for the 75,000 men to wage a war of coercion.

"This proclamation," says Mr. Stuart, "was carefully withheld from us, although it was in print; and we knew nothing of it until Monday morning when it appeared in the Richmond papers. When I saw it at breakfast, I thought it must be a mischievous hoax; for I could not believe Lincoln guilty of such duplicity. Firmly believing it was a forgery, I wrote a telegram at the breakfast table of the Exchange Hotel and sent it to Seward asking him if it was genuine. Before Seward's reply was received, the Fredericksburg train came in, bringing the Washington papers, containing the proclamation."

The other confirmation of Colonel Baldwin's hypothesis was presented a few weeks after the end of the war, in a curious interview with a personal friend and apologist of Seward. The first volume of my life of Jackson had been published in London, in which I characterized the shameless lie told by Seward to the commissioners from Montgomery, through Judge Campbell, touching the evacuation of Sumter. This friend and apologist of Seward said that I was unjust to him, because when he promised the evacuation, he designed and thought himself able to fulfill it; but between the making and breaking of the pledge, a total change of policy had been forced upon the administration against Mr. Seward's advice, by Thaddeus Stevens and the radical governors.

Seward, abolitionist and knave as he was, still retained enough of the statesman-like traditions of the better days of the republic to know that coercion was unlawful and that a war between the states was, of course, the annihilation of the Union. It suited his partisan and selfish designs to talk of an "irrepressible conflict" and to pretend contempt for "effeminate slavocrats;" but he had sense enough to know that the South would make a desperate defense of her rights, and would be a most formidable adversary, if pushed to the wall.

Hence, Mr. Seward, with General Scott, had advised a temporizing policy towards the Montgomery government without violence, and Mr. Lincoln had acceded to their policy. Hence, the promises to Judge Campbell.

Meantime, the radical governors came down, "having great wrath," to terrorize the administration. They spoke in this strain: "Seward cries perpetually that we must not do this, and that, for fear war should result. Seward is shortsighted. War is precisely the thing we should desire. Our party interests have everything to lose by a peaceable settlement of this trouble and everything to gain by collision. For a generation we have been 'the outs;' now at last we are 'the ins.' While in opposition, it was very well to prate of Constitution and of rights; but now we are the government, and mean to continue so; and our interest is to have a strong and centralized government. It is high time now that the government were revolutionized and consolidated, and these irksome 'States' rights' wiped out. We need a strong government to dispense much wealth and power to its adherents; we want permanently high tariffs to make the South tributary to the North; and now these Southern fellows are giving us precisely the opportunity we want to do all this, and shall Seward sing his silly song of the necessity of avoiding war? War is the very thing we should hail!

"The Southern men are rash, and now profoundly irritated. Our plan should be, by some artifice, to provoke them to seem to strike the first blow. Then we shall have a pretext with which to unite the now divided North and make them fly to arms. The Southerners are a braggart, but a cowardly and effeminate set of bullies; we shall easily whip them in three months.

"But this short war will be, if we are wise, our sufficient occasion. We will use it to destroy slavery and thus permanently cripple the South. And that is the stronghold of all these ideas of 'limited government' and 'rights of the people.'

"Crush the South by abolishing slavery and we shall have all we want – a consolidated government, an indefinite party ascendancy and ability to lay on such tariffs and taxes as we please, and aggrandize ourselves and our section!"

These, Mr. Seward's apologist declared to me, were the reasons which together with their predictions and threats of popular rage converted Lincoln from the policy of Seward to that of Stevens. Hence the former was compelled to break his promise through Judge Campbell and to assist in the malignant stratagem by which the South Carolinians were constrained "to fire on the flag." The diabolical success of the artifice is well known.

The importance of this narrative is that it unmasks the true authors and nature of the bloody war through which we have passed. We see that the Radicals provoked it not to preserve but to destroy the Union. It demonstrates, effectually, that Virginia and the Border States were acting with better faith to preserve the Union than was Lincoln's Cabinet.

> *Colonel Baldwin showed Lincoln conclusively that it was not free-soil, evil as that was, which really endangered the Union but coercion.*

Colonel Baldwin showed Lincoln conclusively that it was not free-soil, evil as that was, which really endangered the Union but coercion. He showed him that if coercion were relinquished, Virginia and the Border States stood pledged to labor with him for the restoration of Union and would assuredly be able to effect it. Eight slave-holding Border States with seventeen hireling States would certainly have wielded sufficient moral and material weight in the cause of what Lincoln professed to believe, the clear truth and right to reassure and win back the seven little seceded States, or, if they became hostile to restrain them.

But coercion arraigned fifteen against seventeen in mutually destructive war. Lincoln acknowledged the conclusiveness of this reasoning in the agony of remorse and perplexity, in the writhing and tearing of hair of which Colonel Baldwin was witness.

But what was the decisive weight that turned the scale against peace, and right, and patriotism? It was the interest of a sectional tariff! His single objection, both to the wise advice of Colonel Baldwin and Mr. Stuart, was: "Then what would become of my tariffs?" He was shrewd enough to see that the just and liberal free-trade policy proposed by the Montgomery Government would speedily build up, by the help of the magnificent Southern staples, a beneficent foreign commerce through Confederate ports; that the Northern people, whose lawless and mercenary character he understood, could never be restrained from smuggling across the long open frontier of the Confederacy; that thus the whole country would become habituated to the benefits of free-trade, so that when the schism was healed (as he knew it would be healed in a few years by the policy of Virginia), it would be too late to restore the iniquitous system of sectional plunder by tariffs, which his section so much craved.

Hence, when Virginia offered him a safe way to preserve the Union, he preferred to destroy the Union and preserve his tariffs. The war was conceived in duplicity and brought forth in iniquity.

The calculated treason of Lincoln's Radical advisers is yet more glaring. When their own chosen leader, Seward, avowed that there was no need for war, they deliberately and malignantly practiced to produce war, for the purpose of overthrowing the Constitution and the Union, to rear their own greedy faction upon the ruins. This war, with all its crimes and miseries, was proximately concocted in Washington City by Northern men with malice purpose.

The principle, on which the war was waged by the North, was simply this: That men may rightfully be compelled to submit to, and support, a government that they do not want; and that resistance, on their part, makes them traitors and criminals. No principle, that is possible to be named, can be more self-evidently false than this; or more self-evidently fatal to all political freedom. Yet it triumphed in the field, and is now assumed to be established. If it really be established, the number of slaves, instead of having been diminished by the war, has been greatly increased; for a man, thus subjected to a government that he does not want, is a slave. And there is no difference, in principle – but only in degree – between political and chattel slavery. The former, no less than the latter, denies a man's ownership of himself and the products of his labor; and asserts that other men may own him, and dispose of him and his property, for their uses, and at their pleasure.

Lysander Spooner

WAR BETWEEN THE STATES – FOUGHT FOR STATES RIGHTS OR SLAVERY?

By Toni Turk, Commanding General of the Military Order of the Stars and Bars

The Battle of Sharpsburg (Antietam) was fought on September 17, 1862. Less than a week later, on September 22nd, Abraham Lincoln announced that he would emancipate all slaves in states that did not return to the Union. States in the Union were not affected. Therein lays a story.

The first Blacks brought to the American colonies, whether in the North or in the South, were dealt with as indentured servants with the same status as White indentured servants; however, within a generation, the status of Blacks devolved into slavery. The decision for slavery was a flawed economic decision. As the Northern colonies moved towards industrialization, slavery became less profitable there. As Northern slave holders analyzed their bottom line, the vast majority elected the option of selling their slaves to Southern slaveholders rather than taking the moral alternative of manumission. As a consequence some of the busiest slave ports were located in the North.

Eventually, as soils depleted in the old South, a similar discovery of the economic liability of slavery was faced in that region. To the extent possible, slaves were relocated to new lands farther west. By the time of the War Between the States, slavery in general was a quandary for slave holders. Generations of economic investment with diminishing economic benefit was their reality.

Coincidental to this plight was the private venture of repatriation of ex-slaves to Liberia beginning in 1820, a solution later much favored by Lincoln.

The British Empire abolished slavery in 1834 – with compensation. Then in 1835, the British and French entered a pact to abolish the slave trade.

Had American statesmen in the generation prior to the War Between the States been able to advance a compensated emancipation plan, it would likely have received a warm embrace from Southern slaveholders as a way to recover their economic investment in a failed system – just as their Northern counterparts had done several generations earlier. The failure of compensated emancipation to gain traction rested in part on the resistance of Northern taxpayers to fund it.

Slavery was a Southern moral albatross. Lincoln strategized a way to hang it around their neck. Realizing that for emancipation to be viewed as a moral act rather than a desperate one, Lincoln needed a contest of arms that could not be viewed as a Southern victory. Sharpsburg was that engagement. Strictly speaking, it was not a clear tactical victory for the North, but it proved to be an immensely strategic one. Lincoln's balancing act was not to offend slave holding states not in secession, while making it complicated for the British and French to continue support for a system that they had renounced.

Slavery is a guilt shared by both North and South. Ironically, the treasure expended in waging the War Between the States exceeded that which would have been required to compensate emancipation. Even left alone, slavery was an economically doomed system that would have inevitably crumbled. The blood and treasure expended form 1861 to 1865 and the subsequent ravages of Reconstruction and Jim Crow could all have been avoided. Emancipated compensation would have improved conditions for both Whites and Blacks much earlier than the hundred years' hiatus that occurred.

I supported President Lincoln. I believed his war policy would be the only way to save the country, but I see my mistake. I visited Washington a few weeks ago, and I saw the corruption of the present administration – and so long as Abraham Lincoln and his Cabinet are in power, so long will war continue. And for what? For the preservation of the Constitution and the Union? No, but for the sake of politicians and government contractors.

J.P. Morgan

Was the Confederate Soldier a Rebel?

By W.R. Hammond. *Southern Historical Society Papers, Volume 28,* page 247.

In *The Green Bag* for December, 1899, and January, 1900, this question is answered by Bushrod C. Washington. His argument is clean-cut, strong and convincing. His conclusion agrees with that of all southern men and must be the verdict of posterity. He argues the legality of the course pursued by the seceding States, and plainly shows that the people of that section were not Rebels. No fair minded man can read history and deny the correctness of his conclusions. Quoting from unquestioned authority, he makes plain, beyond all controversy, that there was never any intention of any of the States in the original compact of union to give up the reserved rights not expressly delegated to the general government under the Constitution. His argument, than which nothing can be clearer, is, that the North broke the compact and that the South, for that reason and that alone, sought to withdraw.

Candid men must admit that the compact was broken by the North. Admitting this, they must justify the South in the course taken by her people. The union was a union of political societies upon an agreed basis, and that basis was the Constitution. Hamilton, as quoted by Mr. Washington, expresses this clearly, "If a number of political societies enter a larger political society, the laws which the latter may enact pursuant to the powers entrusted to it by its Constitution must necessarily be supreme over those societies. But it will not follow from this that the acts of the larger which are not pursuant to its constituted powers but which are invasions of the residuary authorities of the smaller societies will become the supreme law of the land.

These will be merely acts of usurpation, and will deserve to be treated as such."

That Congress committed these acts of usurpation cannot now be denied, and the election of Mr. Lincoln in 1860, upon a platform pledged to override the Constitution, as interpreted by the Supreme Court of the United States, was an unmistakable expression on the part of the people of the North of a determination to disregard the solemn mandates of that instrument which formed the bond of union between the sections. This determination once formed and expressed, the South had the legal right to withdraw from the compact, quietly and peaceably, as she did.

This right was clearly recognized by Mr. Webster, who was a statesman of much larger caliber than Harriet Beecher Stowe. He did not hesitate to say, at the same time that Uncle Tom's Cabin made its appearance, that if the northern States willfully and deliberately refused to carry into effect that part of the Constitution which protected the Southern people in the possession of their property, and Congress refused to provide a remedy, the South would no longer be bound to observe the compact. "A bargain," he said, "cannot be broken on one side and bind the other side."

These words were not heeded at the North, but the false and scandalous utterances of Mrs. Stowe were allowed to drown them, and a flood of fanaticism swept over that section of our country, which culminated in the crime of 1860, and the terrible tragedy of the Civil War. There are men in the North who know this now,

> *A bargain cannot be broken on one side and bind the other side.*

and there will be more of them as the years go by; and the verdict of posterity will be that if the sober and statesmanlike utterances of Daniel Webster had been allowed to prevail at the North, instead of the fanatical words of Harriet Beecher Stowe, who was no more worthy of comparison with that broad-minded and patriotic American than a rush light to the sun; the horrors of the war might have been averted, and the problem of being rid of the evil of slavery, which bound the white man to one end of the chain and the black man to the other, would have been wrought out by the consent of the South itself, on peaceful lines, under the Constitution and not in violation of it.

That this would have been done before the close of the nineteenth century, if antagonisms had not been roused and kept alive by the fostering of the abolition sentiment at the North, and that the condition of the black man would be much better today than it is, can scarcely admit of a doubt. Abolition should have been the result of growth, not of revolution; and might have been wrought out patiently by means of the Constitution, and should not have been brought about in bitter spite of it.

In the second series of Appleton's Popular Library, published in 1852, is an essay published from The *London Times*, in which the author reviews Uncle Tom's Cabin and predicts the evils that were liable to result from the book. He is no sympathizer with slavery, but shows that he was opposed to it with all his might. I will close this review by quoting a part of what he says. Let us remember that these words were written by an Englishman nearly fifty years ago:

"And be it stated to the credit of the slave owners of the South that they are fully alive to the danger of the portentous struggle, and have of late years shown no indisposition to help in their own emancipation as well in that of the slave, provided they may only escape the dire catastrophe we speak of. It is certain that a large class of slave owners in the South are most desirous to relieve their soil of the stain and inconvenience of slavery, if the tremendous step can be taken with safety to all parties concerned in the act of liberation. The efforts made in the South to improve the condition of the slave show at least that humanity is not dead in the bosom of the proprietors. Mrs. Stowe has certainly not done justice to this branch of the subject. Horrors in connection with slavery – itself a horror – unquestionably exist, but all accounts – save her own and those of writers actuated by her extreme views – concur in describing the general condition of the Southern slave as one of comparative happiness and comfort such as many a free man in the United Kingdom might regard with envy."

One authority on this point is too important to be overlooked. In the year 1842, a Scotch weaver named William Thompson traveled through the Southern States. He supported himself on his way by manual labor; he mixed with the humblest classes black and white, and on his return home, he published an account of his journeyings. He had quitted Scotland a sworn hater of slave proprietors, but he confessed that experience had modified his views on this subject to a considerable degree.

He had witnessed slavery in most of the slave-holding States; he had lived for weeks among Negroes in cotton plantations, and he asserted that he had never beheld one-fifth of the real suffering that he had seen among the laboring poor in England. Nay, more, he declared: "That the members of the same family of Negroes are not so much scattered as those of workingmen in Scotland, whose necessities compel them to separate at an age when the American slave is running about gathering health and strength."

Ten years have not increased the hardships of the Southern slave. During that period colonization has come to his relief; education has, legally or illegally, found its way into his cabin, and Christianity has added spiritual consolations to his allowed, admitted physical enjoyments.

It has been justly said that to these men of the South who have done their best for the Negro under the institution of slavery must we look for any great effort in favor of emancipation and they who are best acquainted with the progress of events in those parts declare that at this moment "there are powerful and irresistible influences at work in a large part of the slave States tending toward the abolition of slavery within these boundaries."

We can well believe it. The world is working its way toward liberty, and the blacks will not be left behind in the onward march. Since the adoption of the American Constitution, seven states have voluntarily abolished slavery. When that Constitution was proclaimed, there was scarcely a free black in the country. According to the last census, the free blacks amount to 418,173, and of these 233,691 are blacks of the South, liberated by their owners and not by the force of law. We cannot shut our eyes to these facts. Neither can we deny that, desirable as Negro emancipation may be in the United States, abolition must be the result of growth, not of revolution; must be patiently wrought out by means of the American Constitution, and not in bitter spite of it.

America cannot for any time resist the enlightened spirit of our age, and it is manifestly her interest to adapt her institutions to its temper. That she will eventually do so if she be not a divided household – if the South be not goaded to illiberality by the North – if public writers deal with the matter in the spirit of conciliation, justice, charity and truth, we will not permit ourselves to doubt.

That she is alive to the necessities of the age is manifest from the circumstances that, for the last four years, she has been very busy preparing the way for emancipation by a method that has not failed in older countries to remove national trouble almost as intolerable as that of slavery itself.

Origin of the Late War

By R. M. T. Hunter. *Southern Historical Society Papers, Volume 1*, Pages 1-13.

The late civil war which raged in the United States has been very generally attributed to the abolition of slavery as its cause. When we consider how deeply the institutions of southern society and the operations of southern industry were founded in slavery, we must admit that this was cause enough to have produced such a result. But great and wide as was that cause in its far-reaching effects, a close study of the history of the times will bring us to the conclusion that it was the fear of a mischief far more extensive and deeper even than this which drove cool and reflecting minds in the South to believe that it was better to make the death struggle at once than submit tamely to what was inevitable, unless its coming could be averted by force. Men, too old to be driven blindly by passion; women, whose gentle and kindly instincts were deeply impressed by the horrors of war, and young men, with fortune and position yet to be won in an open and inviting field, if peace could be maintained so as to secure the opportunities of liberty and fair treatment, united in the common cause and determined to make a holocaust of all that was dear to them on the altars of war sooner than submit without resentment to the loss of liberty, honor, and property by a cruel abuse of power and a breach of plighted faith on the part of those who had professed to enter with them into a union of justice and fraternal affection.

When this Union was originally formed, the United States embraced too many degrees of latitude and longitude, and too many varieties of climate and production, to make it practicable to establish and administer justly one common government which should take charge of all the interests of society. To the wise men who were entrusted with the formation of that union and common government it was obvious enough that each separate society should be entrusted with the management of its own peculiar interests and that the united government should take charge only of those interests which were common and general. To enforce this necessary distinction, it was provided that all powers, not specially granted, should be reserved to the people and the States, and a list of the granted powers was carefully and specifically made.

But two parties soon arose in regard to these limitations. Those who wielded the powers thus granted became interested to remove these limitations as far as possible, whilst the minority, who belonged to the governed rather than the governing party early learned to regard these limitations as the best and surest defenses against the abuses and oppression of a despotic majority.

A tendency soon became manifest in the non-slaveholding portion of the union to constitute themselves into that governing party. Endowed with the greater share of power in the commencement, that preponderance was increased by the course of events. The famous northwestern ordinance, to which the old Virginia fathers were driven by their abhorrence of slavery, without looking too closely to its probable consequences, made the predominance of the non-slaveholding section in the government irresistible. The abolition of the slave-trade, after a time, by the Constitution and the northwestern ordinance, left the growing superiority of that section not even doubtful.

But the acquisition of Louisiana made another order of growth in political power possible as between the two sections. The bare possibility of such a result kindled a violent opposition in some portions of the non-slaveholding section. In New England it was particularly angry, and there sprung up for the first time in the history of our government audible threats of separation. The "land hunger" of the Anglo-Saxon race, as Theodore Parker calls it, soon quieted the opposition to the acquisition of territory, but a far more bitter strife arose as to the equal rights of the two sections to settle the vacant territory of the Union and grow possibly part passu in power.

So fierce was the strife, and so loud its tumult, that for the first time it broke upon Mr. Jefferson's ear like "a fire bell in the night." The contest between the two sections over the limitations in the Constitution upon the governing party under it began with the commencement of its history, and ended only, as I shall presently show, with the revolution which destroyed the old form and established the despotism of a majority of numbers. It is in the history of this contest we must look for the true causes of the war, and the use made of the victory by the winning party will show the object and nature of that contest.

> ***The contest between the two sections over the limitations in the Constitution upon the governing party under it began with the commencement of its history, and ended only, as I shall presently show, with the revolution which destroyed the old form and established the despotism of a majority of numbers.***

When it became obvious that the only protection of the rights of the minority against the encroachments of the majority was to be found in the limitations upon the power of the governing party, a death struggle arose between the two parties over the constitutional restraints upon this power.

The struggle between the two parties commenced at the beginning of the government. These were respectively led by Hamilton and Jefferson, the one with an avowed preference for monarchy, the other the great apostle of democracy – men of signal abilities and each conscious of what would be the consequence of complete and perfect victory on either side.

The party of power showed a constant tendency to draw all important subjects of jurisdiction within the vortex of Federal control, and an equally persevering effort on the other to limit that control to the strict necessities of a common government. A great leader, who came into the contest and figured in it until it was well nigh ended, used to say that in all good governments there existed a tax-consuming and a tax-paying party, between whom a constant conflict existed, and in the history of that conflict the history of party strife would be found to consist; but when the first acquired complete supremacy the nature, if not the form of the government – if it was originally republican – was sure to change.

> *In all good governments there existed a tax-consuming and a tax-paying party, between whom a constant conflict existed, and in the history of that conflict the history of party strife would be found to consist; but when the first acquired complete supremacy, the nature, if not the form of the government – if it was originally republican – was sure to change.*

The leaders of the States rights party, aware of this tendency, as the contest went on, became more and more anxious to preserve their constitutional defenses, and loudly proclaimed the danger of yielding them up. Time and again they proclaimed that the worst of all governments was that of a majority of numbers with absolute and unrestricted powers. Despotism of all sorts was bad, but the despotism of a majority of numbers in a democratic form of government was the worst of all – particularly was that the case in regard to slavery, as was often asserted.

In February, 1790, when two abolition petitions, one of them signed by Dr. Franklin, were presented to Congress, that body "resolved that Congress had no authority to interfere in the emancipation of slaves, or even the treatment of them within any of the States, it remaining with the several States alone to provide any regulations therein which humanity or true policy may require." Congress thus clearly declared its view of its power over the subject. Congress was petitioned to do all in its power to discourage slavery, of which a Massachusetts man, in an able history of the long contest, has said: "Congress could not move a hair's breadth towards discouraging it, either lawfully or honestly. The powers of Congress being defined and nominated by the Constitution which framed the government, all it could do in regard to any specific subject was to act upon it, if within its province, and if otherwise, 'to touch not, taste not, handle not.'" (Lunt's Origin of the Late War, p. 25.)

In the debate upon the subject, one Southern gentleman objected to the commitment of these memorials as containing "unconstitutional requests," and said "he feared the commitment would be a very alarming circumstance to the Southern States; for if it was to engage Congress in an unconstitutional measure, it would be considered an interference with their rights, making them uneasy under the government, and causing them to lament that they had ever put additional power into their hands." Another declared "that the States would never have entered into the confederacy unless their property had been guaranteed to them, and that we look upon this measure as an attack upon the palladium of our property" – meaning the Constitution. Another said if he was to hold these slaves in eternal bondage he would feel no uneasiness on account of the present menace, "because he would rely upon the virtue of Congress that they would not exercise any unconstitutional authority." This same historian well says "the impression made upon the southern members of Congress at the earliest period is also significant.

Although evidently considering it of no practical importance, they yet clearly made it known they regarded such action as in violation of the Constitution, and that without the guaranty for their rights of property in slaves, permitted by that instrument, the States which they represented would not have assented to it, and hence the plan for the Union must have failed. No one can doubt that if they had deemed the guaranty afforded insufficient, they could have obtained pledges of a still more precise character, either then or at a later period, since the object of the Union was one of paramount interest to all. But neither they nor their northern compatriots entertained any question of the fidelity of their successors to engagements so solemnly undertaken both express and implied." (Lunt, p. 27.)

The history of this transaction shows how early the South was taught to look to the Constitution for the defenses of their rights in regard to slavery, how fully, too, and clearly the Congress admitted the existence of these defenses, and that the South disregarded the unauthorized menace of these "anarchic Quakers," as Carlisle calls them, because they "relied upon the virtue of Congress that they would not exercise any unconstitutional authority." Their property in slaves was guaranteed by the Constitution; they felt authorized to say so by a solemn declaration of Congress made at the time; and they had too much confidence in the northern majority, who were soon to control that body, to believe that directly or indirectly they would impair or destroy a right so solemnly guaranteed.

To have anticipated such an attack upon their property and peace, would have been to suppose that they had been made the easy victims of a perfidy, which, under all the circumstances, under all the traditions of common sufferings and exertions, was characterized by a wealth of deception which would have excited the envy even of a Carthagenian. Especially would that be the case if the deceit was to be covered up by a constant course of perjury on the part of the officials of the government, who were to be sworn as a qualification for office to support the Constitution which contained that pledge.

How justly our fathers relied upon that instrument to protect their rights, subsequent history has shown. Nothing could be more clearly established than the right on one side to reclaim fugitive slaves, and the obligation on the other to return them – an obligation which surely ought to have rested lightly enough on those who brought them here and sold them. Nor is it easy to see how the remorse for having sold them could be relieved by inveigling them away from those who had bought them.

But so it was, that during the existence of slavery there was an ever-living contest between the slave and the free states on this very subject; the former seeking to enforce, and the latter to evade the constitutional obligation for the return of fugitive slaves. Long before the secession of the slave States, it had become almost impossible, without the assistance of armed forces, to reclaim a fugitive slave openly in the free States. Lunt, p. 320, says: "At length fourteen of the sixteen free States had provided statutes which rendered any attempt to execute the fugitive slave act so difficult as to be practically impossible, and placed each of those States in an attitude of virtual resistance to the laws of the United States."

When Mr. Toombs, in the Senate of the United States, during the session in which he withdrew from that body, referred to these laws and taxed the free states with their violations of constitutional obligation, in evidence of which he produced these statutes, it was pitiful to hear the excuses by which the representatives of these states sought to squirm out of the difficulty – a difficulty for which the executives of Ohio and Iowa would scarcely have cared to apologize, if it be true, as doubtless it is, as Lunt states, that "at a somewhat later period those officers refused to surrender to justice persons charged with participation in the John Brown raid" – see note, page 320. At the era of secession, the Constitution had not only ceased to be a palladium for these rights of secession, but was hardly recognized to be binding at all.

> *It had become manifest that the South had no protection for its rights but the Constitution, nor could it hope to avail itself of that protection without an increase of power in the government.*

If, then, this instrument was to be relied upon by the slave States to protect them, it was only in the event that they could arm themselves with enough political power to enforce its provisions. So obvious had this become by 1819-20, when the State of Missouri was struggling for admission as a slave state, that the slave states at that time solemnly asserted their right to settle the unoccupied and unappropriated territory of the United States with their slave property, under the protection of its laws – a right which was as vehemently denied by the free States. So bitter and fierce was this contest, that its agitations shook the very foundations of American society.

It was settled for a time by a compromise excluding slavery from the United States Territories north of a line 36° 30' north latitude, and admitting it south of that line. Even this line left the South in a condition of hopeless inferiority, which was but little helped by the acquisition of a portion of Texas as a slave state. When the vast territory obtained from Mexico at the close of the war was organized, the Missouri compromise line was set aside, and the non-intervention principle was adopted, by which it became between the sections a mere question of the ability to colonize – a question in regard to which there could scarcely be a doubt, with the superior resources in wealth and population of the free states.

It had become manifest that the South had no protection for its rights but the Constitution, nor could it hope to avail itself of that protection without an increase of power in the government. Its hopes for acquiring that were daily becoming less, whilst sectional animosities were constantly becoming more angry and bitter.

A party had sprung up which proclaimed the constitution to be "an agreement with death and a covenant with hell." This party was daily becoming stronger and more dangerous in spirit. It began at first by taking part in the contests between Whigs and Democrats, and grew upon the agitations in Congress and the newspaper press. This war of petitions for abolition was commenced by John Quincy Adams in 1831, when he presented a petition from Pennsylvania for the abolition of slavery in the District of Columbia, but at the same time declared that he could not vote for it. He who was so denounced when he left the Federal party, on account of its disunion tendencies, and joined the Democratic under Mr. Jefferson, became the "old man eloquent" when he fanned the smoldering spark of sectional division with the burning breath of hate and anger which was yet to burst out in flames and consume the house with the fire whose initial spark he consented to bear and apply to the family dwelling, ever nursing the fire until the building was fairly ablaze.

And what was now, in 1860, the worth of the reliance which kept the South quiet in 1790, because it "relied upon the virtue of Congress that it would exercise no unconstitutional authority?" In regard to the right to recapture fugitive slaves, it was at that time obviously a dead better. The free states had violated that obligation by their personal liberty statutes, which were consonant with the general spirit of their people. The abolition party, which denounced the constitution as a "covenant with death and an agreement with hell," was fast growing in power and influence in the free states, and threatened to become the most powerful political organization within their borders.

Massachusetts had adopted resolutions by her legislature, with the assent of her governor – if his message represented his opinions – resolutions which were denounced at the time as being of a disunion character. Her senator, Bates, presented them in silence, and Colonel King, of Alabama, regretted that a proposition should come from Massachusetts to dissolve the Union. (See Lunt's Origin of the War, 128-9).

All hope of acquiring any additional political strength by the South to defend their rights was gone. The free states had announced their determination to exclude slavery from the territories of the United States, and they had the strength to do it, if they believed, as they affected to do, that the Constitution was no obstacle in their path. The right of growth was thus denied to the power of the slaveholding states, and with the state of feeling then existing and cherished, they had nothing to expect but to be dwarfed and oppressed, judging of the future by the past.

Indeed, an armed invasion of Virginia had been just made by John Brown, with the avowed purpose of exciting servile insurrection, and although suppressed by the United States and state forces, it excited no such outburst of horror and denunciation at the North as it might reasonably be expected to have done. On the contrary, he seemed to have been considered more as a martyr perishing in a great and holy cause, than a criminal seeking to excite a servile war, whose victims were to be women and children. "The tolling of bells and the firing of minute guns upon the occasion of Brown's funeral; the meeting houses draped in mourning, as for a hero; the prayers offered, the sermons and discourses pronounced in his honor, as for a saint — all are of a date too recent and too familiarly known to require more than this passing allusion." (Lunt, 328). Was there anything in all this calculated to discourage such attempts for the future? On the contrary, would it not be apt to stir up still more deeply excited minds, and the next attempt would probably have caused much more suffering. To expect that the attempt to cast a lighted match into a powder magazine would fail more than once would be chimerical indeed.

In considering the value of his defenses under the Constitution, a Southern man could not well forget that Mr. Seward, the leader of the party in power, had not only declared the conflict between freedom and slavery to be "irrepressible," but had affirmed there was a higher law than the Constitution, to which the latter must yield,

> *Could the minority rely upon the Constitution to protect any of their rights, if it suited the passions or the interests of the majority to invade them?*

or that the famous Helper book, endorsed and recommended generally by the Republican members of Congress, declared that "our own banner is inscribed: "no cooperation with slaveholders in politics; no fellowship with them in religion; no affiliation with them in society; no recognition of pro-slavery men, except as ruffians, outlaws and criminals." Again: "we are determined to abolish slavery at all hazards." With such a history of the administration of the Constitution by the party in power, there was no very pleasant outlook for the slaveholder in the future. Had he any hope from amendments?

That no effort to save the Union should be spared, Mr. Crittenden, of Kentucky, introduced certain resolutions proposing amendments to the Constitution, which would have saved the Union and which received every Southern vote except the South Carolina senators, who had withdrawn. They proposed to adopt, in effect, the Missouri compromise line, to prohibit Congress from abolishing the slave trade between the states, or slavery in places where the United States had exclusive jurisdiction, or in the District of Columbia, without the consent of Maryland and of the slaveholders, and proposed a more effectual provision for the recovery of fugitive slaves. For these, a substitute was offered by Mr. Clark, of New Hampshire, declaring, amongst other things, that the provisions of the Constitution are ample for the preservation of the Union, and the resolutions of Mr. Crittenden were voted down, and the substitute adopted by a united vote of the Republicans. Says Lunt: "The vote of the Republican members of the Senate was a blank denial of the necessity of compromise, and showed, of course, that they had deliberately made up their minds to refuse any negotiation." (Lunt's Origin of the War, p. 411).

The adoption of Mr. Crittenden's resolutions, it was said by Mr. Douglass, would have saved every Southern State except South Carolina. Undoubtedly such would have been the effect of a general agreement upon these resolutions between the two sections. But did the Republicans desire it? It would seem not from the postscript to Mr. Chandler's letter to Governor Blair: "Some of the manufacturing states think that a fight would be awful. Without a little bloodletting, this Union will not, in my opinion, be worth a curse." This was from a senator from Michigan, a man of much influence in his party.

Virginia, not yet giving up her hope of preserving the Union, interposed to call "a peace conference." Resolutions were adopted by this body, composed of able and eminent men from the different states, very similar to Mr. Crittenden's, which met with no better success. Under these circumstances what were the slaveholding states to do? In 1790 they kept quiet, because they "relied upon the virtue of Congress that they would do nothing without constitutional authority." Was such a faith any longer rational? Had not the conduct of the free states proved that the guarantees of the Constitution upon the subject of slavery were no longer of the slightest avail to them? Upon that subject the majority in Congress, who were from these states, assumed whatever power they wanted.

Could the minority rely upon the constitution to protect any of their rights, if it suited the passions or the interests of the majority to invade them? Our government was fast being revolutionized, and becoming one of a despotic majority of numbers; the limitations of a written constitution fast proving themselves to be without the defense of the political power to enforce them. Had the South the slightest hope of attaining any increase of that power? It had proved itself unable to do this in the past: what was the hope for the future? Lunt (p. 363) says with justice: "That it is impossible to regard the proceedings of the Chicago convention in any other light than as equivalent to a proclamation of absolutely hostile purposes against the Southern section of the country. They were not, technically, a declaration of war, to be conducted by arms, simply because they proposed only to use the pacific force of superior numbers, in order to deprive the minority of its rights under the Constitution." (Lunt's Origin of the War, p. 362).

Indeed, one of its resolutions was amended so as to declare: "When, in the course of human events, it becomes necessary for one people to dissolve the political bands which have connected them with one another, and to assume among the powers of the earth the separate and equal station to which the laws of nature and nature's God entitle them, a decent respect to the opinions of mankind requires that they should declare the cause which impelled them to the separation." This amendment was introduced by a Pennsylvanian (Lunt, p. 358), and passed unanimously by the convention. (Ibid).

To what did this look but secession and separation? Did it not argue the consciousness of a purpose to drive the South to those extremities? What else could the South do but separate, if possible, from the majority which ruled the government, and were animated by such feelings? Mr. Webster, the great apostle of Union in 1851, had said: "I do not hesitate to say and repeat, that if the Northern States refuse willfully or deliberately to carry into effect that part of the Constitution which respects the restoration of fugitive slaves, the South would no longer be bound to keep the compact. A bargain broken on one side is broken on all sides." (Lunt, p. 321).

Had not the precise case occurred? Had not the North deliberately and persistently refused to carry into effect that part of the Constitution? Was the South bound any longer to keep the compact, according to this high authority? In this opinion of Mr. Webster, Mr. Jefferson undoubtedly concurred. Says Lunt, p. 203: "Mr. Jefferson took a different view of the subject, and it is proper to give his opinion as stated by Mr. John Q. Adams (who appears to have agreed with him) in his eulogy on Mr. Madison. Mr. Adams said: 'Concurring in the doctrines that the separate states have a right to interpose in cases of palpable infractions of the Constitution by the government of the United States, and that the alien and sedition acts presented a case of such infraction, Mr. Jefferson considered them as absolutely null and void, and thought the state legislatures competent, not only to declare but to make them so, to resist their execution within their respective borders by physical force, and to secede from the Union, rather than to submit to them, if attempted to be carried into execution by force.'"

On the 2nd of March, 1861, Mr. Greeley declared: "We have repeatedly said, and we once more insist, that the great principle embodied by Jefferson in the Declaration of Independence, 'that governments derive their just powers from the consent of the governed,' is sound and just, and that if the Slave States, the Cotton States, or the Gulf States only, choose to form an independent nation, they have a moral right to do so." (Lunt, p. 388-9).

Is it strange that those states concurred in this opinion? They believed that the government was now in hands which were fast converting it into one of a majority of numbers with unlimited powers. Did the South enter into any such Union as that? Had not her leaders constantly declared that in their opinion this was the worst of all forms of government, and if she was willing to stake life, liberty and property on the effort to escape it, did she not thereby demonstrate the earnestness of her conviction of her right to escape, and that her faith had been plighted to a very different instrument, by which she refused any longer to be bound to those who were seeking under its name to destroy the rights which it guaranteed to her, and force her to subserve the purposes of those who were seeking to ruin and degrade her own citizens, her men, women and children.

Who drove the South to these extremities? The very men who accuse her of treason. When she accepted the contest, to which she was thus virtually invited on terms of contumelious threat and reproach, she was threatened with being wiped out and annihilated by the superior resources of her antagonist, with whom it was vain and foolish to contend, so unequal were the strength and resources of the two parties. It is true that the South parted in bitterness, but it was in sadness of spirit also. She did not wish – certainly, Virginia did not desire it – if she could maintain her rights within the Union.

Probably few men foresaw the extent or the bitterness of the war. Surely it was a mighty contest to have been waged by two parties of such unequal strength in numbers and resources, with such a promise of success to the weaker for nearly four years, and doubtless there were periods during that time when those who provoked that trial by battle regretted that they had done so.

> *"There is no longer any room for hope. We must fight. I repeat it, Sir, – we must fight. An appeal to arms and to the God of battles is all that is left us."*

The South at last fell from physical exhaustion – the want of food, clothes, and the munitions of war; she yielded to no superiority of valor or of skill, but to the mere avoirdupois of numbers. Physically, she was unable to stand up under such a weight of human beings, gathered from wherever they could be called by appeals to their passions or bought by a promise to supply their necessities.

It is said that after the battle of the Second Cold Harbor, where Grant so foolishly assailed Lee in his lines, and where his dead was piled in thousands after his unsuccessful attack, the northern leaders were ready to have proposed peace, but were prevented by some favorable news from the southwest. They did not propose peace except upon terms of unconditional submission. The South being forced to accept those terms to obtain it, the North was not afraid to avow its purposes and carry them out. Slavery was abolished without compensation, and slaves were awarded equal rights with their masters in the government.

It was the fear of these results which drove the South into the war. Experience proved that this fear was reasonable. The war was alleged as the excuse for such proceedings; but can any man doubt that the North would have done the same thing if all constitutional restraints upon the power of the majority had been peaceably removed. To submit peaceably to the unlimited power of the majority was plainly to submit to these consequences or any other action which this majority may strongly desire to take.

It is sought to be excused, I know, by assuming that these things were done with the assent of the South. That these constitutional amendments represent the well considered opinion of any respectable party in the South, there is none so infatuated as to believe. They were accepted as the terms of the conqueror, and so let them be considered by all who desire to know the true history of their origin.

To introduce hostile and conflicting statements in the formation of the public opinion, by which the action of the South was to be regulated, might, indeed, weaken and injure that section, but how it would help or benefit the North is yet to be seen, if it should so turn out. I think I have shown that the South had good reason to believe that the North meditated the infliction of these things, and that there was but little hope of finding any defense against them in the Constitution.

The alacrity with which she put these designs into execution so soon as our conquest enabled her to do so, proves that we did not suspect her wrongfully. The

South had either to acquiesce in this oppression tamely and submissively, or fight to avert it. According to Mr. Webster, she had the constitutional right to do this; according to Mr. Greeley, she had the moral right to do this. She fought to avert these injuries, and because she was unwilling to remain under the government of a majority with unlimited powers. What this latter change threatens remains to be seen.

Congress has already undertaken by her civil rights bill to regulate social intercourse amongst her people in the states. Will Congress undertake to prescribe fast days, enforce temperance and take charge of the police laws of the states and the towns? These are questions which posterity must answer. Will they have no other remedy against this despotism but to substitute for it the one-man power. They at least will be in no doubt as to the causes, and history will be equally clear as to what parties forced it upon us.

"There is no longer any room for hope. We must fight. I repeat it, Sir – we must fight. An appeal to arms and to the God of battles is all that is left us." So said and thought Patrick Henry, in reply to the British exactions upon the colonies. So thought, too, the people of the Confederate States, and they did fight. They waged a war for which history has no parallel against such odds in resources and numbers. Borne down by odds, against which it was almost vain to contend, we were bound to submit, and they have taken from us that which, in my opinion, it will be found

Not enriches them,
But leaves us poor indeed

Had the South permitted her property, her constitutional rights and her liberties to be surreptitiously taken from her without resistance and made no moan, would she not have lost her honor with them? If the alternative were between such a loss and armed resistance, is it surprising that she preferred the latter?

★ ★ ★

STATES RIGHTS

By Jefferson Davis. *The Southern Historical Society Papers, Volume 14*, pages 408-409.

Among the less-informed persons at the North there exists an opinion that the Negro slave at the South was a mere chattel, having neither rights nor immunities protected by law or public opinion. Southern men knew such was not the case, and others desiring to know could readily learn the fact. On that error the lauded story of Uncle Tom's Cabin was founded, but it is strange that a utilitarian and shrewd people did not ask why a slave, especially valuable, was the object of privation and abuse? Had it been a horse they would have been better able to judge and would most probably have rejected the story for its improbability. Many attempts have been made to evade and misrepresent the exhaustive opinion of Chief Justice Taney in the Dred Scott case, but it remains unanswered.

From the statement in regard to Fort Sumter, a child might suppose that a foreign army had attacked the United States – certainly could not learn that the State of South Carolina was merely seeking possession of a fort on her own soil, and claiming that her grant of the site had become void.

The tyrant's plea of necessity to excuse despotic usurpation is offered for the unconstitutional act of emancipation and the poor resort to prejudice is invoked in the use of the epithet "rebellion" – a word inapplicable to states generally and most especially so to the sovereign members of a voluntary union. But, alas for their ancient prestige, they have even lost the plural reference they had in the Constitution, and seem so small to this utilizing tuition as to be described by the neutral pronoun "it!" Such language would be appropriate to an imperial Government, which in absorbing territories required the subjected inhabitants to swear allegiance to it.

Ignorance and artifice have combined so to misrepresent the matter of official oaths in the United States that it may be well to give the question more than a passing notice. When the "sovereign, independent States of America" formed a constitutional compact of union it was provided in the sixth article thereof that the officers "of the United States and of the several states shall be bound by oath or affirmation to support this Constitution," and by the law of June 1, 1789, the form of the required oath was prescribed as follows: "I, AB ,.do solemnly swear or affirm (as the case may be) that I will support the Constitution of the United States."

That was the oath. The obligation was to support the Constitution. It created no new obligation, for the citizen already owed allegiance to his respective state and through her to the Union of which she was a member. The conclusion is unavoidable that those who did not support, but did not violate the Constitution, were they who broke their official oaths. The General government had only the powers delegated to it by the states. The power to coerce a state was not given but emphatically refused. Therefore, to invade a state, to overthrow its government by force of arms, was a palpable violation of the Constitution, which officers had sworn to support, and thus to levy war against states which the Federal officers claimed to be, notwithstanding their ordinances of secession, still in the Union, was the treason defined in the third section of the third article of the Constitution, the only treason recognized by the fundamental law of the United States.

'When our forefathers assumed for the several states they represented a separate and equal station among the powers of the earth, the central idea around which their political institutions were grouped was that sovereignty belonged to the people, inherent and inalienable; therefore; that governments were their agents, instituted to secure their rights, and deriving their just powers from the consent of the governed, whence they draw the corollary that whenever any form of government becomes destructive of these ends it is the right of the people to alter or abolish it,' etc. What was meant by the word "people" in this connection is manifest from the circumstances. It could only authoritatively refer to the distinct communities who, each for itself, joined in the declaration and in the concurrent act of separation from the government of Great Britain.

By all that is revered in the memory of our Revolutionary sires, and sacred in the principles they established, let not the children of the United States be taught that our Federal government is sovereign; that our sires, after having, by a long and bloody war, won community-independence, used the power, not for the end sought, but to transfer their allegiance, and by oath or otherwise bind their posterity to be the subjects of another government, from which they could only free themselves by force of arms.

To subscribe, send an email to:
thestainlessbanner@gmail.com
or visit our website at www.thestainlessbanner.com

Subscription is free.

The Stainless Banner

An e-zine dedicated to the armies of the Confederacy

Volume 3, Issue 10
November, 2012

WE'RE THE BOYS WHO WENT AROUND MCCLELLAN

The battle of "Seven Pines," or "Fair Oaks," had been fought with no result. The temporary success of the Confederates early in the engagement, had been more than counter balanced by the reverses they sustained on the second day, and the two armies lay passively watching each other in front of Richmond. At this time, the cavalry of Lee's army was commanded by General J.E.B. Stuart, and this restless officer conceived the idea of flanking the right wing of the Federal army near Ashland and moving around to the rear, to cross the Chickahominy River at a place called Sycamore Ford in New Kent County, march over to the James River, and return to the Confederate lines near Deep Bottom, in Henrico County. In carrying out this plan, Stuart would completely encircle the army of General McClellan. At the time of this movement, the writer was adjutant of the 9th Virginia cavalry.

When the orders were issued from headquarters directing the several commands destined to form the expedition to prepare three days' rations, and the ordnance officers to issue sixty rounds of ammunition to each man, I remember the surmises and conjectures as to our destination. The officers and men were in high spirits in anticipation of a fight, and when the bugles rang out "Boots and Saddles," every man was ready. The men left behind in camp were bewailing their luck, and those forming the detail for the expedition were elated at the prospect of some excitement. "Good-bye, boys; we are going to help old Jack drive the Yanks into the Potomac," I heard one of them shout to those left behind.

On the afternoon of June 12th, we went out to the Brooke turnpike, preparatory to the march. The cavalry column was the 9th

What's Inside:	
Stuart's Original Report	9
Around McClellan	20
Seven Days Battle	32
Death of Latane	38
Jeb Stuart	40

www.thestainlessbanner.com

Virginia, commanded by Colonel W.H.F. Lee, the 1st Virginia, led by Colonel Fitz Lee, and the Jeff Davis Legion, under Colonel W.H.F. Martin. A section of the Stuart Horse Artillery, commanded by Captain Pelham, accompanied the expedition. The whole numbered twelve hundred men. They passed in bivouac in the vicinity of Ashland, and orders were issued enforcing strict silence and forbidding the use of fires, as the success of the expedition would depend upon secrecy and celerity.

On the following morning, at the break of dawn, the troopers were mounted and the march was begun without a bugle blast, and the column headed direct for Hanover Court House, distant about two hours' ride. Here we had the first sight of the enemy. A scouting party of the 5th U.S. Cavalry was in the village, but speedily decamped when our troops were ascertained to be Confederates. One prisoner was taken after a hot chase across country. We now moved rapidly to Hawes's Shop, where a Federal picket was surprised and captured without firing a shot. Hardly had the prisoners been disarmed and turned over to the provost guard when the Confederate advance was driven in upon the main body by a squadron of Federal cavalry, sent out from Old Church to ascertain by reconnaissance whether the report of a Confederate advance was true or false.

General Stuart at once ordered Colonel W.H.F. Lee, commanding the regiment leading the column, to throw forward a squadron to meet the enemy. Colonel Lee directed Captain Swann, chief of the leading squadron of his regiment, to charge with the saber. Swann moved off at a trot, and, turning a corner of the road, saw the enemy's squadron about two hundred yards in front of him. The order to charge was given, and the men dashed forward in fine style. The onset was so sudden that the Federal cavalry broke and scattered in confusion. The latter had a start of barely two hundred yards, but the Confederate yell that broke upon the air lent them wings, and only a few fell into our hands. The rest made their escape after a chase of a mile and a half.

Now the road became very narrow, and the brush on either side was a place so favorable for an ambuscade that Captain Swann deemed it prudent to draw rein and sound the bugle to recall his men. Stuart, who had been marching steadily with main body of the Confederate column, soon arrived at the front, and the advance guard, which I had all along commanded, was directed to move forward again. I at once dismounted the men, and pushed forward up a hill in my front. Just beyond the hill, I ran into a force of Federal cavalry drawn up in column of fours, ready to charge. Just as my advance-guard was about to run into him, I heard their commanding officer give the order to charge. I fell back and immediately noticed General Stuart in the presence of the enemy.

Captain Latane, commanding a squadron of the 9th Virginia, was directed to move forward and clear the road. He moved up the hill at a trot, and when in sight of the enemy in the road gave the command to charge, and with a yell the men rushed forward. At the top of the hill,

> *"Good-bye, boys; we are going to help old Jack drive the Yanks into the Potomac."*

simultaneously with Latane's order to charge, a company of Federal cavalry, deployed as skirmishers in the woods on the right of the road, were stampeded, and rushed back into the woods to make good their retreat to their friends. The head of Latane's squadron, then just fairly up the hill, was in the line of their retreat and was separated from the rest of the squadron, cut off by the rush of the Federals, and borne along with them up the road toward the enemy.

I was riding at the side of Latane, and just at the time when the Federal company rushed back into the road Captain Latane fell from his horse, shot dead. Though rush of the Federals separated myself and six of the leading files of the squadron from our friends, and we were borne along by the flying Federals. Although the Federal cavalry both in front and rear were in full retreat, our situation was perilous in the extreme. Soon we were pushed by foes in our rear into the ranks of those in our front, and a series of hand-to-hand combats ensued. To shoot or to cut us down was the aim of every Federal as he neared us, but we did what we could to defend ourselves. Every one of my comrades was shot or cut down, and I alone escaped unhurt. After having been borne along by the retreating enemy for perhaps a quarter of a mile, I leaped my horse over the fence into the field and so got away.

Now came the rush of the Confederate column, sweeping the road clear. At this point my regiment was relieved by the 1st Virginia, and Colonel Lee continued the pursuit. The Federals did not attempt to make a stand until they reached Old Church. Here their officers called a halt, and made an attempt to rally to defend their camp. Fitz Lee soon swept them out and burned their camp. They made no other attempt to stand, and we heard no more of them as an organized body, but many prisoners were taken as we passed along. We had surprised them, taken them in detail, and far outnumbered them at all points. The Federal forces, as we afterward learned, were commanded by General Philip St. George Cooke, father-in-law to General Stuart, to whom the latter sent a polite message. The casualties in this skirmish were slight – one man killed on each side, and about fifteen or twenty wounded on the Confederate side, mostly saber cuts.

We halted for a short time at Old Church, and the people of the neighborhood, hearing of our arrival, came out to greet us and wish us Godspeed. They did not come empty-handed, but brought whatever they could snatch up on the spur of the moment, rightly supposing that anything to allay hunger or thirst would be acceptable to us. Some of the ladies brought bouquets, and presented them to the officers as they marched along. One of these was given to General Stuart, who, always gallant, vowed to preserve it and take it into Richmond. He kept his promise.

We were soon far in rear of McClellan's army, which lay directly between us and Richmond. It was thought probable that the Federal cavalry was concentrating in our rear to cut off our retreat. We kept straight on, by Smith's store, through New Kent County to Tunstall's station, on the York River Railroad. I had been in charge of the Confederate advance guard up to the time when Colonel Fitz Lee came to the front, with the 1st Virginia relieved the 9th of that duty.

Well down in New Kent County, General Stuart sent for me again to the front. Hurrying on, I soon reached the head of the column where I found the general, and was directed by him to take thirty men as an advance guard, and to precede the column by about half a mile. Further, I was directed to halt at the road running from the mills to the White House long enough to cut the telegraph wire on that road; thence to proceed to Tunstall's station on the York River Railroad, at which place, the prisoners had informed the general, a company of Federal infantry was posted. At Tunstall's station I was directed to charge the infantry, disperse or capture them, cut the telegraph, and obstruct the railroad.

Here was our point of danger. Once across the railroad, we were comparatively safe. But in possession of the railroad, with its rolling-stock, the enemy could easily throw troops along its line to any given point. However, no timely information had been furnished to the Federal general. We moved with such celerity that we carried with us the first news of our arrival.

Pushing forward at a trot, and picking up straggling prisoners every few hundred yards, the advance-guard at length reached the telegraph road. At this point we overtook an ordnance wagon, heavily loaded with canteens and Colt's revolvers. The horses had stalled in a mud-hole, and the driver, cutting them out from the wagon, made his escape. The sergeant in charge stood his ground and was captured.

Here was a prize indeed, as in those days we were poorly armed. In order to save time, a man furnished with an ax was sent to cut the telegraph wire, while the rest of the party was engaged in rifling the wagon. While these operations were in progress, a body of Federal cavalry, suddenly turning a bend in the road, made their appearance. As soon as the Federal officer in command saw us he called a halt, and, standing still in the road, seemed at a loss to know what to do. His men drew their sabers, as if about to charge, but they did not come on. By this time the telegraph had been out and the wagon disposed of. Our men were hastily mounted and formed into column of fours, with drawn sabers, ready for any emergency. There we stood, eyeing each other, about two hundred yards apart, until the head of the main Confederate column came in sight. The Federals retreated down the road leading to the White House.

> *The men were weary and hungry, and the horses almost exhausted by the long fast and severe exercise.*

One man of the Federal party was sent back along the road to Tunstall's station, now only about half a mile off. I supposed, of course, that this messenger was sent to warn the Federal troops at Tunstall's of our approach. I was, however, afterward informed that he galloped through Tunstall's but never stopped, and when some one called to him, "What's to pay?" he dashed along, calling out, at the top of his voice, "Hell's to pay!"

The road now being clear, we marched on briskly, and arriving we charged down upon it with a yell. We could see the enemy scattered about the building and lounging around before we charged them. The greater part scattered for cover and were pursued by our people. I pushed straight for the station-house, where I found the captain of the company of infantry, with thirteen of his men, standing in front of the building, but with no arms in their hands.

Only one of them seemed disposed to show fight. He ran to the platform where the muskets were stacked, and, seizing one of them, began to load. Before he could ram his cartridge home, a sweep of the saber, in close proximity to his head, made him throw down his gun, and, jumping into a ditch, he dodged under the bridge over the railroad and made his escape. I had no time to pursue him; but, turning to look after the others, met the captain, who, sword in hand, advanced and surrendered himself and his company as prisoners of war. I then proceeded to obstruct the railroad.

To do this effectually, I caused a tree to be cut down which was standing on the side of the road. It fell across the railroad. In addition to this, I placed across the tracks an oak-sill about a foot square and fourteen feet long. I had barely time to do this before a train from the direction of Richmond came thundering down. At this time General Stuart, with the main body, arrived at the station. The engine driver of the coming train, probably seeing the obstructions on the track and a large force of cavalry there, suspected danger, and, being a plucky fellow, put on all steam, and came rushing down. The engine, striking the obstructions, knocked them out of the way and pressed on without accident. General Stuart had dismounted a number of his men, and posted them on a high bank overlooking a cut in the road, just below the station, through which the train was about to pass. They threw in a close and effective fire upon the passing train, loaded with troops. Many of these were killed and wounded.

It was now the second night since leaving camp, and haversacks with which we started from camp had long since been emptied. The march had been so rapid that there was little opportunity of foraging for man or beast. Except a little bread and meat, brought out to the column by the country people as we passed along, we had had nothing since daybreak. The men were weary and hungry, and the horses almost exhausted by the long fast and severe exercise. As soon as a proper disposition had been made of the prisoners and of the captured horses and mules, the column moved on. Down through New Kent County, to a place called New Baltimore, we marched as rapidly as our condition would permit. I was still in the command of the advance-guard, marching some distance ahead of the column, and had orders to halt at this point and await the coming up of the main body. Fortunately, an enterprising Yankee had established a store here, to catch the trade of all persons passing from McClellan's army to his base of supplies at the White House. He had crackers, cheese, canned fruits, sardines, and many other dainties dear to the cavalryman; and in the brief hour spent with him, we, of the advance, were made new men. I fear little was left to cheer and to invigorate those in the rear. The main body arriving, "forward" was the order-straight down through New Kent to Sycamore Ford on the Chickahominy.

A beautiful full moon lighted our way and cast weird shadows across our path. Expecting each moment to meet the enemy, every bush in the distance looked like a sentinel, and every jagged tree bending over the road like a vedette. Marching all night, we arrived at the ford between daybreak and sunrise; and here our real troubles began. To our chagrin, we found the stream swollen by recent rains almost out of its banks, and running like a torrent. No man or horse could get over without swimming, and it happened that the entrance to the ford on our side was below the point at which we had to come out on the other side. Therefore, we had to swim against the current. Owing to the mud, it was not practicable for any number of horses to approach the river at any point except by the road leading to the ford.

We therefore tried it there for two long hours. The 9th Cavalry made the trial. After repeated efforts to swim the horses over we give up, for we had crossed over only seventy-five men and horses in two hours. While we were trying to reach the opposite bank Stuart came up, and, finding the crossing at this point impracticable, rode off to find another farther down the river. At a point about one mile below, known as Forge Bridge, he succeeded in throwing across one branch of the river a bridge strong enough to bear the artillery, and upon which the men, having been dismounted, could walk. Here the approach on our side was higher up stream than the point at which we would come out on the other side. So the horses were formed into a column of fours, pushed into the water, and, swimming down stream, they easily landed on the other side. After a few horses had been crossed in this manner we found no difficulty, the others following on quite readily.

The column was now upon an island formed by the two branches of the Chickahominy, and to reach the mainland it was necessary to cross the other branch of that river. This was, however, accomplished, but with some difficulty. The ford at this crossing was at that time very deep, and the river out of its banks and overflowing the flats to the depth of about two feet for at least a half-mile. At this place the limber of a caisson stuck fast in the mud, and we left it.

On leaving the river, General Stuart directed me to take charge of the rearguard, and, when all had crossed, to burn the bridge. In accordance with these orders, I directed the men to collect piles of fence rails, heap them on the bridge, and set them afire. By my orders the horses had been led some distance back from the river into the brush, where they were concealed from view. The men were lounging about on the ground when the bridge fell in. I was seated under a tree on the bank of the river, and at the moment that the hissing of the burning timbers of the bridge let me know that it had fallen into the water, a rifle-shot rang out from the other side, and the whistling bullet cut off a small limb over my head, which fell into my lap. The shot was probably fired by some scout who had been following us, but who was afraid to fire until the bridge was gone. With a thankful heart for his bad aim, I, at once, withdrew the men and pushed on after the column. When I came to the ford, I found it necessary to swim the horses a short distance, it having been deepened by the crossing of such a large body of horse. Soon the column was in sight, and the march across Charles City County to the James River was made as vigorously as the jaded horses were able to stand. The men, though weary and hungry; were in fine spirits, and jubilant over the successful crossing of the Chickahominy. About sunset we neared the James, at the plantation of Colonel Wilcox. Here we rested for about two hours, having marched into a field of clover, where the horses ate their all. In the twilight, fires were lighted to cook the rations just brought in by our foragers.

We were now twenty-five miles from Richmond, on the "James River Road." Had the enemy been aware of our position, it would have been easy for him to throw a force between us and Richmond, and so cut us off. But the Federal General was not well served by his scouts, nor did his cavalry furnish him with accurate information of our movements. Relying upon the mistakes of the enemy, Stuart resolved to march straight on into Richmond by the River road on which we now lay. To accomplish this with the greater safety, it was necessary for him to march at once. Accordingly, I was ordered to take the advance guard and move out. As soon as the cravings of hunger were appeased, sleep took possession of us. Although in the saddle and in motion, and aware that the safety of the expedition depended on great vigilance in case the enemy should be encountered, it was hard to keep awake. I was constantly falling asleep, and awaking with a start when almost off my horse. This was the condition of every man in the column. Not one had closed his eyes in sleep for forty-eight hours.

The full moon lighted us on our way as we passed along the River road, and frequently the windings of this road brought us near to and in sight of the James River, where lay the enemy's fleet. In the gray twilight of the dawn of Sunday, we passed the Double Gates, Strawberry Plains, and Tilghman's gate in succession. At Tilghman's, we could see the masts of the Fleet, not far off. Happily for us, the banks were high, and I imagine they had no lookout in the rigging, and we passed by unobserved. The sight of the enemy's fleet had aroused us somewhat, when "Who goes there!" rang out on the stillness of the early morning. The challenger proved to be a vedette of the 10th Virginia Cavalry, commanded by Colonel J. Lucius Davis, who was picketing that road. Soon I was shaking hands with Colonel Davis and receiving his congratulations. Then we crossed the stream by the jug factory, up toward "New Market" heights, by the drill-house, and about a mile beyond we called halt for a little rest and food. From this point the several regiments were dismissed to their respective camps.

We lost one man killed and a few wounded, and no prisoners. The most important result was the confidence the men had gained in themselves and in their leaders. The country rang out with praises of the men who had raided entirely around General McClellan's powerful army, bringing prisoners and plunder from under his very nose. The Southern papers were filled with accounts of the expedition, none accurate, and most of them marvelous.

By W.T. Robins, Colonel, C.S.A, Battles and Leaders of the Civil War, Volume 2, starting on page 271.

Do You Have a Book on the War You Are Longing to Publish?

THE STAINLESS BANNER PUBLISHING COMPANY
A Full-Service Small Press Dedicated to the
Preservation of Southern Heritage and History

History ✯ Memoir ✯ Biography ✯ Novel ✯ Alternate History

What does The Stainless Banner Publishing Company offer you?

- ✯ Professional editing
- ✯ Dynamic covers
- ✯ Hard cover or paperback
- ✯ Competitively priced books
- ✯ High royalties paid monthly
- ✯ Your book on Amazon, Barnes & Noble and many other outlets
- ✯ Free advertising

What is my investment?

Nothing! **The Stainless Banner Publishing Company** is a full service small press. Once your book is published, your only investment will be for any inventory you wish to keep on hand.

Where do I send my query letter?

No need for an agent or a query letter. Just send an email telling us about your book and include the first three chapters in the body of the email. You will receive a prompt reply, usually within 48 hours. Send your email to books@thestainlessbanner.com

The Stainless Banner Publishing Company
www.thestainlessbanner.com

Jeb Stuart's Original Report of the Seven Days Battles

Headquarters Cavalry Brigade
Near Richmond, Virginia, July 14, 1862.
Colonel R. H. CHILTON,
Assistant Adjutant General,
Headquarters – Department of Virginia.

Colonel:

In compliance with the orders of the commanding general I have the honor to submit the following report of the operations of my command from June 26 to July 10, embracing the series of battles with the Federal forces before Richmond:

The part assigned to my command is set forth in General Orders, No. 75 (confidential), of June 6, and I beg leave to congratulate the commanding general upon the signal fulfillment by our army of what was planned in that order of battle, so much so that the order itself affords a very correct history of the battle.

My command on the morning of the 26th ultimo consisted of First Virginia Cavalry, Colonel Fitzhugh Lee; Third Virginia Cavalry, Colonel T.F. Goode; Fourth Virginia Cavalry, Captain Chamberlayne; Fifth Virginia Cavalry, Colonel T. L. Rosser; Ninth Virginia Cavalry, Colonel W.H.F. Lee; Tenth Virginia Cavalry, Colonel J. Lucius Davis; Cobb Legion Cavalry, Colonel T.R.R. Cobb; Jeff. Davis Legion, Lieutenant Colonel W. T. Martin; Stuart Horse Artillery, Captain John Pelham; a squadron of Hampton Legion Cavalry, Captain Scrivener [Screven] (attached to Fifth Virginia); three companies First North Carolina Cavalry, Lieutenant Colonel [James B.] Gordon.

The Third Virginia Cavalry was directed to observe the Charles City road; the Fifth Virginia and detachment First North Carolina Cavalry to watch the enemy's movements toward James River, and notify the commander nearest at hand of any attempt of the enemy to move across from White Oak Swamp to the James, and to harass and delay him en route till our forces could fall upon him.

The Tenth Virginia Cavalry was placed in reserve on the Nine-mile road.

With the remainder of my command, including the Horse Artillery, I marched late on the 25th, without baggage, equipped

> *It is proper to remark here that the commanding general had, on the occasion of my late expedition to the Pamunkey, imparted to me his design of bringing Jackson down upon the enemy's right flank and rear, and directed that I should examine the country with reference to its practicability for such a move.*

in light marching order and three days' rations in haversacks, and crossing Jackson's line of march after he had encamped, so as not to interrupt his progress, placed myself on his left flank, near Ashland.

It is proper to remark here that the commanding general had, on the occasion of my late expedition to the Pamunkey, imparted to me his design of bringing Jackson down upon the enemy's right flank and rear, and directed that I should examine the country with reference to its practicability for such a move. I therefore had studied the features of the country very thoroughly, and knew exactly how to conform my movements to Jackson's route. As that part of my former mission was confidential I made no mention of it in my former report, but it is not, I presume, out of place to remark here that the information obtained then and reported to him verbally convinced the commanding general that the enemy had no defensive works with reference to attack from that direction, the right bank of the Totopotomoy being unoccupied; that his forces were not disposed so as successfully to meet such an attack, and that the natural features of the country were favorable to such a descent.

General Jackson was placed in possession of all these facts. Having bivouacked near Ashland for the night, on the morning of the 26th – the Jeff. Davis Legion and Fourth Virginia Cavalry having joined me here from an advanced position of observation on South Anna, which effectually screened Jackson's movements from the enemy – my command swept down upon Jackson's left.

Extending its observations as far as the Pamunkey River road, passing Taliaferro's Mill, where the enemy had a strong picket, which fled at our approach, I reached General Jackson s line of march at the cross-roads at Dr. Shelton's in advance of his column. From Taliaferro's Mill to this point there was constant skirmishing between the enemy's pickets and my advance guard, Colonel Lee's (Company D, sharpshooters) First Virginia Cavalry, displaying the same courage and address which has already distinguished it on many occasions, killing and wounding several of the enemy without suffering any loss.

> *I continued my march by way of Beulah Church, taking several prisoners enroute to Cold Harbor, where I found General Jackson.*

At Dr. Shelton's I awaited the arrival of General Jackson, sending a squadron in advance (Captain Irving, First Virginia Cavalry) to seize and hold the bridge at the Totopotomoy. The enemy, anticipating us, had torn up the bridge and held the opposite bank and obstructed the road, without, however, making any determined stand. Captain W.W. Blackford, Corps of Engineers, assigned to duty with my command, set about repairing the bridge, and in half an hour, with the details furnished him, the bridge was ready.

Passing Pole Green Church, General Jackson's march led directly toward the crossing of Beaver Dam Creek, opposite Richardson's. Reaching that point, he bivouacked for the night and I disposed my command on both his flanks and rear, with five squadrons on picket, looking well toward Cold Harbor and Old Church.

About sundown the enemy made his appearance near Jackson's flank, on the Old Church road, but a few rounds of shell put him to flight, and my pickets on that road were not disturbed during the night.

The next morning, General Jackson moving directly across Beaver Dam, I took a circuitous route to turn that stream, turning down, first, the Old Church road, both aiming for Old Cold Harbor, and directing my march so as to cover his left flank, he having formed at Beaver Dam a junction with the divisions which marched by way of Mechanicsville.

All day we were skirmishing with, killing and capturing, small detachments of the enemy's cavalry, mostly the Lancers, Colonel Rush. Passing Bethesda Church, I sent the Blakely gun, of the Horse Artillery, and a portion of my command, under Colonel Martin, off to the left to see if any force was about Old Church. Colonel Martin found nothing but some flying cavalry, and I continued my march by way of Beulah Church, taking several prisoners enroute to Cold Harbor, where I found General Jackson. He directed me to take position on his left in reserve.

I kept a squadron in observation down the Old Church road, on the Dispatch road, and made dispositions for action whenever opportunity might offer. Owing, however, to the nature of the ground, the position of the enemy in a wood, and the steadiness of our own troops, the cavalry proper had no hand-to-hand conflict with the enemy, though subject to the severe ordeal of a raking artillery fire from guns beyond its reach.

Vedettes placed on our left kept me advised of the enemy's operations, and about 5:00 or 6:00 p.m. a movement of artillery was observed and reported on the road from Grapevine Bridge. The only artillery under my command being Pelham's Stuart Horse Artillery, the 12-pounder Blakely and Napoleon were ordered forward to meet this bold effort to damage our left flank.

The Blakely was disabled at the first fire, the enemy opening simultaneously eight pieces, proving afterward to be Weed's and Tidball's batteries. Then ensued one of the most gallant and heroic feats of the war. The Napoleon gun, solitary and alone, received the fire of those batteries, concealed in the pines on a ridge commanding its ground, yet not a man quailed, and the noble captain directing the fire himself with a coolness and intrepidity only equaled by his previous brilliant career. The enemy's fire sensibly slackened under the determined fire of this Napoleon, which clung to its ground with unflinching tenacity. I had an opportunity of calling General Jackson's attention to the heroic conduct of the officers and men of this piece, and later he, by his personal efforts, reinforced it with several batteries of rifle pieces, which, firing, advanced en échelon about dark and drove the enemy from his last foothold on the right.

I received information that General D.H. Hill was pursuing the enemy down that road at the point of the bayonet. Expecting a general rout, I immediately joined my cavalry and dashed down the road leading by Dr. Tyler's to its intersection with the White House road, about three miles. It was quite dark, but no evidence of retreat or other movement could be detected on that road, so, leaving a squadron for observation at that point, I returned to Cold Harbor with the main body late at night.

Early in the morning that squadron was so burdened with prisoners, mostly of the Regular Army – among others Major Delozier Davidson, commanding Fourth U.S. Infantry – that I had to reinforce it.

Being sent for by the general commanding at his headquarters, at New Cold Harbor, I galloped up, leaving my command prepared for instant service. I received from the commanding general instructions to strike for the York River Railroad at the nearest point, so as to cut the enemy's line of communication with the York and intercept his retreat. General Ewell's division (infantry) was put in motion for the same object, and Colonel Lee, of the Ninth, with his regiment, preceded him as advance guard, finding en route two fine rifle pieces of artillery abandoned by the enemy.

With the main body of cavalry I pursued a parallel route, and arriving near Dispatch, passed the head of General Ewell's column, and pushing a squadron of Cobb Legion Cavalry rapidly forward, surprised and routed a squadron of the enemy's cavalry, they leaving in their hurried departure the ground strewn with carbines and pistols. They fled in the direction of Bottom's Bridge.

I directed the immediate tearing up of the track and cutting the wire, which was done in a very few minutes, and the result reported to General Ewell and to the commanding general.

General Ewell decided to await further orders at Dispatch. I determined to push boldly down the White House road, resolved to find what force was in that direction and, if possible, rout it. A train of forage wagons with a few cavalry as escort was captured before proceeding far and farther down several sutler's establishments.

The prominent points on the roads were picketed by cavalry, all of which fled at our approach, and long before the column of cavalry had reached half-way to the White House the fleeing pickets had heralded the approach of what no doubt appeared to their affrighted minds to be the whole Army of the Valley, and from the valley of the Pamunkey a dense cloud of smoke revealed the fact of the flight and destruction in the path of a stampeded foe.

All accounts agreed that Generals Stoneman and Emory, with a large command of cavalry, infantry, and artillery, had gone in the direction of the White House, where Casey was said to be in command. I found no resistance till I reached Tunstall's Station; here I found a vacated field work and captured a cavalry flag near it. This work, as well as the evidences of recent encampments along the line of railroad, showed that one of the great results anticipated from my late expedition – the detaching a large force to protect the enemy's line of communication---had been accomplished.

At the crossing of Black Creek near this place the enemy had a squadron drawn up on the farther bank in line of battle and what appeared to be artillery on a commanding height beyond. He had destroyed the bridge over this difficult stream, whose abrupt banks and miry bed presented a serious obstacle to our progress. The artillery was ordered up to the front and a few well-directed rounds of shell dispersed the squadron, as well as disclosed in a scrambling race an adroitly-formed ambuscade of dismounted men on the banks of the stream, and produced no reply from what was supposed to be artillery.

A small party of dismounted men under the daring Captain Farley soon gained the farther bank and scoured the woods beyond, while the ever-ready and indefatigable Blackford set to work to repair the crossing. It was dark, however, before it could be finished, and we slept on our arms till morning, finding ample corn for our jaded horses at Tunstall's Station.

The conflagration raged fearfully at the White House during the entire night, while explosions of shells rent the air. I was informed that 5,000 men held the place. Early next morning I moved cautiously down, catching the scattered fugitives of the day before as we advanced, till, coming in plain view of the White House at a distance of a quarter of a mile, a large gunboat was discovered lying at the landing.

I took the precaution to leave the main body about 2 miles behind, and proceeded to this point with a small party and one piece of artillery. Colonel W.H.F. Lee, the proprietor of this once beautiful estate, low in ashes and desolation, described the ground and pointed out all the localities to me, so that I was convinced that a few bold sharpshooters could compel the gunboat to leave.

I accordingly ordered down about seventy-five, partly of First Virginia Cavalry (Litchfield's Company D), and partly Jeff. Davis Legion and Fourth Virginia Cavalry. They were deployed in pairs, with intervals of forty paces, and were armed with rifle carbines. They advanced boldly on this monster, so terrible to our fancy, and a body of sharpshooters was sent ashore from the boat to meet them.

Quite a determined engagement of skirmishers ensued, but our gallant men never faltered in their determination to expose this Yankee buggaboo called gunboat. To save time, however, I ordered up the howitzer, a few shells from which, fired with great accuracy and bursting directly over her decks, caused an instantaneous withdrawal of sharpshooters and precipitate flight under full headway of steam down the river. The howitzer gave chase at a gallop, the more to cause the apprehension of being cut off below than of really effecting anything. The gunboat never returned.

The command was now entirely out of rations and the horses without forage, and I had relied on the enemy at the White House to supply me with these essentials I was not disappointed, in spite of their efforts to destroy everything. Provisions and delicacies of every description lay in heaps, and the men regaled themselves on the fruits of the tropics as well as the substantials of the land. Large quantities of forage were left also.

An opportunity was here offered for observing the deceitfulness of the enemy's pretended reverence for everything associated with the name of Washington, for the dwelling-house was burned to the ground, and not a vestige left except what told of desolation and vandalism.

Nine large barges loaded with stores were on fire as we approached; immense

> *During the morning I received a note from the commanding general directing me to watch closely any movement of the enemy in my direction and to communicate what my impressions were in regard to his designs.*

numbers of tents, wagons, cars in long trains loaded and five locomotives, a number of forges, quantities of every species of quartermaster's stores and property, making a total of many millions of dollars – all more or less destroyed.

During the morning I received a note from the commanding general directing me to watch closely any movement of the enemy in my direction, and to communicate what my impressions were in regard to his designs. I replied that there was no evidence of a retreat of the main body from the position before Richmond down the Williamsburg roads, and that I had no doubt the enemy since his defeat was endeavoring to reach the James as a new base, being compelled to surrender his connection with the York.

If the Federal people can be convinced that this was a part of McClellan's plan, that it was in his original design for Jackson to turn his right flank and our generals to force him from his strongholds, they certainly can never forgive him for the millions of public treasure that his superb strategy cost the nation. He had no alternative left, and, possessed with the information that his retreat was not progressing toward the York, the commanding general knew as well as McClellan himself that he must seek the only outlet left.

It took the remainder of Sunday to ration my command and complete the destruction of some property I was apprehensive the enemy might return and remove, but I sent that day a regiment (First Virginia Cavalry, Colonel Fitz Lee) across to observe the enemy's movements from Bottom's Bridge to Forge Bridges.

On Monday I moved my whole command in the same direction, except one squadron (Cobb Legion), which was left at the White House. Colonel Lee, First Virginia Cavalry, was stationed near Long Bridge, and the remainder near Forge Bridge. The former reported the enemy's pickets visible on the other side, and at the latter place I observed a force of infantry and two pieces of artillery. The Napoleon was left with Colonel Lee, but it was disabled at the first shot. The Blakely being disabled at Cold Harbor left me with only 12-pounder howitzers (one section being present). Captain Pelham engaged the enemy across the Chickahominy with these, and after a spirited duel against one rifle piece and one howitzer the enemy was driven from his position with the loss of two men and two horses killed, we escaping unhurt. The infantry abandoned their knapsacks in their hurry to depart. I tried in vain to ascertain by scouts the enemy's force beyond, and it being now nearly dark, we bivouacked again.

During the entire day Colonel Lee, of the First, as also the main body, captured many prisoners, but none seemed to know anything of the operations of the army. One was a topographical engineer.

At 3.30 a.m. next morning, I received a dispatch from Colonel Chilton, the hour of his writing being omitted, stating that the enemy had been headed off at the intersection of the Long Bridge and Charles City roads and that his destination seemed for the present fixed, and expressing the commanding general's desire for me to cross the Chickahominy and co-operate with the forces on that side, suggesting Grapevine Bridge as the most suitable point. I asked the courier when it was written. He replied at 9:00 p.m., which point of time was after the heavy firing in the direction of White Oak Swamp Bridge had ceased, and I believe, therefore, that the status of the enemy referred to was subsequent to the heavy firing. I therefore started at once for Bottom's Bridge, eleven miles distant, pushing on rapidly myself.

Arriving at Bottom's Bridge I found our troops had passed down. Galloping on to White Oak Swamp Bridge I found many on the march and saw at once that from the lack of firing in front and the rapid rate of march the only way I could cooperate with the main body was by retracing my steps (fortunately the head of my column had not passed Bottom's Bridge) and crossing at the Forge Bridge to come up again on Jackson's left. I wrote a note to General Jackson to apprise him of this intention and hurried back to carry it out.

I found upon reaching Forge Bridge a party of Hunford's Second Virginia Cavalry, who informed me of the route taken by Jackson's column, and pushed on to join him, fording the river.

Passing Nance's shop about sundown, it was dark before we reached Rock's house, near which we stampeded the enemy's picket without giving it time to destroy a bridge further than to pull off the planks. I aimed for Haxall's Landing, but soon after leaving Rock's encountered picket fires, and a little way beyond saw the light of a considerable encampment. There was no other recourse left but to halt for the night, after a day's march of forty-two miles.

As it was very dark very little could be seen of the country around, but I had previously detached Captain Blackford to notify General Jackson of my position and find where he was. He returned during the night, having found our troops, but could not locate General Jackson's hue. I ascertained also that a battle had been raging for some time and ceased about an hour after I reached this point.

My arrival could not have been more fortunately timed, for, arriving after dark, its ponderous march, with the rolling artillery, must have impressed the enemy's cavalry, watching the approaches to their rear, with the idea of an immense army about to cut off their retreat, and contributed to cause that sudden collapse and stampede that soon after occurred, leaving us in possession of Malvern Hill, which the enemy might have held next day much to our detriment.

It is a remarkable fact worthy of the commanding general's notice that in taking the position I did in rear of Turkey Creek I acted entirely from my own judgment, but was much gratified the next day on receiving his note to find that his orders were to the same effect, though failing to reach me till next morning, after its execution.

Early next morning I received orders from General Jackson, unless you had otherwise directed, to take position near his left. Not yet apprised of the enemy's move in the night I proceeded to execute this order, and having halted the column near Gatewood's, where Colonels Rosser, Baker, and Goode, with their respective regiments joined my command, I went forward to reconnoiter.

Meeting with General Jackson, we rode together to Dr. Poindexter's, where we met Major Meade and Lieutenant Samuel R.

> *Early next morning I received orders from General Jackson, unless I had been otherwise directed, to take position near his left.*

Johnston, of the Engineers, who had just made, in the drenching rain, a personal examination of the enemy's position and found it abandoned.

I galloped back to my command and put it in motion for Haxall's, hoping there to intercept the enemy's column. The Jeff. Davis Legion preceded and soon reached the river road in rear of Turkey Creek, capturing scores of the discomfited and demoralized foe at every turn – wagons, tents, arms, and knapsacks abandoned, and the general drift of accounts given by the prisoners spoke eloquently of the slaughter and rout that will make Malvern Hill memorable in history.

Colonel Martin dashed off with a few men toward Haxall's, and in plain view of the monitor captured one of her crew on shore and marched back several other prisoners; the very boldness of the move apparently transfixing the enemy's guns.

Appreciating the importance of knowing the enemy's position with reference to Shirley I endeavored to gain the fork of roads near that point, but it was strongly defended by two regiments of infantry – a prisoner captured near by said Sickles' brigade. The indications were plain, however, that the enemy had gone below that point.

The day was consumed in collecting prisoners and arms back toward Malvern Hill, the road from which was thoroughly blockaded, and in harassing the enemy's rear, which, in spite of his good position, was very effectually done by Colonel Martin with one of Pelham's howitzers, causing marked havoc and confusion in his ranks.

I also reconnoitered in the direction of Charles City Courthouse, with the view to fall on his flanks if still in motion. The result of the last was to the effect that at 10:00 a.m. no part of his forces had reached Charles City Courthouse. I therefore sent down that night a howitzer toward Westover, under Captain Pelham, supported by Irving's squadron First Virginia Cavalry, with orders to reach the immediate vicinity of the river road below, so as to shell it if the enemy attempted to retreat that night.

A squadron (Cobb Legion) was left near Shirley and the main body bivouacked contiguous to oat fields, of necessity our sole dependence for forage since leaving the White House, but the regiments were warned that the pursuit might be resumed at any moment during the night should Captain Pelham's reconnaissance apprise us of a continuance of the retreat.

During the night Captain Pelham wrote to me that the enemy had taken position between Shirley and Westover, nearer the latter, and described the locality, the nature of Herring Creek, on the enemy's right, and indicated the advantage to be gained by taking possession with artillery of Evelington Heights – a plateau commanding completely the enemy's encampment.

I forwarded his report to the commanding general through General Jackson, and proceeded at once to the ground with my command, except one regiment (the Ninth Virginia Cavalry, Colonel W.H.F. Lee), which was ordered down the road by Nance's shop, and thence across toward Charles City Courthouse, so as to extend my left and keep a lookout toward Forge Bridge, by which route I was liable to be attacked in flank and rear by Stoneman, should he endeavor to form a junction by land with McClellan.

I found Evelington Heights easily gained. A squadron in possession vacated without much hesitation, retreating up the road, the only route by which it could reach Westover, owing to the impassability of Herring Creek below Roland's Mill.

Colonel Martin was sent around farther to the left and the howitzer brought into action in the river road to fire upon the enemy's camp below. Judging from the great commotion and excitement caused below it must have had considerable effect.

We soon had prisoners from various corps and divisions, and from their statements, as well as those of citizens, I learned that the enemy's main body was there, but much reduced and demoralized. I kept the commanding general apprised of my movements, and I soon learned from him that Longstreet and Jackson were enroute to my support. I held the ground from about 9:00 a.m. till 2:00 p.m., when the enemy had contrived to get one battery into position on this side the creek. The fire was, however, kept up until a body of infantry was found approaching by our right flank.

I had no apprehension, however, as I felt sure Longstreet was near by, and although Pelham reported but two rounds of ammunition left, I held out, knowing how important it was to hold the ground till Longstreet arrived.

The enemy's infantry advanced and the battery kept up its fire. I just then learned that Longstreet had taken the wrong road and was then at Nance's shop, six or seven miles off. Pelham fired his last round, and the sharpshooters, strongly posted in the skirt of woods bordering the plateau, exhausted every cartridge, but had at last to retire; not, however, without teaching many a foeman the bitter lesson of death.

My command had been so cut off from sources of supply and so constantly engaged with the enemy that the abundant supply which it began with on June 26 was entirely exhausted. I kept pickets at Bradley's store that night, and remained with my command on the west side of the creek, near Phillips' farm. General Longstreet came up late in the evening; he had been led by his guide out of his proper route.

The next day, July 4, General Jackson's command drove in the enemy's advanced pickets. I pointed out the position of the enemy, now occupying, apparently in force, the plateau from which I shelled their camp the day before, and showed him the routes by which the plateau could be reached to the left, and submitted my plan for dispossessing the enemy and attacking his camp. This was subsequently laid before the commanding general. The enemy's position had been well reconnoitered by Blackford, of the Engineers, the day before from a close view, and farther on this day (July 4), demonstrating that his position was strong, difficult to reach except with rifle cannon, and completely flanked by gunboats; all which were powerful arguments, and no doubt had their due weight with the commanding general against renewing an attack thus far of unbroken successes against a stronghold where the enemy had been re-enforced beyond a doubt. The operations of my own command extended farther to the left, except one regiment (Cobb Legion Cavalry) which was directed to follow up the enemy's rear on the river road, and First North Carolina Cavalry, which remained in reserve near Phillips' farm.

The remainder of July 4 and 5 were spent in reconnoitering and watching the river.

On the afternoon of the 5th, Colonel S.D. Lee, of the artillery, reported to me with a battery of rifle guns, Squires' Washington Artillery, to which I added Pelham's Blakely, which had just returned from Richmond, for attacking transports on the river below the Federal forces. The point selected was Wilcox's Landing, which was reached after dark. The only transport which passed during the night was fired into with evident damage, but she kept on.

On the 6th the battery was augmented by two rifle pieces of Rogers' battery, and proceeded to Wayne Oaks, lower down the river.

During that night and next day (7th) the batteries commanded the river, seriously damaging several transports and compelling the crews from two to take to their small boats for the opposite shore, leaving one boat sinking. The batteries were subject to incessant firing from the gunboats, which invariably convoyed the transports, but Colonel Lee, whose report is very interesting, says no damage was done to the batteries, demonstrating, as was done at the White House, that gunboats are not so dangerous as is generally supposed.

On the afternoon of the 7th, the batteries returned to their camps, the men being much exhausted from loss of rest and continuous exertion.

During the 6th, 7th, and 8th, the enemy persistently annoyed our pickets on the river road below Westover, and with all arms of service tried to compel us to retire from that position. Colonel Rosser, commanding Fifth Virginia Cavalry, was present in charge of the post, and inspired his men with such determined resistance – arranging them so as to resist to best advantage – that the enemy failed in the effort within three-quarters of a mile of his main body and in his rear.

At sundown on the 8th, it being decided to withdraw our forces from before the enemy's position, the cavalry covered the withdrawal of the infantry, and prevented the enemy having any knowledge of the movement.

At daylight on the 9th the cavalry proceeded above Turkey Island Creek with the view to establish a line of cavalry outposts from the vicinity of Shirley across by Nance's shop to the Chickahominy.

On the 10th a portion of the cavalry was left on this duty, and the remainder, by direction of the commanding general, marched to a reserve camp.

I regret that the very extended field of operations of the cavalry has made this report necessarily long. During the whole period it will be observed that my command was in contact with the enemy. No opportunity occurred, however, for an overwhelming charge; a circumstance resulting first from the nature of the positions successively taken by the enemy in woods or behind swamps and ditches, he taking care to change position under cover of night, the distance being so short – only fifteen miles – as to be marched in one night. Added to this was the uncertainty of whether the enemy would attempt the passage of the Chickahominy where I awaited him, or under cover of a demonstration toward Chaffin's Bluff he would gain the James. The country being obscurely wooded and swampy his facilities for effecting the latter were great.

The portion of the cavalry operating under my instructions on the Richmond side of the Chickahominy was under the command at first of Colonel Rosser and afterward of Colonel Lawrence S. Baker, First North Carolina Cavalry. The latter made a gallant charge on the 30th ultimo at

Willis' Church with his and a portion of Colonel Goode's command but were repulsed with some loss. Their reports, enclosed, will give particulars of their operations.

Major Crumpier was mortally wounded and Captain Ruffin taken prisoner. For other casualties you are respectfully referred to Colonel Baker's report. During the series of engagements in which the portion of the brigade with me participated very few casualties occurred, notwithstanding frequent exposure to the enemy's fire.

During the whole period the officers and men exhibited that devotion to duty, thorough discipline, and efficiency which characterize regular troops, and claim at my hands the highest measure of praise and grateful acknowledgment.

Colonels. T.R.R. Cobb, Fitz Lee, W.H.F. Lee, and Lieutenant Col. W.T. Martin, under my immediate command, were frequently entrusted with distinct isolated commands and displayed that zeal and ability which entitle them to favorable notice and give evidence of capacity for higher trusts.

Captain John Pelham, of the Horse Artillery, displayed much signal ability as an artillerist, such heroic example and devotion in danger, and indomitable energy under difficulties in the movement of his battery, that, reluctant as I am at the chance of losing such a valuable limb from the brigade, I feel bound to ask for his promotion, with the remark that in either cavalry or artillery no field grade is too high for his merit and capacity. The officers and men of that battery emulated the example of their captain, and did justice to the reputation already won.

Captain William W. Blackford, of the Engineers, assigned to duty with me the day before the battles, was always in advance, obtaining valuable information of the enemy's strength, movements, and position, locating routes, and making hurried but accurate sketches. He is bold in reconnaissance, fearless in danger, and remarkably cool and correct in judgment. His services are invaluable to the advance guard of an army.

Captain J. Hardeman Stuart, Signal Corps, was particularly active and fearless in the transmission of orders at Cold Harbor, and deserves my special thanks for his gallant conduct.

Captain Norman R. Fitzhugh, assistant adjutant-general, chief of staff, though but recently promoted from the ranks, gave evidence of those rare qualities, united with personal gallantry, which constitute a capable and efficient adjutant-general.

Captain Heros von Borcke, assistant adjutant-general, was ever present, fearless and untiring in the zealous discharge of the duties assigned him.

Major Samuel Hardin Hairston, quartermaster, and Major Dabney Ball, commissary of subsistence, were prevented by their duties of office from participating in the dangers of the conflict, but are entitled to my thanks for the thorough discharge of their duties.

The following officers attached to my staff deserve honorable mention in this report for their valuable services: Captain Redmond Burke; Lieutenant John Esten Cooke, ordnance officer; Lieutenant J.T.W. Hairston, Lieutenant, Jones R. Christian, Third Virginia Cavalry; Lieutenant Chiswell Dabney, aide; CaptainS.W.D. Farley and W.E. Towles, volunteer aides, they having contributed their full share to whatever success was achieved by the brigade.

My escort did good service. Private Frank Stringfellow, Fourth Virginia Cavalry, was particularly conspicuous for gallantry and efficiency at Cold Harbor. The majority of the Hanover Company (G), Fourth Virginia Cavalry, possessing invaluable merits as guides, was distributed as such among the various generals. First Lieutenant D.A. Timberlake accompanied me, and from his intimate acquaintance with the country, as well as his personal bravery, was an indispensable aid to my march. His deeds of individual prowess in Hanover place him high among partisan warriors, and enabled us to know exactly the enemy's position and strength near Atlee's Station.

The reports of other commanders have not been received; should they be sent in subsequently they will be forwarded.

My command captured several thousand prisoners and arms, the precise number it being impossible to ascertain. The detachment of cavalry left at the White House secured much valuable public property, enumerated already.

I have the honor to be, colonel, your obedient servant,
J. E. B. STUART,
Brigadier-General

STUART RIDES AROUND MCCLELLAN

By John Esten Cooke. *Wearing of the Gray,* pages 165-181.

Those who that went with Stuart on his famous Ride around McClellan in the summer of 1862, just before the bloody battles of the Chickahominy, will ever forget the fun, the frolic, the romance – and the peril too – of that fine journey? Thinking of the gay ride now, when a century seems to have swept between that epoch and the present, I recall every particular, live over every emotion. Once more I hear the ringing laugh of Stuart and see the keen flash of the blue eyes under the black feather of the prince of cavaliers!

If the reader will follow me, he shall see what took place on this rapid ride, witness some incidents of this first and king of raids. The record will be that of an eyewitness, and the personal prominence of the writer must be excused as inseparable from the narrative.

I need not dwell upon the situation in June, 1862. All the world knows that, at that time, McClellan had advanced with his magnificent army of 156,000 men to the banks of the Chickahominy and pushing across had fought on the last day of May the bloody but indecisive battle of Seven Pines.

On the right it was a Confederate, on the left a Federal success; and General McClellan drew back, marshaled his great lines, darkening both the northern and southern banks of the Chickahominy and prepared for a more decisive blow at the Confederate capital, whose spires were in sight.

Before him, however, lay the Southern army, commanded now by Lee, who had succeeded Johnston, wounded at Seven Pines. The moment was favorable for a heavy attack by Lee. Jackson had just driven before him the combined forces of Shields and Fremont, and on the bloody field of Port Republic ended the great campaign of the Valley at one blow. The veterans of his command could now be concentrated on the banks of the Chickahominy against McClellan; a combined advance of the forces under Lee and Jackson might save the capital.

But how should the attack be made? In council of war, General Stuart told me he proposed an assault upon General McClellan's left wing from the direction of James River, to cut him off from that base. But this suggestion was not adopted; the defenses were regarded as too strong. It was considered a better plan to attack the Federal army on the north bank of the Chickahominy, drive it from its works and try the issue in the fields around Cold Harbor.

The great point was to ascertain if this was practicable, and especially to find what defenses, if any, the enemy had to guard the approach to their right wing. If these were slight, the attack could be made with fair prospects of success. Jackson could sweep around while Lee assailed the lines near Mechanicsville; then one combined assault would probably defeat the Federal force.

To find the character of the enemy's works beyond the stream – his positions and movements – General Stuart was directed to take a portion of his cavalry, advance as far as Old Church, if practicable, and then be guided by circumstances. Such were the orders with which Stuart set out about moonrise on the night, I think, of June __, upon this dangerous expedition.

As the young cavalier mounted his horse on that moonlight night, he was a gallant figure to look at. The gray coat buttoned to the chin; the light French saber balanced by the pistol in its black holster; the cavalry boots above the knee, and the brown hat with its black plume floating above the bearded features, the brilliant eyes, and the huge moustache, which curled with laughter at the slightest provocation – these made Stuart the perfect picture of a gay cavalier, and the spirited horse he rode seemed to feel that he carried one whose motto was to "do or die."

I chanced to be his sole companion as he galloped over the broad field near his headquarters, and the glance of the blue eyes of Stuart at that moment was as brilliant as the lightning itself. Catching up with his column of about 1,500 horsemen and two pieces of horse-artillery under Colonels William H.F. Lee, Fitz Lee, and William T. Martin, of Mississippi – a cavalier as brave as ever drew saber – Stuart pushed on northward as if going to join Jackson and reaching the vicinity of Taylorsville, near Hanover Junction, went that night into bivouac.

He embraced the opportunity, after midnight, of riding with Colonel W.H.F. Lee to "Hickory Hill," the residence of Colonel Williams Wickham – afterward

> *In a council of war, General Stuart told me he proposed an assault upon General McClellan's left wing from the direction of James River, to cut him off from that base.*

General Wickham – who had been recently wounded and paroled. Here he went to sleep in his chair after talking with Colonel Wickham, narrowly escaped capture from the enemy rear, and returning before daylight, advanced with his column straight upon Hanover Courthouse.

Have you ever visited this picturesque spot, reader? We looked upon it on that day of June – upon its old brick courthouse, where Patrick Henry made his famous speech against the parsons, its ancient tavern, its modest roofs, the whole surrounded by the fertile fields waving with golden grain – all this we looked at with unusual interest. For in this little bird's nest, lost as it were in a sea of rippling wheat and waving foliage, some Yankee cavalry had taken up their abode; their horses stood ready saddled in the street, and this dark mass we now gazed at furtively from behind a wooden knoll, in rear of which Stuart's column was drawn up ready to move at the word.

Before he gave the signal, the General dispatched Colonel Fitz Lee round to the right, to flank and cut off the party. But all at once the scouts in front were descried by the enemy; shots resounded; and seeing that his presence was discovered, Stuart gave the word, and swept at a thundering gallop down the hill. The startled blue birds, as we used to call our Northern friends, did not wait; the squadron on picket at the courthouse, numbering some one hundred and fifty men, hastily got to horse then presto! They disappeared in a dense cloud of dust from which echo some parting salutes from their carbines.

Stuart pressed on rapidly took the road to Old Church and near a place called Hawes' Shop, in a thickly wooded spot, was suddenly charged himself. It did not amount to much and seemed rather an attempt at reconnaissance. A Federal officer at the head of a detachment came on at full gallop, very nearly ran into the head of our column and then seeing the dense mass of gray coats, fired his pistol, wheeled short about and went back at full speed with his detachment.

Stuart had given, in his ringing voice, the order: "Form fours! Draw saber! Charge!" and now the Confederate people pursued at headlong speed uttering shouts and yells sufficiently loud to awaken the seven sleepers! The men were evidently exhilarated by the chase, the enemy just keeping near enough to make an occasional shot practicable.

A considerable number of the Federal cavalrymen were overtaken and captured, and these proved to belong to the company in which Colonel Fitz Lee had formerly been a lieutenant. I could not refrain from laughter at the pleasure which Colonel Fitz – whose motto should be "toujours gai" – seemed to take in inquiring after his old cronies. "Was Brown alive? Where was Jones? and Was Robinson sergeant still?" Fitz never stopped until he found out everything. The prisoners laughed as they recognized him. Altogether, reader, the interview was the most friendly imaginable.

The gay chase continued until we reached the Tottapotamoy, a sluggish stream, dragging its muddy waters slowly between rush clad banks, beneath drooping trees and this was crossed by a small rustic bridge. The line of the stream was entirely undefended by works. The enemy's right wing was unprotected. Stuart had accomplished the object of his expedition and afterward piloted Jackson over this very same road.

But to continue the narrative of his movements. The picket at the bridge had been quickly driven in and disappeared at a gallop. On the high ground beyond, Colonel W.H.F. Lee, who had taken the front, encountered the enemy. The force appeared to be about a regiment, and they were drawn up in line of battle in the fields to receive our attack. It came without delay. Placing himself at the head of his horsemen, Colonel Lee swept forward at the *pas de charge* and with shouts the two lines came together.

The shock was heavy, and the enemy – a portion of the old United States Regulars, commanded by Captain Royal – stood their ground bravely, meeting the attack with the sabre. Swords clashed, pistols and carbines banged, yells, shouts, cheers resounded; then the Federal line was seen to give back and take to headlong flight. They were pursued with ardor and the men were wild with this – to many of them – their first fight.

But soon after all joy disappeared from their faces, at sight of a spectacle which greeted them. Captain Latane of the Essex cavalry had been mortally wounded in the charge, and as the men of his company saw him lying bloody before them, many a bearded face was wet with tears. The scene at his grave afterward became the subject of Mr. Washington's picture, *The Burial of Latane*, and in his general order after the expedition, Stuart called upon his command to take for their watchword in the future "Avenge Latane!"

Captain Royal, the Federal commander, had also been badly wounded and many of his force killed. I remember passing a Dutch cavalryman who was writhing with a bullet through the breast and biting and tearing up the ground. He called for water, and I directed a servant at a house near by to bring him some. The last I saw of him, a destitute cavalryman was taking off his spurs as he was dying. War is a hard trade.

Fitz Lee immediately pressed on and burst into the camp near Old Church, where large supplies of boots, pistols, liquors, and other commodities were found. These were speedily appropriated by the men and the tents were set on fire amid loud shouts. The spectacle was animating; but a report having got abroad that one of the tents contained powder, the vicinity thereof was evacuated in almost less than no time.

We were now at Old Church, where Stuart was to be guided in his further movements by circumstances. I looked at him; he was evidently reflecting. In a moment he turned round to me and said: "Tell Fitz Lee to come along, I'm going to move on with my column." These words terminated my doubt, and I understood in an instant that the General had decided on the bold and hazardous plan of passing entirely round McClellan's army.

"I think the quicker we move now the better," I said, with a laugh.

"Right," was Stuart's reply; "tell the column to move on at a trot."

So at a rapid trot the column moved.

The gayest portion of the raid now began. From this moment it was neck or nothing, do or die. We had one chance of escape against ten of capture or destruction.

Stuart had decided upon his course with that rapidity, good judgment, and decision, which were the real secrets of his splendid efficiency as a leader of cavalry, in which capacity I believe that he has never been surpassed, either in the late war or any other. He was now in the very heart of the enemy's citadel, with their enormous masses upon every side. He had driven in their advanced force, passed within sight of the white tents of General McClellan's headquarters, burned their camps and ascertained all that he wished.

How was he to return? He could not cross the Pamunkey and make a circuit back; he had no pontoons. He could not return over the route by which he had advanced. As events afterward showed, the alarm had been given and an overpowering force of infantry, cavalry, and artillery had been rapidly moved in that direction to intercept the daring raider. Capture stared him in the face, on both of these routes – across the Pamunkey, or back as he came. He must find some other loophole of escape.

Such was the dangerous posture of affairs and such was the important problem which Stuart decided in five minutes. He determined to make the complete circuit of McClellan's army; and crossing the Chickahominy below Long Bridge, re-enter the Confederate lines from Charles City. If on his way, he encountered cavalry he intended to fight it. If a heavy force of infantry barred his way, he would elude, or cut a path through it. If driven to the wall and debarred from escape he did not mean to surrender.

A few days afterward I said to him: "That was a tight place at the river, General. If the enemy had come down on us, you would have been compelled to have surrendered."

"No," was his reply; "one other course was left."

"What was that?"

"To die game."

And I know that such was his intention. When a commander means to die game rather than surrender he is a dangerous adversary.

From Old Church onward it was terra incognita. What force of the enemy barred the road was a question of the utmost interest, but adventure of some description might be safely counted on. In about twenty-four hours I, for one, expected either to be laughing with my friends within the Southern lines, or dead, or captured. Which of these three results would follow seemed largely to depend upon the chapter of accidents.

At a steady trot now, with drawn sabers and carbines ready, the cavalry followed by the horse-artillery, which was not used during the whole expedition, approached Tunstall's Station on the York River railroad, the enemy's direct line of communication with his base of supplies at the White House.

Everywhere the ride was crowded with incident. The scouting and flanking parties constantly picked up stragglers and overhauled unsuspecting wagons filled with the most tempting stores. In this manner, a wagon, stocked with champagne and every variety of wines, belonging to a General of the Federal army, fell prey to the thirsty graybacks.

Still they pressed on. Every moment an attack was expected in front or rear. Colonel William T. Martin commanded the latter.

"Tell Colonel Martin," Stuart said to me, "to have his artillery ready and look out for an attack at any moment."

I had delivered the message and was riding to the front again, when suddenly a loud cry arose of Yankees in the rear! Every saber flashed, fours were formed, the men wheeled about, when all at once a stunning

> *"To die game."*

roar of laughter ran along the line; it was a canard. The column moved up again with its flanking parties well out. The men composing the latter were, many of them from the region, and for the first time for months saw their mothers and sisters.

These went quite wild at sight of their sons and brothers. They laughed and cried, and on the appearance of the long gray column instead of the familiar blue coats of the Federal cavalry, they clapped their hands and fell into ecstasies of delight. One young lady was seen to throw her arms around a brother she had not before met for a long time, bursting into alternate sobs and laughter.

The column was now skirting the Pamunkey, and a detachment hurried off to seize and burn two or three transports lying in the river. Soon a dense cloud rose from them, the flames soared up, and the column pushed on. Everywhere were seen the traces of flight – for the alarm of "hornets in the hive" was given. Wagons had turned over and were abandoned – from others the excellent army stores had been hastily thrown. This writer got a fine red blanket and an excellent pair of cavalry pantaloons for which he still owes the United States. Other things lay about in tempting array, but we were approaching Tunstall's, where the column would doubtless make a charge and to load down a weary horse was injudicious.

The advance guard was now in sight of the railroad. There was no question about the affair before us. The column must cut through whatever force guarded the railroad. To reach the lower Chickahominy, the guard here must be overpowered. Now was the time to use the artillery, and every effort was made to hurry it forward. But alas! it had got into a tremendous mud hole and the wheels were buried to the axle. The horses were lashed and jumped almost breaking the traces. The drivers swore; the harness cracked, but the guns did not move.

> *Suddenly in the midst of the tumult was heard the shrill whistle of a train coming from the direction of the Chickahominy. Stuart quickly drew up his men in a line on the side of the road, and he had no sooner done so than the train came slowly round a wooded bend and bore down.*

"Gat! Lieutenant," said a sergeant of Dutch origin to the brave Lieutenant McGregor, "it can't be done. But just put that keg on the gun, Lieutenant," pointing, as he spoke, to a keg of whiskey in an ambulance, the spoil of the Federal camp, "and tell the men they can have it if they only pull through!" McGregor laughed and the keg was quickly perched on the gun.

Then took place an exhibition of Herculean muscularity which would have delighted Guy Livingston. With eyes fixed ardently upon the keg, the powerful cannoneers waded into the mud hole up to their knees, seized the wheels of gun and caisson loaded down with ammunition, and just simply lifted the whole out and put them on firm ground. The piece whirled on – the keg had been dismounted – the cannoneers reveled in the spoils they had earned.

Tunstall's was now nearly in sight, and that good fellow Captain Frayser, afterward Stuart's signal officer, came back and reported one or two companies of infantry at the railroad. Their commander had politely beckoned to him as he reconnoitered, exclaiming in wheedling accents, full of Teutonic blandishment, "Koom yay!"

But this cordial invitation was disregarded! Frayser galloped back and reported, and the ringing voice of Stuart ordered "Form platoons! Draw saber! Charge!" At the word, the sabers flashed, a thundering shout arose and sweeping on in column of platoons, the gray people fell upon their blue adversaries, gobbling them up, almost without a shot. It was here that my friend Major F— got the hideous little wooden pipe he used to smoke afterward. He had been smoking a meerschaum when the order to charge was given; and in the rush of the horsemen, dropped and lost it. He now wished to smoke and seeing that the captain of the Federal infantry had just filled his pipe, leaned down from the saddle and politely requested him to surrender it.

"I want to smoke!" growled the Federal captain.

"So do I," retorted Major F—.

"This pipe is my property," said the captain.

"Oh! What a mistake!" responded the major politely, as he gently took the small affair and inserted it between his lips. Anything more hideous than the carved head upon it I never saw.

The men swarmed upon the railroad. Quick axes were applied to the telegraph poles, which crashed down, and Redmond Burke went in command of a detachment to burn a small bridge on the railroad near.

Suddenly in the midst of the tumult was heard the shrill whistle of a train coming from the direction of the Chickahominy. Stuart quickly drew up his men in a line on the side of the road, and he had no sooner done so than the train came slowly round a wooded bend and bore down. When within 200 yards, it was ordered to halt, but the command was not obeyed. The engineer crowded on all steam. The train rushed on, and then a thundering volley was opened upon the "flats" containing officers and men.

> *So Stuart pushed on.*

The engineer was shot by Captain Farley, of Stuart's staff, and a number of the soldiers were wounded. The rest threw themselves upon their faces; the train rushed headlong by like some frightened monster bent upon escape, and in an instant it had disappeared.

Stuart then reflected for a single moment. The question was, should he go back and attack the White House, where enormous stores were piled up? It was tempting, and he afterwards told me he could scarcely resist it. But a considerable force of infantry was posted there. The firing had doubtless given them the alarm, and the attempt was too hazardous. The best thing for that gray column was to set their faces toward home, and keep moving, well closed up both day and night, for the lower Chickahominy.

So Stuart pushed on. Beyond the railroad appeared a world of wagons, loaded with grain and coffee standing in the road abandoned. Quick work was made of them. They were all set on fire, and their contents destroyed. From the horse trough of one, I rescued a small volume bearing on the flyleaf the name of a young lady of Williamsburg. I think it was a volume of poems – poetic wagon drivers!

These wagons were only the vaunt couriers – the advance guard – of the main body. In a field beyond the stream, thirty acres were covered with them. They were all burned. The roar of the soaring flames was like the sound of a forest on fire. How they roared and crackled! The sky overhead, when night had descended, was bloody looking in the glare.

Meanwhile the main column had moved on, and I was riding after it when I heard the voice of Stuart in the darkness exclaiming with strange agitation: "Who is here?"

"I am," I answered.

He recognized my voice, he exclaimed: "Good! Where is Rooney Lee?"

"I think he has moved on, General."

"Do you know it?" came in the same agitated tone.

"No, but I believe it."

"Will you swear to it? I must know! He may take the wrong road, and the column will get separated!"

"I will ascertain if he is in front."

"Well, do so; but take care. You will be captured!"

I told the General I would gallop on forever till I found him, but I had not gone 200 yards in the darkness when hoof strokes in front were heard, and I ordered: "Halt! who goes there?"

"Courier from Colonel William Lee."

"Is he in front?"

"About a mile, sir."

"Good!" exclaimed the voice of Stuart, who had galloped up. I never heard in human accents such an expression of relief. If the reader of this has ever commanded cavalry, moving at night in an enemy's country, he will understand why Stuart drew that long, deep breath, and uttered that brief word, "Good!" Once separated from the main column and lost, good-by then to Colonel Lee!

Pushing on by large hospitals which were not interfered with, we reached at midnight the three or four houses known as Talleysville and here a halt was ordered to rest men and horses, and permit the artillery to come up. This pause was fatal to a sutler's store from which the owners had fled. It was remorselessly ransacked and the edibles consumed.

This historian ate in succession figs, beef-tongue, pickle, candy, tomato catsup, preserves, lemons, cakes, sausages, molasses, crackers and canned meats. In presence of these attractive commodities the spirits of many rose.

Those who in the morning had made me laugh by saying, "General Stuart is going to get his command destroyed. This movement is mad," now regarded Stuart as the first of men. The raid, as a feat of splendor and judicious daring, which could not fail in terminating successfully. Such is the difference in the views of the military machine, unfed and fed.

In an hour the column moved again. Meanwhile a little incident had happened which still makes me laugh. There was a lady living some miles off in the enemy's line whom I wished to visit, but I could not obtain the General's consent.

"It is certain capture," he said. "Send her a note by some citizen, say Dr. H ; he lives near here."

This I determined to do and set off at a gallop through the moonlight for the house, some half a mile distant, looking out for the scouting parties which were probably prowling on our flanks. Reaching the lonely house, outside the pickets, I dismounted, knocked at the front door, then the back, but received no answer. All at once, however, a dark figure was seen gliding beneath the trees, and this figure cautiously approached. I recognized the Doctor and called to him whereupon he quickly approached, and said, "I thought you were a Yankee!" and greeting me cordially, led the way into the house.

Here I wrote my note and entrusted it to him for delivery taking one from him to his wife, within our lines. In half an hour I rode away, but before doing so asked for some water, which was brought from the well by a sleepy, sullen, and insolent negro.

This incident was fruitful of woes to Dr. H! A month or two afterwards I met him looking as thin and white as a ghost.

"What is the matter?" I said.

"The matter is," he replied, with a melancholy laugh, "that I have been starving for three weeks in Fortress Monroe on your account. Do you remember that servant who brought you the water that night on Stuart's raid?"

"Perfectly."

"Well, the very next day he went over to the Yankee picket and told them that I had entertained Confederate officers and given you all information which enabled you to get off safely. In consequence I was arrested, carried to Old Point, and am just out!"

I rejoined the column at Talleysville just as it began to move on the road to Forge Bridge. The highway lay before us, white in the unclouded splendor of the moon. The critical moment was yet to come. Our safety was to turn apparently on a throw of the dice, rattled in the hand of Chance.

> *There was only one man who never desponded or bated one jot or tittle of the heart of hope. That was Stuart. I had never been with him in a tight place before, but from that moment, I felt convinced that he was one of those men who rise under pressure.*

The exhaustion of the march now began to tell on the men. Whole companies went to sleep in the saddle and Stuart himself was no exception. He had thrown one knee over the pommel of his saddle, folded his arms, dropped the bridle, and, chin on breast, his plumed hat drooping over his forehead was sound asleep. His surefooted horse moved steadily, but the form of the General tottered from side to side and for miles I held him erect by the arm.

The column thus moved on during the remainder of the night, the wary advance guard encountering no enemies and giving no alarm. At the first streak of dawn, the Chickahominy was in sight, and Stuart was spurring forward to the ford.

It was impassable! The heavy rains had so swollen the waters that the crossing was utterly impracticable! Here we were within a few miles of McClellan's army, with an enraged enemy rushing on our track to make us rue the day we had circumvented them and inflicted on them such injury and insult; here we were with a swollen and impassable stream directly in our front – the angry waters roaring around the half-submerged trunks of the trees – and expecting every instant to hear the crack of carbines from the rearguard indicating the enemy's approach! The situation was not pleasing.

I certainly thought that the enemy would be upon us in about an hour, and death or capture would be the sure alternative. This view was general. I found that cool and resolute officer, Colonel

William H.F. Lee, on the river's bank. He had just attempted to swim the river, and nearly drowned his horse among the tangled roots and snags. I said to him: "What do you think of the situation, Colonel?"

"Well, Captain," was the reply, in the speaker's habitual tone of cheerful courtesy, "I think we are caught."

The men evidently shared this sentiment. The scene upon the river's bank was curious and under other circumstances would have been laughable. The men lay about in every attitude, half-overcome with sleep, but holding their bridles and ready to mount at the first alarm. Others sat their horses asleep, with drooping shoulders. Some gnawed crackers; others ate figs, or smoked, or yawned. Things looked blue and that color was figuratively spread over every countenance.

When this writer assumed a gay expression of countenance, laughed, and told the men it was all right, they looked at him as sane men regard a lunatic! The general conviction evidently was that all right was the very last phrase by which to describe the situation.

There was only one man who never desponded, or bated one jot or tittle of the heart of hope. That was Stuart. I had never been with him in a tight place before, but from that moment I felt convinced that he was one of those men who rise under pressure. He was aroused, strung for the hard struggle before him, and resolute to do or die; but he was not excited. All I noticed in his bearing to attract attention was a peculiar fashion of twisting his beard, certain proof with him of surrounding peril. Otherwise he was cool and looked dangerous.

He said a few words to Colonel Lee, found the ford impassable and then ordering his column to move on, galloped down the stream to a spot where an old bridge had formerly stood. Reaching this point, a strong rearguard was thrown out, the artillery placed in position, and Stuart set to work vigorously to rebuild the bridge, determined to bring out his guns or die trying.

> *Standing in the boat beneath, Stuart worked with the men, and as the planks thundered down and the bridge steadily advanced, the gay voice of the General was heard humming a song.*

The bridge had been destroyed, but the stone abutments remained some thirty or forty feet only apart, for the river here ran deep and narrow between steep banks. Between these stone sentinels, facing each other, was an aching void which it was necessary to fill.

Stuart gave his personal superintendence to the work, he and his staff laboring with the men. A skiff was procured; this was affixed by a rope to a tree, in the mid-current, just above the abutments, and thus a movable pier was secured in the middle of the stream. An old barn was then hastily torn to pieces and robbed of its timbers. These were stretched down to the boat and up to the opposite abutment, and a foot-bridge was thus ready. Large numbers of the men immediately unsaddled their horses, took their equipments over, and then returning, drove or rode their horses into the stream, and swam them over. In this manner a considerable number crossed; but the process was much too slow. There, besides, was the artillery, which Stuart had no intention of leaving. A regular bridge must be built without a moment's delay, and to this work Stuart now applied himself with ardor.

Heavier blows resounded from the old barn. Huge timbers approached, borne on brawny shoulders and descending into the boat, anchored in the middle of the stream, the men lifted them across. They were just long enough; the ends rested on the abutments, and immediately thick planks were hurried forward and laid crosswise, forming a secure footway for the cavalry and artillery horses.

Standing in the boat beneath, Stuart worked with the men, and as the planks thundered down, and the bridge steadily advanced, the gay voice of the General was heard humming a song. He was singing carelessly, although at every instant an overpowering force of the enemy was looked for and a heavy attack upon the disordered cavalry.

At last the bridge was finished. The artillery crossed amid hurrahs from the men, and then Stuart slowly moved his cavalry across the shaky footway. A little beyond was another arm of the river, which was, however, fordable, as I ascertained and reported to the General; the water just deep enough to swim a small horse; and through this, as through the interminable sloughs of the swamp beyond, the head of the column moved.

The prisoners, who were numerous, had been marched over in advance of everything, and these were now mounted on mules, of which several hundred had been cut from the captured wagons and brought along. They were started under an escort across the ford and into the swamp beyond. Here, mounted often two on a mule, they had a disagreeable time; the mules constantly falling in the treacherous mud holes, and rolling their riders in the ooze. When a third swamp appeared before them, one of the Federal prisoners exclaimed, with tremendous indignation, "How many d—d Chicken-hominies are there, I wonder, in this infernal country!"

> **Stuart had thus eluded his pursuers and was over the Chickahominy in the hospitable county of Charles City.**

The rearguard, under Colonel W.H.F. Lee, had meanwhile moved down steadily from the high ground and defiled across the bridge. The hoofs clattered on the hasty structure, the head of the column was turned toward the ford beyond, the last squadron had just passed, and the bridge was being destroyed, when shots resounded on the opposite bank of the stream and Colonel Rush thundered down with his "lancers" to the bank.

He was exactly ten minutes too late. Stuart was over with his artillery, and the swollen stream barred the way, even if Colonel Rush thought it prudent to "knock up against" the 1,500 crack cavalry of Stuart. His men banged away at Colonel Lee and a parting salute whizzed through the trees as the gray column slowly disappeared.

A lady of New Kent afterwards told me that Colonel Rush stopped at her house on his return, looking weary, broken down, and out of humor. When she asked him if he had "caught Stuart," he replied, "No, he has gone at the back door. I only saw his rearguard as it passed into the swamp."

Stuart had thus eluded his pursuers and was over the Chickahominy in the hospitable county of Charles City. The gentlemen of the county, we afterwards heard, had been electrified by the rumor that Stuart was down at the river trying to get across and had built a hasty bridge for us lower down. We were over, however, and reaching Mr. C –'s, the General and his staff lay down on a carpet spread on the grass in the June sunshine and went to sleep. This was Sunday. I had not slept since Friday night, except by snatches in the saddle, and in going on to Richmond afterwards fell asleep every few minutes on horseback.

Two hours of slumber, however, made Stuart as fresh as a lark; and having eaten Mr. C – very nearly out of house and home, we pushed on all day.

At night the column stopped, and I thought the General would stop too; but he said, "I am going to Richmond tonight; would you like to ride with me?"

I was obliged to decline; my horse was worn out. Stuart set out by himself, rode all night, and before daylight had passed over the thirty miles. An hour afterwards, General Lee and the President knew the result of his expedition. The cavalry returned on the same day, moving slowly in front of the gunboats, which fired upon them; but no harm was done. Richmond was reached and amid an ovation from delighted friends, we all went to sleep.

Such was Stuart's ride around McClellan's army in those summer days of 1862. The men who went with him look back to it as the most romantic and adventurous incident of the war. It was not indeed so much a military expedition as a raid of romance – a "scout" of Stuart's with fifteen hundred horsemen!

It was the conception of a bold and brilliant mind and the execution was as fearless. "That was the most dangerous of all my expeditions," the General said to me long afterwards; "if I had not succeeded in crossing the Chickahominy, I would have been ruined, as there was no way of getting out."

The Emperor Napoleon, a good soldier, took this view of it; when tracing out on the map Stuart's route from Taylorsville by Old Church to the lower Chickahominy, he characterized the movement as that of a cavalry officer of the first distinction.

This criticism was only just, and the raid will live in history for three reasons:

1. It taught the enemy "the trick," and showed them the meaning of the words "cavalry raid." What General Kilpatrick, Sheridan, and others afterwards effected, was the work of the pupil following the master.

2. It was on a magnificent arena, to which the eyes of the whole world were attracted at the time.

3. In consequence of the information which Stuart furnished, General Lee, a fortnight afterwards, attacked and defeated General McClellan.

These circumstances give a very great interest to all the incidents of the movement. I hope the reader has not been wearied by my minute record of them. To the old soldiers of Stuart there is a melancholy pleasure in recalling the gay scenes amid which he moved, the exploits which he performed, the hard work he did. He is gone; but even in memory it is something to again follow his feather.

JEB STUART AT THE SEVEN DAYS BATTLES

By Major H.B. McClellan. *The Life and Campaigns of Major General JEB Stuart.*

In order that the movements of the Confederate cavalry during the Seven Days Battles around Richmond may be understood, it is necessary to relate the operations of those divisions of Lee's army with which it was in immediate connection.

The order of battle issued by General Lee on the 24th of June assumed that Jackson's command would be able to reach the vicinity of the Central Railroad on the 25th and be in position to turn the enemy's right flank early on the 26th. Jackson's march was, however, delayed to such an extent that he only reached the vicinity of Ashland on the night of the 25th. Here he was joined by Stuart with the 1st, 4th, and 9th regiments of Virginia Cavalry, the Cobb Georgia Legion, the Jeff Davis Legion, and the Stuart Horse Artillery. The 3rd and 5th regiments of Virginia Cavalry, the Hampton Legion, and the 1st North Carolina Cavalry were stationed on the right of the Confederate army, observing the country between the White Oak Swamp and the James River. The 10th Virginia Cavalry was held in reserve on the Nine Mile Road.

The positions held by the Federal army on the 25th of June were nearly the same as at the time of Stuart's reconnaissance. The three divisions of the 5th Corps under General Fitz John Porter occupied the north bank of the Chickahominy. Taylor's brigade of Franklin's Corps, which had constituted the extreme right at Mechanicsville, was withdrawn on the 19th and replaced by McCall's division of the 5th Corps. This appears to have been the only change on the Federal right wing since the 15th of June. The remainder of McClellan's forces extended south of the Chickahominy to the White Oak Swamp. Generals Stoneman and Emory observed the country north of Mechanicsville towards Atlee's Station and Hanover Courthouse with cavalry.

The official reports do not show what cavalry was under General Stoneman's command, but on the night of the 25th, he was reinforced by two regiments of infantry from Morell's division – the 18th Massachusetts and the 17th New York. Colonel H.S. Lansing, of the 17th New York, states that the cavalry force under General Stoneman consisted of two regiments and a light battery.

Leaving Ashland early on the 26th, Jackson pursued the Ashcake Road and crossed the Central Railroad about 10:00 a.m. Here, he met the first Federal cavalry

> *General Lee's plan of battle had contemplated an attack upon the enemy's positions in the vicinity of Mechanicsville at an early hour on the 26th, in which Jackson was to play the all-important part of turning the Federal right in their strong position on Beaver Dam Creek.*

picket or scout. Stuart covered his left flank by the march of his column and by scouts as far north as Hanover Courthouse. At Taliaferro's Mill, Stuart encountered a cavalry picket, which retired before him, skirmishing by way of Dr. Shelton's to the Totopotomoy.

A part of Stuart's command scouted the road past Enon Church to Hawes' Shop. At Dr. Shelton's, Stuart awaited the arrival of Jackson's column, having sent one squadron to seize the bridge over the Totopotomoy. The enemy, however, had burned this bridge and held the opposite bank until the arrival of the Texas brigade of Whiting's division, whose skirmishers crossed the stream and drove them away. The bridge was rebuilt, and Jackson's march was continued. His divisions rested for the night in the vicinity of Pole Green Church and Hundley's Corner, his left still covered by Stuart's cavalry.

General Lee's plan of battle had contemplated an attack upon the enemy's positions in the vicinity of Mechanicsville at an early hour on the 26th, in which Jackson was to play the all-important part of turning the Federal right in their strong position on Beaver Dam Creek. But as we have already seen, Jackson's march had been unexpectedly delayed and although nothing had been heard from him since early in the day, General A.P. Hill, at 3:00 p.m. crossed the Chickahominy at Meadow Bridge with five of his brigades and drove the enemy back upon their impregnable line on Beaver Dam Creek. This movement uncovered the bridge at Mechanicsville for Longstreet, who had been waiting since early in the morning for an opportunity to cross.

The Federal position on Beaver Dam was to be approached only by two roads; the one leading from Mechanicsville towards Cold Harbor. Field's, Archer's, and Anderson's brigades, of A.P. Hill's division, attacked the upper position, while Pender's brigade, aided by Ripley's of D.H. Hill's division, assailed the lower. Neither effort was attended with success and after sustaining heavy losses, the Confederate lines withdrew, at 9:00, from the unavailing contest.

Early the next morning the attack was renewed but without more favorable results. After two hours of fighting, the Federal troops were withdrawn to take position at Gaines' Mill. This movement was the necessary result of the march of Jackson's command, which now rendered the position at Beaver Dam untenable. It seems from General Trimble's report that Jackson might have reached this same point on the previous evening, and that it was within his power to have rendered efficient aid to the troops which were there engaged.

The Federal line on Beaver Dam was held, mainly, by two brigades of McCall's division, who, protected by their works, inflicted upon their assailants a loss probably ten times as great as they themselves suffered. The withdrawal of McCall on the morning of the 27th under fire, and in the presence of a superior force of the enemy, was conducted in a manner worthy of praise.

During the morning of the 27th, Longstreet and A.P. Hill moved down the Chickahominy towards Gaines' Mill, while D.H. Hill moved by way of Bethesda Church to Cold Harbor. Jackson crossed Beaver Dam Creek early in the morning, and advanced to Walnut Grove Church; then bearing to his left moved on Cold Harbor. Finding his road obstructed, he was compelled to make a still wider *détour* to the left, which threw him in the rear of D.H. Hill.

Meantime Stuart had covered the left of Jackson's march, and having thoroughly scoured the country toward the Pamunkey as far as Old Church, had advanced by way of Beulah Church and had taken position on Jackson's left in readiness to intercept the enemy should he attempt to retreat to the Pamunkey by way of Old Cold Harbor. The battle at Gaines' Mill was opened by A.P. Hill at about 2:30 p.m., and soon extended from right to left along the whole Confederate line.

On Jackson's line there was no opportunity to use artillery during the earlier part of the battle. Stuart was the first to find a suitable position. Observing, late in the evening, a movement of the enemy's artillery on the road from Grapevine Bridge, two of Pelham's guns, a twelve-pounder Blakely and a Napoleon, were ordered forward to meet it. The Blakely gun was disabled at the first fire, leaving the Napoleon to encounter alone the two batteries to which it was opposed. Pelham maintained the unequal contest with the same courage which subsequently, at Fredericksburg, called forth the praise of Lee and Jackson. By the personal efforts of General Jackson, whose attention was called to the position occupied by Pelham, he was reinforced by the batteries of Brockenborough, Carrington, and Courtney.

The design of the Federal commander was not yet manifest, and it was still deemed possible that he might attempt to retreat toward the Pamunkey River. When the Federal lines had been forced at Gaines' Mill and Cold Harbor, Stuart proceeded three miles still further to his left, to intercept any movement in that direction; but finding no evidences of a retreat, he returned the same night to Cold Harbor. Early the next morning, the 28th, General Ewell's division was sent down the Chickahominy to Dispatch Station, and the 9th Virginia Cavalry constituted his advance-guard. With his main body Stuart pursued a parallel route to the left and pushing ahead of Ewell's column surprised a squadron of the enemy's cavalry at Dispatch Station. The enemy retreated in the direction of Bottom's Bridge. Ewell remained at Dispatch Station during the rest of the day, and on the 29th moved to Bottom's Bridge. On the following day he rejoined his corps.

After Ewell had taken position at Dispatch Station on the 28th, Stuart determined to advance toward the White House. General Stoneman and General Emory had retired in that direction, and had occupied Tunstall's Station on the evening of the 27th, stationing pickets on the roads towards Dispatch Station. Stuart advanced to Tunstall's Station. Here he found that a fieldwork commanding the approaches to the station had been constructed since his recent visit on the 13th, which gave proof by its presence that one of the results desired in his late reconnaissance had been accomplished, and that a considerable force of the enemy had been detached to guard his communications.

Immediately beyond Tunstall's Station the enemy had destroyed the bridge over Black Creek, and there awaited Stuart's advance with cavalry and artillery posted on the hills beyond. The fire of Stuart's guns dispersed the cavalry, and Captain Farley, having gained the opposite bank with a few dismounted men, drove off the sharpshooters who commanded the bridge. Captain Blackford at once proceeded to rebuild, but it was after dark before a practicable crossing could be made.

Meantime Stoneman had sent his infantry to the White House, where, with all the infantry of General Casey's command, it was received on board transports and gunboats and moved down the river. At dark the evacuation of the White House Landing was completed. So far as their hasty departure permitted, the government property was destroyed by the Federal troops, and, last of all, the torch was applied to the home of Colonel W.H.F. Lee. It is but just to General Casey to state that he says in his report that this last act was performed without his knowledge and against his express orders.

Early the next morning, the 29th, Stuart moved cautiously toward the White House. He had reason to think that it was held by a considerable force of the enemy; nothing, however, was in sight but a Federal gunboat, the *Marblehead*, which occupied a threatening position in the river. Imagination had clothed the gunboat with marvelous terrors, and at this stage of the war there was nothing which inspired more of fear than the screech of its enormous shells. Stuart determined to illustrate to his command its real character. Leaving his main body about two miles in the rear, he advanced with seventy-five men selected from the 1st and 4th Virginia Cavalry and the Jeff Davis Legion.

These men were armed with rifle carbines. Deployed in pairs, with intervals of forty paces, they advanced across the open ground to attack the boat, from which a party of sharpshooters was promptly sent on shore to meet them. A lively skirmish ensued, during which Stuart brought up one of Pelham's howitzers and placed it in position to command the gunboat. Pelham's shells were soon exploding directly over her decks. To this fire she was unable to reply; for while her guns might throw shot far inland, they could not be brought to command that point of the bank where the howitzer was posted. The skirmishers were soon withdrawn to the boat, and under a full head of steam she disappeared down the river, followed as far as was practicable by the impudent and tormenting howitzer.

> *During the 30th, Stuart moved his command to Long and Forge bridges, and at the latter place, he bivouacked that night.*

Although the destruction of Federal property at the White House had been great, it was by no means complete, and sufficient remained to supply both men and horses of Stuart's command. Having sent Colonel Fitz Lee with the 1st Virginia Cavalry to observe the Chickahominy from Bottom's Bridge to Forge Bridge, Stuart remained at the White House for the rest of the day. The information which he had been able to send to General Lee was of importance, for it was demonstrated that the enemy had abandoned his base on the Pamunkey and was seeking a new one on the James.

Late in the afternoon of the 29th, Magruder engaged the enemy at Savage Station. Jackson's route lay to the flank and rear of this position, but he was unable to participate in the battle, being delayed by the necessity of rebuilding Grapevine Bridge, which the enemy had destroyed on his retreat. He succeeded in crossing the

Chickahominy during the night, and by noon on the 30th had advanced to White Oak Swamp.

On the 29th, a reconnaissance was made on the Charles City Road by five companies of the 1st North Carolina Cavalry and the 3rd Virginia Cavalry under the command of Colonel L.S. Baker of the 1st North Carolina. The enemy's cavalry was discovered on the Quaker Road, and a charge, the 1st North Carolina leading, drove it back to Willis' Church. Here the head of the column was greeted by a fire of artillery and infantry, and Colonel Baker was forced to retire, having sustained a loss of sixty-three in killed, wounded and missing. His charge had led him unwittingly into the presence of a large force of infantry.

During the 30th, Stuart moved his command to Long and Forge bridges, and at the latter place he bivouacked that night. At half past three the next morning, July 1, he received orders to cross the Chickahominy at Grapevine Bridge and connect with Jackson. He moved at once up the Chickahominy but on reaching Bottom's Bridge discovered that the army had passed on to the south, and that the only practicable way for him to connect with Jackson was to retrace his steps and cross at one of the lower fords.

Turning the head of his column about, he returned to Forge Bridge, where he found the 2nd Virginia Cavalry, Colonel T.T. Munford, which at that time belonged to the Valley cavalry and had accompanied Jackson's command. Fording the river at this point, Stuart pressed on past Nance's shop to Rock's house, near which he encountered a picket, which he pursued until within sight of the camp-fires of a large body of the enemy. Here he encamped for the night.

While Stuart was thus occupied, Longstreet and A.P. Hill had fought the bloody battle of Frayser's Farm, or Glendale, on the afternoon of the 30th. Could Jackson have participated in this battle the result must have been fatal to the Federal army. He had reached White Oak Swamp at midday, but found the bridge destroyed and the passage disputed by a large force of infantry and artillery. After sending Munford's regiment of cavalry across, Jackson decided that the passage was impracticable. The Reverend Dr. Dabney seems to be of the opinion that this was, perhaps, the sole occasion on which the great Stonewall did not accomplish all that lay within his power. On the afternoon of the next day, July 1, Jackson, D.H. Hill, Huger, and Magruder fought the battle of Malvern Hill.

Early on July 2, Stuart took position at Gatewood's on Jackson's left; but as soon as it was known that the enemy had abandoned the position at Malvern Hill, Stuart started down the river to ascertain his location. Lieutenant Colonel W.T. Martin, of the Jeff Davis Legion, was sent in advance. To his command the 4th Virginia Cavalry had been temporarily added because all of the field officers of that regiment were disabled. When opposite Haxall's, Colonel Martin and a few of his men proceeded to the river bank, where, in full sight of the *Monitor* and the *Galena*, which were lying in the river not one hundred yards distant, he captured a sailor belonging to the *Monitor*, drove off thirty mules from the open field, and, scouring the adjacent woods, retired in safety with 150 prisoners. Privates Volney Metcalf and William Barnard are especially mentioned by Colonel Martin for boldness in this affair.

At the Cross Roads near Shirley, Stuart

found the rearguard of the enemy in such force that he was unable to move it. He spent the remainder of the day in collecting prisoners toward Malvern Hill, and in reconnoitering toward Charles City Courthouse.

Having ascertained that the enemy had not moved in that direction, Captain Pelham was sent with one howitzer and Irving's squadron, of the 1st Virginia Cavalry, with orders to take position in the vicinity of Westover, and shell the enemy should he attempt to move down the river road during the night. Pelham discovered the position of the Federal army at Westover and informed Stuart of the advantages which might possibly be gained by occupying Evelington Heights, a plateau which commanded the enemy's encampment.

Pelham's report was received during the night, and Stuart at once moved his command, as it suggested, having forwarded the information to the commanding general, through General Jackson, and occupied the heights at about 9:00 in the morning of the 3rd. He had been informed that Longstreet and Jackson were moving to his support, and believing that Longstreet was close at hand, he opened with Pelham's howitzer on the Federal camps on the plain below. Artillery and infantry were moved to confront him, but he maintained his ground until nearly 2:00 in the afternoon, when, having exhausted his ammunition, and having learned that Longstreet had advanced no further than Nance's Shop, he withdrew.

THE DEATH OF CAPTAIN LATANE

By William Campbell, *Southern Historical Society Papers, Volume 29*, pages 97-90.

At your request, I undertake, after an intervention of more than thirty-four years, to write (from memory) my recollections of Stuart's famous ride around McClellan's army in the early summer of 1862; and also of the death of Captain William Latane, of the Essex Light Dragoons, who fell in a charge made by his squadron upon the enemy near the "Old Church," in Hanover County, Virginia.

Captain Latane, a son of Henry Waring and Susan Allen Latane, was born at The Meadow on January 16, 1833, and grew to man's estate surrounded by home influences not inferior to any in Virginia.

After receiving such training as the surrounding educational institutions could afford, he began the study of medicine at the University of Virginia in October, 1851. Here, he remained until the following summer, not offering for graduation.

In the fall of 1852, for some unexplained reason, he did not return to the University, but transferred the scene of his studies to the Richmond Medical College where he graduated in the spring of 1853. The following winter he spent in Philadelphia, taking a post-graduate course at one of the medical schools of that city, and also attending the hospital practice of the city.

On returning home in the spring of 1854, he located at The Meadow, and at once became a candidate for the practice of medicine. Here he remained until the breaking out of the war, not only attending to his practice – which soon became extensive, in consequence of his doing a large amount of charity practice among the poor around him – but giving successful attention to his large farm. In the management of the labor on this farm, he was "without any thought of it on his part," thus receiving preliminary training for the handling of large bodies of soldiers when the clash of arms should come upon his loved country. This would surely have been realized, had not his young life been snatched so suddenly away.

Early in 1861, when Mr. Lincoln made his call for troops to put down what he termed the rebellion, there was a rush to arms all over Virginia, and soon a cavalry company, called the Essex Light Dragoons, was formed, electing as their officers Dr. R.S. Cauthorn, captain; William L. Waring, first lieutenant; William A. Oliver, second lieutenant, and William Latane, third lieutenant. The company was soon mustered into the Confederate service for one year.

In the spring of 1862 it became necessary to re-enlist the men and reorganize the company, and in this reorganization, by common consent, William Latane was made captain. It was about this time that your writer made the acquaintance of his captain. I found him a man of small stature and quiet demeanor, but quick to perceive the wrong and very assertive in his opposition to it.

He commanded the confidence of his men by his evenhanded justice to all, and at the same time he brooked no disorder. Soon after the reorganization he was ordered to report with his company at Hick's Hill, near Fredericksburg, to become one of the constituent companies of the Ninth Virginia Cavalry, of which W.H.F. Lee, a son of General R.E. Lee, was colonel; R.L.T. Beale, lieutenant-colonel, and Thomas Waller, major. The Essex Light Dragoons became Company F of that famous regiment, and in the years that followed few of the recruits knew the company by its original name.

> *He commanded the confidence of his men by his evenhanded justice to all, and at the same time, he brooked no disorder.*

The month of service around Fredericksburg amounted to little except picket and drill duty, but McClellan's landing on the Peninsula and his march on Richmond made it necessary for us to retire to the lines around that city. Our regiment found a camp near Young's Millpond, and not far from the Brook Turnpike, occupying a position on the extreme left of the army defending Richmond. Nothing of special interest occurred during the following month other than the usual routine camp life.

But on Thursday, June 12th, came orders to prepare three days' rations and hold ourselves ready to march at a moment's notice. There was naturally suppressed excitement and speculation as to what we were to do or where we were to go, but no news came, and we could only indulge in speculation as to our destination.

About 1:00 p.m., the regimental bugler sounded "saddle up," which was caught up by the company buglers and soon the camp was in commotion. "To horse" was soon sounded, and through the whole camp could be heard the command of the officers, "Fall in, men."

Companies were formed and our regiment marched out of camp to participate in the most memorable and daring raid that was made during the war. We marched in the direction of Hanover Courthouse and went into camp after dark having marched some fifteen miles. Early dawn on the following morning found us in the saddle, the Ninth Virginia in the front, and our squadron, composed of the Mercer Cavalry, of Spotsylvania, and our company being in the front of the regiment, the Mercer being in advance.

Captain Crutchfield being absent, Captain Latane commanded the squadron, and, of course, rode in front, immediately in the rear of Colonel Lee and staff. Our march proceeded via Hanover Courthouse and on toward the Old Church. The first indication of an enemy we saw was the bringing in of a Yankee by one of our scouts. Soon thereafter Captain Latane rode to the rear and ordered four of his own company to advance to the front and form the first set of fours.

This had scarcely been accomplished before Colonel Lee ordered Captain Latane to throw out four flankers, two on either side, and four members of his company were at once ordered to proceed, two to the right and the others to the left and march a little in advance of the regiment.
Your writer was one of those on the left.

Moving forward, not seeing an enemy or supposing one to be near, I suddenly heard the command to charge and then the clash of arms with rapid pistol shots. Riding rapidly towards the firing, I found our squadron occupying the road and two companies of the Fifth United States Regulars attempting to form in a field near at hand, and Lieutenant Oliver urging his men to charge them. This was promptly done, and the enemy driven to the woods.

Just before reaching the timber, I overtook Lieutenant McLane, of the Federals, and he, seeing the utter futility of resistance, surrendered. As I was taking him to the rear, I met Colonel Lee and was told by him of the death of Captain Latane. He ordered me to turn my prisoner over to the guard, and then go and look after my captain. I soon found his body surrounded by some half a dozen of his men, one of whom was his brother, John – who was afterwards elected a lieutenant in the company, and the following year he too sealed his devotion to his country with his life.

Another of those present was S.W. Mitchell, a sergeant in the company, and, I wish to add, as gallant a spirit as ever did battle for a country. Mitchell, being the stoutest man present, was selected to bear the body from the field. He having mounted his horse, we tenderly raised the body and placed it in front of him. John Latane then mounted his horse, and he and Mitchell passed to the rear, while the rest of us hurried to join our command on its perilous journey.

I wish I could write my feelings as I looked upon the form of him who but a few moments before was the embodiment of life and duty. I wish I could describe to you the beautiful half Arabian horse that he rode, The Colonel, and how splendidly he sat him, but I fully realize that I am not equal to the task.

John R. Thompson, in his beautiful poem, "The Burial of Latane," and William D. Washington, in his painting of the same name, have, by pen and brush, so enshrined the name of Latane in the hearts of the people of our Southland that it will endure as long as men are admired for the devotion to duty and for risking their lives upon the perilous edge of battle in defense of homes and country.

I can only add that the glorious Stuart continued to ride grandly on his way, the Ninth Virginia still holding the post of honor at the front. Passing the Old Church, we hastened on toward the York River Railroad. Soon it was crossed and night came on but not halting.

On we marched into the county of New Kent. All that long night was spent in the saddle, pushing our way toward the lower Chickahominy, which we reached in the early morning, only to find that the bridge over which we intended to cross had been burned. But General Stuart was equal to the emergency. He soon had his rear guarded and the men swimming their horses over, while others were tearing down an old barn out of which a temporary bridge was constructed. On this the artillery and the few horses that remained were taken over.

The bridge was again burned in order to prevent pursuit. Again there was an all-night march, as we hurried up through the county of James City and on to Richmond, which we reached about midday on Sunday, June 15th, and went back to our camp that afternoon.

We brought back many trophies of our raid, consisting of several hundred prisoners and as many horses. But these went little way towards compensating the Essex Light Dragoons for the loss they had sustained in the death of their gallant captain.

As the years have crept on and I have called back to memory one incident after another of the deeds of daring and scenes of danger through which the cavalry of the Army of Northern Virginia passed in the four years of conflict, I recall none more splendidly conceived, more dashingly executed, and showing more favorable results than Stuart's raid around McClellan at Richmond.

A Sketch of Jeb Stuart

By John Esten Cook. *Wearing of the Gray.*

This sketch, may it please the reader, will not contain any historic events. Not a single piece of artillery will roar in it. Not a single volley of musketry will sound. No life will be lost from the very beginning to the end of it. It aims only to draw a familiar outline of a famous personage as he worked his work in the early months of the war, and the muse of comedy, not tragedy, will hold the pen. For that brutal thing called war contains much of comedy; the warp and woof of the fabric is of strangely mingled threads – blood and merriment, tears and laughter follow each other, and are mixed in a manner quite bewildering! Today it is the bright side of the tapestry I look at – my aim is to sketch some little trifling scenes "upon the outpost."

To do so, it will be necessary to go back to the early years of the late war, and to its first arena, the country between Manassas and the Potomac. Let us, therefore, leave the present year, 1866, of which many persons are weary, and return to 1861, of which many never grow tired talking – 1861, with its joy, its laughter, its inexperience, and its confiding simplicity, when everybody thought that the big battle on the shores of Bull's Run had terminated the war at one blow.

At that time, the present writer was attached to Beauregard's or Johnston's Army of the Potomac, and had gone with the advance force of the army, after Manassas, to the little village of Vienna – General Bonham commanding the detachment of a brigade or so. Here we duly waited for an enemy who did not come; watched his mysterious balloons hovering above the trees, and regularly turned out whenever one picket fired into another. This was tiresome, and one day in August I mounted my horse and set forward toward Fairfax Courthouse, intent on visiting that gay cavalry man, Colonel Jeb Stuart, who had been put in command of the front toward Annandale. A pleasant ride through the summer woods brought me to the picturesque little village; and at a small mansion about a mile east of the town, I came upon the cavalry headquarters.

The last time I had seen the gay young Colonel he was stretched upon his red blanket under a great oak by the roadside, holding audience with a group of country people around him, honest folks who came to ascertain by what unheard – of cruelty they were prevented from passing through his pickets to their homes. The laughing, bantering air of the young commandant of the outpost that day had amused me much. I well remembered now his keen eye, and curling moustache, and cavalry humor – thus it was a good companion whom I was about to visit, not a stiff and silent personage, weighed down with official business. Whether this anticipation was realized or not, the reader will discover.

The little house in which Colonel Jeb Stuart had taken up his residence, was embowered in foliage. I approached it through a whole squadron of horses, picketed to the boughs; and in front of the

> *Let me sketch him as he then appeared – the man who was to become so famous as the chief of cavalry of General Lee's army; who was to inaugurate with the hand of a master a whole new system of cavalry tactics – to invent the raid which his opponents were to imitate with such good results – and to fall, after a hundred hot fights in which no bullet ever touched him, near the scene of his first great "ride" around the army of McClellan.*

portico a new blood-red battle flag, with its blue St. Andrew's cross and white stars, rippled in the wind. Bugles sounded, spurs clashed, sabers rattled, as couriers or officers, scouts or escorts of prisoners came and went; huge-bearded cavalrymen awaited orders or the reply to dispatches – and from within came song and laughter from the young commander.

Let me sketch him as he then appeared – the man who was to become so famous as the chief of cavalry of General Lee's army; who was to inaugurate with the hand of a master, a whole new system of cavalry tactics – to invent the raid which his opponents were to imitate with such good results – and to fall, after a hundred hot fights in which no bullet ever touched him, near the scene of his first great ride around the army of McClellan.

As he rose to meet me, I took in at a glance every detail of his appearance. His low athletic figure was clad in an old blue undress coat of the United States Army, brown velveteen pantaloons worn white by rubbing against the saddle, high cavalry boots with small brass spurs, a gray waistcoat, and carelessly tied cravat. On the table at his side lay a Zouave cap, covered with a white Havelock – an article then very popular – and beside this two huge yellow leathern gauntlets, reaching nearly to the elbow, lay ready for use. Around his waist, Stuart wore a black leather belt, from which depended on the right a holster containing his revolver, and on the left a light, keen saber, of French pattern, with a basket hilt. The figure thus was that of a man every inch a soldier, and the face was in keeping with the rest. The broad and lofty forehead – one of the finest I have ever seen – was bronzed by sun and wind; the eyes were clear, piercing, and of an intense and dazzling blue; the nose prominent, with large and mobile nostrils, and the mouth was completely covered by a heavy brown moustache, which swept down and mingled with a huge beard of the same tint, reaching to his breast. Such was the figure of the young commandant, as he appeared that day, in the midst of the ring of bugles and the clatter of arms, there in the center of his web upon the outpost. It was the soldier ready for work at any instant; prepared to mount at the sound of the trumpet and lead his squadrons in person, like the hardy, gallant man-at-arms he was.

After friendly greetings and dinner on the lid of a camp-chest, where that gay and good companion, Captain Tiernan Brien, did the honors, as second in command, Stuart proposed that we should ride into Fairfax Courthouse and see a lady prisoner of his there. When this announcement of a lady prisoner drew forth some expressions of astonishment, he explained with a laugh that the lady in question had been captured a few days before in suspicious proximity to the Confederate lines, which she appeared to be reconnoitering; and that she was a friend of the other faction was proved by the circumstance that when captured she was riding a Federal Colonel's horse, with army saddle, holsters, and equipments complete. While on a little reconnaissance, all by herself, in this guise she had fallen into Stuart's net and had been conducted to his headquarters; assigned by him to the care of a lady resident at the Courthouse, until he received orders in relation to her from the army headquarters – and this lady we were now about to visit.

We set out for the village, Stuart riding his favorite Skylark, that good sorrel which had carried him through all the scouting of the Valley and was captured afterwards near Sharpsburg. This horse was of extraordinary toughness, and I remember one day his master said to me, "Ride as hard as you choose, you can't tire Skylark." On this occasion, the good steed was in full feather and as I am not composing a majestic historic narrative, it will be permitted me to note that his equipments were a plain McClellan tree, upon which a red blanket was confined by a gaily colored surcingle: a bridle with single head-stall, light curb bit, and single rein. Mounted upon his sorrel, Stuart was thoroughly the cavalry man, and he went on at a rapid gallop, humming a song as he rode.

We found the lady prisoner at a hospitable house of the village, and there was little in her appearance or manner to indicate the poor captive, nor did she exhibit any freezing terror, as the romance writers say, at sight of the young *militaire*. At that time some amusing opinions of the Southerners were prevalent at the North. The rebels were looked upon pretty much as monsters of a weird and horrible character-a sort of anthropophagi, Cyclops-eyed, and with heads that "did grow beneath their shoulders." Short rations, it was popularly supposed, compelled them to devour the bodies of their enemies; and to fall into their bloody clutch was worse than death. This view of the subject, however, plainly did not possess the captive here. Her fears, if she had ever had any of the terrible gray people, were quite dissipated; and she received us with a nonchalant smile and great indifference.

I shall not give the fair dame's name, nor even venture to describe her person, or conjecture her age – further than to say that her face was handsome and laughing, her age about twenty-five or thirty.

The scene which followed was a little comedy, whose gay particulars it is easier to recall than to describe. It was a veritable crossing of swords on the arena of qit, and I am not sure that the lady did not get the better of it. Her tone of *badinage* was even more than a match for the gay young officer's – and of *badinage* he was a master – but he was doubtless restrained on the occasion by that perfect good breeding and courtesy which uniformly marked his demeanor to the sex, and his fair adversary had him at a disadvantage. She certainly allowed her wit and humor to flash like a Damascus blade; and, with a gay laugh, denounced the rebels as perfect wretches for coercing her movements. Why, she would like to know, was she ever arrested? She had only ridden out on a short pleasure excursion from Alexandria and now demanded to be permitted to return thither.

"Why was she riding a Federal officer's horse?" Why, simply because he was one of her friends. If the Colonel would please let her return through his pickets she would not tell anybody anything – upon her word!

The Colonel in question was smiling – probably at the idea of allowing anything on two feet to pass through his pickets to the enemy. But the impossibility of permitting this was not the burden of his reply. With that odd laughter of the eye always visible in him when thoroughly amused, he opposed the lady's return on the ground that he would miss her society. This he could not think of, and it was not friendly in her to contemplate leaving him for ever so soon after making his acquaintance! Then she was losing other pleasant things. There was Richmond – she would see all the sights of the Confederate capital. Then an agreeable trip by way of Old Point would restore her to her friends.

Reply of the lady extremely vivacious: She did not wish to see the Confederate capital! She wished to go back to Alexandria! Straight! She was not anxious to get away from *him*, for he had treated her with the very greatest courtesy, and she should always regard him as her friend. But she wanted to go back to Alexandria, through the pickets – straight!

That the statement of her friendly regard for the young Colonel was unaffected, the fair captive afterwards proved. When in due course of time she was sent by orders from army headquarters to Richmond, and thence *via* Old Point to Washington, she wrote and published an account of her adventures, in which she denounced the Confederate officials everywhere, including those at the center of Rebeldom, as ruffians, monsters, and tyrants of the deepest dye, but excepted from this sweeping characterization the youthful Colonel of cavalry, who was the author of all her woes. So far from complaining of him, she extolled his kindness, courtesy, and uniform care of her comfort, declaring that he was the noblest gentleman she had ever known.

There was indeed about Colonel Jeb Stuart, as about Major-General Stuart, a smiling air of courtesy and gallantry, which made friends for him among the fair sex, even when they were enemies; and Bayard himself could not have exhibited toward them more respect and consideration than he did uniformly. He must have had serious doubts in regard to the errand of his fair prisoner, so near the Confederate lines, but he treated her with the greatest consideration; and when he left her, the bow he made was as low as to the finest lady in the land.

To subscribe, send an email to:
thestainlessbanner@gmail.com
or visit our website at www.thestainlessbanner.com

Subscription is free.

The Stainless Banner

An e-zine dedicated to the armies of the Confederacy

Volume 3, Issue 11
December, 2012

THE RELIGIOUS CHARACTER OF STONEWALL JACKSON

It is not an accident that in the impressive exercise with which you open this building, there is a place assigned for the religious character of him whose name is here to abide.

It is not only that any study of his character and career would be incomplete, but that it would be wholly unphilosophical and untruthful, without a statement of that which lay so effectively in his heart and covered so entirely all that we know of him. It was Thomas Carlyle who said, "A man's religion is the chief fact with regard to him." And more than of any man of renown of modern times, it is true of Jackson, that his religion was the man himself. It was not only that he was a religious man, but he was that rare man among men to whom religion was everything.

It is a remarkable fact that Oliver Cromwell, the great Puritan protector, of whom Thackeray spoke as "our great king," whose whole career has been the study of historians and critics, is in our day receiving a final study in his personal religion. Eminent critics are telling us that the campaigns will be the study and admiration of military schools for centuries to come. However true that may be, of this we are sure, the religion of Stonewall Jackson will be the chief and most effective way into the secret springs of the character and career of the strange man, who as the years go by is rising into the ranks of the great soldier-saints of history – Saint Louis of France, Gustavus Aldolphus of Sweden, Oliver Cromwell of England, and Stonewall Jackson of America.

What's Inside:	
We Remember!	9
The Day Lee Cried	11
Winter 1864	13
Born in a Barn	26

www.thestainlessbanner.com

Page 1

In this brief address I am to make today; in the hurried sketch I am to attempt of the inner springs of life and power in the story of Stonewall Jackson, I cannot be unmindful of the laws of heredity and the strong inbred qualities that came in the blood of a stalwart race. Nor can I forget the discipline of the hard life of his childhood, a homeless orphan boy drifting from place to place, and in the tenderest of years of youth, unprotected and exposed, seeking his bread as he could find it.

Certainly, I must not fail to recall that a mother of piety and love left him a little child of seven years, with nothing of religious instruction, no mother's knee at which to say his childhood's prayer, nothing to gentle and refine, nothing to restrain and guide him into an upright manhood, save the one unfading memory of that mother's love and parting blessing.

Running away from a harsh and unloving home, with an older brother boating on the Ohio, camping in hunger and cold, riding an uncle's horses on a race-course, attempting the rude work of a country constable in the mountains of West Virginia, there was absolutely no instruction, no counsel, and no ruling authority in all the young years of growth and formation.

It is marvelous indeed that out of such a youth, he came with purity and integrity, truthful, honest, modest and writing in rude characters that first brave maxim of life, "You may be whatever you resolve to be." I can find no mark of conscious religious sentiment in all this; though I see plainly the directing hand of a Divine providence fitting for a short life as rare and disciplined within, as it was brilliant and heroic without.

The thoughts of religion began to stir in his heart under the influence of a pious friend at West Point, and were felt with some power, when a young lieutenant at Fort Hamilton, he was, of his own desire, baptized into the Christian faith, by a Episcopal clergyman. They were moving effectually upon heart and conscience, when in the City of Mexico, applauded and promoted for conspicuous bravery, with a rare candor and open-heartedness he sought instruction from a bishop of the Catholic Church, of whom he was accustomed to speak with the most sincere respect.

The truths of the religion of Christ found a deep and abiding place in his heart, in the more quiet and regulated conditions of his first years in

> *It is not one truth or another about God, or one feature of our Christian religion rather than another that became real and dominant to him; but God, God Himself, the living, personal and present God, became the one transcendent fact that dwelt in all his thoughts and possessed his whole being.*

Lexington, when under the ministry of the venerable Presbyterian pastor, Dr. William S. White, he made a public confession of his personal faith in Christ. Acknowledging his ignorance of religions, he came with the entire candor and simplicity to be taught as a little child. The truths he heard were not wholly clear to him, and some things he agonized with an honesty and courage that were most admirable in the sincere seeker after truth. Only through the long process of study, reflection and prayer, he was led into a clear vision of the great essential truths of evangelical religion. As they came out, like stars fixed in the firmament of his upward gaze, he bowed his head and his heart and gave them their rightful authority over all his manhood.

The inspired Psalmist declares of the wicked man, "God is not in all his thoughts." The supreme fact that in the character of Jackson was that far beyond any man of whom we read, "God was in all his thoughts."

It is not one truth or another about God, or one feature of our Christian religion, rather than another, that became real and dominant to him; but God, God Himself, the living, personal and present God, became the one transcendent fact that dwelt in all his thoughts and possessed his whole being. It was not God only as the surpassingly glorious subject of reflection, or as a living and working and revealing himself in nature and in history, nor as partially known by Hebrew prophets in the childhood of humanity; but God revealed in Christ, the God of law and love, whose law is love, and whose love leads back to law.

I am careful to say this, that I may also say, the supreme thought of God gave unity to his religion and unity to his life. As it went down into the hidden nature within, it possessed the whole man with unwonted power and made him one and the same, a man of God within and without. Unto a personal and present God, he gave the undivided faith of his heart. He acknowledged his supreme authority as maker and redeemer over part of his being and every breath of his life, and to that authority, he bowed his will implicitly. "He came nearer putting God in God's place," said Dr. Stiles, "than any man we have ever known." And in this he put himself in the one rightful place to which man belongs, the humblest and the most majestic, the strongest, the safest, and the happiest man can ever occupy.

It gave simplicity and directness and personal humility in an uncommon degree. All things were viewed in the light of the supreme fact of God. All things were referred to it. All things were submitted to the rulings of that fact. It covered all other facts, all other truth, it ruled all action, it answered all questions of duty and made all his life and service one and simple forever.

How inevitably came his humility. He owed all to God, all that he was, all he had attained, all he had accomplished in classroom or on battlefield, and unto Him belong all the praise and the glory: "God has given us a brilliant victory at Harper's Ferry today," he wrote from the field; "Our Heavenly Father blesses us exceedingly."

On his campped in the Wilderness

hospital, when I read General Lee's magnanimous note congratulating him on the victory Jackson had won at Chancellorsville, he replied with emotion, "General Lee is very kind to me, but he should give the glory to God!"

How unquestionable was his dependence! As he lifted his hand in the morning twilight, riding down to the field of Fredericksburg, he said, "I trust our God will give us a great victory today, Captain!"

How immediately came his obedience. A friend in Lexington asked whether would obey, if the Lord bade him leave the home he loved and all that it contained, and go on some mission to Africa. He rose and with intense feeling and prompt decision declared, "I would go without my hat." And asked if it were required of him to give up the activity and happiness of life, the exquisite happiness of energy and lie on a bed of pain, he said, "I would lie there a thousand years without a murmur, if I knew it to be the will of my heavenly Father!"

I have been accustomed to recall two notable things in the religion of Jackson: his belief in the providence of a present *God*, ruling and directing in wisdom, power and goodness in all the affairs of men; and his consequent belief in the right and power of prayer, to Him whose ears are always open to the cry of his children, and who is ready to hear and answer above all that his children can ask or think. He was as all who knew who were at all in touch with his daily life, a man of prayer; humble, truthful, confident prayer, from which he came as the saint comes, with unspeakable joy in his heart and serenity in all his face and bearing.

It is an old jest, that the Puritan could scarcely be said to enjoy his religion; but if Jackson were in any sense a puritan, his personal happiness was unbroken and abiding. The performance of duty was not hard, because the fear of the Lord he loved and served was the only fear he knew. There was no asceticism in his life, because there was no gloom in his heart. "I do rejoice," he said, "to walk in the love of God."

> *"He came nearer putting God in God's place," said Dr. Stiles, "than any man we have ever known."*

There were no lacking those who neither knew nor understood the character of Jackson, nor had the most remote conception of truth and power of his religion. If it appeared to any that sternness and rigidity marred his character, it was only because in such rare degree among men he lived and acted from deep conviction of duty and that was strange to us. Whatever was remarkable about his personal bearing, and was sometimes criticized or ridiculed, was due to the absolute possession of him the great things of religion had taken.

These were the things that were the strong iron of his blood; they were the constant inspiration of his gentler, simpler life in his Lexington home, and as well as the animating power of his matchless campaigns that have given him undying fame.

His patriotism was a duty to God. His obedience to the State that called him to the field was made clear and plain to him, as obedience to God. All soldierly duty was rendered as a service to his God. He loved and reverenced the Sabbath day with great ardor; yet on a Sabbath morning, he came from his knees in his happy home, turned away from the services of the sanctuary he loved, and buckling on his sword, took command of the Cadet corps on yonder parade grounds, and rang out clear and sharp, his first command in the Civil War, "Battalion, march!" He went without fear, without regret, without selfish ambition, to the unknown fortunes of war. Whatever was the marvelous development of soldierly qualities, of brilliant generalship, whatever the story of campaign and victory, from which he never asked a furlough, and from which he never returned, he was the same devout and single-hearted servant of the living God.

Capable of anger and indignation in high degree, he had cultivated a self-control that gave him a self-mastery that was sometimes marvelous. An officer of rank came on Sunday afternoon to the little office building at Moss Neck to urge his personal application for a leave of absence. He violated the guard and entered General Jackson's private apartment without announcement. Never had I seen General Jackson so surprised and then so angry. His face flushed, his form grew erect, his hands were clenched behind his back, and he quivered with the tremendous effort at self-control. And no word was permitted to pass his lips until his passion was entirely mastered, when he quietly explained wherein the unfortunate colonel was violating all rules and all propriety and sent him to his quarters, the most thoroughly whipped man I ever saw.

Having strong attachment to the church of which he was a member and positive convictions concerning what he thought was a member and positive convictions concerning what he thought was true and right, he was yet generous and catholic in his esteem of all other churches and had sincere respect for the views of others. Ruling himself with a severe discipline in things he deemed right, he was never censorious or dictatorial. He worshipped in all churches alike with devoutness and comfort. He encouraged the chaplains of all churches, Protestant and Catholic. A Protestant and a Presbyterian of Presbyterians, he obtained the appointment of a Catholic priest to a chaplaincy.

In nothing perhaps was the reality and power of his own religion so evident as in his interest in the religious welfare of others. With an unwearying diligence, he conducted his Sunday-school for colored people. Visiting at Beverly, of his own volition, he gathered the village people to instruct them himself in the truths of religion.

He was profoundly interested in the work of the army chaplains and used all his great influence and opportunity to sustain them. He was accustomed to make individual friends the subject of his earnest and continued prayer. He once came walking to the camp of the Rockbridge Artillery, asking for a certain corporal and leaving a package for him in his absence. It was a matter of intense curiosity in the camp, as perhaps containing some handsome gift or unexpected promotion for the corporal; who, when he returned to camp, found the package to contain religious tracts for distribution among his comrades.

Not all devoid of humor was the earnest, reticent man. His fondness for General J.E.B. Stuart was very great and the humor and frolic of that genial and splendid cavalryman was a source of unbounded delight.

Dr. George Junkin, President of Washington College, and father of the first Mrs. Jackson, went back to Pennsylvania at the opening of the war and wrote a vigorous book on the errors into which he believed the South had fallen. He forwarded a copy of his book, under a flag of truce, from General Hooker's headquarters to General Lee's. It came to us about the time of the battle of Fredericksburg, and when I opened the package and told our General its title, "Political Heresies," he said with a grim smile, "I expect it is well named, Captain; that's just what the book contains, political heresies."

I remember two young girls in a mansion on the Rappahannock were with great earnestness asking for locks of his hair. Blushing like a girl himself, he pleaded that they had so much more hair than he had, then that he had gray hairs, and their friends would think he was an old man. They protested that he had no gray hair and was not an old man, when he said, "Why, don't you know the boys call me Old Jack?"

The stern warrior was one of the gentlest of men. He had the tenderest affection for little children. Little Janie Corbin was a pleasure and delight to him in the afternoons of his days of office toil at Moss Neck, as she folded papers and cut lines of soldiers and paraded them on his table. He heard from me of her death with an outburst of tears and a convulsed frame.

> *All soldierly duty was rendered as a service to his God.*

It was complained by one of his distinguished generals of division, in a severe paper, that ladies, mothers, wives and daughters had invaded the vicinity of our camps and were diverting officers and men from military duty. When that paper was read to him, Jackson rose and paced the room impatiently and to the request that he would order the ladies to retire, he said, "I will do no such thing. I am glad my people can have their friends with them. I wish my wife could come to see me."

No one who ever entered his house or obtained access to his office at his corps headquarters can forget the marked courtesy with which he was received. His attention was the same to his guest whether he was the General Commanding or a private soldier. Your hat was taken by his own hands and his own black stool from the mess hall of the Institute must be your seat while you were his guest.

Are these the things that mark the gentleman? Are purity and truth, modesty and courtesy the things by which we know him? These things he had, not by conventionality, but as the constant expression of a gentle nature and the fruit of religious principle.

An English gentleman of rank and of large touch with polite society, at the end of a week's sojourn, spent chiefly in General Jackson's room, said, "He is a revelation to me; Jackson is the best informed soldier I have met in America and as perfect a gentlemen as I have ever known!"

How surpassingly fitting it seems that the two Virginia heroes of our civil war should meet again and find their resting place in tombs so near; in this retired place among the strong mountains of the state they loved so well! How unlike they were in many things, in origin, in culture, in family tradition, in the conventionalities of society and in the knowledge of the world! How much alike they were in unselfish devotion to the same cause, in true and simple piety, and in the generous honor that each paid to the other! They, who set one over against the other and study to give either one the greater glory of his campaign or that, do an unworthy violence to their spirit and are rebuked in the presence of their silent tombs. Two lofty peaks, they stand on fame's eternal camping ground, each giving unfading glory to the other.

How happy and hopeful it is that here the young man from Virginia, from mountains and lowlands alike, are to be gathered in growing numbers and to be trained for life under the pervading inspiration of names and stories, than which none in all history are more true and effulgent in all things pure and lovely and of good report. If any young man shall go out from the institutions of Lexington to anything if life that is corrupt, or unmanly, or forget of the honor of Virginia, he will do so against the example and the appeal of Robert E. Lee and Stonewall Jackson.

Ten years of faithful toil Jackson gave to the Virginia Military Institute with difficulties that have not always been well understood. Through uncounted years to come his great name will rest upon this building as a benediction! The memory of the soldier and his campaigns and victories will abide in this hall and the spirit of the honest and God-fearing Christian gentleman will come back to speak forever of that fear of God which is the beginning of wisdom and of that simple and humble faith which is the sure and only way to enduring honor and exaltation.

In the lowly building at Guinea's Station, where he lay suffering, failing, dreaming, passing away, he spoke of a grave "in Lexington, in the valley of Virginia."   And then his thoughts so easily passed to another rest and other shades.

By James Power Smith, D.D.
Captain and aide-de-camp to Jackson

Do You Have a Book on the War You Are Longing to Publish?

THE STAINLESS BANNER PUBLISHING COMPANY
A Full-Service Small Press Dedicated to the
Preservation of Southern Heritage and History

History ✶ Memoir ✶ Biography ✶ Novel ✶ Alternate History

What does **The Stainless Banner Publishing Company** offer you?

- ✶ Professional editing
- ✶ Dynamic covers
- ✶ Hard cover or paperback
- ✶ Competitively priced books
- ✶ High royalties paid monthly
- ✶ Your book on Amazon, Barnes & Noble and many other outlets
- ✶ Free advertising

What is my investment?

Nothing! **The Stainless Banner Publishing Company** is a full service small press. Once your book is published, your only investment will be for any inventory you wish to keep on hand.

Where do I send my query letter?

No need for an agent or a query letter. Just send an email telling us about your book and include the first three chapters in the body of the email. You will receive a prompt reply, usually within 48 hours. Send your email to books@thestainlessbanner.com

The Stainless Banner Publishing Company
www.thestainlessbanner.com

The Stainless Banner
December 2012

We Remember!

By Brother Len Patterson, Chaplain, Army of Trans-Mississippi, Sons of the Confederate Veterans.

Even the most casual reading of the Bible tells us that God is very much interested in genealogy. In the very beginning of the Scriptures, Genesis 5, we are given the lineage from Adam to Noah and his sons. And the first nine chapters of First Chronicles is nothing but genealogy, and it doesn't end there. Then of course, the New Testament begins with the "generation" or genealogy of Jesus Christ. It seems that God considers knowing our forefathers (and mothers) very important. And, I submit that if it's important to God, it should be important to us.

As the Apostle Paul sat in the dungeon of the Mamertine Prison in Rome awaiting his execution, he penned his final letter to his young protégé, Timothy, who was then leader of the Christians in Ephesus. In this letter, we call "Second Timothy," following his usual opening salutation, Paul begins chapter one, verse five, by speaking in remembrance of Timothy's mother and grandmother. Then in Chapter three, verses fourteen and fifteen, he tells the young preacher to continue in the things he learned as a child.

Earlier in his ministry, Paul wrote to the Church at Ephesus itself, and in Ephesians 6:2, admonished them to, "Honour thy father and mother (which is the first commandment with promise;)" We should notice this is a command, not a suggestion or request. And, what is the promise for obedience to this command? The next verse reads, "That it may be well with thee, and thou mayest live long on the earth."

So, we are not only taught by Scripture to remember our forefathers and fore mothers with due respect, but commanded to honor them as well. It has been said, and rightly so, "If we do not honor our ancestors, we have no right to expect honor from our descendants."

Since the beginning of human history, countries have gone to war with each other for one reason or another. And, at such times its citizens have been called upon to bare arms, most commonly the country's young men. These young men did not go to war for glory, conquest, or profit. They answered their countries call because it was their duty to serve. In America in 1861, after forcing the issue at Fort Sumter, Abraham Lincoln asked the Northern states, usually referred to as the "Union," to furnish him with seventy-five thousand troops to put down, what he referred to as "the rebellion." The Union then invaded the

> *If we do not honor our ancestors, we have no right to expect honor from our descendants.*

Southern states of the Confederacy, which had legally and peacefully succeeded from the United States which they had every lawful and constitutional right to do.

It must be realized that in 1861 each state of the United States was autonomous, independent, self-governing, and had it's own military forces. On July 2, 1776, as our country was being founded, the Continental Congress passed a resolution stating firmly that, "these United Colonies are, and of right ought to be, free and independent States." In 1812, thirty-six years later, as the States again went to war with England, Connecticut Governor, Roger Grizwold, declared his state's militia would not serve in the war against Britain. He was not being rebellious or unpatriotic, nor was he considered a traitor. Connecticut, as would any other free and independent state, had the lawful and Constitutional right to make their own decision and abstain from serving.

It must also be remembered people at that time were citizens of their state, not the country. Therefore, in 1861, when the Union invaded the newly formed independent states of the Confederacy: Texas, Louisiana, Mississippi, Alabama, Georgia, Tennessee, Virginia, and the rest called upon their young men to come to their defense. And they did. By the thousands, the young men of the South left their homes, families, and fields and came to the aid of their home state. Of course they fought in unity with citizen soldiers of other states to form the Confederate Army.

These men, some no more than boys, fought long and hard, and fought well. They did not fight for glory or reward, or for any ideology. They fought because they saw it as their duty to serve their state. They were out-gunned, out-manned, under fed, under supplied and faced an unrelenting and merciless enemy. For four years they suffered hardship and deprivation and died by the thousands in defense of their homeland. They were brave! They were heroic! They were the sons of the South. Our fathers!

But today, I hear their blood screaming from a thousand hills and a hundred battlefields, "Where is our respect? Where is our honor? Where is the remembrance for our sacrifice and service?" Who remembers their brave hearts and courage? What of the states that called them from their peaceful homes and families to the horrors of the battlefield. What of Texas, Louisiana, Mississippi, Alabama, and the rest? Do they care? Do the very states they fought for, and many died for, remember their devotion to duty and their unselfish patriotism?

I also hear the mournful cries of mothers who lost their sons, wives who lost their husbands, sisters who lost their brothers and children who lost their fathers. In many cases they only know their loved ones answered their states call to arms and never came home. What of their solace? Where are there condolences? How can they be comforted? Who remembers their pain and suffering? What of their descendants? Do they remember the unselfish and determined stand of those to whom they owe their very existence? Do they remember their courage and determination? Do they care?

Who cares about people who are long dead? How does our remembrance help them? Who cares about a country that only lasted a few years a century and a half ago? What difference could that possibly make now? Who cares about a heritage that is only a faint shadow of what it once was? What does that have to do with life today? Who cares about all that? We have a simple answer. "We do!" We care about our heritage, our forefathers, their cause, the country they fought for, and the flags they fought under. Yes, we care.

We care for the memory of those who answered their states call to duty and valiantly fought in defense of their homeland? We remember their diligence, sacrifice, patriotism and honor? We remember them and proclaim, "I am the proud descendant of a brave and noble Confederate Soldier." We! The Sons of Confederate Veterans. We! The Military Order of the Stars and Bars. We! The United Daughters of the Confederacy. We! The Order of Confederate Rose. We! Who come together to memorialize their lives and service. We care, and we remember!

THE DAY ROBERT E. LEE CRIED

By Curt Steger

I was there the day Robert E. Lee cried.
It was Palm Sunday in 1865.
I saw him don his finest uniform when the morning broke,
And I heard the last artillery sounds and I saw the smoke.

I rode the last ride with him to meet his foe.
He was Master Lee to us and we loved him so.
I saw him dismount Traveller at the McLean house and go in,
And I thought of the battles this man and horse had fought to win.

There was a countenance about him as he waited for Grant to arrive
That confirmed within my soul that he was the noblest man alive.
He was beaten but not defeated as he sat there and waited
For that dreaded moment to occur that both he and The South hated.

He rose erect when the two foes in the parlor there did meet,
And it was Lee that day that taught me that there can be dignity in defeat.
I watched him sign his honorable name and then be the first to depart.
I saw his last duty done – I saw his heavy heart.

He stopped on the porch at the top of the steps to look across at his last battlefield,
And I thought of this man's character and how it would be needed to help our country heal.
I saw the salutes from his former foes gathered there in the yard,
And I thought of how his will to battle them had made their last four years so hard.

I knew I had just seen one of those rare moments in history
When more eyes fell on the man vanquished than on the man in victory.
We rode back down that lane together to a different world,
And I watched how his Army of Northern Virginia around him swirled.

Then I saw the saddest thing that ever a man's eyes could meet.
I saw the tears of this great General roll down and off his cheek.
A sight I knew that I would never forget—the day I saw grown men cry;
Half starved and shoeless, yet still proud men lifting hats as he rode by.

Their devotion to him still in their eyes as each reached out to touch
This famous rider they had fought for—this one they loved so much.
Then this man of character forever immortalized himself in Southern lore
When he spoke to those gallant men there,

"I have done the best I could for you. My heart is to full too say more."
And when our Cause was no more, I heard the last rebel yell
And I thought of our great General of which one day the world will tell
How one man alone furled the standards never lowered in defeat

How his army never vanquished duty bound laid down its arms
No more battles to fight for him—no more buglers' sound of retreat.
It is to the future generations I ask forgiveness for being unable to describe
The sadness I felt deep inside me on the day Robert E. Lee cried.

THE WINTER OF GROWING DESPAIR

By Douglas Southall Freeman from his book, *R.E. LEE*, pages 525-544.

Anxious as had been the months of November and December, 1864, there had been some hours when Lee could think of other things than troop movements, and after the repulse of the attack on Fort Fisher, while Sherman waited in Savannah, there came a brief respite. Fortunately, Lee's health continued good, though he looked much older. He rarely showed any sign of his gnawing anxiety and was ceaselessly active.

After he had completed his inspection of the lines on the right, early in November, he had returned to Petersburg, and then, about November 25, had decided to move headquarters farther toward the right, whither Grant seemed perpetually to be extending his flank in his efforts to reach the Southside Railroad.

Lee had wished to remain in a tent, where his visitors would be no disturbance to others, but Mrs. Lee and Walter Taylor between them had prevailed on him to accept the invitation of the Turnbull family and to establish himself at their home, Edge Hill, which was about two miles west of Petersburg, on a healthy, convenient, and accessible site.

"After locating the General and my associates of the staff," Taylor wrote at the time, "I concluded that I would have to occupy one of the miserable little back rooms, but the gentleman of the house suggested that I should take the parlor. I think that the General was pleased with his room, and on entering mine he remarked: 'Ah, you are finely fixed. Couldn't you find any other room? 'No,' I replied, 'but this will do. I can make myself tolerably comfortable here.' He was struck dumb with amazement at my impudence, and soon vanished."

The quarters were, indeed, the best Lee had ever had during his campaigning, a fact which probably caused him some inward twinges, for Taylor believed the General was never so well satisfied as when he was living like a Spartan. But if the rooms were pleasant, headquarters fare was of the scantiest. When General Ewell visited him, Lee insisted that his guest have his lunch, which consisted of two cold sweet potatoes. An Irish M. P. who came to Edge Hill remarked to Mrs. Pryor, who had furnished him a room, because Lee could not, "You should have seen Uncle Robert's dinner today, madam! He had two biscuits and he gave me one!" Another day the Irishman reported, "What a glorious dinner today, madam! Somebody sent Uncle Robert a box of sardines."

Lee's chief recreation, during many unoccupied hours, was entertaining children. The Federals had ceased bombarding the town during November, but on the 28th they had opened fire again, deliberately, Lee thought, for, as he remarked to Reverend Henry C. Clay, "whenever a house was set on fire, we saw the fire of the enemy increased and converging on that point." Lee was distressed that the bursting of the shells kept his young friends from playing in the city streets and occasionally he would send in a wagon and bring them out to headquarters, where they could frolic, free of danger.

One day he was riding back to town with a party of youngsters, when a young guest began to whip the mules to make them go faster. "Don't do that, my little child," he said. The girl forgot after a few minutes and struck the beasts a second time. "Anne," he said, sternly but sadly, "you must not do that again. My conscience is not entirely at ease about using these animals for this extra service, for they are half fed, as we all are."

Once he spent a few moments playing with a child who was sick abed, and on a Sunday when he entered a crowded chapel and found a plainly dressed little miss vainly looking for a seat, he escorted her to the pew that was reserved for him, and had her remain at his side during the service.

"Yesterday afternoon," he wrote Mrs. Lee in January, "three little girls walked into my room, each with a small basket. The eldest carried some fresh eggs, laid by her own hens; the second, some pickles made by her mother; the third, some popcorn grown in her garden. . . . I have not had so pleasant a visit for a long time. I fortunately was able to fill their baskets with apples, which distressed poor Bryan, and I begged them to bring me nothing but kisses and to keep the eggs, corn, etc., for themselves."

Another recreation, though rare, was that of reading a new book. He lingered affectionately, no doubt, over the life-story of that admiring old friend and chief from whom, in April, 1861, he had found it so difficult to part. "I have put in the bag General Scott's autobiography, which I thought you might like to read," he wrote Mrs. Lee late in the winter. "The General, of course, stands out prominently, and does not hide his light under a bushel, but appears the bold, sagacious, truthful man he is."

His rides into Petersburg and his visits to Richmond were a comfort to him, of course. In Petersburg he called often on Mrs. A.P. Hill, the young and lovely wife of the commander of the Third Corps. "General Lee comes very frequently to see me," Mrs. Hill wrote her mother, all in a breath, "he is the best and greatest man on earth, brought me the last time some delicious apples." When the General "went home" to Richmond, it was always to find Mrs. Lee more and more a cripple, though she was interested in everything and kept her needles busy knitting socks for the soldiers.

The capital was more crowded than ever, dejected and negligently dilapidated. Sometimes, from the sad seniors, Lee would turn away to the children. "I don't want to see you," he would say half in jest and half in reproof, "you are too gloomy and despondent; where is ---?" and would name the little girl of the family.

The young belles of the town were much in doubt whether it was proper to have dances at so dark a time, and a committee of them asked his advice, with the assurance that if he disapproved, they would not dance a single step. "Why, of course, my dear child," he answered. "My boys need to be heartened up when they get their furloughs. Go on, look your prettiest, and be just as nice to them as ever you can be!"

At Christmas time, when Savannah had fallen and the fate of Wilmington seemed to hang by a thread, he went to Richmond to see his family, but he could not stay more than a day. On his return he learned that some of his friends had sent him a saddle of mutton to brighten the mess at Edge Hill. It went astray, however, and never reached him. "If the soldiers get it," he said, simply, in report its non-arrival, "I shall be content. I can do very well without it. In fact, I should rather they should have it than I."

At a Yuletide dinner in Petersburg he was no little embarrassed because he wanted to save his portion of turkey so that he could carry it to one of his staff officers who, as he explained, had been very ill and had "nothing to eat but corn bread and sweet-potato coffee." When a barrel of turkeys arrived for himself and his staff he ordered his fowl sent to the hospital, and he announced his purpose in such a tone that the other officers sadly repacked the barrel and sent all its contents to the scanty mess of the convalescents.

He discouraged personal gifts as far as he could. Daily, for three months after he came to Petersburg, the wife of Judge W.W. Crump, a distinguished jurist of Richmond, sent fresh bread to his mess by a special messenger. "Although it is very delicious," he felt constrained to write her, "I must beg you to cease sending it. I cannot consent to tax you so heavily. In these times no one can supply their families and furnish the Army, too. We have plenty to eat and our appetites are so good that they do not require tempting."

There was much to be done, during this period and throughout the winter, in maintaining the morale of the officers, for many of them now regarded the Southern cause as lost. In their private speculations as to how long the Confederacy could survive, few affirmed that even the Army of Northern Virginia could resist the enemy longer than July, 1865. Carelessness increased, drinking became worse. Lee had constantly to be stirring some of his subordinates to vigilance.

Occasionally he would snub a man whom he thought had been needlessly absent; sometimes he would present his coldest mien to those he found loafing at headquarters; if an officer seemed to be too dainty about his food, Lee would chaff him with exaggerated attention. When a man complained of injustice on the part of his superiors, Lee would urge him to his full duty and not to fear the consequences. The sick or captured officer was always his special care. To the family of wounded or captured men, as well as to the kin of one who was killed, he was quick to send his condolence. Furloughs he had to decline, even in a case so pathetic as that of General Pendleton, who wished to go home to baptize the posthumous child of his son, Sandie, who had been killed in Early's Valley campaign.

So far as circumstances permitted, Lee continued to give encouragement and to administer rebukes by tactful suggestion. Riding out one day in

January, 1865, he inquired of General John B. Gordon and of General Heth concerning the progress being made on two heavy redoubts that were under construction on Hatcher's Run. Gordon assured him that his fort was nearly finished. Heth said, with some embarrassment, "I think the fort on my side of the run also about finished, sir."

Lee decided to go with them and to see for himself. Gordon's works were in the conditions described. On Heth's front, the digging had scarcely begun.

"General," said Lee, "you say this fort is about finished?"

"I must have misunderstood my engineers, sir."

"But you did not speak of your engineers. You spoke of the fort as nearly completed."

Heth was riding a very spirited horse that had been presented to his wife, and in his humiliation he must have tugged at the reins, for the animal began to prance about excitedly.

"General," said Lee, in his blandest manner, "doesn't Mrs. Heth ride that horse occasionally?"

"Yes, sir."

"Well, General, you know that I am very much interested in Mrs. Heth's safety. I fear that horse is too nervous for her to ride without danger, and I suggest that in order to make him more quiet, you ride him at least once a day to this fort." That was all, but it was a rebuke that sank into the heart of Heth.

One evening the General came upon a group of young officers who were working over a bit of mathematics, drinking all the while from two tin cups that were replenished at the mouth of a jug that had a guilty, bibulous look. Lee solved their problem for them and went his way with no reference to their refreshment; but the next morning, when one of the group began to recount a very curious dream of the night, Lee quietly observed, "That is not at all remarkable. When young gentlemen discuss at midnight mathematical problems, the unknown quantities of which are a stone jug and two tin cups, they may expect to have strange dreams."

He was always careful, however, never to rebuke an officer of rank in the presence of others. On a tour of inspection with Gordon he found some earthworks that had been very badly located, and he said so in plain words. Turning, however, he noticed some young officers within earshot; so he added audibly, "But these works were laid out by skilled engineers, who probably know their business better than we do."

If Lee had on occasion to admonish officers during that dreadful winter, he likewise had occasion to be grateful to some of them, and to none more than to Brigadier General Archibald Gracie for an act of a kind that Lee best appreciated. One day in November he had been on the lines with Gracie, who commanded a brigade in Johnson's division. Being perhaps unfamiliar with the deadliness of the sharpshooting on that part of the front, Lee carelessly stood up on the parapet. Gracie, without a word, instantly interposed his body between that of Lee and the enemy. Both were pulled back over the works before either was hit, but Lee never forgot the spirit Gracie exhibited. A few weeks later, on December 3, Gracie was killed by a fragment of shrapnel on a point on the fortifications where there was not supposed to be any danger.

Through these anxious days, as always, Lee's reliance was on a Power which, as he wrote Mr. Davis at the time he recalled Rodes' division from the Valley, "will cause all things to work together for our good." Again he told Mrs. Lee: "I pray daily and almost hourly to our Heavenly Father to come to the relief of you and our afflicted country. I know He will order all things for our good, and we must be content."

The type of his prayer book having become too small for his vision, he mentioned the fact one day to Mrs. Churchill J. Gibson, wife of the rector of Grace Church, Petersburg, and said that he intended to give it to some soldier. He remarked, as he spoke, that the volume was the one he had used during the Mexican War. Mrs. Gibson at once offered to give him several new copies of the prayer book in exchange for so interesting a memento. Lee gladly agreed and distributed the new books through one of his chaplains to men who asked for them. In each copy he inscribed a line of presentation.

Yet, faithfully as he used his new book of devotion, with the humility that marked his every act, he doubted if his own prayers would avail. In an exchange of letters with General Pendleton, during the autumn, when Pendleton explained that he had omitted to say grace at the General's table because he did not know his chief had asked him to do so, Lee said "[I] am deeply obliged to you for your fervent prayers in my behalf. No one stands in greater need of them. My feeble petitions I dare hardly hope will be answered."

A nation's prayers, and not an individual's only, were needed as January, 1865, passed. Hourly along the line of thirty-five miles from the Williamsburg road to the unstable right flank on Hatcher's Run the pickets kept their rifles barking, and the sharpshooters watched the embrasures on the reddish-yellow parapet across the fields.

Nightly the fuse of each bomb could be traced, like giants' fireworks, from the mouth of the mortar through the high trajectory and back again to earth. Never was there silence, never a day without casualties; yet from the time of the raid on Belfield in December, 1864, until February, 1865, there was no large action on the Richmond-Petersburg front, largely because of the condition of the roads.

Elsewhere, calamity followed on the heels of disaster. Before the middle of January it became apparent that Sherman would soon start his advance from Savannah toward Charleston. There were only scattered forces to oppose him. Kershaw's old brigade, now under General James Conner, was immediately ordered to Charleston. Cavalry was much needed there. After conference with President Davis, Lee dispatched Butler's division to the Palmetto State and authorized General Hampton to go thither, also, in the hope that his great reputation in South Carolina would bring new volunteers to the colors.

In retrospect, Lee regarded this as the great mistake he made during the campaign, because it crippled him in dealing with subsequent Federal operations against his right flank. When the movement was ordered there seemed no alternative to it, unless Sherman was to be permitted to advance unhindered up the coast.

Before Butler's cavalry could get under way for South Carolina, a great Federal fleet again appeared off Wilmington, convoying an infantry force on transports. This time there was no delay and no experimentation with powder boats. Under the direction of General Alfred H. Terry, the troops were thrown ashore and on January 13, a bombardment of Fort Fisher wwas begun. Before the early winter's sun had set, two days later, the Union flag was flying over the shattered works, and the last port of the Confederacy was closed.

With Wilmington lost and Sherman about to march northward, the alarm in Richmond grew into a frenzy. Davis was blamed, as the executive of a waning cause always is, both for what he had done and what he failed to accomplish. Some of those who had been so insistent on a rigid interpretation of the Federal Constitution in 1860 now began to clamor for a dictator. Lee was to be the man. The President must step aside and place all power in the hands of the one person who had the genius to save the South. Longstreet had hinted at something of the sort in December, and Lee had ignored it. To his mind, the very suggestion was abhorrent and a reflection on his loyalty as a soldier and a citizen. So far as the record shows, nobody ever presumed to mention the subject to him personally. At length, as a sort of desperate compromise with Congress, the President consented to the appointment of a general-in-chief.

As it happened, the nearest approach to an open break between General Lee and the President had occurred only a few days before. Late in January or early in February there was an exchange of correspondence regarding the destruction of tobacco in the warehouses of Richmond, to prevent its falling into the hands of the enemy. President Davis telegraphed Lee, in effect, "Rumor said to be based on orders given by you create concern and obstruct necessary legislation. Come over. I wish to have your views on the subject."

Lee replied in cipher, which Davis had employed, that it was difficult for him to leave Petersburg. "Send me the measures," his telegram concluded, "and I will send you my views." This made Davis very angry. He replied at some length, and ended: "Rest assured I will not ask your views in answer to measures. Your counsels are no longer wanted in this matter." Lee received this in silence when it was decoded, then quietly ordered his horse, rode to the railroad and took the train to Richmond. When he returned he said nothing of what had passed between the President and himself. Evidently, however, all misunderstandings were cleared up for, on February 6, Davis named Lee to the newly created office.

The appointment came just at the time when the negotiations for peace at the so-called Hampton Roads conference had failed and when the Federals were active on Lee's right flank. He had his hands full, more than full, and was under no illusions as to what he could do in general command. He wrote characteristically:

"I know I am indebted entirely to your indulgence and kind consideration for this honorable position. I must beg of you to continue these same feelings to me in the future and allow me to refer to you at all times for counsel and advice. I cannot otherwise hope to be of service to you or the country. If I can relieve you from a portion of the constant labor and anxiety which now presses upon you, and maintain a harmonious action between the great armies, I shall be more than compensated for the addition to my present burdens. I must, however, rely upon the several commanders for the conduct of the military operations with which they are charged, and hold them responsible. In the event of their neglect or failure I must ask for their removal."

He did not attempt to do more than he indicated in this letter and he did not consider that his appointment conferred the right to assign generals to command armies. "I can only employ such troops and officers," he said, "as may be placed at my disposal by the War Department. Those withheld or relieved from service are not at my disposal."

It was all he could do to watch Grant, to conserve the strength of his dwindling army, and to combat the dark forces of hunger and disintegration that had long been at work. In December the shortage of provisions had become more acute than ever. No salt meat was available in the depots, and none was arriving from the South. In the emergency the navy lent the army 1500 barrels of salt beef and pork, but the commissary general confessed himself desperate, and a special secret report to Congress bore out his dark view of the South's resources.

In January heavy rains temporarily broke down transportation on the Piedmont Railroad, which linked Lee's army with the western Carolinas. About the same time floods cut off supplies from the upper valley of the James River. Lee then had only two days' rations for his men and already had scoured clean the country within reach of his foragers.

In this crisis, at the instance of the War Department, which fell back in every emergency on the magic of his name and on the compelling power of his appeal, he asked the people to contribute food for the army. Almost before he could ascertain whether appreciable results would follow this call, he had to march a heavy column to the extreme right to meet new Federal demonstrations on Hatcher's Run. This was on February 5, the eve of the very worst weather of a bad winter.

The military results were negligible, but for three night and three days a large part of the Confederate forces had to remain in line of battle, with no meat and little food of any sort. The suffering of the men so deeply aroused Lee that he broke over the usual restraint he displayed in dealing with the civil authorities. "If some change is not made and the commissary department reorganized, I apprehend dire results," he wrote the Secretary of War. "The physical strength of the men, if their courage survives, must fail under this treatment."

He did not demand the resignation of the grumbling Northrop, the commissary general of subsistence, but his reference to the necessity of a change was not lost on President Davis. "This is too sad to be patiently considered," Davis endorsed on Lee's dispatch, "and cannot have occurred without criminal neglect or gross incapacity." Within a few days, Northrop was quietly relieved of duty and was succeeded by Brigadier General I.M. St. John, who had much distinguished himself by his diligent management of the mining and nitre bureau.

St. John was most reluctant to take the post, but he immediately organized a system by which supplies were to be collected from the farmers, hauled to the railroad and dispatched directly to the army, without being handled through central depots. Lee welcomed the change, and was encouraged by it to believe that if communications could be maintained, the army would be better fed. The people, he reasoned, "have simply to choose whether they will contribute such . . . stores as they can possibly spare to support an army that has borne and done so much in their behalf, or retain those stores to maintain the army of the enemy engaged in their subjugation."

This view was at once made the basis of an ingenious appeal for food, addressed to the people of Virginia by a special committee of Richmond ministers and other citizens. A plan was outlined by which a farmer could ration a soldier for six months – much as money was raised in America during the war with Germany to feed Belgian babies and Armenian orphans.

Anxiously, agonizingly, Lee awaited the response of the people. When he was asked early in March for an appraisal of the military situation, he postulated everything, in his reply, on transportation and on the willingness of the people to make further sacrifices. "Unless the men and animals can be subsisted," he said, "the army cannot be kept together, and our present lines must be abandoned. Nor can it be moved to any other position where it can operate to advantage without provisions to enable it to move in a body… Everything, in my opinion, has depended and still depends upon the disposition and feelings of the people. Their representatives can best decide how they will bear the difficulties and sufferings of their condition and how they will respond to the demands which the public safety requires."

The representatives of Virginia in the Congress were brought together to answer Lee's question. He was present and told them of lengthened lines and thinning forces, of the privations the soldiers had to meet, and of the scarcity of food for them and for the horses. The Virginians replied that the people of the state, with loyalty and devotion, would meet any new demand made on them, but they seemed to General Lee to content themselves with words and assertions of their faith in their constituents. They proposed nothing; they did nothing. Lee said no more – the facts were warning enough – but he went from the building and made his way to his residence with distress and indignation battling in his heart.

When dinner was over, Custis (Lee's oldest son) sat down by the fire to smoke a cigar and to read the news, but Lee paced the floor restlessly. "He was so much engrossed in his own thoughts," wrote a silent young observer, years afterwards, "that he seemed to be oblivious to the presence of a third person. I watched him closely as he went to the end of the room, turned and tramped back again, with his hands behind him. I saw he was deeply troubled. Never had I seen him look so grave.

"Suddenly he stopped in front of his son and faced him: 'Well, Mr. Custis,' he said, 'I have been up to see the Congress and they do not seem to be able to anything except to eat peanuts and chew tobacco, while my army is starving. I told them the condition the men were in, and that something must be done at once, but I can't get them to do anything, or they are unable to do anything.'… there was some bitterness in his tones….

"The General resumed his promenade, but after a few more turns he again stopped in the same place and resumed: 'Mr. Custis, when this war began I was opposed to it, bitterly opposed to it, and I told these people that unless every man should do his whole duty, they would repent it; and now' (he paused slightly as if to give emphasis to his words) 'they will repent.'"

It was on this visit to Richmond, or on another about the same time, that he was chatting with a group of gentlemen at the President's house when one of them said: "Cheer up, General, we have done a good work for you today. The legislature has passed a bill to raise an additional 15,000 men for you." Lee, who had been very silent and thoughtful, bowed his acknowledgments. "Yes," he said, "passing resolutions is kindly meant, but getting the men is another matter." He hesitated for a moment, and his eyes flashed. "Yet," he went on, "if I had 15,000 fresh troops, things would look very different."

Outraged as Lee was by the apparent incapacity of Congress, he warmly encouraged General St. John to do his utmost in applying the same methods of direct appeal the new commissary general had used with notable success in collecting nitre; but as Lee sought to find food for his men he saw new military difficulties added to those of transportation, weather, distress, and growing public despair.

The danger of the destruction of all lines of communication with the South and the occupation of the only territory from which he was now drawing supplies were daily brought nearer and nearer. "The perils and privations of the troops," in the opinion of an observant colonel who saw him often, "were never absent from his thought."

Bound up, now as always, with subsistence for the men was the old, tragic question of provender for the horses during a winter when there was no pasturage. It was the experience of 1862-63 and 1863-64 more poignantly repeated. Many of the army wagons were used, during most periods of quiet, to collect food and bring it to the railroad. When the army was in a country that had not been stripped of food the wagons could gather enough to make up for the deficiencies of the regular supply from the depots. Now, the territory around Petersburg having been swept of the last provisions, such horses as were not too feeble and too ill-fed to be sent out, had to be used at a long distance from the army, in North Carolina and in western Virginia.

Those that remained had then to be fed at places where inability to employ them in foraging made the army wholly dependent on what came by railroad. The familiar "vicious circle" thus was rounded more speedily than ever, and the mobility of the army and its range of vision were hourly decreased. There was danger that the troops might remain where they were until, in a literal sense, they had no horses to move their trains. Yet Lee could not circumvent this by an early departure from Petersburg, because the mud was so heavy the teams could not pull the wagons. He had to wait until the roads were better, even if he had to risk immobility then.

What was true of the wagon trains applied also, of course, to the artillery. The horses had to be taken from most of the guns and scattered throughout the countryside, at a distance from the line, in order to keep them alive. As late as March 20 it had not been possible to call in even the animals of the horse artillery. Many commands had to be consolidated and reorganized because there were not enough horses for all the batteries.

The cavalry suffered with the wagon train and with the artillery. No substantial force could be kept close to the infantry. When Butler's division was sent off, the horses were subsisted in North Carolina. Two other divisions were scattered in small units because supplies could not be transported to the places where the troops should have been concentrated. At the time of the operations against Hatcher's Run, in February, W.H.F. Lee's division had to be brought forty miles, by roundabout roads, from Stony Creek, where forage was being delivered.

Early in March, when it was necessary to send out cavalry on a forced reconnaissance, five days elapsed before Fitz Lee could get his men together and start after the enemy. Rooney Lee, called up at a critical hour, had to be returned to Stony Creek on the very day Lee thought that Grant's cavalry was being heavily reinforced.

Before the end of the winter Lee was uncertain whether he would be able to maintain even a small cavalry force around Richmond. There was virtually nothing he could do to maintain the arm of service on which he had to depend not only for early information of the enemy's movements but also for the protection of his communications and for the safety of his right flank from a sudden turning movement. He urged the government to new endeavor in procuring horses, and when it was reported that animals could not be had for lack of money, he frankly advocated the seizure of cotton and tobacco, their sale for gold and the purchase of horses with this medium. Nothing coming of this, he was compelled to extemporize new tactics.

Infantry were to be stationed as close as possible to any point whence the enemy was expected to start a raid, and were then to be moved rapidly to support the thin cavalry that might be thrown forward – a scheme that seems to have been proposed by Longstreet. This meant, of course, that the defensive line had to be weakened, and the danger of a break increased by this detachment of infantry, even when troopers who had no horses were put in the trenches. It was a grim plight for an army that once had boasted a Stuart and stout squadrons of faultlessly mounted boys who had mocked the awkward cavalry of McClellan as they had ridden around his army.

Desertion continued to sap the man-power of the army. After Christmas, when the winter chill entered into doubting hearts, and every mail told the Georgia and Carolina troops of the enemy's nearer approach to their homes, more and more men slipped off in the darkness. Desertions between February 15 and March 18 numbered 2934, nearly 8 per cent of the effective strength of the army. From Pickett's division alone, a command that had won the plaudits of the world, 512 soldiers deserted about the middle of March, during the progress of a single move.

There was suspicion that men from different brigades were communicating with one another and were arranging rendezvous. When they left, taking their arms with them, they usually went home, but not a few of the weaker-spirited joined the enemy. From one division, a good one at that, 178 were reported to have "gone over into the Union," in the language of the trenches. Conditions became so bad that when it was necessary to move one of Pickett's brigades through Richmond, Longstreet's adjutant general did not think it safe to let men wait long in the streets.

The reasons for this wastage in an army that had been distinguished for nothing more than for its morale were all too apparent – hunger, delayed pay, the growing despair of the public mind, and, perhaps more than anything else, woeful letters from wives and families telling of danger or privation at home. Lee noted with much distress that the largest number of desertions were among the North Carolina regiments, which previously had fought as valiantly as any troops in his command. The army was melting away faster than was the snow.

Lee had been able to do little about subsistence and the supply of horses, but desertion and the conditions it brought about were military problems. He faced them. After offering amnesty, he had to enforce very sternly the law for the execution of deserters who were recaptured, and when clemency was shown in a case where a court-martial had decreed the death penalty, he telegraphed: "Hundreds of men are deserting nightly, and I cannot keep the army together unless examples are made of such cases." He sent a large detachment to western North Carolina to bring back deserters, and he felt compelled to take from his insecurely held trenches a whole brigade to guard the crossings of the Roanoke River.

The articles of war on desertion and the regulations forbidding any man to propose such a course, even in jest, were read throughout the army for three days. Longstreet issued an order in which he announced that he would recommend for commission with the proposed Negro regiments any man who thwarted the attempt of another soldier to desert.

Despair had not entered every heart. If hundreds deserted, there were thousands who had resolved that neither hunger nor cold, neither danger nor the bad example of feebler spirits could induce them to leave "Marse Robert." Many of them "came to look upon the cause as General Lee's cause, and they fought for it because they loved him. To them he represented cause, country and all."

The soldiers' letters of this dark period present a hundred contrasts. One Marylander wrote in January: "There are a good many of us who believe this shooting match has been carried on long enough. A government that has run out of rations can't expect to do much more fighting, and to keep on in a reckless and wanton expenditure of human life. Our rations are all the way from a pint to a quart of cornmeal a day, and occasionally a piece of bacon large enough to grease your palate."

A young North Carolinian, in precisely the opposite mood, expressed his regret that the people of his state were despairing because of the loss of Fort Fisher. "If some of them could come up here," he wrote, "and catch the good spirits of the soldiers, I think they would feel better."

Lee understood the fears of the faint-hearted as much as he valued the courage of those who, knowing the cause to be hopeless, determined to sustain it to the end. In his appeals to all his men, he spoke now as a father to his sons. The little that he could do for their comfort, he did with warm affection.

One winter's day, as he and his staff were riding along, he met four private soldiers plodding through the mud toward the lines. Stopping, he asked where they were going, and when they explained that they had been to Petersburg and were afraid they would not reach their posts before roll-call, he had some of his officers take the men up behind them on their horses and carry them to the trenches.

When a sergeant of the fine old Fourth South Carolina came to Lee's headquarters and asked for transportation on a furlough he had earned, the General was distressed that the railroad pass had not been issued with the furlough. "They ought to have given you transportation without putting you to this trouble," he said. He was accessible to all his men, even to the cooks. When one Negro attendant presented himself at Edge Hill, Lee had him admitted.

"General Lee," the man began, "I been wantin' to see you for a long time. I's a soldier."

"Ah," Lee answered, "to what army do you belong – to the Union army or to the Southern army?"

"Oh, General, I belong to your army."

"Well, have you been shot?"

"No, sah, I ain't been shot yet."

"How is that?" Lee inquired. "Nearly all of our men get shot."

"Why, General, I ain't been shot 'case I stays back whar de generals stay."

Desperate as were the times, Lee found delight in that answer and repeated it more than once to his lieutenants.

Of the men to whom the heart of Lee went out, the wounded always came first. One day he was journeying over to Richmond for an interview with the President. As the train neared the city a crippled soldier got up and struggled to put on his overcoat. Nobody in the crowded car did anything to help him. Observing this Lee rose and assisted the veteran.

After he had seen that he could not count on the employment of the reserves, Lee had exerted himself in the early winter to organize the local defense troops in Richmond, but he found it progressively more difficult to get them out as the weather grew worse.

He had sought, also, to retain the Negro laborers over the Christmas holidays. Although he had previously had a low opinion of the fighting quality of Negro troops, he saw now that the South must use them, if possible. After the beginning of 1865 he declared himself for their enlistment, coupled with a system of "gradual and general emancipation." Congress hesitated and debated long, but at last, on March 13, the President signed a bill to bring Negroes into the ranks, though without any pledge of emancipation, such as Lee had considered necessary to the success of the new policy.

Bad as was the law, Lee undertook at once to set up a proper organization for the Negro troops. While Congress had argued, Virginia had acted in providing for the enrollment of Negroes, slave and free, in the military service. On March 24 Lee applied for the maximum number allowable under the statute of the commonwealth. "The services of these men," he said, "are now necessary to enable us to oppose the enemy."

He urged on his lieutenants new economy of force and he strengthened his lines against sudden attack. Personal appeals were made to returned prisoners of war to waive the usual furlough and to rejoin their commands; all able-bodied men were taken from the bureaus; all "leaves" for officers were suspended; new combat rules and revised marching instructions were issued to meet changed conditions.

All that Lee had learned in nearly four years of war, all that his quiet energy inspired, all that his associates could suggest or his official superiors devise – all was thrown into a last effort to organize and strengthen the thin, shivering, hungry Army of Northern Virginia for the last grapple with the well-fed, well-clad, ever-increasing host that crowded the countryside opposite Lee's lines.

✯ ✯ ✯

In all my perplexities and distresses, the Bible has never failed to give me light and strength.

Robert E. Lee

BORN IN A BARN

By Elaine Bridge.

"Close that door! Were you born in a barn?"

Well…yes, He was. He was born in a manger of hay, at least, likely surrounded by animals. And He's been leaving doors open behind Him ever since.

Many more years ago than her youngest daughter is old, my friend Terry had an encounter with God. While I don't know the details, I do know that Jesus came knocking on her heart's door one day, and she let Him in. Not only did He never leave, He left the door open behind Him, and countless others have followed His lead into her heart and life.

I joined that crowd when she began hosting a mid-week fellowship and Bible study gathering several years ago and literally opened the door to her home for a group of us who have been faithfully meeting there ever since. For years upon years now, she's rushed home from her job on a Thursday to start a meal for the lot of us and then sits on her porch and waits for us all to arrive. Enveloped in hugs and love when we do, we're ushered inside for a time of food and friendly banter. When the meal has moved from plates to waists, we move the dishes to the kitchen, break out the coffee cups, pens and worksheets and study the night's lesson in one accord. The evening ends as it began, with laughter on the front porch as the crowd gradually dissipates into the cars that fill the driveway and heads for home.

As the last guest leaves and her door is shut on the night, Terry's heart and life are still open and available. Because she freely shares her struggles and foibles, friends and acquaintances feel comfortable to be real about theirs around her and come calling on the phone or crowding a campfire in her backyard as together they search for the keys to dealing with the problems they face. More often than not, her visitors find Jesus in their discussions with her, whether they open a conversation with Him on their own at that point yet or not. They are at least introduced to the Master Locksmith who can open any door that is closed to them, whether it be a physical door of opportunity in their lives, or the entryway into the heart of another person that was closed over an event that happened in the past. And it begins by responding to the pounding of their own hearts, and letting Him in.

I have a fear of facing a locked door on my car. Because the key contains a computer chip it is expensive to duplicate, and I haven't gone to the expense of doing so again after the first

> *He has come to you with a simple message from your Father's heart: The door is open. The porch light is on. Hurry Home.*

spare was lost on a snowboarding hill and the second simply disappeared somewhere in the house. As a result, I never leave my car casually anymore, but stop to make sure I'm holding the key in my hand before I exit the vehicle. Even so, I still remember the horror of accidentally dropping my lone means of entry into the trunk one day just as the lid to the thing came slamming down. And many of us live restricted lives because we're similarly afraid of somehow dropping the ball with God, and hearing the Gates of Heaven slam shut against us forever as a result.

Yet God doesn't want us to live our lives in fear. Knowing that we were without the means ourselves to purchase our passage into His Presence, He sent His Son to rescue us. Jesus was born in a barn in Bethlehem for the sole purpose of reopening the door of communication between fallen man and the God Who still desperately loved His creation, despite their rejection and sin. With His death, He made restitution for us all and reconciliation a precious possibility as Heaven's portals swung wide open once more.

From the night of His birth, while yet in His mother's womb as she and Joseph arrived at a crowded inn, He's been knocking on doors that are closed to Him, seeking a place to be born anew. If this Christmas you feel strangely moved, perhaps it's because He has come to you with a simple message from your Father's heart: The door is open. The porch light is on. Hurry Home.

"For it is through Him that we …now have…(access)…to the Father…"
(Ephesians 2:18 AMP)

To subscribe, send an email to: thestainlessbanner@gmail.com or visit our website at www.thestainlessbanner.com

Subscription is free.

CPSIA information can be obtained at www.ICGtesting.com
Printed in the USA
BVOW06s0103070415

394942BV00005B/29/P

9 781627 520102